Dingyü Xue
Modeling and Simulation with Simulink®

Also of Interest

MATLAB® Programming
Mathematical Problem Solutions
Dingyü Xue, 2020
ISBN 978-3-11-066356-3, e-ISBN (PDF) 978-3-11-066695-3,
e-ISBN (EPUB) 978-3-11-066370-9

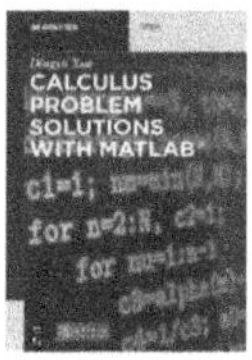

Calculus Problem Solutions With MATLAB®
Dingyü Xue, 2020
ISBN 978-3-11-066362-4, e-ISBN (PDF) 978-3-11-066697-7,
e-ISBN (EPUB) 978-3-11-066375-4

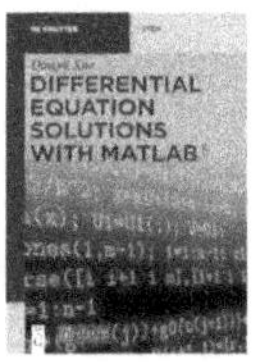

Differential Equation Solutions with MATLAB®
Dingyü Xue, 2020
ISBN 978-3-11-067524-5, e-ISBN (PDF) 978-3-11-067525-2,
e-ISBN (EPUB) 978-3-11-067531-3

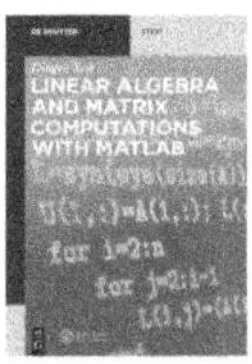

Linear Algebra and Matrix Computations With MATLAB®
Dingyü Xue, 2020
ISBN 978-3-11-066363-1, e-ISBN (PDF) 978-3-11-066699-1,
e-ISBN (EPUB) 978-3-11-066371-6

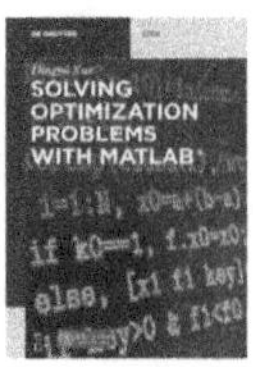

Solving Optimization Problems with MATLAB®
Dingyü Xue, 2020
ISBN 978-3-11-066364-8, e-ISBN (PDF) 978-3-11-066701-1,
e-ISBN (EPUB) 978-3-11-066369-3

Dingyü Xue

Modeling and Simulation with Simulink®

For Engineering and Information Systems

DE GRUYTER

清華大學出版社
TSINGHUA UNIVERSITY PRESS

Author
Prof. Dingyü Xue
School of Information Science and Engineering
Northeastern University
Wenhua Road 3rd Street
110819 Shenyang
China
xuedingyu@mail.neu.edu.cn

ISBN 978-3-11-073904-6
e-ISBN (PDF) 978-3-11-073495-9
e-ISBN (EPUB) 978-3-11-073499-7

Library of Congress Control Number: 2021949680

Bibliographic information published by the Deutsche Nationalbibliothek
The Deutsche Nationalbibliothek lists this publication in the Deutsche Nationalbibliografie; detailed bibliographic data are available on the Internet at http://dnb.dnb.de.

Cover image: Dingyü Xue
Typesetting: VTeX UAB, Lithuania
Printing and binding: CPI books GmbH, Leck

www.degruyter.com

Preface

Scientific computing is commonly and inevitably encountered in course learning, scientific research and engineering practice for each scientific and engineering students and researchers. For the students and researchers in the disciplines, which are not pure mathematics, it is usually not a wise thing to learn thoroughly low-level details of related mathematical problems, and also it is not a simple thing to find solutions of complicated problems by hand. It is an effective way to tackle scientific problems, in a high efficiency, accurate and creative manner with the most advanced computer tools. This method is especially useful in satisfying the needs for those in the area of science and engineering.

The author has made some effort to this goal to address directly the solution methods for various branches in mathematics in a single book. Such a book, entitled "MATLAB based solutions to advanced applied mathematics," was published first in 2004 by Tsinghua University Press. Several new editions are published afterward, in 2015, the second edition in English by CRC Press and in 2018, the 4th Edition in Chinese were published. Based on the latest Chinese edition, a brand new MOOC project was released in 2018,[1] and received significant attention. The number of the registered students is around 14,000, in the first round of the MOOC course, and tens of thousands in later rounds. The textbook has been cited by tens of thousands times by journal papers, books and degree theses. The English version of the MOOC is now also available.[2]

The author has over 30 years of extensive experience of using MATLAB in scientific research and education. Significant amount of materials and first-hand knowledge have been accumulated. These cannot be covered in a single book. A series entitled "Professor Xue Dingyü's Lecture Hall" of such works are scheduled with Tsinghua University Press, and the English editions are included in the DG STEM with De Gruyter. These books are intended to provide systematic, extensive and deep explorations in scientific computing skills with the use of MATLAB and related tools. The author wants to express his sincere gratitude to his supervisor, Professor Derek Atherton of Sussex University, who first brought him into the paradise of MATLAB.

The MATLAB series is not a simple revision of the existing books. With decades of experience and material accumulation, the idea of "revisiting" is adopted in authoring the series, in contrast to other mathematics and other MATLAB-rich books. The viewpoint of an engineering professor is established and concentrations are made on solving various applied mathematical problems with tools. Many innovative skills and general-purpose solvers are provided to solve problems with MATLAB, whereas not possible by any other existing solvers, so as to better illustrate the applications

1 MOOC address: https://www.icourse163.org/learn/NEU-1002660001
2 MOOC address: https://www.icourse163.org/course/NEU1-1458222170

https://doi.org/10.1515/9783110734959-201

of computer tools in solving mathematical problems in every mathematics branch. It also helps the readers to broaden their viewpoints in solving scientific computing, and even find innovative solutions by themselves to scientific computing, which cannot be solved by any other existing methods.

The first title in the MATLAB series: "MATLAB Programming," can be used as an entry-level textbook or reference book to MATLAB programming, so as to establish a solid foundation and deep understanding for the application of MATLAB in scientific computing. Each subsequent volumes tries to cover a branch or topic in mathematical courses. Bearing in mind the "computational thinking" in authoring the series, deep understanding and explorations are made for each mathematics branch involved. These MATLAB books are suitable for the readers who have already learned the related mathematical courses, and revisit the courses to learn how to solve the problems by the use of computer tools. It can also be used as a companion in synchronizing the learning of related mathematics courses, and view the course from a different angle so that the readers may expand their knowledge in learning the related courses, so as to better learn, understand and practice the materials in the courses.

This book is the sixth one in the MATLAB series. Systematic presentations are made on introducing various simulation methods with MATLAB and Simulink. Commonly used blocks are introduced in Simulink, with basic manipulations of the interface. Then the block diagram modeling methods for various differential equation are studied. For linear systems, a unified simulation framework for linear control systems is proposed, and the simulation of transfer functions with nonzero initial conditions is studied. For nonlinear systems, the universal static nonlinear element is constructed, and various approximate simulation methods are introduced. Advanced topics such as subsystem and its masking, command based modeling and simulation methods. In particular, the combination of Simulink and numerical optimization techniques are explored, and many conventionally unsolvable problems can be studied. In this book, application problems such as multimedia information manipulation and processing, modeling and simulation of engineering and non-engineering problems are studied.

At the time that the books are published, the author wishes to express his sincere gratitude to his wife, Professor Yang Jun. Her love and selfless care over the decades have provided the author immense power, which supported the authors' academic research, teaching and writing.

January 2022 Xue Dingyü

Contents

1 Simulink and system simulation

1.1 An introduction to system simulation techniques

Systems are the integrated wholes composed of interrelated and interacting entities with specific functions. There are various of ways to classify systems. Normally they can be classified into engineering systems or non-engineering systems. Engineering systems are the whole composed of interacting components such that certain system objectives can be achieved. For instance, motor drive systems are composed of an actuating component, a power transfer component and a signal measurement component, so as to control the motor speed or position, among other objectives.

The coverage of nonengineering systems is much wider. From universe to the microworld, any integrated whole can also be regarded as a system, since there are interrelated and interacting relationships. To study quantitatively the system behaviors, its characteristics and internal relationships should be extracted to establish the system models.

System simulation is the study to investigate the system behaviors based on the mathematical models and physical behaviors of real systems. In particular, it refers the study of system behaviors with computers. The fundamental contents of computer simulation includes systems, models, algorithms, programming, and the display, analysis and validation of the results.

System simulation technology is becoming the third methodology to investigate and change the world, after theoretical study and scientific experiment[41]. A significant amount of work, which cannot be carried out by ordinary theoretical research and experiment, are completed by system simulation methods. The work such as those in hazardous situations, and those with high cost and time consuming are more suitable to be carried out with computer simulation methods, so as to achieve reliable research results. System simulation technology has been applied successfully in all the fields in science and engineering.

The most popular MATLAB language and Simulink environment are used as the mainline to introduce system simulation methods in this book. In Section 1.2, the concept of mathematical modeling and physical modeling are introduced, with example demonstrations. The stages in system simulation development are studied in Section 1.3, with concentrations on mathematical software and simulation languages. The development of MATLAB and Simulink are also presented. The structures of the book and other essential issues are presented in Section 1.4.

https://doi.org/10.1515/9783110734959-001

1.2 Mathematical modeling versus physical modeling

There are many ways to classify modeling and simulation methods. One way is classify them as block diagram based and equation based methods. The key point is whether the simulation model is built upon block diagrams or upon equations. Another classifying method is mathematical modeling and physical modeling methods. The key point is whether the simulation system is built upon mathematical expressions or based on the assembly of physical blocks.

Mathematical expressions are used to depict the dynamical behaviors of the systems in mathematical modeling, and the model is used to describe the quantities of the variables in the systems. It is the basis in the analysis and design of the systems. Based on the characteristics or mathematical tools in describing the behaviors of the systems, the systems can further be classified as continuous, discrete-time, discrete event and hybrid systems. It can also be classified as linear and nonlinear; time invariant and time-varying; lumped parameter and distributed parameter; deterministic and stochastic systems.

With the rapid development and extensive application of computer technology, it seems that the utilization of mathematical models are more and more adopted. The entire system can be expressed by formulas with ordinary differential equations (ODEs) or difference equations. With mathematical models, numerical algorithms for differential and difference equations can be applied to find the numerical solutions of system behaviors. To build mathematical models, solid background and knowledge are needed in the specific field. In this section, examples are presented to demonstrate the meaning and differences of mathematical and physical modeling, and also to point out their advantages and disadvantages.

Example 1.1. Consider the simple electric circuit shown in Figure 1.1. Build up the mathematical model of the circuit.

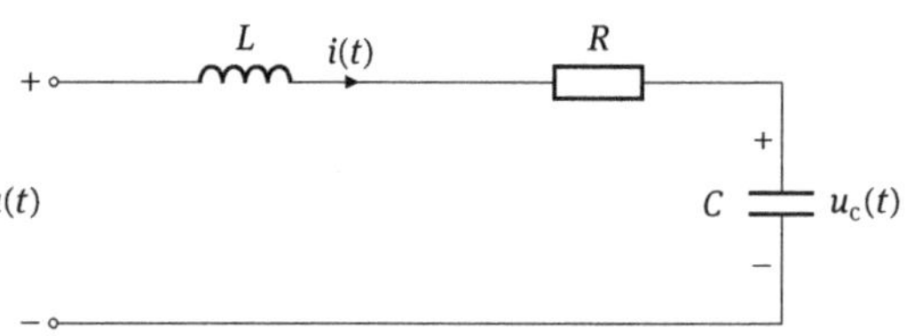

Figure 1.1: Series circuit of resistor, conductor and inductor.

Solutions. From the circuit, the dynamic relationship between the given voltage $u(t)$ and the voltage $u_c(t)$ across the capacitor C. The former signal is often known as the input signal, and the latter is the output of the system. The current signal $i(t)$ is the intermediate signal in the circuit. It is known from circuit theory that the voltage across the inductor L and the loop current $i(t)$ satisfy the following differential equation:

$$u_l(t) = L\frac{di(t)}{dt}. \tag{1.2.1}$$

With the well-known Kirchhoff's law, the following equation regarding the input and output signals can be established as

$$u(t) = Ri(t) + L\frac{di(t)}{dt} + u_c(t). \tag{1.2.2}$$

Besides, the voltage $u_c(t)$ across the capacitor and the loop current $i(t)$ satisfy the following differential equation:

$$i(t) = C\frac{du_c(t)}{dt}. \tag{1.2.3}$$

Substituting (1.2.3) into (1.2.2), the differential equation of the entire circuit can be written as

$$u(t) = RC\frac{du_c(t)}{dt} + LC\frac{d^2u_c(t)}{dt^2} + u_c(t). \tag{1.2.4}$$

Besides, if $u_c(t)$ and $i(t)$ are selected as the states of the circuit, the following state space equation (the concept of state space equation will be presented later) can be written as

$$\begin{cases} \dfrac{du_c(t)}{dt} = \dfrac{1}{C}i(t), \\ \dfrac{di(t)}{dt} = \dfrac{1}{L}u(t) - \dfrac{R}{L}i(t) + \dfrac{1}{L}u_c(t). \end{cases} \tag{1.2.5}$$

With the mathematical expressions, simulation languages and tools can be used to describe them, and work out simulation results.

It can be seen from the example that specific knowledge in the related disciplines are needed to write out the mathematical models of the systems. If the mathematical modeling methods in the fields are not quite solid, it might be hard to work out mathematical models. Therefore, physical modeling methods emerged, and more and more widely accepted by the researchers. The so-called physical modeling is to build the simulation models with provided components using building block methods. For instance, if the user is not able to write out the mathematical model for the circuit in Example 1.1, a software supporting physical modeling is needed to connect the circuit with blocks for resistors, inductors and capacitors, according to the circuit. The remaining parts are left to the physical modeling software, so as to generate the simulation model automatically for the circuit, so as to complete the simulation behaviors of the circuits.

In the field of electric and electronic circuits, Spice (simulation program with integrated circuit emphasis) language is the most widely used simulation system. The language is the exemplary software in electronic circuits and the *de facto* world standard in the field.[67] The components for the resistors, capacitors, inductors, semiconductors, independent and control sources are provided in the language, such that the electric circuit can be established in the same way as building real circuits. The simulation model can then be generated automatically. There is no need for the users to write out the mathematical models of the circuits by hands. Spice language appears in 1972 and various versions and variations follow such as PSpice. A simple example is demonstrated next for physical modeling of an electric circuit.

Example 1.2. If $R = 10\,\Omega$, $C = 15\,\mu\text{F}$, $L = 20\,\text{mH}$ and $u(t) = 8\,\text{V}$, use Spice language to write out physical models of the circuit in Figure 1.1.

Solutions. Of course, the simulation of the DC circuit is useless for this particular example. Here, Spice language in physical modeling is illustrated. To use Spice language in modeling, the nodes should be labeled, as shown in Figure 1.2. With the labels, the start and end nodes for each branch can be adopted to write out the following program for the circuit.

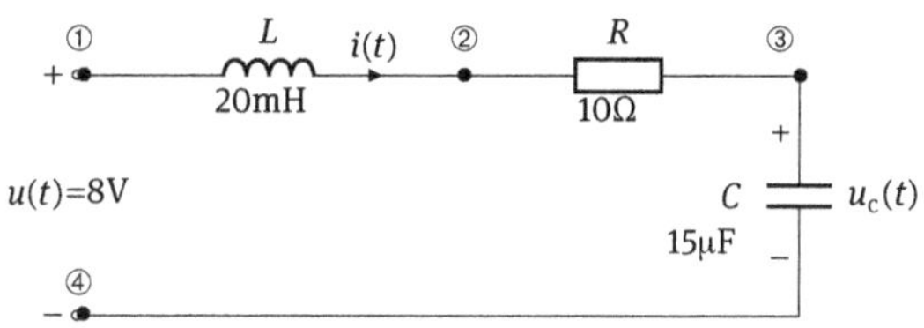

Figure 1.2: Modified R-L-C circuit.

```
* this is the comment: a Spice model file name: c1mspice.cir
* The format in: component name, start and end nodes, values and units
VIN 1 4 8.0volt
L 1 2 20mH
R 2 3 10ohm
C 3 4 15uF
.END
```

It can be seen from the Spice source code that in describing complicated circuits, there is no need to write out the differential equation models for the circuits. The Spice simulation program determines the variations in the signals according to the connections of the components in the circuit automatically. This is the major difference between physical and mathematical modelings. Note that, this is only a demonstration, and it does not imply that the Spice modeling method should be mastered by the readers of this book. It can be seen from the examples that physical modeling and simulation have equally their own benefit.

1.3 Introduction to the development of simulation tools

The earlier development of computer simulation techniques experienced the stages such as analog simulation in 1940s, mixed analog and digital simulation in 1950s, simulation languages emerged since mid 1960s. The object oriented simulation technique was initiated in the earlier 1980s and brought new technology into system simulation. With the development and popularity of digital computers, many dedicated simulation languages emerged and developed in the last 50 years, such as DYMOLA, CSMP, ACSL, SIMNON, MATLAB/Simulink, MatrixX/System Build, Modelica and so on. With the dominance of the MATLAB and Simulink tools, many of the above mentioned languages extinguished. MATLAB and Simulink have become the top selected computer language and tool in simulation applications. In this section, the development stages of simulation languages and tools will be briefly summarized, with some of the tools commented.

1.3.1 Analog and mixed computer simulation

Analog simulation is a simulation technique based on similarity principles. For instance, for different physical systems, such that the mathematical models of the temperature variations in a furnace, mechanical and speed relationship in motor drive systems, the relationship between voltage and current in electric circuits are the same, and they can all be described by first-order differential equations. Based on similarity principles, real electronic devices can be used in the simulation process. This is the stage of analog simulation.[55] Operational amplifiers are the essential components in analog simulation. With operational amplifiers, operational units such as sums and integrators can be constructed. On analog computers, these units can be connected to establish simulation system.[60] Analog simulation can be classified as block diagram based simulation, since the mathematical model of differential equations are represented by fundamental blocks such as sums and integrators.

The precisions of analog simulation are relatively low and depend heavily on the computer hardware. For linear components, the error is around 0.1 to 0.2 %, while for nonlinear ones, 0.3 to 0.5 %. For linear ODEs, the error may reach 5 %, while for nonlinear ODEs, the error may reach as high as 10 to 20 %.[55] Besides, the application field is rather narrow, and cannot be used to handle simulation problems of complicated systems.

The first block diagram based simulation language appeared in 1955, and the name was DAS (digital analog simulator).[24] The language is digital implementation of analog simulation, where the components and linking methods on analog computers are represented on digital computers, and translated into machine code, so as to perform simulation. Later, various mixed analog-digital simulation tools appeared, and a good survey was made in [11].

1.3.2 Simulation languages on digital computers

In the 1960s, great support to simulation was provided by digital computers and computer languages such as FORTRAN, ALGOL and COBOL. Different dedicated simulation programs appeared. In 1965, the first digital simulation language named MIMIC was proposed,[53] together with some subsequent languages and programs, such as the DSL language.[62] In 1967, the developers of various simulation programs were gathered by the Simulation Councils Inc. (SCi) to fully investigate the MIMIC simulation language, and worked out a new standard for continuous system simulation language, known as the CSSL standard.[66] In fact, the term "digital simulation" was proposed in contrast to "analog simulation," a more precise phrase is "numerical simulation." Therefore, the term "numerical simulation" will be used in the subsequent parts of this book.

The CSMP (continuous system modeling program) languages was proposed based on the CSSL standard, and became an influential language for continuous system.[59, 27] The entire modeling and simulation program is divided into three modules in CSMP language, that is, `INITIAL`, `DYNAMIC` and `TERMINAL`, with 45 components provided. The programming was made more standardized and simpler. The major task of CSMP language is modeling and simulation of continuous systems. Besides, digital electronic devices such as logic gates, flip flops are provided such that digital electronic circuits can also be studied. Furthermore, automatical sorting of components are made possible such that the execution orders of components are independent of their places in the source code.

Meanwhile, the other popular simulation languages and programs are CSS, ACSL,[50] ESL, μSIL[37] and SIMNON series of programs.[4]

CSMP-C simulation language developed by Professor Jihu Ma in Shenyang Institution of Automation, Academia Sinica is a representative simulation language in China. The CADCSC program[72] developed by the team and lead by Professor Jingqing Han can be used in the simulation and design of control systems.

ACSL language (Advanced Continuous Simulation Language) is a representative and widely used simulation language, proposed by Mitchell and Gauthier Associate.[50] ACSL was developed on the CSSL standard. Before the popularity of Simulink, ACSL was the top selected mainstream simulation language in control systems studies.

A program in ACSL language is required to describe the system with the proposed components. The syntaxes of ACSL is very similar to Fortran language. An example in ACSL language is given next.

Example 1.3. The well-known van der Pol equation is given by[68]

$$y''(t) + \mu(y^2(t) - 1)y'(t) + y(t) = 0,$$

with $\mu = 1$, and the known initial conditions $y(0) = 2.5$ and $y'(0) = 3$. The interested simulation range is $t \in (0, 15)$. Observe the ACSL description in ACSL language.

Solutions. State space models can be directly described in ACSL, and the current ODE given above cannot be expressed directly. Therefore, it should be converted into state space models first. For instance, select state variables $y_1(t) = y'(t)$ and $y_2(t) = y(t)$, therefore, the van der Pol equation can be converted into the following state space model as

$$\begin{cases} y_1'(t) = (1 - y_2^2(t))y_1(t) - y_2(t), \\ y_2'(t) = y_1(t). \end{cases}$$

Therefore, the following ACSL program can be written for the system.

```
PROGRAM VAN DER POL EQUATION
CINTERVAL CINT=0.01
CONSTANT  Y1C=3.0, Y2C=2.5, TSTP=15.0
   Y1=INTEG((1-Y2**2)*Y1-Y2, Y1C)
   Y2=INTEG(Y1, Y2C)
TERMT (T.GE.TSTP)
END
```

In the program, the communication step-size `CINT` is set to 0.01. The state variables are integrated by the function `INTEG()`, with the initial values described by constants `Y1C` and `Y2C`. The stop time `TSTP` for simulation is set to 15. In the program, the variable `T` is the actual time in simulation. The compiler of ACSL can be used and the link to the ACSL library can be made to create an executable file. When executing the file, a prompt `ACSL>` is provided automatically so that further commands can be made to control the simulation process under the prompt.

```
ACSL> PREPAR T, Y1, Y2
ACSL> START
ACSL> PLOT Y1, Y2
```

In the above commands, ACSL is informed to reserve the three variables `T`, `Y1` and `Y2`. Then simulation process in invoked, and finally the phase plane trajectory, that is, the relationship between `Y1` and `Y2` can be displayed.

1.3.3 Object-oriented simulation language

New methodology and ideas in system simulation are fueled by the emersion of Simulink. With the tool, simulation model can be drawn easily using graphical methods, such that the modeling and simulation study of complicated systems can easily be implemented. The development of Simulink will be further discussed later.

Another active representative object-oriented simulation simulation language is Modelica[49, 65], and originated with the collaborative work performed in European universities. The language is a physical modeling one, and still widely used in simulation.

1.3.4 The development of MATLAB and Simulink

MATLAB language provides a stimulative force in solving scientific computing problems, and especially in the study of control systems. In 1978, Professor Cleve Moler, the director of the Computer Science Department at the University of New Mexico thought that the then advanced EISPACK and LINPACK packages were not good enough in dealing with numerical linear algebra problems. He conceived and implemented an interactive computer language, and he named it MATLAB (Matrix Laboratory). At the very beginning, it was a free tool. A control engineer, Jack Little, rewrote MATLAB and developed CTRL-C language, dedicated for solving control related problems.[43]

In 1984, Jack Little and Cleve Moler cofounded a company named The MathWorks Inc., and released the first commercial version of PC-MATLAB at the IEEE Conference on Decision and Control, in Las Vegas, held at the end of 1984. The next year Pro-MATLAB for the Unix system was released by MathWorks Inc.

The debut of MATLAB language just met the period of the rapid development of state space theory in control, where a significant amount of linear algebra problems are to be solved. Therefore, attention from scholars in the control community were attracted, and a dedicated control oriented tool named Control System Toolbox was developed in MATLAB.[40] A significant impact to the control community was received and it soon became the *de facto* standard computer language in control related fields. With the sustainable development of MATLAB, its functions were made stronger and stronger. It should be claimed that, although MATLAB was initiated by the numerical mathematician, it was widely utilized and popularized by the scholars in the control community. At the very beginning, almost all the toolboxes were control related. With its benign development, more and more dedicated toolboxes in various other fields appeared, and MATLAB soon became the top selected computer language in many fields.

In 1990, a graphical modeling and simulation tool named Simulab was released by MathWorks Inc. It can be used in Pro-MATLAB running on OpenWindows on SUN

SPARC, or X-windows on Unix. Such a tool was made available for PC-MATLAB (MATLAB 3.5) in 1992, and renamed Simulink, since the original name was similar to another language Simula, and easy to be mixed. There are two meanings in Simulink—Simu and link, implying that it can be used to link systems and simulation.

Since MATLAB was very popular in the control community at that time, and Simulink modeling is very closely related to block diagram modeling of control systems, it was immediately recognized and accepted by the scholars in the control community. Then widely used tools such as ACSL were abandoned, and Simulink became the universal tool in modeling and simulation.

In 1996, the released MATLAB 5.0 supported more data types, such as multidimensional arrays, structures, cell arrays, classes and objects. Windows programming was fully supported, such that the facilities in MATLAB programming was made more powerful.

Thirty years have passed since the first debut of Simulink. Various versions of Simulink have been released, and some are the variations in the interface. A great amount of blocksets and APPs have been released. Two important variations are considered as milestones and should be recommended:

In 1997, Stateflow was introduced, supporting finite-state machine (FSM) modeling, which is an extension of Simulink, applicable for the event-driven hybrid systems.

In MATLAB R2008a, the concept of multidomain physical modeling was proposed, and a modeling tool named Simscape was released, allowing physical modeling of mechanical, electrical, electronic and control systems in a building block manner. These facilities promoted the Simulink capabilities in engineering systems.

Simulink version 10.0 was released with MATLAB R2019b, with a brand-new appearance, and it made modeling and simulation even easier, while some of the earlier versions are still in use. To balance the needs of users of different versions, concentrations were made in this book upon new versions, where the necessary utilization of old versions are also introduced.

1.4 Main structure and comments of the book

The main structure and contents of the book is briefly introduced in this section to provide study guidelines for the readers. Besides, illustrations are made for the Simulink setting and typesetting of the book.

1.4.1 Main structure of the book

In the many factors in system simulation technology, algorithms and computer program design capabilities are the most important ones. It determines directly whether the prespecified problem can be solved correctly or not. The simulation problem solutions based on the most powerful and reliable computer tools such as MATLAB and Simulink are the highlights of this book.

In Chapter 1, this chapter, system simulation techniques are classified and the concepts of mathematical and physical modeling are illustrated. The development stages in simulation tools are reviewed, and in particular, the progress and major facilities in MATLAB and Simulink are presented.

In Chapter 2, the ways to initiate Simulink environments are illustrated. Then concentrations are made on the main Simulink interface and the commonly used blocks provided in Simulink are addressed. Simple demonstrations are provided for Simulink modeling, such that a good foundation can be established for the readers to further use Simulink modeling and simulation capabilities.

In Chapter 3, the creation and decoration methods for Simulink models are demonstrated, and parameter dialog boxes and controls are illustrated. In this chapter, two main-stream versions of Simulink interface are presented, including the use of toolbars and menu systems. Finally, the use of the tools such as model browser are demonstrated.

In Chapter 4, the modeling and solutions of ordinary differential equations are addressed. The modeling rules and methods in Simulink are illustrated, and the template design method for Simulink is presented. The modeling and simulation methods for various ODEs are presented, including first-order implicit ODEs, high-order explicit ODE sets, implicit ODEs, delay differential equations, switched ODEs and ODEs driven by stochastic processes. Finally, the modeling and solution of difference equations are presented.

In Chapter 5, the modeling and simulation methods for various control systems are illustrated. The linear continuous system modeling for linear systems are addressed first, for multivariable systems, singular systems and transfer functions with nonzero initial conditions. The modeling of discrete systems are also addressed. Then the modeling of static nonlinearities are presented. The simulation mechanism in Simulink such as automatic sorting rules are demonstrated. Approximate analysis methods such as linearization, Padé approximation, concept and elimination of algebraic loops are illustrated.

In Chapter 6, the concepts and construction of subsystems are illustrated. Subsystem extraction and various of special subsystems such as enable and triggered ones are presented. Subsystem masking techniques are presented with MATLAB interface and MATLAB commands. Command-line driven simulation processes and model parameter extraction are demonstrated.

In Chapter 7, programming implementation of Simulink blocks are addressed. M-function block is demonstrated first with restrictions pointed out. The concepts of system functions are presented and concentrations are made on structure and programming of S-functions. Examples are given to show how to design reusable blocks.

Also, with active disturbance rejection control as the example, S-function modeling and applications for systems with complicated algorithms are demonstrated.

In Chapter 8, efforts are proposed on how to solve some application problems with the combined simulation capabilities in Simulink and numerical optimization capabilities provided in MATLAB. Two categories of problems are explored—one is boundary value problem solutions for ODEs, the other is optimal controller design. Different problems are converted into numerical optimization problems and solved directly. In solving ODEs, boundary value problems in delay differential equations and Caputo differential equations are explored. A guideline in APP design is demonstrated, and based on it, the development of two optimal controller design APPs are addressed.

In Chapter 9, multimedia information processing with Simulink is illustrated. Audio information processing is presented first, including filter and dynamic range control. Audio activity detection is also illustrated. Video information processing methods are also discussed, such as color space conversion, Gamma correction, histogram equalization, morphological methods and edge detection.

In Chapter 10, the concepts of multidomain physical modeling with Simulink are presented, and the Simscape tool is briefly discussed. With the tools provided in Simscape, the modeling in electrical, electronic and mechanical systems are illustrated. Low-level mathematical modeling of the systems are no longer necessary with the methods. The building block method is introduced instead to create simulation models for engineering systems.

In Chapter 11, modeling and simulation methods for nonengineering systems are discussed. The pharmacokinetics and pharmacodynamic modeling of anaesthesia are studied. Then simulation problems of discrete event systems are explored with the tools such as Stateflow and SimEvent. A queueing system simulation example is created and demonstrated.

1.4.2 Comments on the typesetting

A significant amount of blocks are provided in Simulink. The block icons in the groups are adjusted manually such that the icons are arranged more reasonably. Also, in the screenshots of some dialog boxes, highlights are made to better illustrate the use of Simulink interfaces.

The results obtained by Simulink are directly related to the setting of Simulink environment. If you find the results different from the one illustrated in the book, the template setting should be noticed. For instance, the command `plot(tout,yout)` may yield an error message under the default template setting in Simulink. Therefore, the template should be modified accordingly to the one illustrated in Figure 4.5. Details in template modification may refer related contents in the book.

1.5 Exercises

1.1 The mathematical model of van der Pol equation is $y''(t)+\mu(y^2(t)-1)y'(t)+y(t) = 0$, and the initial values are $y(0) = 1$ and $y'(0) = -1$. Draw the Simulink model, if you had already some experience using Simulink.

1.2 PID controllers are the most used controller types in control engineering practice. The structure of PID control systems is shown in Figure 1.3. If you had already some experience using Simulink, set up the Simulink model and draw the unit step response of the closed-loop system.

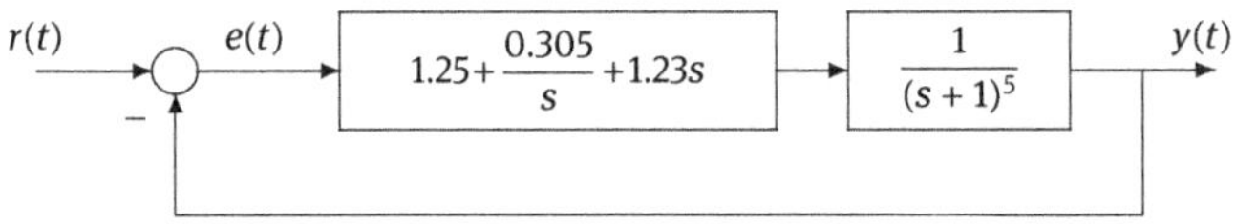

Figure 1.3: PID control system structure.

1.3 A Simulink model is shown in Figure 1.4, with key signals labeled. Write out the corresponding differential equation model for the system.

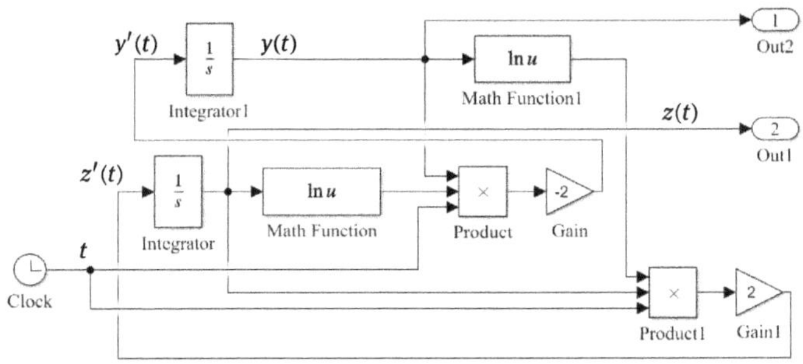

Figure 1.4: Simulink model of a differential equation.

2 Commonly used blocks in Simulink

Simulink is the block diagram based modeling and simulation tool provided in MATLAB. It is also an important tool in solving simulation problems. The major difference between MATLAB simulation and Simulink simulation is that command-line driven simulation is carried out in MATLAB, and generally loop structures can be used to complete simulation process, step-by-step. While in Simulink, the users need to draw the entire system with block diagrams. Then the mechanism in Simulink can be used to complete the simulation process. If one is familiar with MATLAB programming, this does not imply that he is automatically familiar with the Simulink simulation mechanism. Extensive study and practice are needed to use Simulink to complete his work. It is useful for the reader to get familiar with the execution mechanism and modeling methods so as to better solve Simulink modeling and simulation problems.

Compared with traditional command-line driven problem solutions in MATLAB, Simulink interface and utilization methods may depend heavily on the versions of MATLAB. Under different versions, the appearance and facilities of Simulink may be different. The Version 10.0 of MATLAB is mainly used in this book, while meanwhile the use of earlier versions is also presented.

In Section 2.1, the methods to invoke the Simulink environment are introduced, and the main Simulink interface is addressed. There are many ways to invoke the Simulink environment, and display major block groups. A significant amount of blocks and subgroups are provided, and these blocks can be used to build simulation models directly. Then a computer can be used to investigate the behaviors of the complicated systems. In Sections 2.2 to 2.5, the groups are further classified and presented as input and output groups, system groups, mathematical operation groups, signal manipulation groups and subsystem groups. The commonly used blocks among them are discussed. Of course, the introduction here is rather concise, and the target is to let readers to have a first look at the existing Simulink blocks. In the subsequent sections and chapters, more detailed information are provided to some of them. In Section 2.6, some of the existing toolboxes and blocksets are also briefly illustrated.

2.1 Invoke the Simulink environment

To use Simulink for modeling and simulation tasks, the Simulink environment must be invoked first. Two major methods to invoke Simulink are presented here, and the use of Simulink interface and utilization methods are introduced for main-stream versions.

https://doi.org/10.1515/9783110734959-002

2.1.1 Invoke Simulink with interface

Start MATLAB first. The MATLAB toolbar in the main command window is shown in Figure 2.1. Click the **Simulink** button (circled in Figure 2.1), and the Simulink environment can be invoked directly. The main Simulink window is opened, as shown in Figure 2.2, where the main buttons at the lower right portion of the window can be

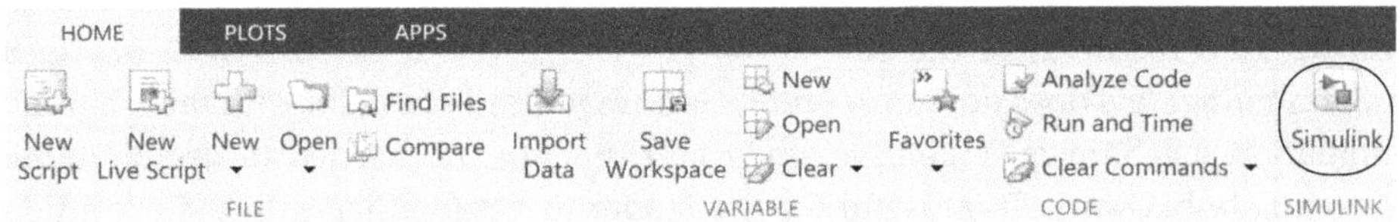

Figure 2.1: MATLAB window toolbar.

Figure 2.2: Initial Simulink interface.

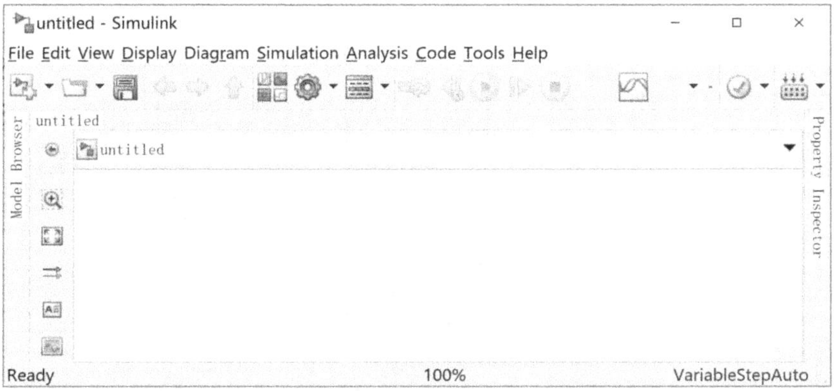

Figure 2.3: The blank model window in MATLAB R2019a or earlier ones.

clicked. The list on the left are the Simulink names in the current MATLAB folder. The suffixes of Simulink model names can be `mdl` and `slx`, with the former, ASCII code ones, usually saved in old versions of Simulink, while the latter, binary files saved by recent versions. These files are usually not readable.

The main region in the Simulink window is the Simulink modeling zone. The column labeled **My Templates** allow the users to design his own Simulink template. For a newly installed Simulink environment, this portion is blank. Details of user designed templates will be presented later.

The lower-right region of the Simulink main window contains icons **Blank Model**, **Blank Library** and so on. The two icons are two important options. If one wants to create a new Simulink model, the icon **Blank Model** should be clicked first. A blank model window in Figure 2.3 will be opened.

The form of the model window may be different in different versions of MATLAB. The one shown in the figure is the window opened in earlier versions of MATLAB. If MATLAB R2019b (with Simulink upgraded to version 10.0) or later versions are used, a blank model window is opened. The menu system is no longer displayed. Instead, a toolbar shown in Figure 2.4 is displayed. Therefore, when different versions are used, please pay attention to the differences in using Simulink environment.

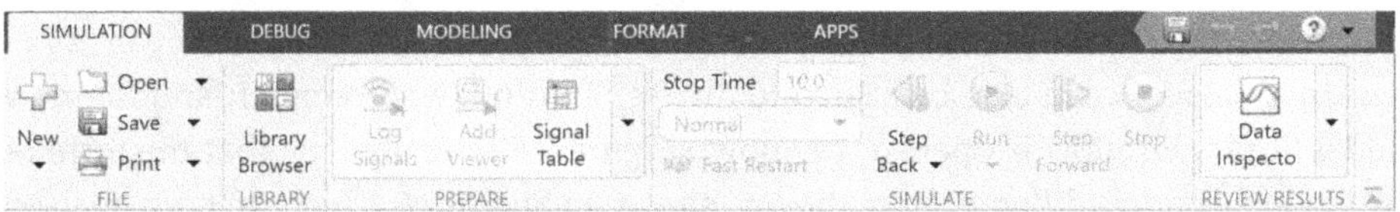

Figure 2.4: Toolbar in version R2019b or later ones.

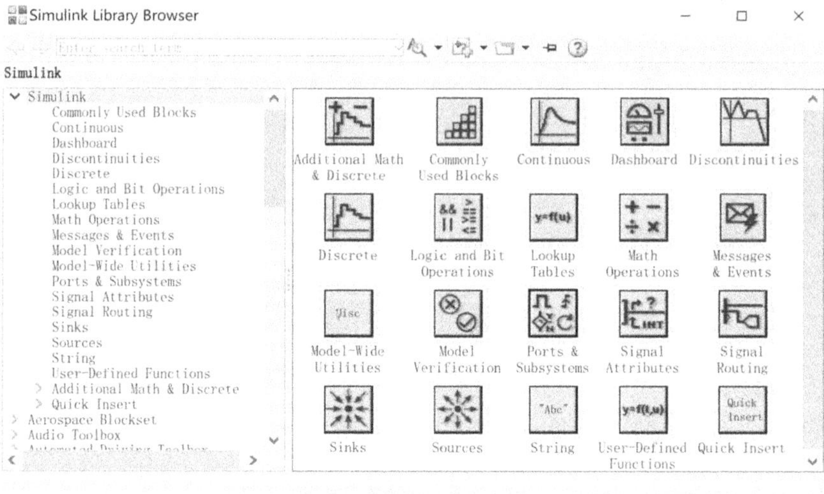

Figure 2.5: Simulink model library (one form).

Click the button in the Simulink model window in Figure 2.3 or click **Library Browser** button in the toolbar shown in Figure 2.4, and the Simulink model library shown in Figure 2.5 is opened. All the blocks in Simulink can be directly visited from the model library. Click the icon in the model library, and a blank Simulink model window is opened.

In fact, with the "**New→Simulink Model**" menu in the main MATLAB window shown in Figure 2.1, the blank Simulink model window can also be opened.

The use of menu system and toolbar will be further illustrated in later sections, and details are not given here.

With the blank model window, the modeling and simulation processes with Simulink can be initiated. Details in system modeling and simulation will be further presented later.

2.1.2 Invoke Simulink with MATLAB commands

Apart from the invoke method discussed above, the MATLAB command `open_system('simulink')` can also be issued to invoke Simulink environment, and the new model library window is opened, as shown Figure 2.6. It is obvious that the form of the model library is different from the one shown in Figure 2.5. They can both be used in Simulink modeling.

It seems that the model library opened in this way is more concise, and easy for teaching and demonstration. Therefore, this type of model library will be used later in this book.

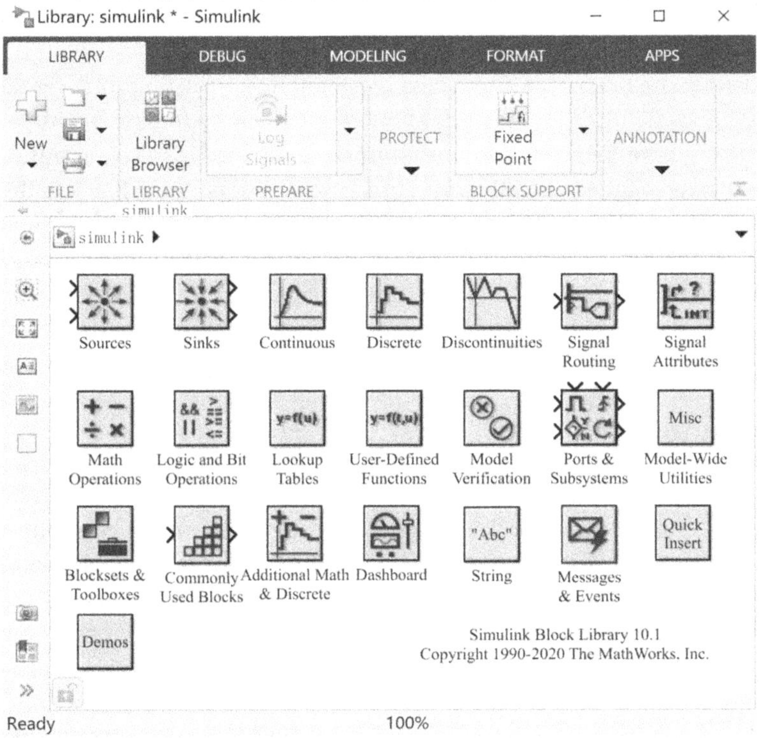

Figure 2.6: Simulink model library (another form).

In Figure 2.6, each icon in the model library corresponds to a lower-level block group. The block groups provided in Simulink include Sources (for input signals), Sinks (for output signals), Continuous, Discrete, Discontinuities, Signal Routing, Signal Attributes, Math Operations, Logic and Bit Operations, Lookup Table, User-Defined Functions, Model Verification, Ports & Subsystems and so on. For different versions, the contents in each group may be different. Here, the groups in MATLAB R2020a are illustrated. The readers of other versions have to find their own blocks themselves, if necessary.

The groups named Dashboard, Demos and Messages & Events are also important. The blocks in them may also be used directly in real modeling processes. It is recommended that the reader may execute the demo models in their own area of interest, and experience the modeling and simulation methods faster.

Blocksets & Toolboxes group provides various toolboxes and Simulink blocks. Details will be presented later.

Another way to invoke Simulink is to issue command `simulink`. The action is the same as the invoke method presented in Section 2.1.1.

2.1.3 Signals and ports

In Simulink modeling, the blocks are connected by signal lines. The Simulink signal lines are oriented, and the direction of signal flow is indicated by the arrows of the lines. Each block may have its input and output ports, and more than one input or output ports are allowed. Signals usually travel from the output port of a block to the input port of another one. The signal lines can be scalar on vector ones. This is quite similar to the triplex cable in daily life, where three scalar wires are grouped together to form one vectorized cable. More than one scalar signal can be transmitted within such a signal line.

2.2 Input and output groups

Double click an icon in the model library, the corresponding group window can be opened. A group may contain some block icons, and even icons for lower-level groups. A significant amount of blocks are provided in Simulink, and it is not possible to address all of them within one book. In this book, classifications of the groups are made, and the commonly groups are further divided into input/output groups, system groups, mathematical operation groups, signal manipulation groups and so on. The commonly used groups and blocks are presented in this and the subsequent sections.

In this section, concentrations are made on input and output groups. Since there are so many blocks in each group, it is not possible to present them all; only the commonly used ones are addressed here, and the users may have to refer to the manuals[58] of other documents for the details of the blocks.

2.2.1 Input signal group

Double click the Sources group icon, and the signal source group is opened, as shown in Figure 2.7. For the convenience of type-setting, the layout of the blocks in this and subsequent groups are adjusted. The commonly used blocks in the source group are:

(1) Constant block. It can be used to generate constant input signals. Double click the block icon, a parameter dialog box shown in Figure 2.8 is shown where numeric values can be filled into the Constant Value edit box. If a numeric vector is specified, vector signal with many constants are generated. The edit box Sample time is used to accept simple time values. If 0 or the default `inf` are specified, continuous signals will be generated.

Normally many blocks may have many tag panes in the parameter dialog box, with the default one labeled as Main, where the most important parameters are provided. The block parameters can be specified in the dialog box. Apart from these parameters, some additional parameters are also allowed. For instance, the pane named Signal

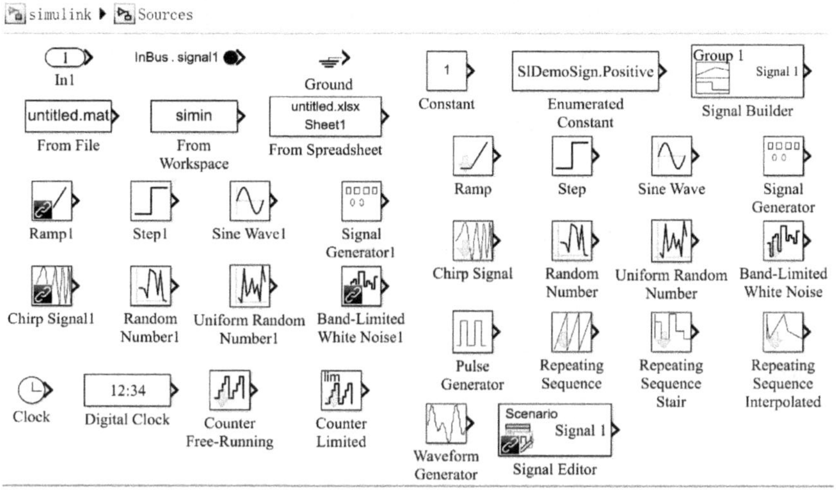

Figure 2.7: Sources group.

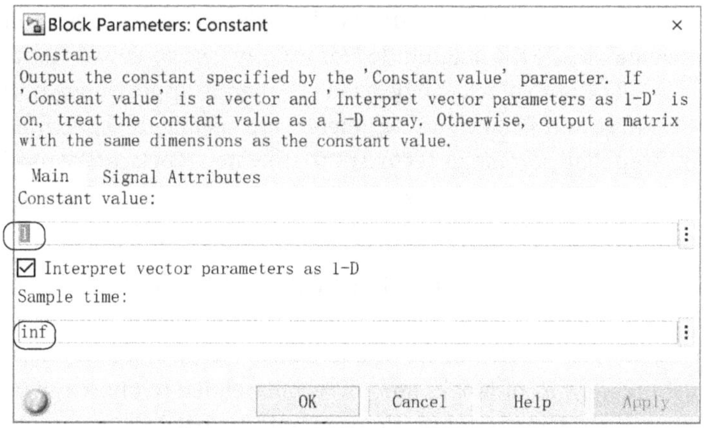

Figure 2.8: Parameter dialog box of Constant block.

Attributes and other panes can be assessed, so that the related parameters can be modified, if necessary.

(2) **Clock block.** It can be used to extract current time t. Double click the block and a parameter dialog box also appears. Due to the page budgets of the book, if not necessary, the screenshot of the parameter dialog box is not given. In the **Display time** check box in the dialog box, and it determines whether the display of time is allowed.

(3) **Step block.** It is used to generate step input signals. The waveform of step signal is shown in Figure 2.9. Several parameters such as **Step time**, **Initial value** and **Final**

value can all be modified directly in the dialog box. Besides, the sample time can also be specified. For discrete signals, it can be modified to an actual nonzero constant. In normal simulation processes, the step time is always assigned as 0, rather than the default 1 here. It should be noticed in real applications.

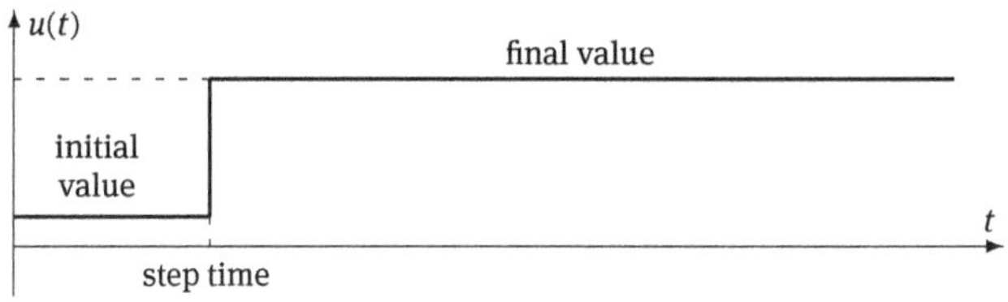

Figure 2.9: Illustration of step input signal.

Example 2.1. Use Simulink to generate a unit impulse signal.

Solutions. Impulse signals cannot be generated by any existing Simulink blocks, nor such signals in the real world. It only exists in theory. The width δ of the ideal unit impulse signal is infinitesimal, while the magnitude M is infinity. The so-called "unit" means that the area $M\delta$ is 1. The Step block can be used to generate approximate unit impulse signals. If Initial value and Final value parameters are specified respectively as 10000 and 0, with the Step time parameter set to 0.0001, the Step block can be used to generate approximately "unit impulse signals."

(4) Commonly used functions as inputs. These include Ramp block (with mathematical form of t), Pulse block, Sine Wave block, Chirp block (with mathematical form $\sin t^2$) and so on.

(5) Random Number generator block is used to generate pseudorandom numbers. For continuous systems, these types of blocks are not recommended to approximate white noise signals. Block Band-Limited White Noise should be used instead, with fixed-step simulation algorithms.

(6) Repeating Sequence block is used to generate periodic signals. The two vectors, Time values and Output values, should be used to describe the sample pairs of the signal. Linear interpolation is used to generate signals within each period. In the subsequent periods, the periodic signals can be generated. Note that the Time values vector should be a strictly increasing vector; otherwise, error messages will be generated. There are two variations of the block: the Repeating Sequence Stair and Repeating Sequence Interpolated blocks.

(7) In1 block is for the import port. If many such blocks are used in one model, they can be numbered automatically. The use of such blocks will be demonstrated later.

(8) Besides, From Workspace, From File and From Spreadsheet blocks are also provided to use MATLAB workspace variables, MATLAB data files or Microsoft Excel

files as input signals in Simulink. The use of these blocks are quite simple and straightforward, and they are not demonstrated here.

Example 2.2. Generate the periodic signal shown in Figure 2.10.

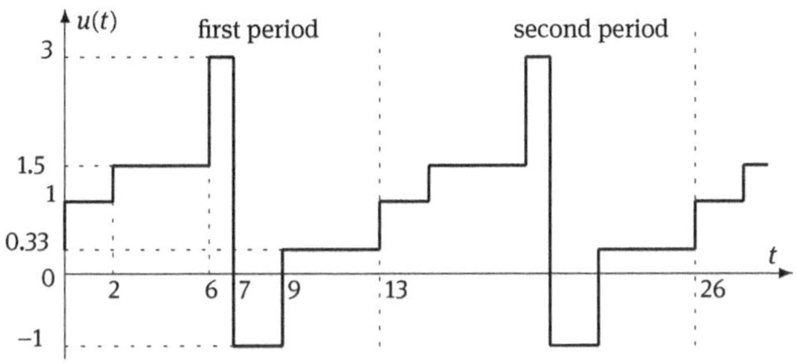

Figure 2.10: Illustration of periodic signals.

Solutions. It is, of course, a periodic signal in stairs. It is quite natural to try the Repeating Sequence Stair block to generate the expected signal directly. Unfortunately, this block only applies to the case where time is equally spaced. Therefore, the Repeating Sequence block should be used instead to generate the expected signal in stairs. For this specific problem, the period of the signal is $T = 13$ seconds. In the first period, the turning points are $(0, 0.33)$, $(0 + \epsilon, 1)$, $(2, 1)$, $(2 + \epsilon, 1.5)$, $(6, 1.5)$, $(6 + \epsilon, 3)$, $(7, 3)$, $(7 + \epsilon, -1)$, $(9, -1)$, $(9 + \epsilon, 0.33)$ and $(13, 0.33)$, where ϵ should be selected as a very small quantity, such as $\epsilon = 10^{-5}$. Therefore, the expected periodic stairs signal can be generated. The following commands can be used to construct the two vectors:

```
>> ee=1e-5;
   t=[0 ee 2 2+ee 6 6+ee 7 7+ee 9 9+ee 13];
   u=[0.33 1 1 1.5 1.5 3 3 -1 -1 0.33 0.33]; plot(t,u)
```

A simulation model shown in Figure 2.11(a) can be established. In the other two output signal observing blocks, Scope and Out1 blocks are used. These blocks are provided in Sinks group, to be presented next.

Double click Repeating Sequence block icon, the parameter dialog box shown in Figure 2.11(b) is displayed. The horizontal and vertical vectors should be specified accordingly in the dialog box. Theoretically, the expected periodic stairs signal can be generated.

To better illustrate modeling process, the model names are deliberately displayed. The shortcut menu item Format → Show Block Name → On can be selected to enable display of the block names. If we do not want to display the names, it can be set to Auto or Off.

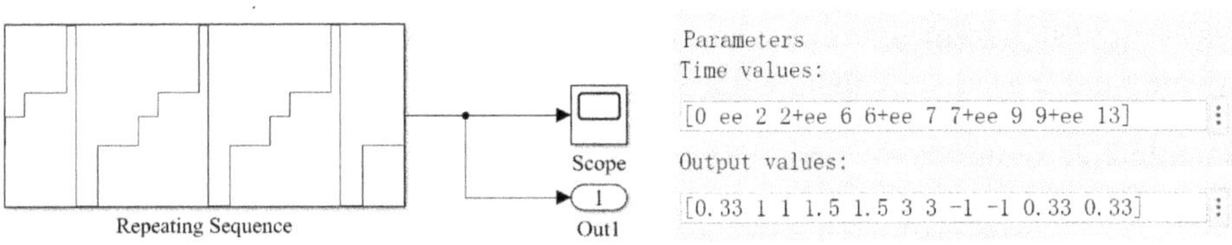

(a) simulation model (file: `c2mperiod.slx`) (b) parameter dialog box

Figure 2.11: Simulation model and parameter setting.

If we want to display or hide block names in an easier manner, the block should be clicked first. Then the three dots ... appear above the block icon. Click the dots, and several simple buttons are displayed, within which the second one, , can be used to switch the block name display status.

If simulation is carried out, even through very tough error tolerances are selected, simulation results in Figure 2.12 can be obtained. Unfortunately, the curve obtained is erroneous, since it is different from the expected one in Figure 2.10.

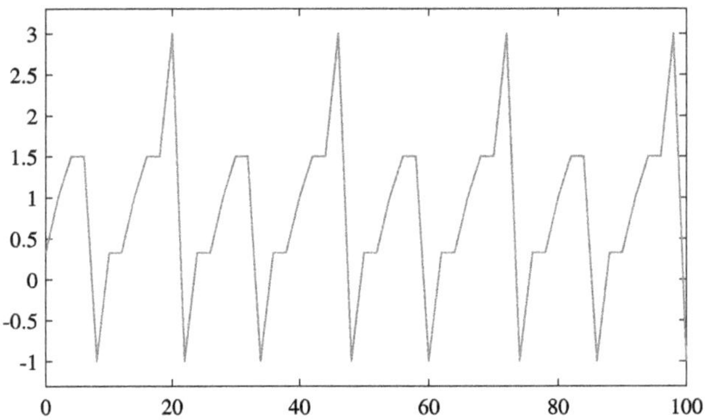

Figure 2.12: Erroneous output signal.

The reason for the error is because the output signal depends only upon time t. No matter what the step-size selected in simulation might be, the results on the selected time instances are exact. While outside the selected points, the precisions using variable step-sizes are not ensured. If there are dynamic elements in the simulation model, such as transfer functions and state space models, the default variable-step algorithms may ensure the correctness of the results. For this specific example, fixed-step simulation algorithms can be used to complete the simulation process, if the step-size selected is small enough. This information will be further discussed in detail later.

2.2.2 Signal builder interface

Signal Builder block provides a visual interface to construct desired waveform of the input signals. The block can be copied into the model window first. Then double click the model, and the design interface shown in Figure 2.13 is displayed. Generally speaking, the interface is suitable to construct piecewise linear input waveforms.

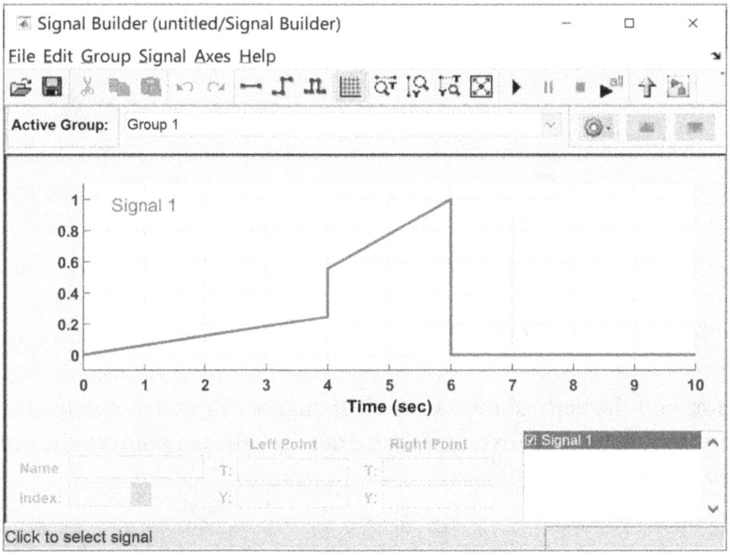

Figure 2.13: Signal Builder interface.

The prototype waveform of the input signal is shown in Figure 2.13. If one wants to insert new turning points on the current curve, then hold the **Shift** key, and click a point on the current curve, and a new turning point is specified. One can drag the point independently to any desired position, so as to construct the expected piecewise linear signal model.

Example 2.3. Construct a piecewise linear signal with the **Signal Builder** block.

Solutions. Open a blank model window, and copy the **Signal Builder** block into the window. Double click the block, and the signal builder interface is opened. Hold the **Shift** key, and click a point on the original waveform, and a new turning point is created, as shown in Figure 2.14.

The turning point can be operated independently such that the waveform in Figure 2.15 can be generated (file: `c2msgb.slx`).

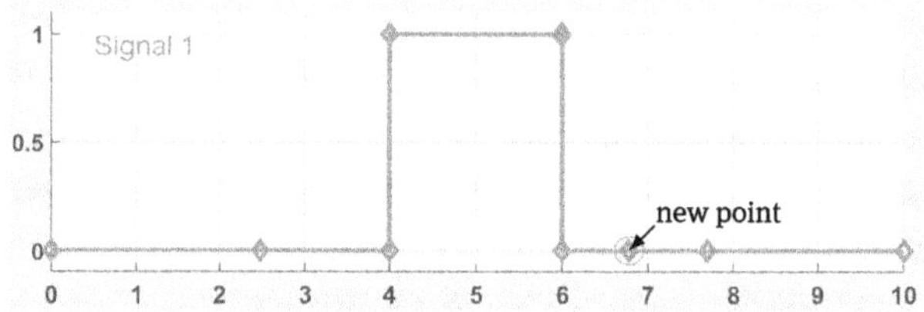

Figure 2.14: Insert a new turning point.

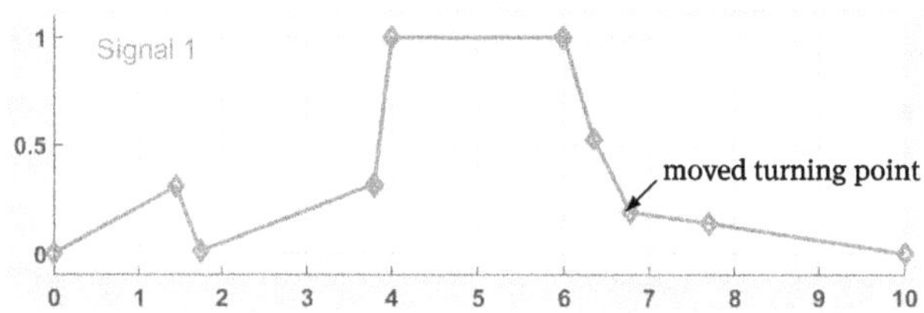

Figure 2.15: Drag the turning point to modify the waveform.

Note that the waveform on the vertical lines cannot be simply dragged to change the shapes. The **Shift** key should be held first, and then drag the selected point to change the shape of the waveform.

2.2.3 Output sinks group

Generally, there should be output blocks in any simulation model to display simulation results in some way. Double click the **Sinks** group icon, and the output sinks group can be opened, as shown in Figure 2.16.

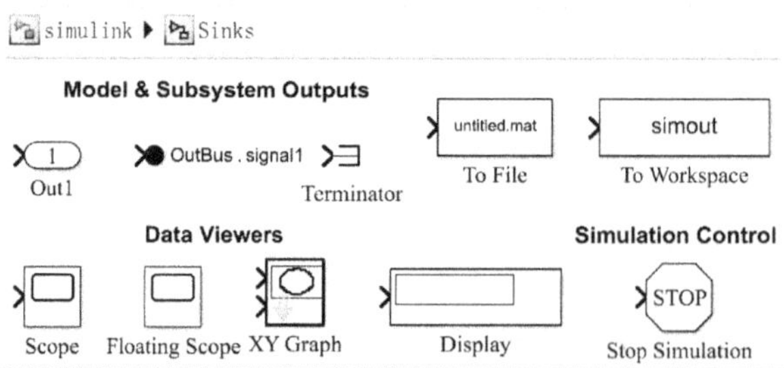

Figure 2.16: Sinks group.

The commonly used blocks in the group are:

(1) **Out1** block, which is the output port, with actions similar to the input ports. The use and facilities will be discussed later.

(2) **Scope** block, which should be connected directly to the signal lines to be observed, so as to display the simulation results in real-time. Besides, variations of the scope block can also be used, such as **Floating Scope** and **XY Graph** blocks. The former one can be placed above a signal line, and the latter can be used to draw the relationship of different signals, such as in phase plane trajectories.

(3) **Terminator** block. In real modeling, if some of the signals are not connected to any other models, they are in a suspending status. Warning messages may appear in the simulation process. To avoid such warnings, the suspended signals should be connected to the terminator blocks.

(4) **Stop Simulation** block. If the input to the block is true, the simulation process can be terminated automatically. This block is rather useful in certain simulation models.

(5) **To File** and **To Workspace** blocks can be used to save the simulation results to data files and MATLAB workspace, respectively.

Example 2.4. Use scope block to display the waveform generated in Example 2.2.

Solutions. Input the two vectors $\boldsymbol{t}$ and $\boldsymbol{u}$ into the MATLAB workspace first, with the following commands. The Simulink model in Figure 2.11(a) can be opened, and the simulation algorithm can be set to fixed-step one, with step-size set to 0.0001. Set also the stop time to 100. Details of the setting will be illustrated later. Save the model into a new name of `c2fperiod1.slx`. To simulate the system, the scope display shown in Figure 2.17 can be obtained. It can be seen that the waveform obtained is the same as the expected one in Figure 2.10.

```
>> ee=1e-5;
   t=[0 ee 2 2+ee 6 6+ee 7 7+ee 9 9+ee 13];
   u=[0.33 1 1 1.5 1.5 3 3 -1 -1 0.33 0.33];
```

Click the ⚙ icon in the scope window, a further dialog box shown in Figure 2.18 is given, which allows the users to set certain properties of the scope block. For instance, the check box **Open at simulation start** can be selected such that each time when simulation process is invoked, and the scope block is opened automatically to display in real-time the simulation results. If the **Logging** pane is clicked, the results are allowed to be returned in different ways.

In the Simulink model, an extra output port signal is also connected to the output signal. This block is left for further use. It also implies that the signals can be connected to several output blocks simultaneously.

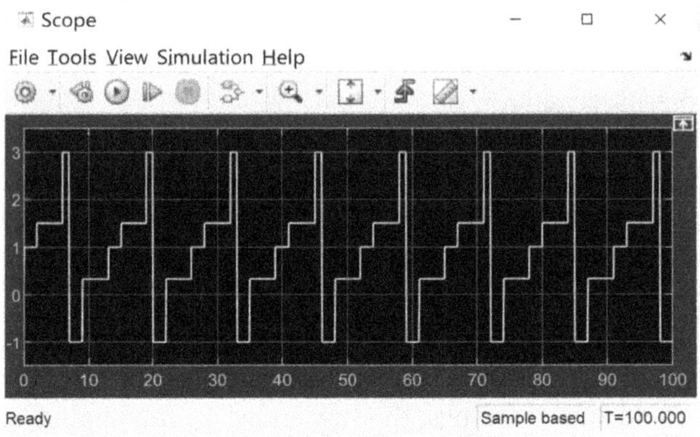

Figure 2.17: Scope display of simulation results.

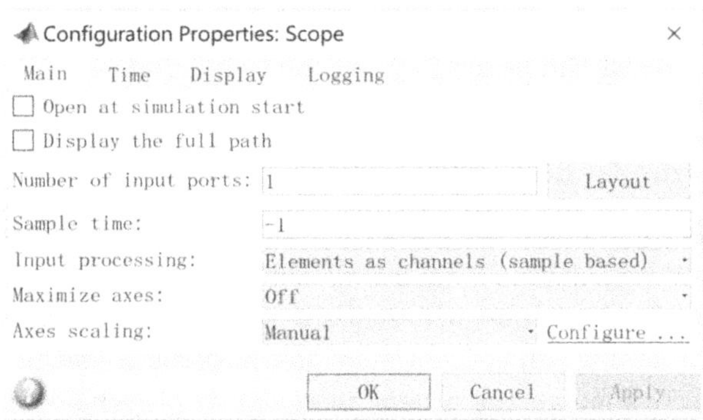

Figure 2.18: Property setting in the scope block.

Apart from genuine scope blocks, "virtual" scopes can be applied to certain interested signals. Specifically, click the right mouse button on a certain signal line, and the shortcut menu shown in Figure 2.19 is displayed. Expand the menu and select the Scope item, and an icon appears above the selected signal line. In the simulation process, double click the icon, and the scope display of the signal is shown.

2.2.4 Dashboard group

In ordinary applications, the simulation results can be visualized with curves. For some engineers, they may prefer to visualize the signals by gauges or indicators. A Dashboard group is provided in Simulink, with elements shown in Figure 2.20. The dashboard blocks can be connected to the interested signals, and the simulation results can be displayed directly.

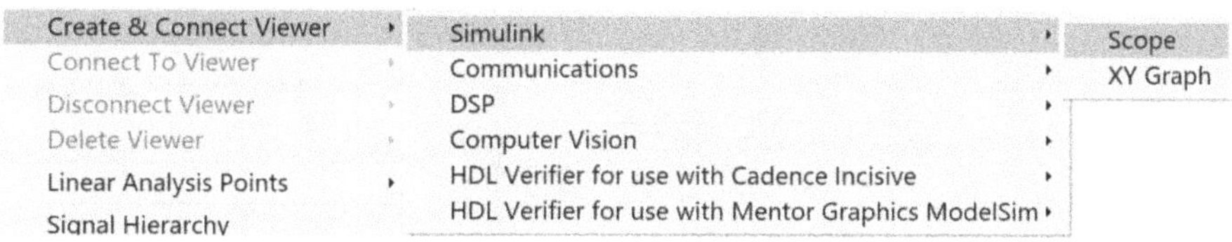

Figure 2.19: Part of the shortcut menu.

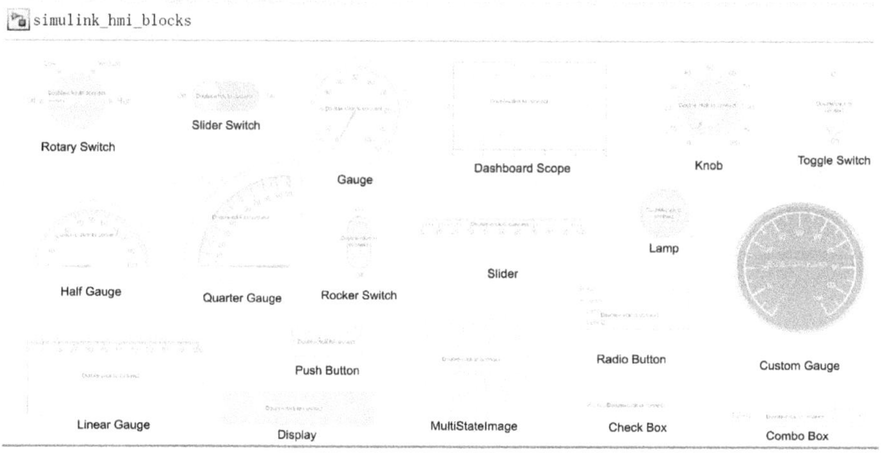

Figure 2.20: Dashboard group.

It can be seen from the group that there are three types of blocks: gauges, switches and dialog box controls. These controls are useful in graphical display of signals, or in the design of dialog boxes.

Note that a label of Double-click to connect is marked beside all of the components. If one wants to visualize certain signals, they should be first connected to the interested signal lines. The specific method of block connection will be demonstrated through examples.

Example 2.5. Use a gauge to display the signal generated in Example 2.2.

Solutions. The scope block can be deleted first, and then copy Gauge block to the model window, as shown in Figure 2.21(a). It can be seen that above the gauge block, an icon is shown. Click the icon, and move it to Repeating Sequence block, the prompt shown in Figure 2.21(b) is given. Select the Repeating Sequence:1 item, where :1 indicate the first port of the block, the associative relationship between the signal and the gauge is set up. Simulating the system, the simulation results are shown in the gauge block.

In real applications, apart from the associative setting, the scales of the gauge block should also be adjusted. For instance, if the signal varies in the interval $(-1, 3)$,

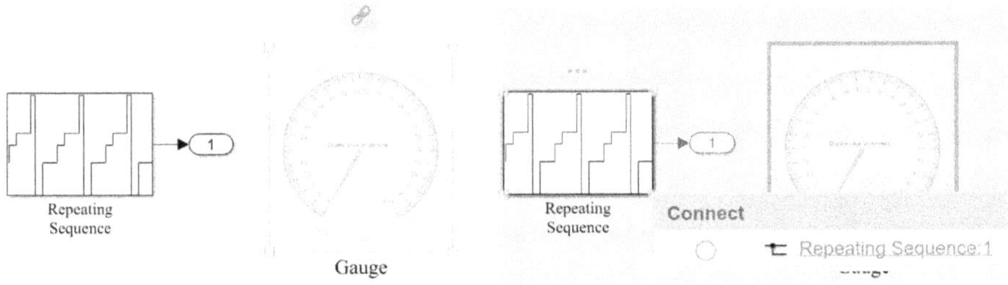

(a) model (file: `c2mperiod2a.slx`) (b) signal association

Figure 2.21: Modeling and gauge connections.

and the default scale is (0, 100), it should be redefined. Double click the gauge block, and the dialog box shown in Figure 2.22 is displayed. Set Minimum and Maximum edit boxes with −1.2 and 3.2, and then directly initiate the simulation process, and the variations of the stairs signal can be visualized on the gauge.

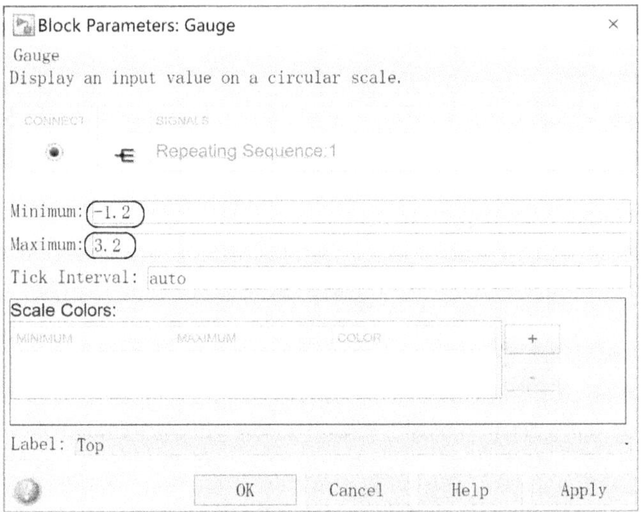

Figure 2.22: Parameter dialog box for the gauge block.

2.3 System groups

The mathematical models of systems are usually mathematical expressions describing the dynamic behaviors of the systems. The models are usually used to analyze and design for the systems. Based on the mathematical tools in describing the models, the systems can also be classified as continuous, discrete and hybrid systems. Also, they can be further classified as linear and nonlinear, time invariant and time-varying, lumped and distributed parameter, deterministic and stochastic systems. Simulink is suitable in describing lumped parameter systems. The corresponding mathematical models are ordinary differential equations and difference equations.

There are three related groups for describing systems. They are Continuous, Discrete and Discontinuities groups, suitable for describing linear continuous and discrete systems, with simple nonlinearities. It can be seen that the blocks in the three groups are far from adequate for complicated modeling. Many more blocks are needed and will be illustrated later.

2.3.1 Linear continuous group

The mathematical foundation of continuous systems is differential equation. Laplace transforms are often used to convert them into complex domain expressions. Double click the Continuous group icon, and the linear continuous group window is opened, as shown in Figure 2.23. The commonly used blocks are:

(1) Some Integrator blocks, which are the essential elements in continuous dynamical systems. Such blocks include Integrator, Integrator Limited, Integrator, Second-Order and Integrator, Second-Order Limited blocks.

(2) Descriptor State-Space and State-Space blocks for describing state space models of systems. These models will be further discussed later.

(3) Transfer Fcn and Zero-Pole blocks are the commonly used transfer function-type models. The mathematical form of a transfer function is given by

$$G(s) = \frac{b_1 s^m + b_2 s^{m-1} + \cdots + b_m s + b_{m+1}}{a_1 s^n + a_2 s^{n-1} + \cdots + a_n s + a_{n+1}}, \tag{2.3.1}$$

while the zero-pole model is the factorized form.

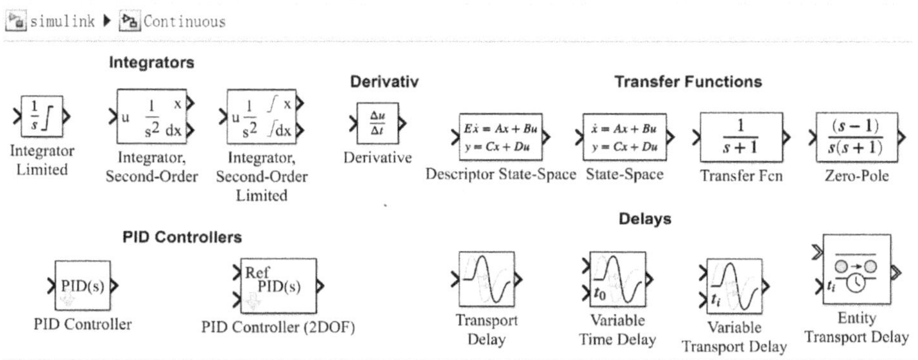

Figure 2.23: Continuous group.

(4) PID Controller and PID Controller(2DOF) blocks are commonly used PID controller blocks.

(5) Various delay blocks, such as Transport Delay and Variable Time Delay blocks and so on. If the input signal to a Transport Delay block is $u(t)$, the output signal is $u(t-T)$, where T is the delay time. The physical meaning is that the block returns the value of the input T seconds ago. If variable delay block is used, the delay time T is varying. The transfer function of the fixed delay term is e^{-Ts}.

(6) Other continuous blocks are provided, including Derivative and First-Order Hold blocks. It is known from Volume II that the accuracy and numerical stability of the numerical differentiation operation is not quite satisfactory and the former block is not recommended. The latter is in fact a discrete block.

Example 2.6. Represented in Simulink, the higher-order transfer function model

$$G(s)=\frac{2s^2+18s+40}{s^6+10s^5+40s^4+82s^3+91s^2+52s+12}.$$

Solutions. To describe a transfer function model, two coefficient vectors, $\boldsymbol{n}$ and $\boldsymbol{d}$, are needed, describing respectively the coefficients of the numerator and denominator polynomials, in the descending order of s. It is quite natural to find that $\boldsymbol{n}$ = `[2,18,40]` and $\boldsymbol{d}$ = `[1,10,40,82,91,52,12]`. Copy the Transfer Fcn block into the model window. Double click the model icon, and the parameter dialog box is opened as shown in Figure 2.24. The two vectors can be specified respectively in the Numerator coefficients and Denominator coefficients edit boxes, such that the transfer function block can be established.

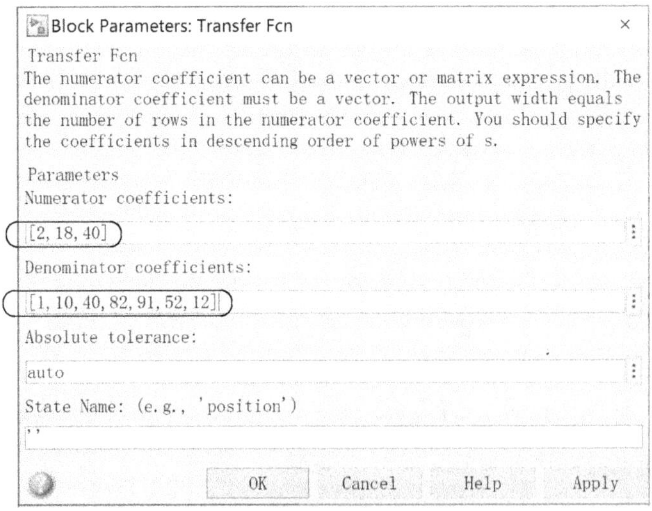

Figure 2.24: Parameter dialog box of a transfer function block.

Example 2.7. Describe the following composite transfer function model[12] in Simulink:

$$G(s) = \frac{1 + \frac{3e^{-s}}{s+1}}{s+2}.$$

Solutions. It can be seen that the complicated transfer function model cannot be described directly with the Transfer Fcn block, since part of the transfer function contains the delay term of e^{-s}, meaning the delay constant of 1. The following manual formulation is needed:

$$G(s) = \frac{1}{s+2}\left(1 + \frac{3e^{-s}}{s+1}\right).$$

It can be seen that in the brackets, the term $3e^{-s}/(s+1)$ is in parallel connection of gain 1. Since it is multiplied with $1/(s+2)$, they can be modeled with a series connection. Therefore, the Simulink model shown in 2.25 can be created directly. In the model, the delay term is modeled directly with the Transport Delay block. Double click the block, and the value of 1 can be specified in the Time delay edit box to describe the delay constant. Apart from these blocks, a Sum block is used to complete parallel connection. This block belongs to the Math group, to be presented later.

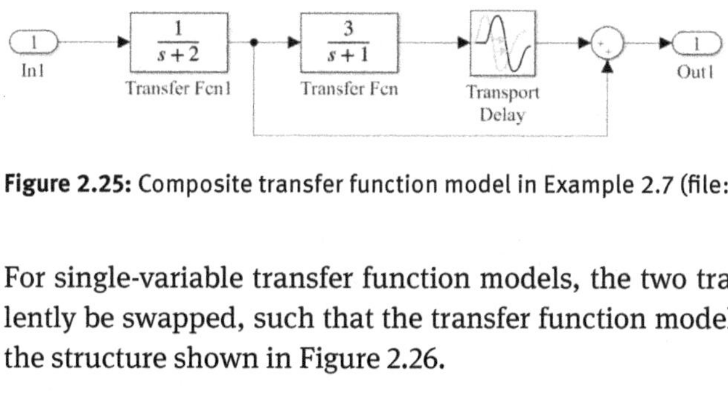

Figure 2.25: Composite transfer function model in Example 2.7 (file: `c2mtf2.slx`).

For single-variable transfer function models, the two transfer functions can equivalently be swapped, such that the transfer function model can also be represented by the structure shown in Figure 2.26.

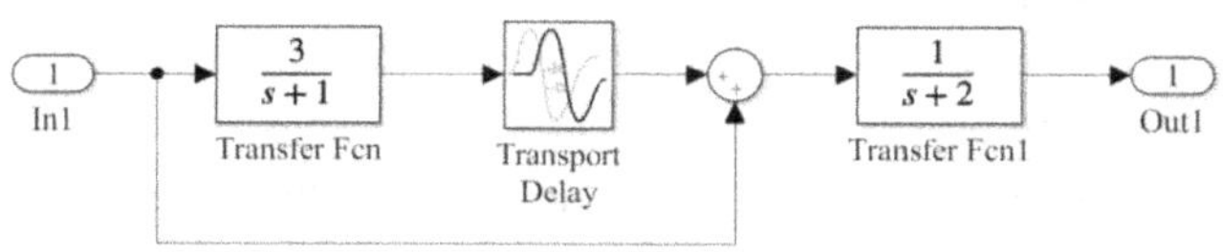

Figure 2.26: Equivalent model after swapping (file: `c2mtf2a.slx`).

2.3.2 Linear discrete group

Discrete blocks are often used in digital systems and computer control systems. The mathematical model is difference equations, and the complex domain model under z transform. Double click the Discrete group icon, and the discrete group in Figure 2.27 is displayed. The commonly used blocks are:

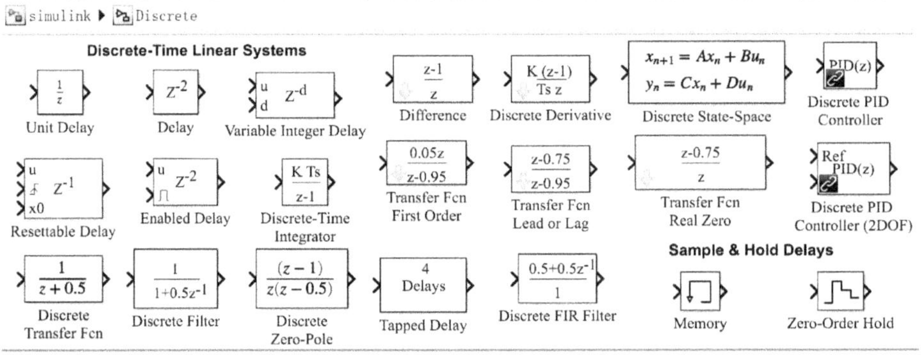

Figure 2.27: Discrete group.

(1) In contrast to the cases in continuous blocks, the discrete group also includes discrete transfer functions, state space models, discrete difference, various delay blocks and discrete PID controller blocks. The descriptions are similar to the cases in continuous systems. The mathematical form of discrete transfer function is

$$G(z) = \frac{b_1 z^m + b_2 z^{m-1} + \cdots + b_m z + b_{m+1}}{a_1 z^n + a_2 z^{n-1} + \cdots + a_n z + a_{n+1}}, \tag{2.3.2}$$

where z is the z transform operator.

Compared with their continuous counterparts, there is one more entry in the discrete parameter dialog box, that is, the sample time.

(2) Memory block. If a signal is connected to the block, the output returns the value of the signal at the last sample. The block can also be used in continuous signals, where the output returns the value of the input signal at the previous step-size.

(3) Zero-Order Hold block, with it, the value of the output signal is kept constant in the entire sample. This is commonly used in real discrete control systems.

(4) Various of delay blocks. Unit Delay is the most important block, returns the input signal at the previous sample. Delay blocks returns the values of the input several samples before. The entry in the parameter dialog box can be specified as the number of steps in delay. Variable Integer Delay block also allows to specify the number of delays by its second input signal.

2.3.3 Discontinuities group

The so-called discontinuous block means that continuous block may fail to describe its characteristics. These blocks are usually nonlinear. Therefore, a special category of blocks are provided to describe the nonlinearities. Double click the **Discontinuities** group icon, and the discontinuous group in Figure 2.28 is opened, with the commonly used blocks:

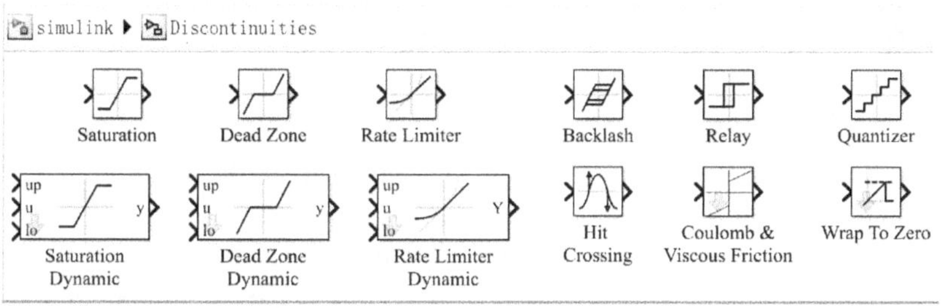

Figure 2.28: Discontinuities group.

(1) **Saturation, Dead Zone** and **Rate Limiter** blocks. The saturation and dead-zone nonlinearities are shown in Figure 2.29. The mathematical forms of these two nonlinearities are

$$y(u)=\begin{cases} u_{\rm m}, & u\leqslant u_{\rm m},\\ u, & u_{\rm m}<u\leqslant u_{\rm M},\\ u_{\rm M}, & u>u_{\rm M},\end{cases}\qquad y(u)=\begin{cases} u-u_{\rm m}, & u\leqslant u_{\rm m},\\ 0, & u_{\rm m}<u\leqslant u_{\rm M},\\ u-u_{\rm M}, & u>u_{\rm M},\end{cases}\tag{2.3.3}$$

where $u_{\rm m}$ and $u_{\rm M}$ are given constants. In the mathematical expressions, $y(u)$ is the output signal of the block and u is the input. The relationship between the input and output is a nonlinear function.

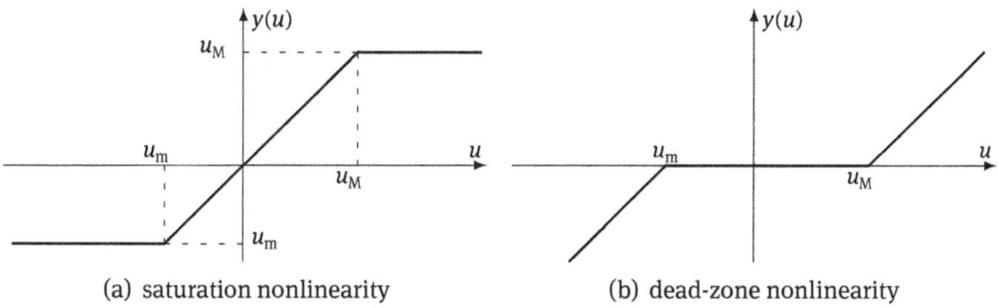

Figure 2.29: Illustrations of saturation and dead-zone nonlinearities.

In the blocks provided in Simulink, u_m and u_M also be variables, with the blocks such as **Saturation Dynamic** and **Dead Zone Dynamic**, where the variable boundaries u_m and u_M are driven by two external input signals. The **Rate Limiter** block ensures that the changing rate of the input signal is kept within a certain range. The variable boundary of the rate is also supported in the **Rate Limiter Dynamic** block.

(2) The **Backlash** block is used the relationship between the input and output signals at the instance the gear is switch from positive rotation to negative, or vice versa. A loop nonlinearity is described in the block.

(3) **Relay** block is also a commonly used loop nonlinearity.

(4) Some other commonly used nonlinearities are supported such as **Quantizer** and **Coulomb & Viscous Friction** blocks.

Example 2.8. It seems that the nonlinearity blocks provided in Figure 2.28 are quite limited. In some earlier works on nonlinear control systems, complicated nonlinearities can also be equivalently constructed by several low-level nonlinearities.[8] An example is shown in Figure 2.30, where the nonlinearity in Figure 2.30(a) can be decomposed of two lower-level nonlinearities shown in Figure 2.30(b), in series connection. Use Simulink to construct the nonlinearity.

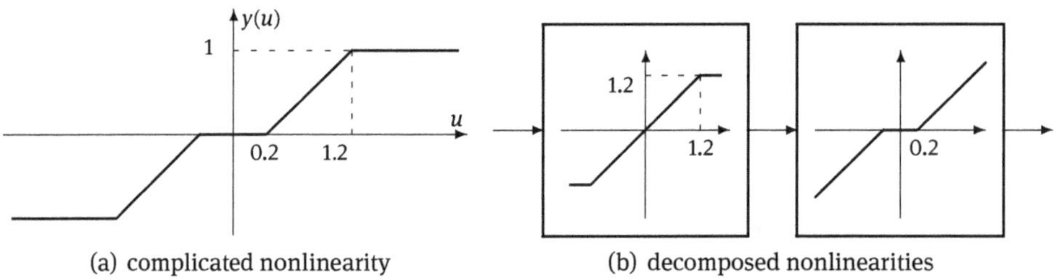

Figure 2.30: Decomposition of complicated nonlinearities.

Solutions. Based on the decomposition method in Figure 2.30, it is not hard to use two nonlinearities, **Saturation** and **Dead Zone**, to construct directly the Simulink model, shown in Figure 2.31, where the input and output are represented respectively by **In1** and **Out1** blocks. The parameters of the nonlinearities can be set with their parameter dialog boxes.

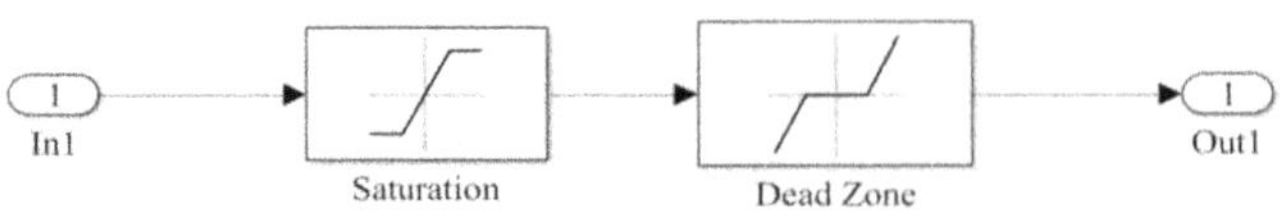

Figure 2.31: Simulink model for the nonlinearity (file: `c2mnl1.slx`).

Of course, the modeling method here is rather complicated, and cannot be validated. Later, a simpler and more straightforward method is presented to build the complicated nonlinearities.

Although this method can be used build nonlinearities, the nonlinearities handled in this way is extremely limited. Skills are usually needed. In later chapters, a universal method to construct any static nonlinearities will be presented.

2.4 Mathematical operations and manipulating groups

Mathematical operations and manipulating blocks are usually used to impose algebraic and logic operations to the signals. Also static and dynamic function evaluations of any complexity are made possible. In this section, the commonly used blocks in the three groups, Math Operations, Bit & Logic Operations and User-Defined Functions, are illustrated.

2.4.1 Mathematical operations group

Here, mathematical operations simply refer to algebraic computing and some simple transcendental functions. Double click Math Operations group icon, and the mathematical operations group is opened, as shown in Figure 2.32. The commonly used blocks in the group include:

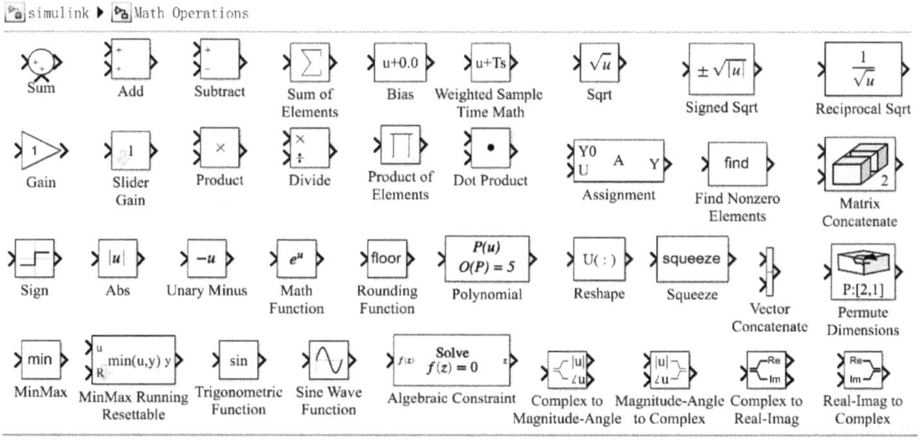

Figure 2.32: Math Operations group.

(1) **Algebraic computing blocks.** The blocks **Add** and **Sum** are used to implement addition to signals. Also **Subtract**, **Product** and **Divide** blocks can be used to carry out algebraic computing on input signals. More complicated operations are made possible with these blocks.

Example 2.9. Use Simulink blocks to implement complicated mathematical operations such as $y = u_1 \frac{u_2}{u_3} u_4 \frac{u_5}{u_6 u_7}$, where u_i are the input signals to the block.

Solutions. The computation can be implemented with the **Product** block. Feed the seven signals accordingly to the **Product** block. Double click the block, and fill in the **List of Signs** edit box the string like **/**//, where "*" means the corresponding input signal is the numerator, while "/" implies the signals are denominators. Therefore, complicated multiply and divide operations of signals can be handled in this way.

(2) **Gain** blocks. If the input signal is u, the output signal is then $y = Ku$, where K is the gain of the block. There are variations in the gain blocks, such as **Slider Gain** block, which allows the user to adjust the gain value using scroll bars. Besides, dot operations are allowed with the block.

(3) The **Math Function** block implements various simple mathematical functions, which can be assigned in the parameter dialog boxes. Double the block, and in the **Function** list box of the dialog box, available mathematical functions are provided, as shown in Figure 2.33.

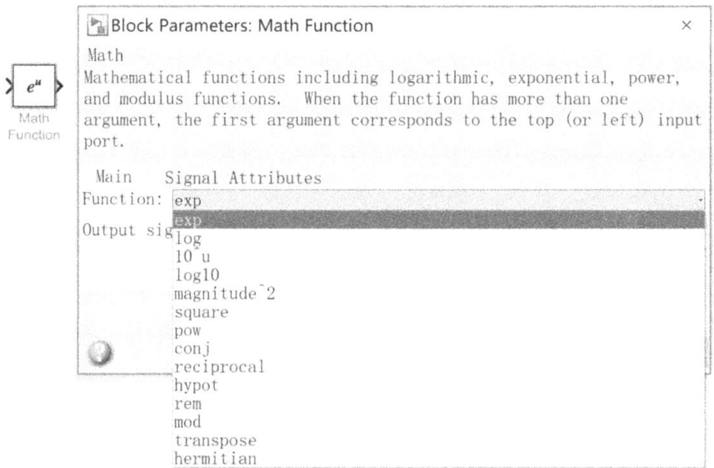

Figure 2.33: Available list of functions in Math Function dialog box.

(4) Other mathematical operation blocks. For instance, the **Abs** block returns the absolute values of the input signal. The **Trigonometric Function** block carries trigonometric operations on the input signal, and the particular operation can be assigned

in the parameter dialog box. The Sign block extracts the signs of the input signal. The Rounding Function block can be used to find integers from its input signals.

Example 2.10. The integral of time weighted absolute error (ITAE) performance index is a commonly used criterion in optimal control. Use Simulink to represent such an integral signal:

$$I(t) = \int_0^t \tau|e(\tau)|\mathrm{d}\tau, \tag{2.4.1}$$

where $e(\cdot)$ is the input signal to the block, t or τ are time and the output $I(t)$ is a function of time t.

Solutions. The time information can be generated with the Clock block in the sources group. Signal $e(t)$ can be provided by the In1 block. It is obvious that $|e(t)|$ can be obtained, by feeding $e(t)$ into the Abs block. Multiplication can be made such that $t|e(t)|$ can be generated. Feed the signal into the Integrator block, the expected ITAE integral can be obtained and the result can be linked to the Out1 block, as shown in Figure 2.34.

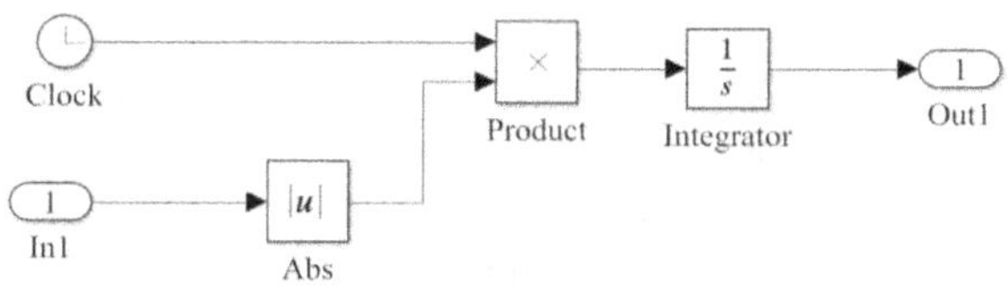

Figure 2.34: Simulink model to compute ITAE integral signal (file: `c2mITAE.slx`).

(5) The Algebraic Constraint block can be used to solve algebraic equations. The use of the block will be demonstrated with examples later.

(6) Other blocks, including those for matrix and vector manipulation, and also complex signal processing. They are not covered in this book.

2.4.2 User-defined functions group

Apart from the simple mathematical operations stated above, complicated function operations can also be implemented with Simulink blocks. Double click the User-Defined Functions group icon, and the group window can be obtained as shown in Figure 2.35. The commonly used blocks in the group are:

(1) The Fcn block can be used to describe static functions. In the parameter dialog box, the input signal to the block is regarded as vector u. If the mathematical expres-

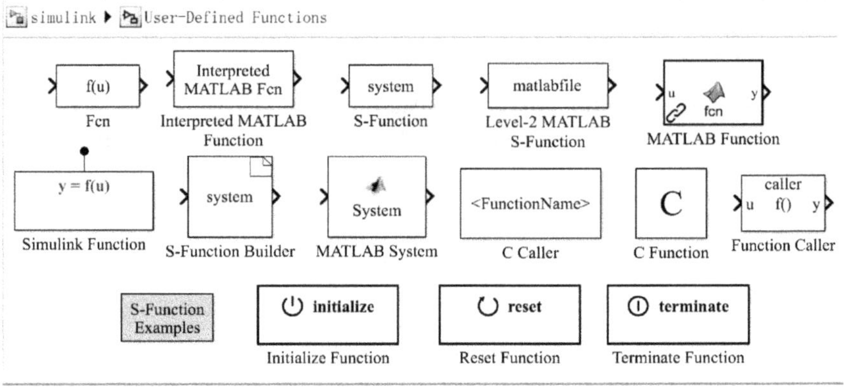

Figure 2.35: User-Defined Functions group.

sion $x_1^2 + x_1 \sin x_2 e^{-0.5x_3x_1}$ is to be described, where $\boldsymbol{x}$ vector signal is the input to the block, one should specify in the dialog box the following expression:

```
u(1)^2+u(1)*sin(u(2))*exp(-0.5*u(3)*u(1))
```

Note that, although the real input signal is $\boldsymbol{x}$, only the notation u is accepted. If x is used, an error message is given.

(2) The Interpreted MATLAB Function block can be used to describe complicated static mathematical expressions. The term static is used, which implies that the output signals are uniquely determined by the current values of the input signals. A MATLAB function can be written, with an input argument of u, and output argument of y. Such a function is used to describe the static functions. If one does not want to create a MATLAB file, the MATLAB Function block can be used, where the static function editor interface is provided in the block, for describing the static function. There is no need to create a real file.

(3) S-Function block. In contrast to the static function blocks, S-function blocks are used to describe dynamic functions. S in S-function stands for system. The mathematical forms of the systems are usually continuous or discrete state space expressions. The relationship between input and output signals are no longer static; they are in dynamic relationships, since apart from input and output signals, dynamic "state" signals are involved. The S-function block requires the user to write his own MATLAB code, according to the S-function format. If we do not want to create a real S-function file, the block MATLAB System can be used, and S-function code can be embedded in the block.

(4) Callback functions blocks. The related blocks are Initialize Function, Terminate Function and Reset Function. They allow Simulink to execute callback functions automatically.

(5) The C Caller block allows the user to embed C code in the block.

2.4.3 Lookup table group

Double click the Lookup Table group icon in the Simulink model browser, and the group is opened, as shown in Figure 2.36. The commonly used blocks in the group are:

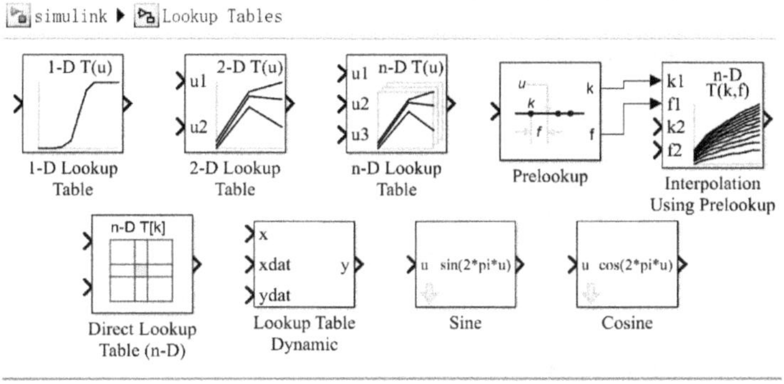

Figure 2.36: Lookup Table group.

(1) **1-D Lookup Table** block. The information of the turning points should be specified. This block will be further specified in Chapter 5. Apart from this block, **2-D Lookup Table** and **n-D Lookup Table** blocks are also provided.

(2) **Prelookup** and **Interpolation Using Prelookup** blocks are used together to lookup table using the index method. The efficiency is higher than other lookup table blocks.

(3) The **Lookup Table Dynamic** block allows the user to use external control signals to describe turning points, so as to implement the lookup table dynamically.

2.4.4 Logic and bit operations group

The Logic and bit operations group is also provided in Simulink, as shown in Figure 2.37. The commonly used blocks are:

(1) The **Logic Operators** block icon shows that it can be used to carry out "and" logic operations. Double click the block, the logic operations can be set to "or," "not," "not and," "nor or" and "exclusive or" in the parameter dialog box.

(2) The **Relational Operators** blocks allows the user to set "greater than," "less than" and other relational operations in parameter dialog box, so as to compare two input signals. If the relation holds, the output is logic 1; otherwise, it returns 0.

(3) The **Combinatorial Logic** block allows the user to describe truth tables and it will be demonstrated later through examples.

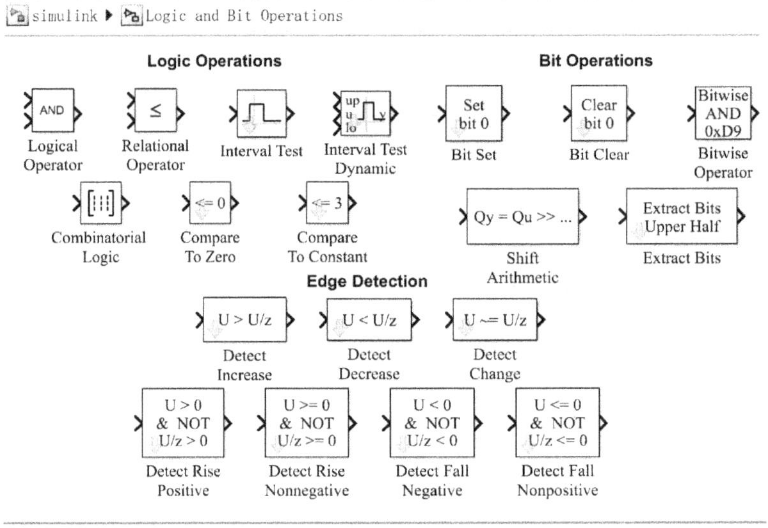

Figure 2.37: Logic & Bit Operations group.

2.5 Signal manipulation groups

Simulink allows the user to convert signals. For instance, data structures of signals can be converted, or several scalar signals can be combined into a vector signal. In this section, signal manipulation groups in Simulink are introduced, while related examples will be given later, not in this section.

2.5.1 Signal routing group

Many signal selection and operation blocks are provided in the signal routing group, as shown in Figure 2.38. Besides the above mentioned conversion blocks, various switch blocks are also provided. Some commonly used blocks are:

(1) Vector conversion blocks. Mux and Demux blocks are a pair of vector conversion blocks. The former can be used to convert several scalar or even vector signals into a vector signal, while the latter converts a vector signal into several scalar or even vector signals. The actual number of input or output ports can be set in the Number of inputs in Mux block or Number of outputs in the Demux block, through parameter dialog boxes. If there are 7 input signals in the Demux block as a vector, and the parameter is set to `[3,4]`, there will be two output ports in the block; one is a vector signal of 3, the other is a vector of 4. The Selector block allows the user to select several signals from the input vector and send them to the output vector. The signals to select can be specified in the Index edit box in its parameter dialog box.

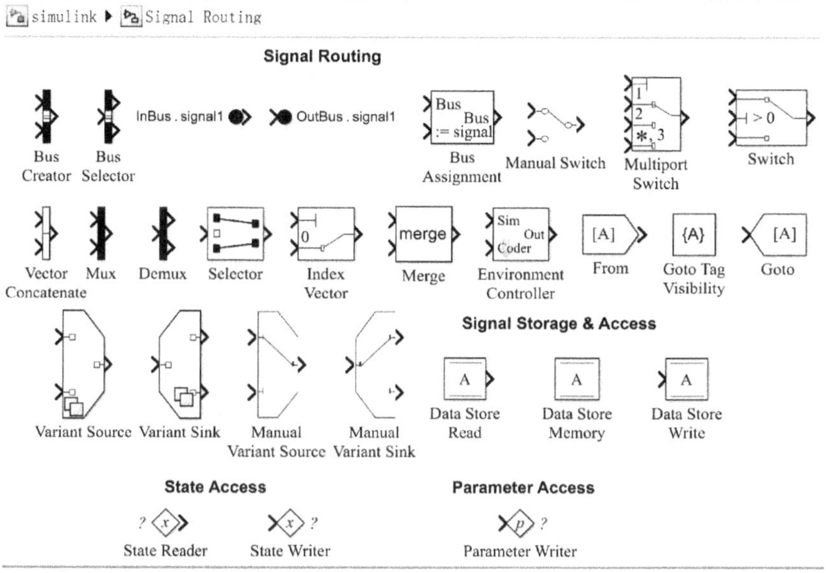

Figure 2.38: Signal Routing group.

(2) Bus signal conversion blocks. Bus signals are very similar to vector ones. The difference is that, in bus signals, signals of different data types are allowed to be grouped together, while vector signals only allow to combine signals of the same data types.

(3) Switch blocks. Switch block is the most widely used one, which has three input ports and one output port. The second input port accepts the control signal. If its value is greater than the Threshold prespecified in the parameter dialog box, the first input signal is passed to the output; otherwise, the third is passed. The Manual Switch block allows the user to change the status of the switch manually during the simulation process. There are two input ports and one output port. At the initial stage of simulation, the first input is passed to the output. If the user double clicked the icon, the first input is broken, and the second input is passed to the output. Double click the icon again, and the status of the switch is altered. In this way, the users may alter the status of the switch manually. More generally, the Multiport Switch block can be used. This block has $n+1$ input ports, and one output port. Among the input signals, the first one is the control signal, whose values can be $1, 2, \ldots, n$. If the control signal value is k, then the $(k+1)$th input signal is passed directly to the output port, while the others are pending.

(4) Storage blocks. In very complicated simulation models, sometimes there might be too many signal line crossings, which looks rather messy. Therefore, some signals in one region can be connected to a Goto block. At the other end, if the signal is needed, it can be connected to a From block. The correspondence of the signals can be specified through Tags in the parameter dialog boxes.

(5) Other blocks are not illustrated here.

2.5.2 Signal attribute group

In simulation models, sometimes the attributes of signals need to be modified. Open the signal attribute group, and the blocks provided are shown in Figure 2.39. The commonly used one is the **Data Type Conversion** block, which allows to convert the input signal from one data type into another.

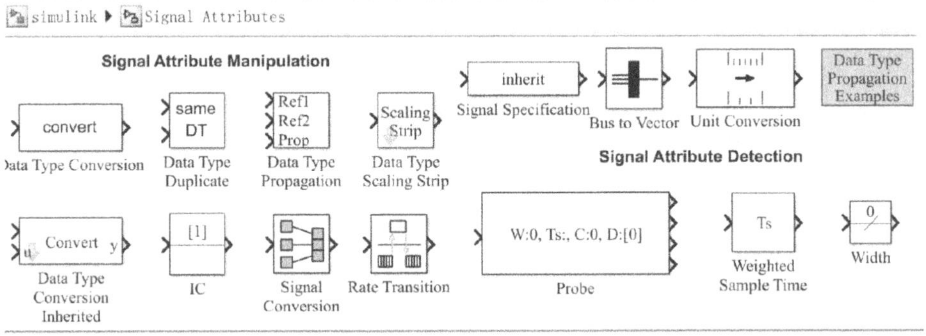

Figure 2.39: Signal Attributes group.

2.5.3 Ports and subsystems group

Ports and subsystems group are shown in Figure 2.40. Subsystem blocks are only briefly presented here, and details can be found in Chapter 5. The commonly used blocks are:

(1) Port blocks, such as **In1**, **Out1** and bus blocks are illustrated earlier.

(2) Various subsystem blocks. Different subsystem blocks are supported in Simulink, for instance, **SubSystem**, **Atomic Subsystem**, **Enabled Subsystem**, **Triggered Subsystem**, **Enabled and Triggered Subsystem** and so on. An enabled subsystem is controlled by an external signal to determine whether the subsystem block is executed or not. The triggered subsystems are controlled by the rising and/or falling edges of the external signal. The **Enable** and **Trigger** signal generator blocks are also provided in the group.

(3) The **Model** block allows the user to embed a Simulink file into another Simulink model. The user may specify the file name in the parameter dialog box. When using the model file embedding, make sure that the input and output ports are matched. Examples will be given later to show how to embed models.

(4) Flow control blocks. These include subsystem blocks such as **For Each Subsystem**, **For Iterator Subsystem**, **Switch Case**, **If** blocks and so on.

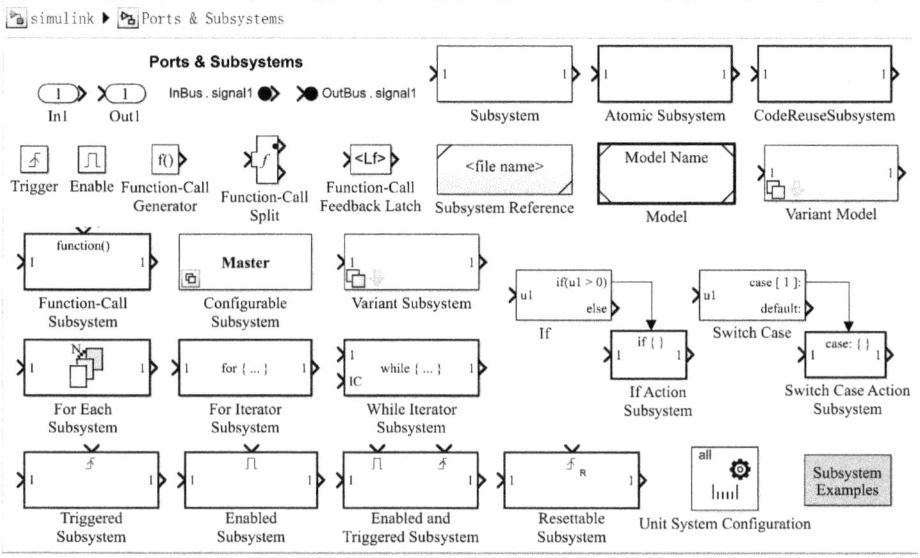

Figure 2.40: Ports & Subsystems group.

2.6 Simulink toolboxes and blocksets

Apart from the standard groups provided in Simulink, Simulink blocks are provided in many toolboxes. The block collections are referred as blocksets. Double click the Blocksets & Toolboxes icon, and all of the blocksets installed are listed as shown in Figure 2.41. On the left portion, there is a blockset list, and on the right, all of the blocks or groups in the selected blockset are displayed. The one shown in the figure is the blockset provided in the Computer Vision Toolbox.

In recent versions of MATLAB, many toolboxes and Simulink blocksets are provided, and some of them are listed below:

(1) Automatic control related blocksets, including Control System Toolbox, Instrument Control Toolbox, Model Predictive Control Toolbox, Fuzzy Logic Toolbox, Motor Control Blockset, Robust Control Toolbox, System Identification Toolbox and so on.

(2) Artificial intelligence and robotics related blocksets, including Robotics Toolbox, ROS Toolbox, Reinforcement Learning, Deep Learning Toolbox and so on.

(3) Multimedia related blocksets, including Data Acquisition Toolbox, Audio System Toolbox, Image Acquisition Toolbox, Computer Vision Toolbox, DSP System Toolbox, Sensor Fusion and Tracking Toolbox, Automated Driving Toolbox and so on.

(4) Physical modeling blocksets, including Simscape and related tools, Aerospace Blockset, Vehicle Dynamics Blockset, Vehicle Network Toolbox, Communication Toolbox and so on.

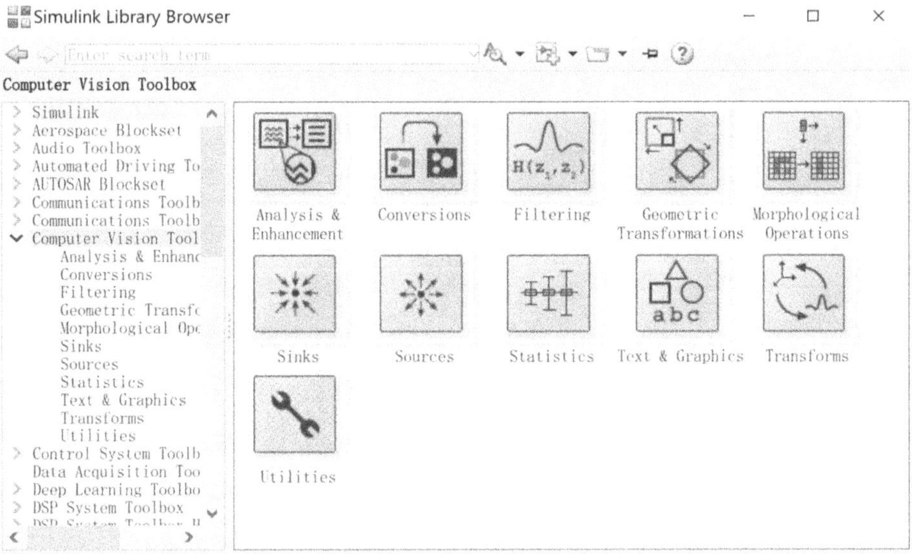

Figure 2.41: Blocksets & Toolboxes window.

Some of the above blocksets and toolboxes will be demonstrated later. If further information is needed for the blocksets, the readers are advised to refer to manuals or other materials, so as to better learn Simulink and its applications.

2.7 Exercises

2.1 Create a blank library window, and copy the commonly used blocks such as Clock, Step, In1, Out1, Scope, Transfer Fcn, Integrator, Transport Delay, PID Controller, Gain, Sum, Product, Mux, Demux, Saturation and Dead Zone into the library window, and save it as the new commonly used block library. The readers may refresh the library later on according to their own needs.

2.2 Consider Constant block. If the vector `[1,2,3]` is filled into its parameter dialog box, and connected to a scope, what is the output signal?

2.3 The default Product block has two input ports, and one output port. Modify the block so that four input signals are accepted.

2.4 If a transfer function model is given below, use Simulink to model the block:

$$G(s) = \frac{s^2 + 4s + 4}{(s+1)(s+2)(s+3)(s+4)}.$$

2.5 If the open-loop transfer function is

$$G(s) = \frac{s^3 + 7s^2 + 24s + 24}{s^4 + 10s^3 + 35s^2 + 50s + 24},$$

and it is known that the system is with the unity negative feedback structure. Create the simulation model for the closed-loop system. Modify the open-loop transfer function block name to New Transfer Function. Draw the step response of the closed-loop system. Besides, use a dashboard block to display the simulation result, and change properly the scales for the output signal.

2.6 Use the scope to show the output signal of the above exercise, and set the scales of the scope block such that y signal is restricted to the interval $[-2, 4]$.

2.7 Open the Simulink model `c2mnl1.slx` in Example 2.8. Add a virtual scope to the output line of the Saturation block, and observe the signal. Add also a virtual scope to the output line of Dead Zone block. Is there any way to display the two signals together on the same virtual scope?

2.8 A state space model with two inputs and two outputs is given below. If the two input ports are driven respectively by signals $u_1(t) = \sin t$ and $u_2(t) = 2\cos 2t$, draw the outputs of the system:

$$\begin{cases} \boldsymbol{x}'(t) = \begin{bmatrix} 2.25 & -5 & -1.25 & -0.5 \\ 2.25 & -4.25 & -1.25 & -0.25 \\ 0.25 & -0.5 & -1.25 & -1 \\ 1.25 & -1.75 & -0.25 & -0.75 \end{bmatrix} \boldsymbol{x}(t) + \begin{bmatrix} 4 & 6 \\ 2 & 4 \\ 2 & 2 \\ 0 & 2 \end{bmatrix} \boldsymbol{u}(t), \\ \boldsymbol{y}(t) = \begin{bmatrix} 0 & 0 & 0 & 1 \\ 0 & 2 & 0 & 2 \end{bmatrix} \boldsymbol{x}(t). \end{cases}$$

2.9 ISE (integral of squared error) performance index is a commonly used criterion in control. The mathematical form is

$$I(t) = \int_0^t e^2(\tau)\mathrm{d}\tau.$$

Use In1 block for the $e(t)$ signal, draw a Simulink model to compute the signal $I(t)$ and connect the signal to block Out1.

2.10 ITSE (integral of time weighted squared error) performance index is a commonly used criterion in control. The mathematical form is

$$I(t) = \int_0^t \tau e^2(\tau)\mathrm{d}\tau.$$

Use **In1** block for the $e(t)$ signal, draw a Simulink model to compute the signal $I(t)$ and connect the signal to block **Out1**.

2.11 The simulation result in Example 2.2 is erroneous. Connect a stable transfer function block to the signal and connect the signal to a scope block. Observe whether the output signal is correct or not.

2.12 Consider the plant model $G(s)$ in Example 2.7. If a PID controller is used, with the model $G_c(s) = 0.998 + 0.333/s + 0.0809s$, draw the closed-loop step response of the system.

2.13 For a unity negative feedback control system with the plant and controller models given by

$$G(s) = \frac{10e^{-20s}}{2s+1}, \quad G_c(s) = 0.6 + \frac{0.008}{s},$$

draw the simulation model of the system, and use a dashboard to show the output signal.

2.14 The relationship of the input and output signals of a multivariable system is given by

$$\begin{bmatrix} y_1(s) \\ y_2(s) \end{bmatrix} = \begin{bmatrix} \dfrac{12.8e^{-s}}{16.7s+1} & \dfrac{-18.9e^{-3s}}{21s+1} \\ \dfrac{6.6e^{-7s}}{10.9s+1} & \dfrac{-19.6e^{-3s}}{14.4s+1} \end{bmatrix} \begin{bmatrix} u_1(s) \\ u_2(s) \end{bmatrix},$$

with $u_1(t) = \sin t$, and $u_2(t)$ is a unit step signal, draw the low-level Simulink model of the system and observe the output signals.

2.15 A block diagram of a DC motor drive system is given in Figure 2.42.[55] Build the Simulink model, and draw the unit step response of the system.

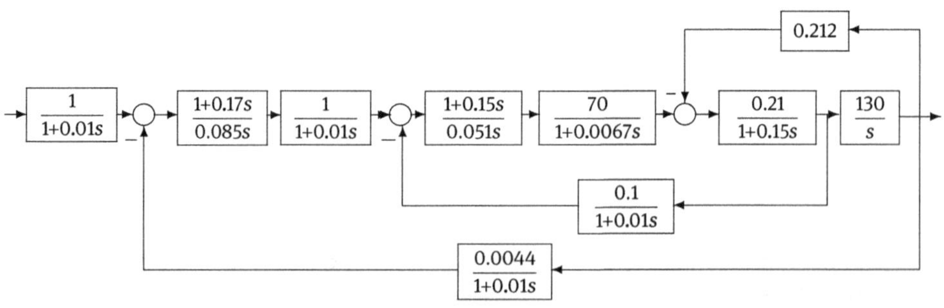

Figure 2.42: Block diagram of a DC motor drive system.

3 Block manipulation and menu systems

The main task in Simulink modeling is to draw the required model in the model window. All of the Simulink models are composed of interconnected blocks. In this chapter, concentrations are made on manipulations of blocks. In Section 3.1, the block input and decoration are presented, including block rotation and flip, font and shadow setting and so on. Examples are used to demonstrate Simulink system modeling process. In Section 3.2, block parameter input and modification are introduced. Concentrations are made on parameter dialog box controls and there use.

In September 2019, MATLAB R2019b was released, and Simulink was upgraded to version 10.0. The appearance, manipulation and execution methods have significant differences with those in old versions since there are significant differences in different mainstream versions in Simulink, in this and subsequent chapters. The two mainstream versions, before and after R2019b, are both presented. In Section 3.3, the menu system in the Simulink environment is illustrated. The output port and simulation parameters are presented, and show system analysis methods. In Section 3.4, the menu system and toolbar in the new versions are presented, including the simulation parameters setting, model debugging and the model format setting are illustrated. The call to APPs is discussed. In Section 3.5, the tools such as the model browser in Simulink are demonstrated. Better ways in model management and manipulation are illustrated. With these tools, Simulink modeling is made much easier.

Some of the manipulation methods are the foundations to be discussed in this book. Later, some key operation procedures will be further demonstrated.

3.1 Simple manipulation and decoration of blocks

Blocks and signal lines are the essential and most important components in Simulink modeling. When establishing Simulink models, the model library and a blank model windows should be opened first. Then the needed blocks can be copied into the model window. Signal lines can then be drawn to set up the Simulink model, and perform simulation. In this section, block copying and processing are illustrated, and the model parameter setting is then discussed. Examples are given to demonstrate Simulink modeling and skills.

3.1.1 Shortcut menus

Right click a block icon, and the shortcut menu can be displayed. Since the menu is too long, it is divided into five portions and is given in Figure 3.1. It can be seen that the general manipulation and rotating facilities are presented. Further manipulation of blocks is presented. The commonly used menu manipulations are presented next.

https://doi.org/10.1515/9783110734959-003

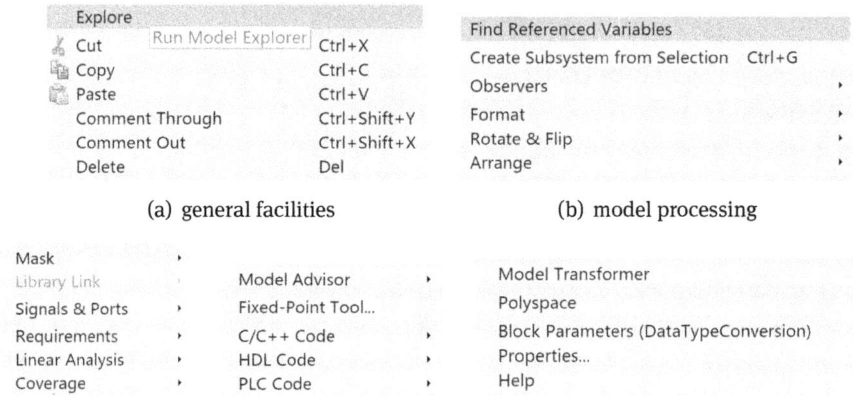

(a) general facilities (b) model processing (c) model mask tools (d) code compiling (e) other processing

Figure 3.1: Shortcut menu of a block.

3.1.2 Copying and moving blocks

To set up a Simulink model, the necessary blocks should be copied from the existing model library to the model window. If one wants to copy the blocks, different methods can be used.

(1) One obvious way is to drag directly the block from one window into another one.

(2) Another way is to select the block to be copied by clicking it with the left mouse button. Hold on Ctrl and press the C key, or right click the block to get the shortcut menu and select Copy menu item, as shown in Figure 3.1(a). The block is then coped into the Microsoft Windows clipboard. Then in the destination window, press Ctrl+V keys, or select Paste item from the shortcut menu, and the block can then be copied.

(3) In the same model window, hold the Ctrl key, then drag the block to make another copy.

(4) If one wants to select all the blocks within a certain range, then drag the mouse button over the range to draw a rectangular shadow, and then all of the blocks in the selected region are selected simultaneously. If the blocks to be selected in the window are not in the same region, hold the Ctrl key, and then click the blocks, and they can be selected.

(5) When the blocks are selected, they can be moved to other places by dragging the mouse, or drag one of the four corners to resize the blocks. These are conventional Microsoft Windows operations, and not presented in detail here.

Example 3.1. A block diagram of a feedback control system is shown in Figure 3.2. Establish a Simulink model for the control system.

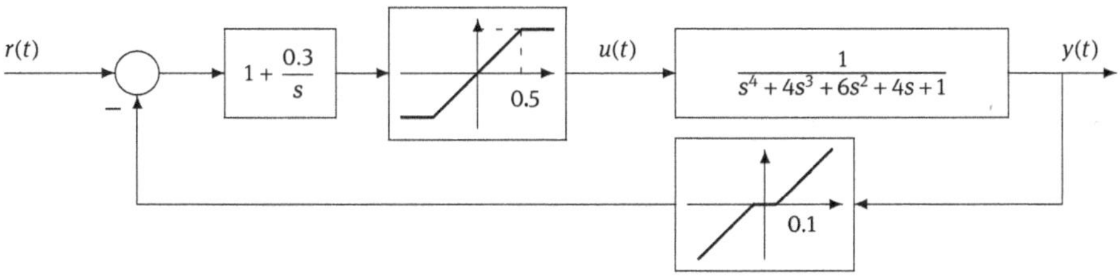

Figure 3.2: Block diagram of a feedback control system.

Solutions. By analyzing the feedback control system model in Figure 3.2, it can be seen that the following blocks are needed to set up the Simulink model:

(1) Transfer function block. The plant $1/(s^4 + 4s^3 + 6s^2 + 4s + 1)$ and controller $1 + 0.3/s$ are all continuous transfer function blocks. Copy the **Transfer Fcn** block from the **Continuous** group to the model window, and modify the parameters of the transfer function models, and keep them the same as the ones in Figure 3.2. Besides, since there are two **Transfer Fcn** blocks, then in the Simulink model, the one with repeated name is automatically changed into **Transfer Fcn1**. It can also be changed into any other names. The users may change their names manually. The specific method is that by clicking the block name such that it is selected, it can be changed manually. Make sure that the block name must be made unique; otherwise, it cannot be accepted.

(2) Nonlinearity blocks. Two nonlinear blocks are needed, **Saturation** and **Dead Zone**. They should be copied directly from the **Discontinuities** group to the model window, and assigned their parameters.

(3) Input and output port blocks. Input signal is the unit step input signal, and **Step** block from the **Sources** group can be selected and copied into the model window. The output signal can be connected both to a scope and output port, where **Scope** and **Out1** blocks from the **Sinks** group can be used.

(4) Sum block. Copy the **Sum** block from **Math Operations** group to the model window.

All of these blocks can be copied into the model window, and the parameters of them can be modified accordingly. The model established is as shown in Figure 3.3. Of course, this model is only a semifinished model, since the models are not corrected to complete the simulation model.

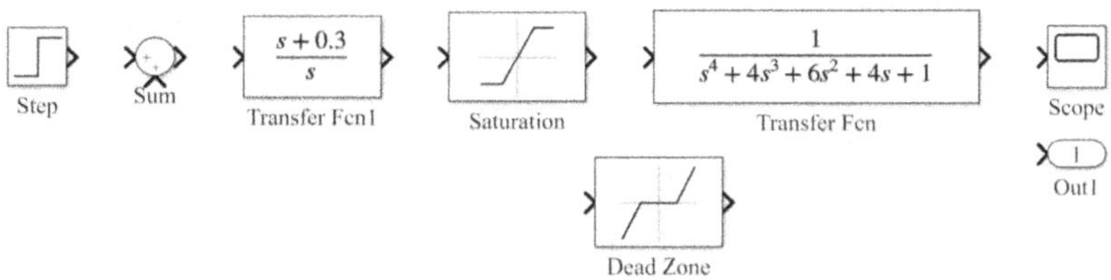

Figure 3.3: All of the expected blocks for the simulation model I (file: `c3mmod1a.slx`).

3.1.3 Font setting in blocks

Font setting and block rotation can be made by selecting the **Format** menu item from the shortcut menu, shown in Figure 3.1(b), and the submenu is shown in Figure 3.4. Select the **Font Style for Selection** menu item, and a standard font setting dialog box is given, so that the user may select the allowed font and sizes. Due to the limitation in page budget, the screenshot is not given.

Figure 3.4: Format menu.

3.1.4 Linking blocks

Linking blocks in Simulink is rather simple and straightforward. Several commonly used linking methods are presented here:

(1) To link two blocks, the simplest way is to click the triangle at the output port of the former block, and drag the mouse to the > port of the latter block. If the button is released, the link between the two blocks is made.

(2) A more convenient way is hold the **Ctrl** key, then click the input block, then click the output block, and then link between the two blocks can be made.

(3) In recent versions, if a block is copied and moved nearer to another block, at the same level of the output port of the former block, a link in shallow color is shown. To confirm the link, click the shallow color line, and the link is made. If one wants to abandon the link, just remove the mouse away.

Example 3.2. Set up a Simulink model for the control system shown in Figure 3.1.

Solutions. With all the blocks copied as shown in Figure 3.1, the following actions can be made to link and complete the simulation model.

Using the methods shown above, the blocks can be linked directly. Of all the methods, method (2) is the most convenient one for this example. For this particular example, hold the **Ctrl** key, then click in turn from **Step** to all other blocks in the forward path, and finally link the feedback path so that the Simulink model shown in Figure 3.5 can be created.

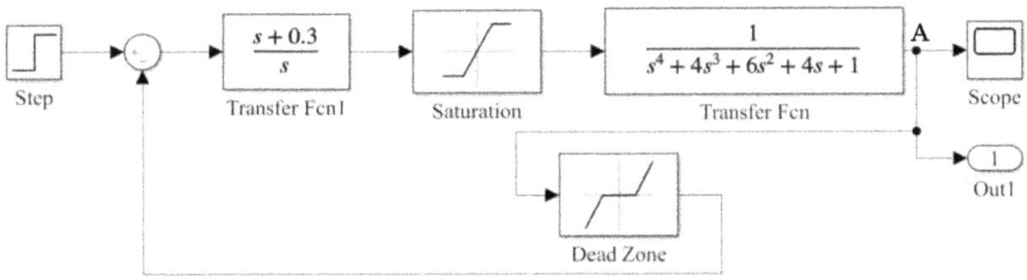

Figure 3.5: The constructed model II (file: `c3mmod1b.slx`).

The link of some blocks are rather complicated in this example. For instance, the extraction of the signal $y(t)$ from Figure 3.5 so as to construct the feedback path. To extract signal from point A, mishandling may occur. Two alternative methods can be adopted to draw the signal line:

(1) Hold the Ctrl key at A, and draw lines regularly.

(2) From the input port of Dead Zone block, draw a line backwardly to point A and release the button, the needed signal line can be drawn automatically.

3.1.5 Rotating and flipping blocks

In normal cases, Simulink blocks have input ports on the left side, and output ports on the right. If one wants to shift the input ports to the top of the block, and output port to the bottom, rotations of the blocks should be made. Select a block, and open the shortcut menu, then select **Rotate & Flip** option. The submenu is obtained as shown in Figure 3.6. One may select **Clockwise**, **Counterclockwise** and **Flip Block** options from it to directly rotate or flip blocks. If the **Flip Block Name** is selected, the model name is moved to the top of the block. Of course, if the model name is selected, one may also drag it to the desired positions.

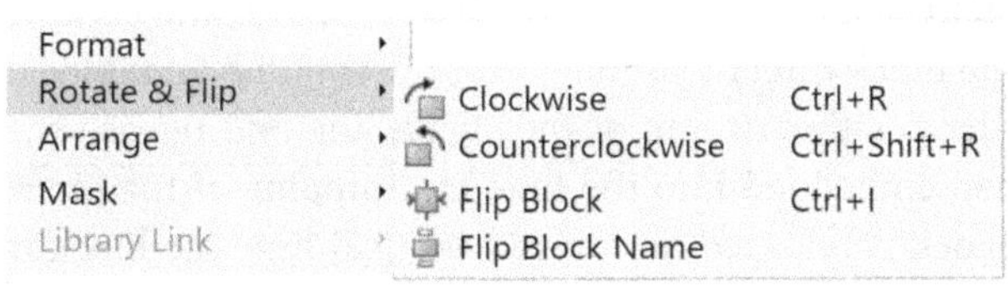

Figure 3.6: Block rotating and flipping submenu.

As an example, consider the **Transfer Fcn** block. The clockwise rotation and flipping of the block are shown respectively as shown in Figure 3.7(a) and (b). When rotating blocks, the model names are also changed accordingly, while when flipping blocks, the positions of the block name is unchanged. If one wants to move the model name to the top, use the **Flip model name** option, and the result is as shown in Figure 3.7(c).

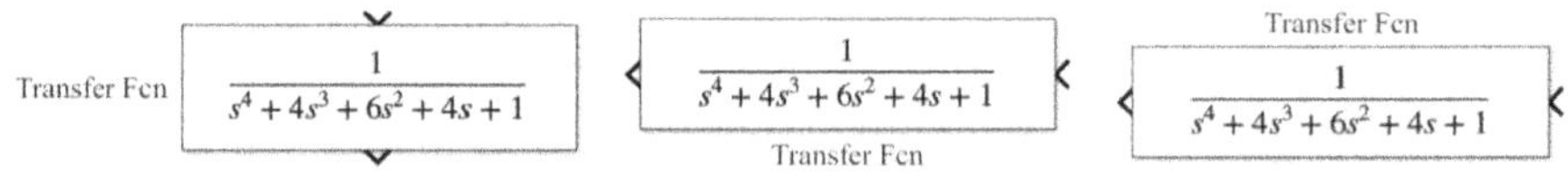

(a) clockwise rotation (b) block flipping (c) model name flipping

Figure 3.7: Rotation and flipping of blocks.

Example 3.3. Although the Simulink model in Figure 3.5 implements faithfully the original nonlinear feedback control system model, the layout of the feedback path looks ugly. This is because the **Dead Zone** block is the same as the standard Simulink blocks; the input port is on the left, while the output port on the right. Therefore, the block can be flipped, so that normal Simulink model can be constructed. Flip the block and draw the correct model.

Solutions. Delete the existing signal lines to and from the **Dead Zone** block. Click the block, open the shortcut menu and select **Rotate & Flip** → **Flip Block** option. The block is then flipped automatically. The signal lines can then be made so that the new model shown in Figure 3.8 can be established.

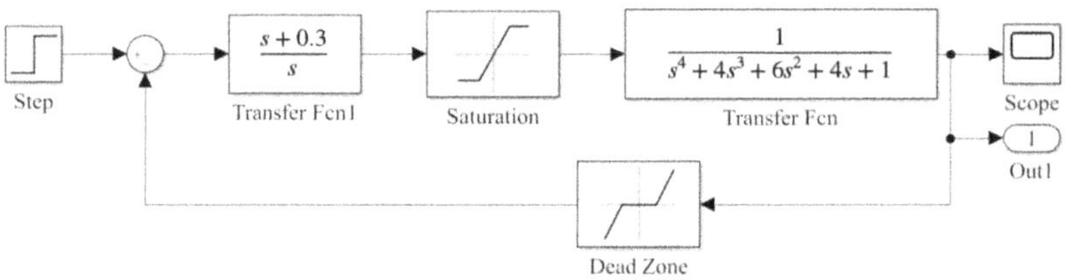

Figure 3.8: Further modify the model (file: `c3mmod1c.slx`).

Compared with the rotating and flipping methods shown above, a better and concise way is described. Link the scope block directly to the – input port of the **Sum** block. Drag the **Dead Zone** block to the position on top of the signal line and release the mouse button. The block is then embedded into the line, and flipping of the block is automatically made. If the block is released on top of a vertical line, the block is embedded and rotated automatically.

3.1.6 Color and shading setting of blocks

The Simulink environment allows the user to modify the color or other properties of the blocks and lines independently. One may change the color of a group of blocks the same, so as to increase the readability of the Simulink model.

If one selects one or a group of blocks from the shortcut menu shown in Figure 3.4, the Foreground Color or Background Color submenus can be chosen, such that a standard color dialog box is shown. It can be used to set the colors of the blocks. In the blocks, the text are referred to as the foreground, while background color refers to the background of the block.

Example 3.4. In industrial control systems, PID controllers are usually connected to saturation nonlinearity as shown in Figure 3.9. Assume that the plant is given in Example 2.7, and the parameters of the PID controller are $K_\mathrm{p} = 0.998$, $K_\mathrm{i} = 0.333$ and $K_\mathrm{d} = 0.0809$. The bounds of the saturation element are $|u(t)| \leqslant 0.5$. Create a Simulink model, and change the controller to blue, and change the color of the plant model to red.

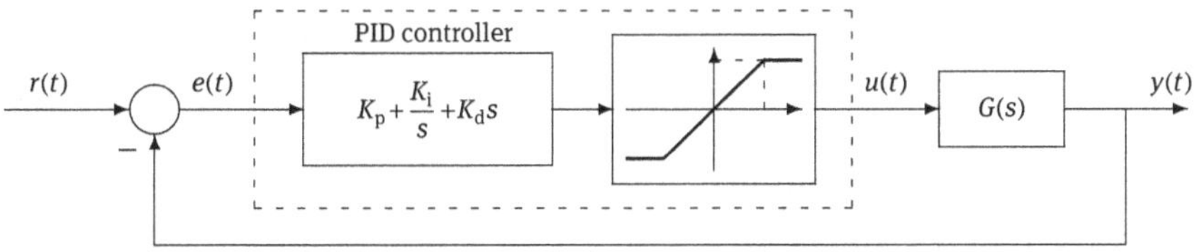

Figure 3.9: Block diagram of a PID control system.

Solutions. Similar to the cases demonstrated earlier, the PID controller can be established with low-level blocks. Also, copy the saturation element, step input, scope, output port and sum block. Copy the plant model in Example 2.7 to the new Simulink model. The new control system model can be constructed as shown in Figure 3.10.

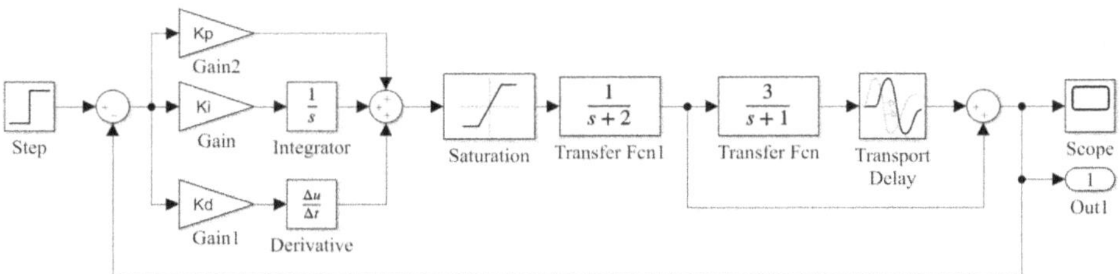

Figure 3.10: Simulink model of a PID control system (file: `c3mpid1.slx`).

Drag the portion of the PID controller, such that all the blocks are selected. With the Format → Background Color submenu item in the shortcut menu, change the color to Cyan, from the standard color setting dialog box. All of the parts in the controller are then set to the color cyan. Similarly, select all of the blocks of the plant model, and set the background to Red. The decorated Simulink model is obtained as shown in

Figure 3.11. It can be seen that the model such decorated is easier to understand and maintain.

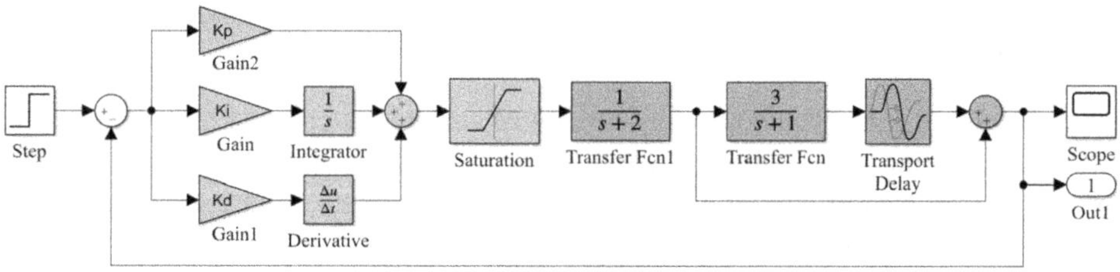

Figure 3.11: Simulink model with color changed (file: `c3mpid2.slx`).

To make the Simulink model look prettier, some blocks can be made striker. For instance, shading effect can be added to the blocks. Select a block, or a group of blocks, then from the Format → Shadow submenu from the shortcut menu in Figure 3.4, shadows can be added to the block. If the item is selected again, shadows can be canceled.

Example 3.5. Add a transfer function block for $G(s) = 1/(s^4 + 4s^3 + 6s^2 + 4s + 1)$ in a model window, and draw shadows to the block. Observe the result.

Solutions. Create a blank Simulink model window, shading effect can be made to some blocks, as shown in Figure 3.12(a). Select the block, and pick the Format → Shadow submenu, and the shadow will added to the transfer function block, as shown in Figure 3.12(b). In the earlier versions, the shadow has rather prominence, where in the new versions, the shadow effect can still be witnessed if observed carefully.

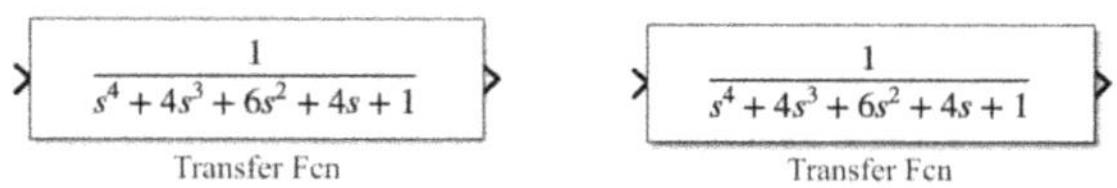

(a) original transfer function (b) block with shadow added

Figure 3.12: Shadow decoration of a block.

3.2 Modifying block parameters

Each Simulink block has its own parameters. If a block with a known mathematical model is designed, such that the block can be used easily, the commonly used parameters can be assigned through the parameter dialog boxes. In this section, parameter dialog boxes and commonly used controls are discussed.

3.2.1 Commonly used controls in dialog boxes

Double click a block, and the parameter dialog box can be opened. Parameter dialog boxes have already been demonstrated in the earlier examples. For instance, in the parameter dialog box of a Constant block shown in Figure 2.8, constant values are allowed to be specified in the edit box of the dialog box. Therefore, edit boxes are the most widely used controls in parameter dialog boxes. The following controls are supported in ordinary Simulink blocks:

(1) Edit boxes. Click the edit box, then the values can be entered into the edit box. The input method is rather straightforward.

(2) Listboxes. Usually a  button appears with the listboxes. In old versions, the button is shown like . The contents in the listbox are then displayed. The needed item can be selected from the list. If only very few options are provided in the listbox, all the options are displayed. The  button may not appear.

(3) Radio buttons. A set of such buttons may be provided as a group. Only one item can be selected from them. The relationship among the options are referred to as mutual exclusive.

(4) Checkboxes. The behavior of checkboxes are very similar to radio buttons. They are often grouped. The difference is that many items can be selected from them simultaneously, or none of them may be chosen.

(5) Scroll bars. Scroll bars are commonly used controls in dialog boxes. Usually, it is similar to edit boxes. The difference is that the values of the controls are adjusted by scroll bars.

(6) Buttons. There is at least one button in a parameter dialog box. It can be used to accept or abandon information, and close the parameter dialog box.

(7) Panes. Sometimes if the parameters to be adjusted are too many, it may be arranged to different pages. Each page is lead by a pane.

Examples are used next to show various controls and their applications in parameter dialog boxes.

Example 3.6. Consider the Slider Gain block in the Math Operations group. Double click the block, and the parameter dialog box is opened as shown in Figure 3.13. The dialog allows the user to adjust the value of the gain with the slider between 0 and 2.

If the user wants to adjust the gain within a different range, the Low and High values should be changed to redefine the range of the sliders.

Example 3.7. The PID Controller block in the Continuous group is a powerful block. The parameter dialog box is a representative one.

Double click the block icon, and the parameter dialog box is displayed, as shown in Figure 3.14. In the dialog box, Proportional and other options are given in edit boxes. The values of the parameters can be filled into the edit boxes directly. The Continuous-time and Discrete-time controls are a pair of radio buttons. Only one of them can

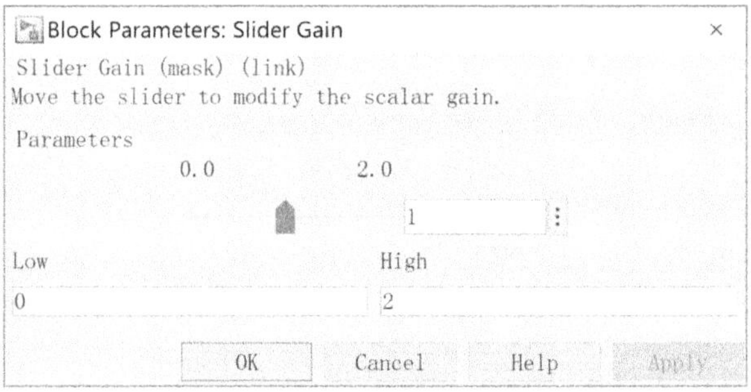

Figure 3.13: Slider Gain parameter dialog box.

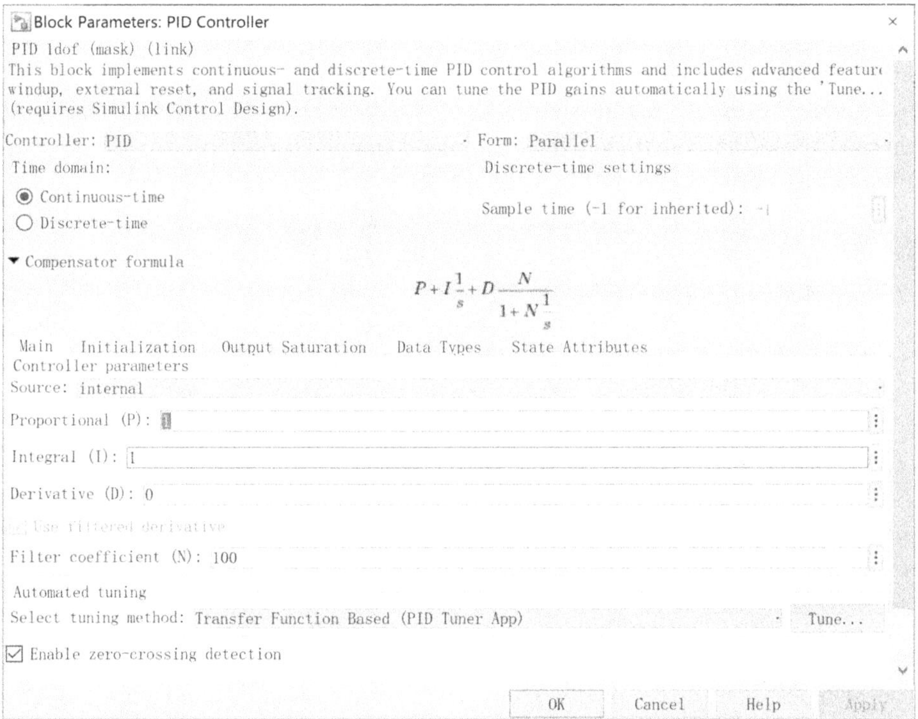

Figure 3.14: PID Controller parameter dialog box.

be chosen at a time to decide if the controller is continuous or discrete. Controller and From controls are listboxes. Different controller types can be selected from the list boxes. **Enable zero-crossing detection** control is a checkbox. The user may decide whether zero-crossing detections are made (the concept of zero-crossing will be given later). The controls **OK** and **Cancel** are push buttons, which indicate whether to accept or ignore the parameter changes in the dialog box.

Main and Initialization panes are the panes in the dialog box. Since there are too many variations in the parameters, they are arranged in different pages. The user may visit each pane to find and assign a particular parameter.

3.2.2 Filling information in dialog boxes

The controls in the parameter dialog box can be filled as specific values or as variable names. However, the variables must be assigned in the MATLAB workspace first. Otherwise, error messages may be given. In Simulink modeling, if the values of some parameters are not given, error messages may be displayed during the simulation process. In this section, the use of parameter dialog boxes are demonstrated.

Example 3.8. Consider again the transfer function block in Example 2.6. Use different methods to input the parameters of the transfer function into the block.

Solutions. In Example 2.6, it was demonstrated that the coefficient vectors are filled directly into the parameter dialog box. If there are two vectors `n1` and `d1` in MATLAB workspace, the following commands can be issued:

```
>> n1=[2,18,40]; d1=[1,10,40,82,91,52,12];
```

then in the parameter dialog box in Figure 2.24, the variable names, `n1` and `d1`, can be specified. Therefore, the transfer function parameters can be correctly assigned to the block, so that it can be used in the model correctly.
If `n1` and `d1` variable names are filled into the parameter dialog box, but the two variables do not exist in MATLAB workspace, in the parameter dialog box, mark is given, as shown in Figure 3.15, indicating that the variables are not available. The user must ensure that the parameters exist in the MATLAB workspace; otherwise, the simulation process cannot proceed. If the parameters are already in the MATLAB workspace, no error message will be given in simulation. If the values in the MATLAB workspace is changed, the new values will be used automatically.

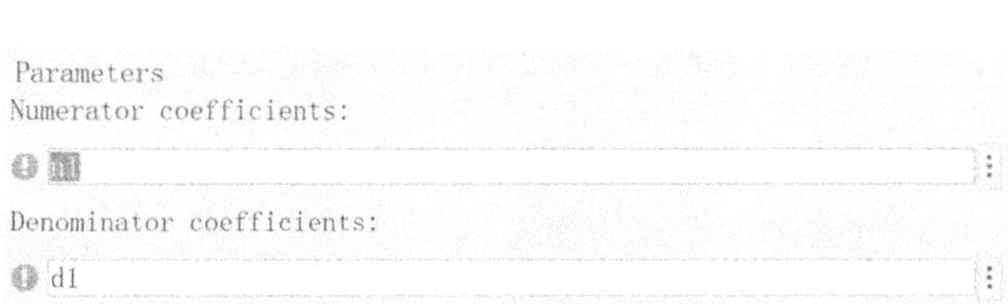

Figure 3.15: Error message display for the parameter dialog box.

Example 3.9. The integrator block is used in the Simulink model shown in Figure 3.11. The integrator block is a relatively representative block in Simulink. It has many variations. Use the parameter dialog box to change the form of such a block.

Solutions. Integrators are the most important component in differential equation modeling and simulation. The integrator block provided in Simulink is very powerful, and it has many variations. Under the default setting, if the input signal to the block is $u(t)$, the output signal $x(t)$ is mathematically expressed as

$$x(t) = x_0 + \int_{t_0}^{t} u(\tau)\mathrm{d}\tau, \tag{3.2.1}$$

where t_0 is the start time instance, x_0 is the initial value of the output signal $x(t)$. Double click the integrator block, and the dialog box shown in Figure 3.16(a) is opened. The initial value x_0 can be specified in the **Initial condition** edit box.

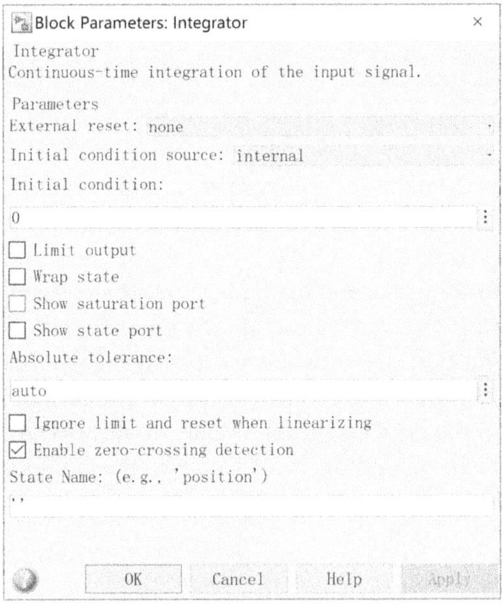

(a) Dialog box of **Integrator** block

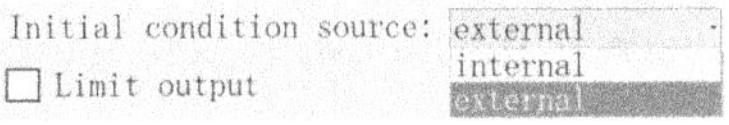

(b) initial value options

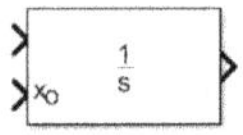

(c) with external initial values

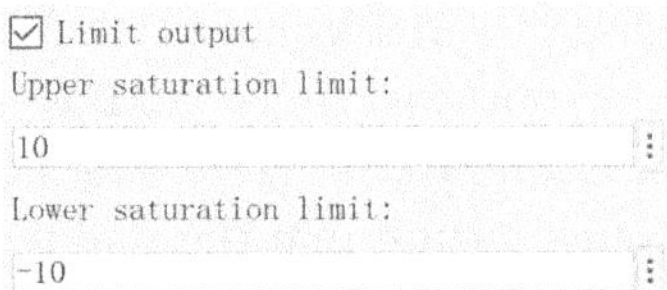

(d) saturation setting

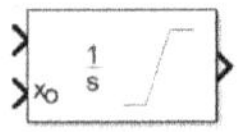

(e) the new integrator icon

Figure 3.16: Options and setting of integrator blocks.

If x_0 is not a given value, and it is specified as an external signal, the listbox named **Initial condition source** should be opened. Select **external** option as shown in Figure 3.16(b). Then the integrator will have one more input port, labeled as x0, as shown in Figure 3.16(c), allowing the user to specify the initial values with an external signal.

A saturation nonlinearity can be appended to the integrator, so as to restrict the range of the output of the block. The check box **Limit output** can be selected. Two edit boxes labeled **Upper saturation limit** and **Lower saturation limit** are then displayed automatically as shown in Figure 3.16(d). It allows the user to specify the upper and lower bounds of the saturation nonlinearity. A saturation curve will appear automatically on the integrator block icon, as shown in Figure 3.16(e).

Of course, the integrator block has more variations. For instance, the facilities such as **reset** and **Ignore limit or reset when linearizing**. They are not discussed here.

3.3 Menu systems in earlier versions of Simulink

As mentioned earlier, the interface of Simulink may be different on different versions. It can be said that MATLAB R2019b is a dividing point in the Simulink interface, since Simulink was updated into version 10.0. For ease of illustration, we refer the versions R2019a and earlier ones as "old versions," and R2019b and later versions as "current versions." To better describe the two major versions, the manipulations of menu systems and interface of the two mainstream versions will be presented in this section and the next one.

The shortcut menu of Simulink is an easy and flexible way to manipulate Simulink. More manipulating actions can be implemented with the menu systems provided in Simulink. In the earlier versions, the main menu in Simulink is shown in Figure 3.17.

Figure 3.17: Main menu system in the earlier versions of Simulink.

3.3.1 File management menus

Many of the menu items in the Simulink model window are conventional Microsoft Windows menus, and some new menu items are also added to Simulink. Therefore, these menu items are relatively easy for Simulink users. Some of the commonly used submenus are discussed here.

The **File** menu in Simulink is shown in Figure 3.18(a), where **New**, **Open** menu items are similar to those in other Microsoft Windows programs.

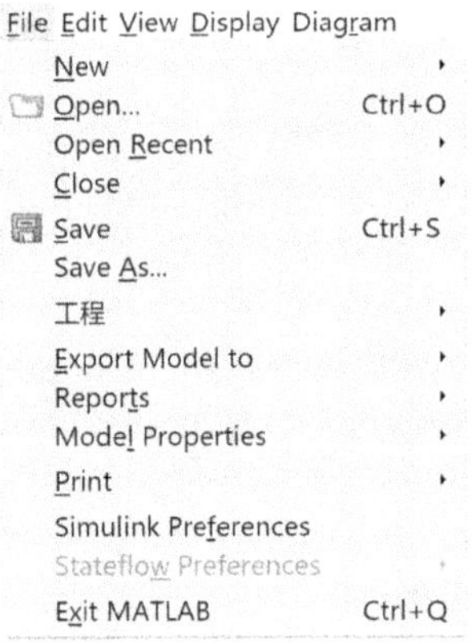

(a) file manipulation menu

(b) output menu

(c) model properties

Figure 3.18: File menu in Simulink.

There is a black triangular sign ▸ after the New menu item, indicating that there are submenu items. Click the triangular sign, and the submenu items are displayed, including Blank Model, Chart (for Stateflow), Library and other options. Click the New menu, and a new blank model window is opened automatically. These operations are similar to other Windows programs. They are not presented further here.

The three different menus are addressed in this section:

(1) Export Model to menu, as shown in Figure 3.18(b), where Previous Version allows the user to save the model into the form in earlier versions of Simulink. The user needs to select a suitable version. Of course, Simulink models can also be saved in other formats.

(2) The Model Properties menu, as shown in Figure 3.18(c), opens a dialog box shown in Figure 3.19. Simulink model information is displayed, such as file created time, modified time and so on. The information displayed is in the Main pane (as shown in Figure 3.19).

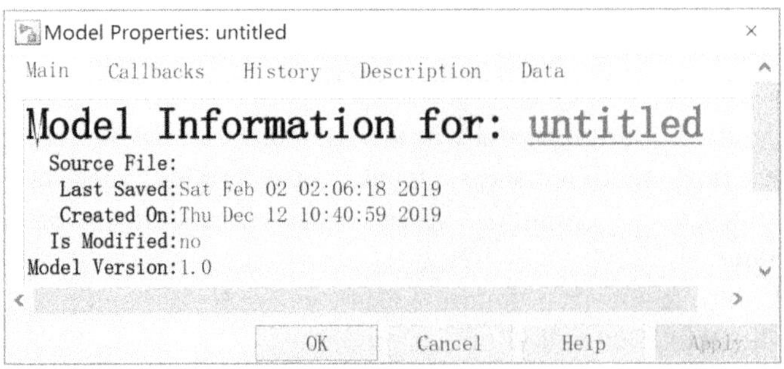

Figure 3.19: Simulink model properties.

If the **Callback** pane is selected, the dialog box in Figure 3.20 is displayed. Some callback functions such as **PreLoadFcn** and **PostLoadFcn** are allowed to be specified.

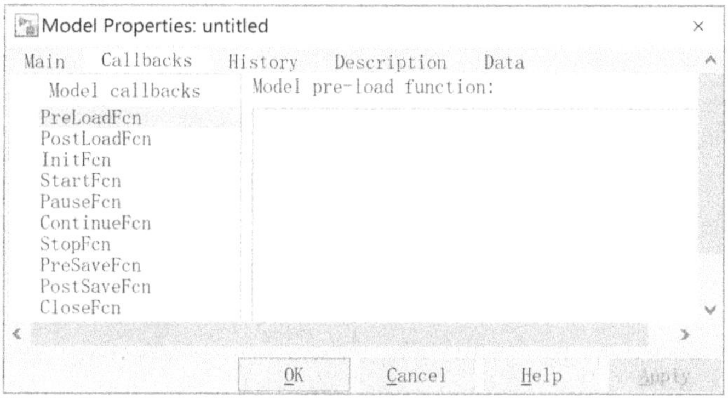

Figure 3.20: Preload callback function setting in Simulink.

Example 3.10. Consider the PID control system model in Example 3.4, where the three parameters in the controller are described by variable names. This means that before running the model, the variables must be specified first in MATLAB workspace; otherwise, the model cannot be correctly executed. Use relevant callback functions to automatically set the variables.

Solutions. The so-called preload means that every time the model is opened, the variables are assigned. Preload variable specification can be implemented with the above mentioned **PreLoadFcn** callback function. Open the dialog box in Figure 3.20, and fill in the following commands in the **Model preload function** edit box:

```
Kp=0.998; Ki=0.333; Kd=0.0809;
```

and save the model. The next time when the model is opened, the above statements are executed first automatically. It is no longer necessary to worry about unassigned variables in the models.

(3) The **Simulink Preferences** menu is mainly used to set the operating environment in Simulink and also the user's own preference. The dialog box shown in Figure 3.21 is opened, allowing the user to make different settings. For instance, the **Background Color** can be used to specify the background of the model window, and the folders of various files can also be assigned.

If the **Editor** item in the listbox on the left in Figure 3.21 is selected, the dialog box is changed to the new one in Figure 3.22, allowing the user to change the settings for the Simulink model editor. For instance, the **Line crossing style** can be set to specify

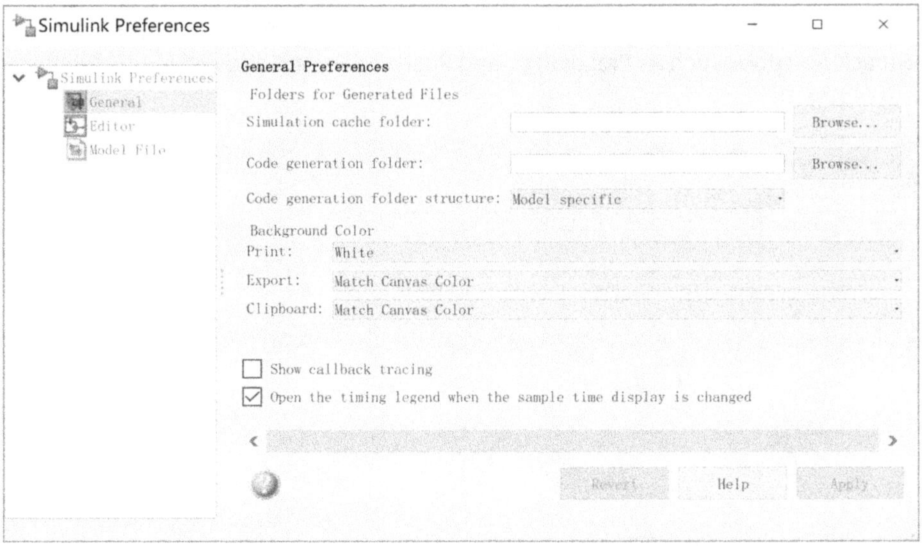

Figure 3.21: Global setting in Simulink.

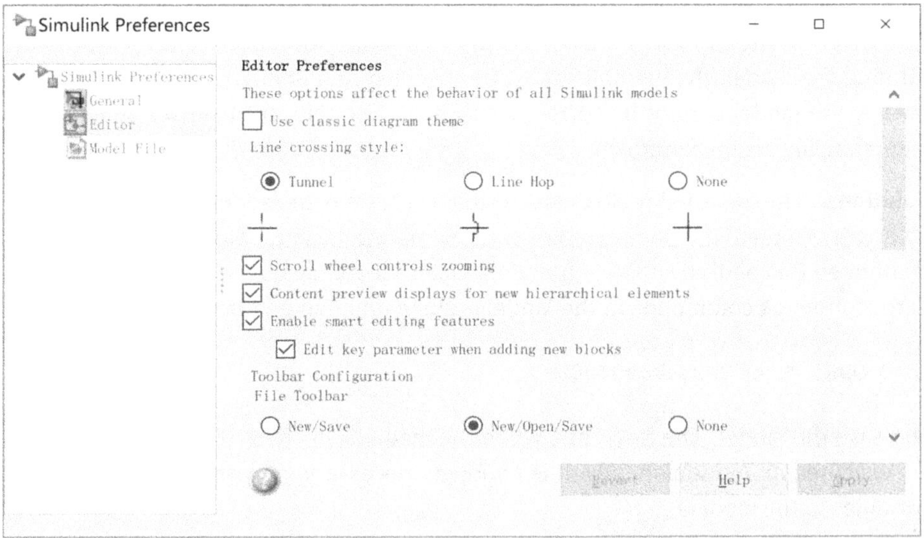

Figure 3.22: Simulink editor setting.

the presentations in signal line crossing, where Tunnel is the default one, and it is not recommended to alter the option. Model zooming methods can also be set, such as with Scroll wheel controls zooming option, and the model window display can be magnified or reduced with the mouse scroll wheel controls. If touch-screen computers are used, gestures can be used to magnify or reduce the models as well.

If the **Model File** item on the left is selected, the dialog box shown in Figure 3.23 is opened, allowing the user to select the default suffix of the models, `slx` or `mdl`. The former is the default one. Other model information can also be specified.

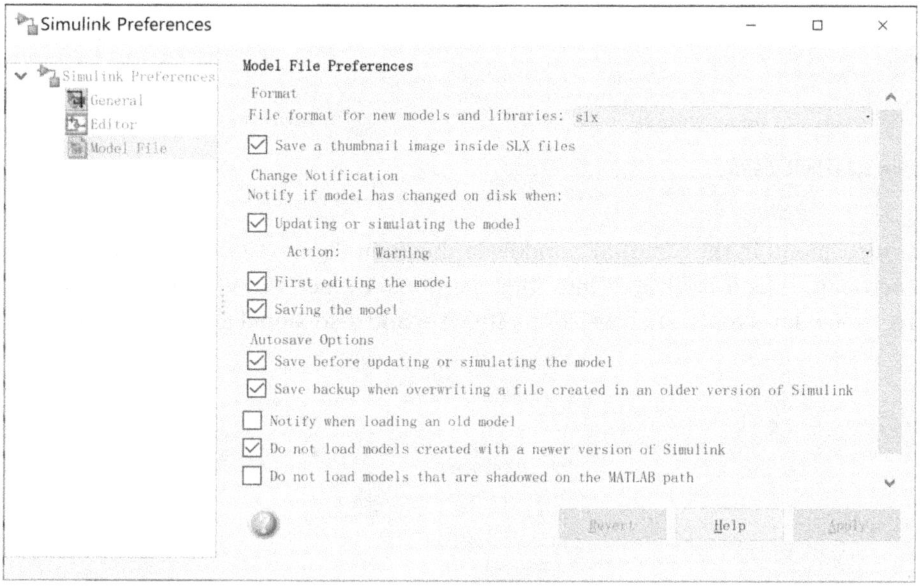

Figure 3.23: Simulink model setting.

3.3.2 Edit, view and display menus

If the **Edit** menu in Simulink model window is selected, the submenu items are displayed in Figure 3.24. We break the menu into two portions, since it is too long. In fact, many items are similar to the ordinary Microsoft Windows ones. Simple operations such as copy and paste are allowed to manipulate blocks. If the **Select All** item is selected, all of the menu items in the model window are selected. The **Find** menu item provides an edit box, allowing the user to search for a block name in the model window.

Copy Current View to Clipboard menu item copies the Simulink model into Windows' clipboard. Two submenu items are provided: one is **Bitmap** and the other is **Metafile**. The so-called metafile memorizes the way how the block diagram is drawn, including the layout of the blocks and how they are connected. The saving system of these two formats are different. The former faithfully memorizes the image information, and the latter is suitable for scaling of the pictures.

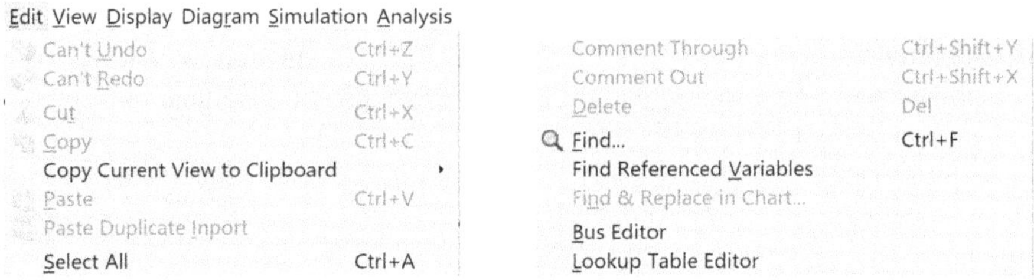

(a) upper portion (b) lower portion

Figure 3.24: Edit menu.

The **View** menu in the Simulink window is shown in Figure 3.25. Again, the menu is rather long, thus it is divided into three portions. In fact, the **View** menu may visit directly low-level tools, such as **Library Browser** and also **Model Explorer** and so on.

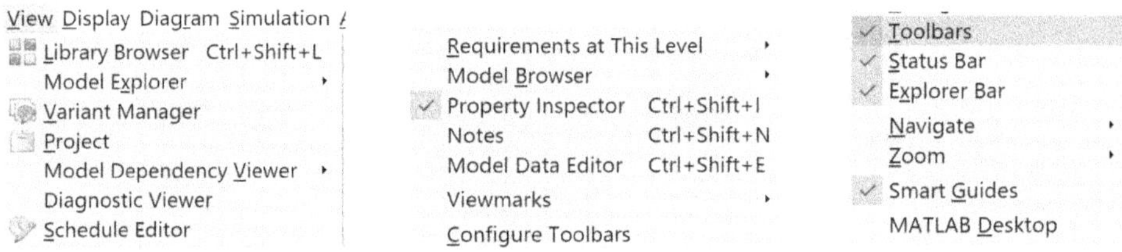

(a) upper portion (b) middle portion (c) lower portion

Figure 3.25: View menu.

The **Display** menu is used to specify the display format. The menu is given in Figure 3.26(a). The **Interface** submenu displays things related to the interfaces of the models. **Library Links** display information of each block and its prototypes and positions. The user is also allowed to break the link between the current block and its original prototype blocks.

Signals & Ports menu is shown in Figure 3.26(b), and the display and manipulation of signals and ports are allowed. The commonly used ones are **Signal Dimensions**, **Wide Nonscalar Lines**, **Port Data Types** and so on. The processing methods make Simulink models easy to maintain and error check.

Example 3.11. Display the port information for the model in Example 3.4.

Solutions. The Simulink model for PID control system in Example 3.9 is an independent model. If the **Display** → **Interface** menu item is selected, the port display is as shown in Figure 3.27. It can be seen from the display that there is no input port in the model, since all the input signals are already described in the model, and there is no extra input port. There is only one output port labeled **Out1**. It is seen on the display

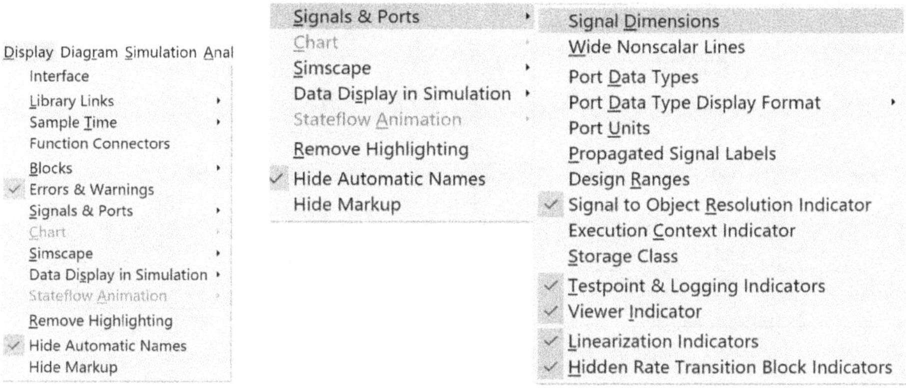

(a) Display menu (b) Signals & ports submenu

Figure 3.26: Display menu and options.

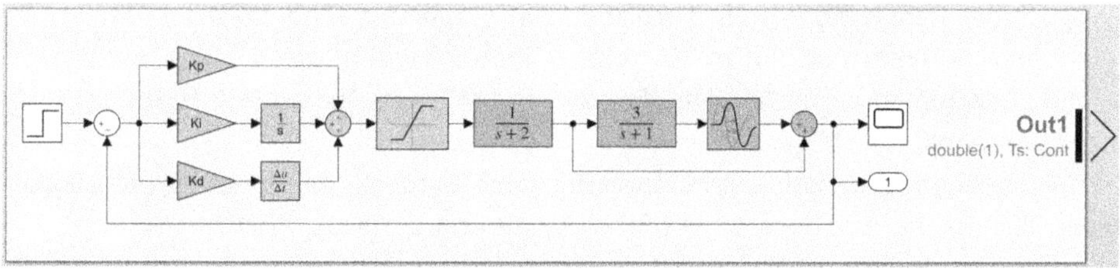

Figure 3.27: Port information display of PID control model.

to the right that the data type of the output port is **double(1)**, where `(1)` indicates that it is a scalar line. **Ts: Cont** indicates that the signal is a continuous one. It can be seen that the information of the ports in the model can be described.

Simple demonstrations on ports and signal line decorations are given here. Some other examples will be given later on these topics.

3.3.3 Model manipulation menus

Model manipulation can be handled with the **Diagram** menu shown in Figure 3.28(a). In fact, some of the **Format** submenu items further displayed in Figure 3.28(b) can be visited directly from the shortcut menu discussed earlier. Then there are slight differences in the menu item arrangement but they are the same in actual use.

Other commonly used items in **Diagram** menu are:

(1) **Refresh Blocks** allow the user to update the information in the model. For instance, in Example 3.8, a warning message was demonstrated, where variables `n1` and `d1` for the transfer function were not available in MATLAB workspace. Even though

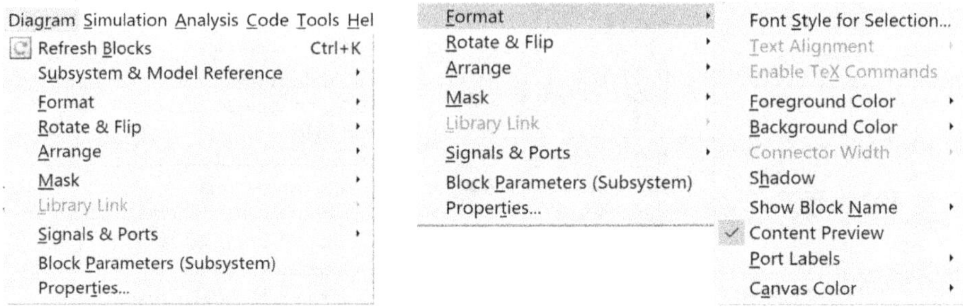

(a) Diagram menu (b) Format submenu

Figure 3.28: Diagram menu.

they are later specified, the warning signal may not vanish immediately. The block can be refreshed to immediately eliminate the warnings.

(2) Subsystem & Model Reference and Mask submenus allow the user to construct subsystems and reusable blocks. The related methods will be presented in detail later.

(3) The submenus Rotate & Flip, Arrange and Mask are the same as the ones in the shortcut menus.

(4) The Signals & Ports submenu is the same as the submenu under the Display menu. The display and decoration of the signals and ports are allowed.

3.3.4 Setting menus for simulation parameters

The Simulation menu is the most important and exclusive one in Simulink. The contents of the menu is shown in Figure 3.29, where Mode lists several supported simulation execution modes, such as Normal, Accelerator and External. The normal mode is the direct and straightforward execution mode in Simulink. The accelerator mode allows the Simulink model to compile, such that the execution is usually faster than the normal mode. The external mode refers that the simulation model can be made as a standalone execution code, without the support of MATLAB. Among these three modes, the normal mode is the default and recommended one. Although the external is faster, it is not possible to maintain. Besides, since the conversion process to the standalone code also needs time, the overall time consumption may not be reduced. If the model is changed, the compiling process is needed to again generate the executable code. Therefore, unless necessary, this mode is not recommended.

The Model Configuration Parameters submenu opens a model parameter setting dialog box, as shown in Figure 3.30. Various simulation parameters can be set in this dialog box. It is an important dialog box, since all the necessary simulation control parameters can be set in this dialog box. Some parameters may significantly impact the modeling and simulation results, so that they must be assigned carefully.

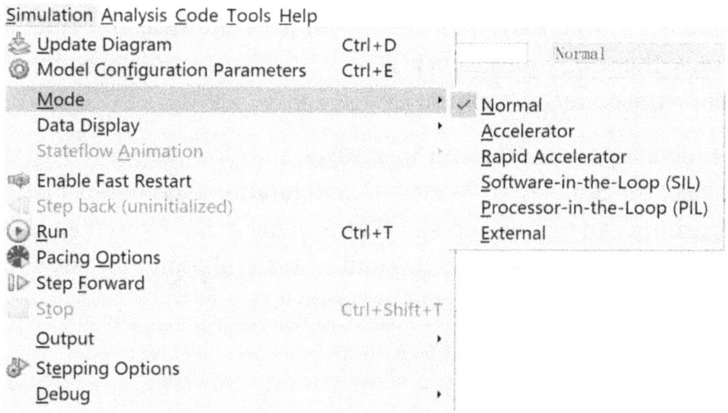

Figure 3.29: Simulation menu.

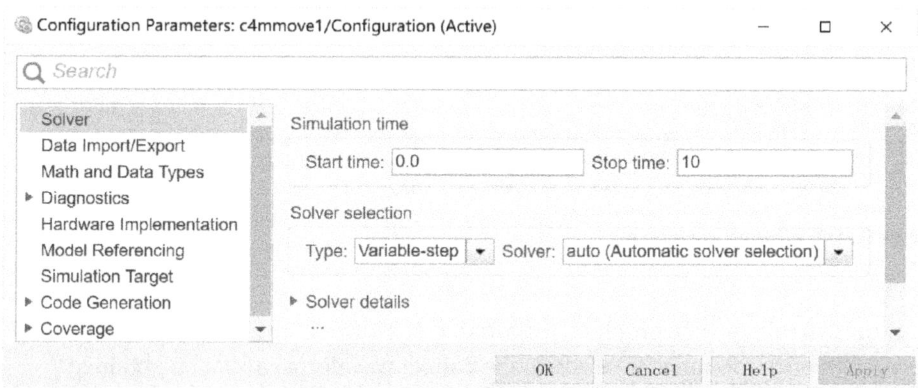

Figure 3.30: Simulation parameter setting dialog box.

Clicking the icon in the Simulink toolbar, the dialog box opened is the same as the **Model Configuration Parameters** submenu.

It can be seen from the parameter setting dialog box in Figure 3.30 that the **Start time**, **Stop time** can be assigned. The **Solver** listbox allows the user to select simulation algorithms. Details of some of the algorithms may refer to Volume V of the book. Algorithms and control parameters will be given later in this book, and show the impact to simulation results.

Solver selection pane provides Type listbox, allowing two types of algorithms: fixed-step and variable-step algorithms. The difference and advantages of these two types are presented in detail in Volume V, and variable-step algorithms are recommended. If the user insists on using a fixed-step algorithm, the user may select the step-size as well.

Example 3.12. In Example 2.2, a periodic staircase signal generator was illustrated. Under the default settings, the wrong waveform is generated. Use a fixed-step simulation algorithm to generate the correct waveform.

Solutions. The model file `c2mperiod.slx` can be resaved into file `c3mperiod.slx`. Open the model and select the **Simulation → Model Configuration Parameters** menu item, the fixed-step algorithm can be selected and set the value of the **Fixed-step size** edit box to 0.0001. The stop time can be set to 100 and start simulation process, and the correct waveform can be generated, as shown in Figure 3.31.

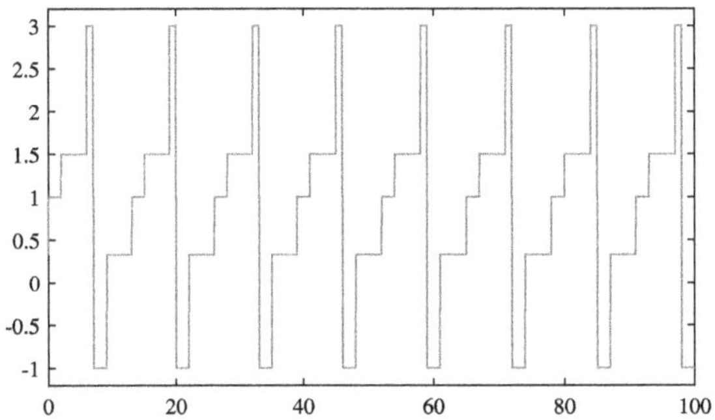

Figure 3.31: The correct waveform.

Note that for this specific problem, if a variable-step algorithm is selected, even though very tough error tolerance is selected, the correct waveform cannot be obtained.

If the ▸ sign to the right of **Solver details** label is clicked, the simulation parameter dialog box is expanded to the next level, as shown in Figure 3.32. In this dialog box, further simulation control parameters such as **Relative tolerance**, **Absolute tolerance** and so on appear. The status of **Zero-crossing options** can also be set. By default, the value of relative error tolerance is `1e-3`, that is, allows one-thousandth relative error. This value is far too large in a real numerical simulation. It should be set to a much smaller value.

Some of the control parameters may have a significant impact on the simulation process and results. Suggestions will be given later on how to select relevant control parameters.

3.3.5 Analysis tool menus

In the Simulink model window, **Analysis** menu shown in Figure 3.33 is provided. It can be seen that some analysis on the model properties can be made. For instance,

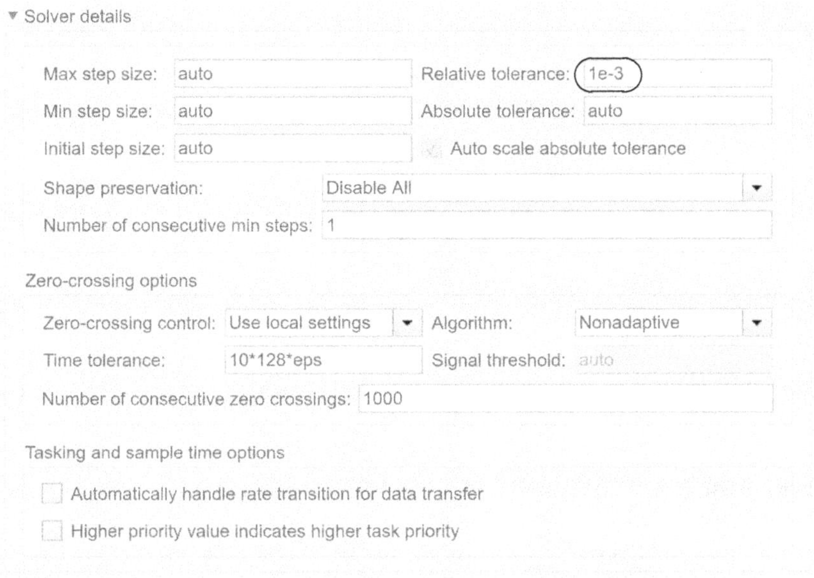

Figure 3.32: Details in the solver setting.

the Model Advisor can be used to analyze the static behaviors of the system. If there are errors or faults in the model, they may be indicated.

Many modeling manipulating APPs such as Performance Tools and Compare To are provided in Simulink. Also, the tools such as Control Design, Parameter Estimation, Sensitivity Analysis, Design Verifier and Response Optimization are also provided. These tools can be used in the analysis and design of control systems.

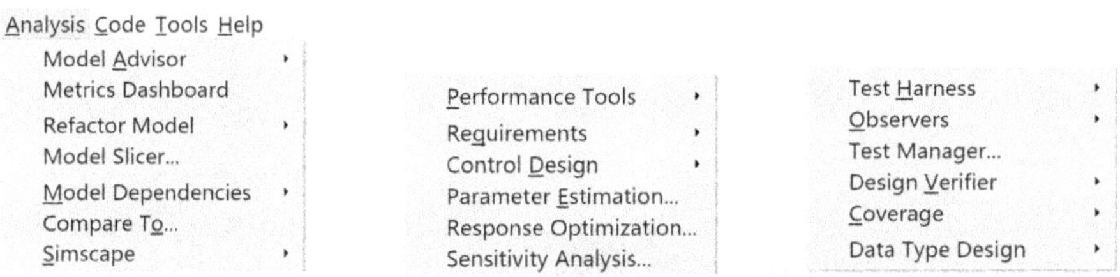

(a) model analysis (b) control system analysis (c) other analysis tools

Figure 3.33: Menus for system analysis and design.

The Code menu in Simulink model window is shown in Figure 3.34(a), which is used to perform code conversion problems for Simulink models.

The Tools menu in the Simulink model window is shown in Figure 3.34(b), where a lot of Simulink programs such as model browser and model advisor aforementioned can be visited. Some tools are to be presented later.

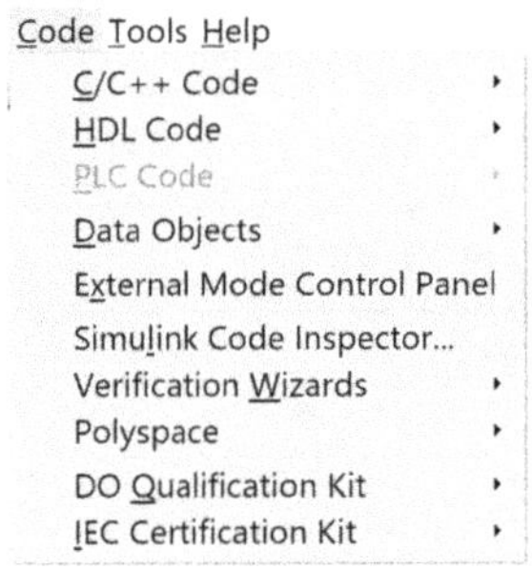

(a) Code menu

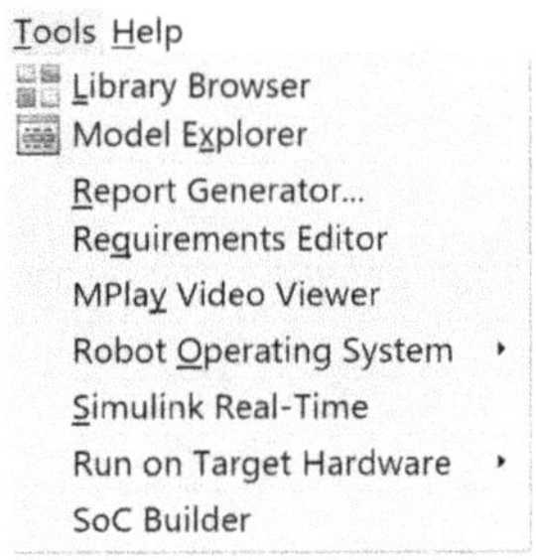

(b) Tools menu

Figure 3.34: Code and Tools menus.

3.4 Toolbar and menu systems in new versions

In the current Simulink versions, the interface provided is no longer a menu system, but the graphical toolbar. Although the actions of the toolbars are similar to the menu system, the operations have some problems. The menu-based operation methods presented in the previous section cannot be used directly.

In fact, as presented earlier, the main toolbar in the Simulink versions are shown in Figure 3.35. Since the toolbar is too wide, it is given here in two rows. In the toolbar, five panes are provided, and the default one is SIMULATION. These panes can also be regarded as the tags. Each pane corresponds to a certain toolbar. In this section, some of the commonly used options in the toolbars are presented.

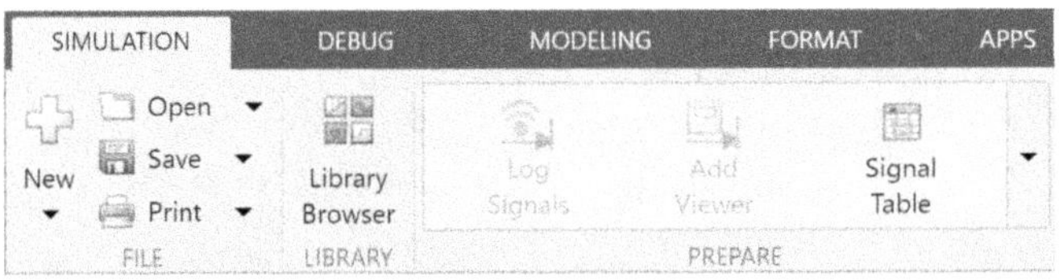

(a) the left one

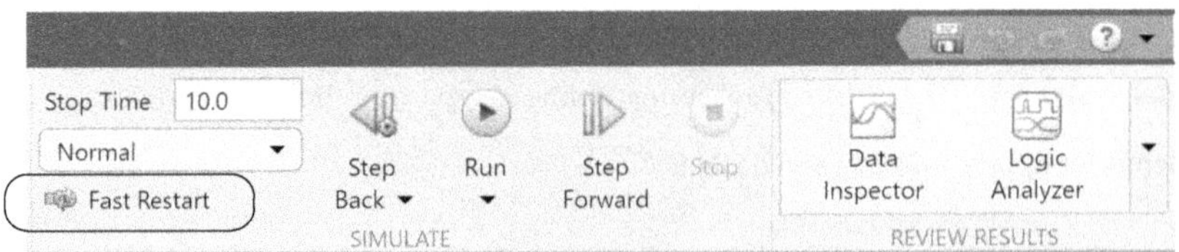

(b) the right one

Figure 3.35: Toolbars in the new versions of Simulink.

3.4.1 Simulation parameters and setting

In the default SIMULATION pane, the options provided appear in different portions of the toolbar for different actions. Each portion has its own name, and appear at the bottom of the toolbar. The following portions are presented:

(1) The left most portion is labeled as FILE, which is quite similar to the File menu in earlier versions. The icons for file operations, such as New, Open, Save and Print are provided. The functions of these buttons are very close to the menu items in the old versions. To the right of these buttons, there is a ▼ sign, indicating that there are lower-level pop-up menus. For instance, click the ▼ sign to the right of the Open button, and a pop-up menu shown in Figure 3.36 is given. The user is allowed to select a model name. If the ▼ sign is not clicked, only press the button and the default action of this button is executed. For instance, click Open may bring in a standard filename dialog box.

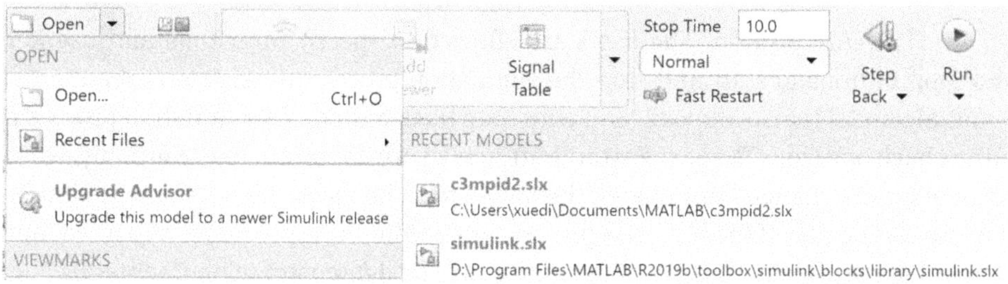

Figure 3.36: Pop-up menu of Open button.

The pop-up menu also allows the user to select Upgrade Advisor, and the Simulink model can be upgraded into the most recent version.

(2) The LIBRARY portion has only one icon, Library Browser, and by selecting it, the block library can be opened directly.

(3) PREPARE portion allows the user to add an observing point, and lists Signal Table. The tool can be used to trace all the signal lines. Next, an example can be used to demonstrate the use of signal table tool.

Example 3.13. List all of the signal lines in the Simulink model in Example 3.4.

Solutions. Command `c3mpid1.slx` can be used to open the Simulink model. Of course, the Open menu in the SIMULATION pane and the listbox shown in Figure 3.36 is displayed. The Simulink model name in Example 3.4 can be selected, and the Simulink model window can be opened automatically. Click the Signal Table icon in the toolbar, and the interface shown in Figure 3.37 is displayed. The users are allowed to assign names to the signal lines in any block. Double click the Name cell of the signal line, and the signal line name can be specified in the cell. If the **Test Point** checkbox of a certain block is checked, a mark ⚲ is added over the signal line.

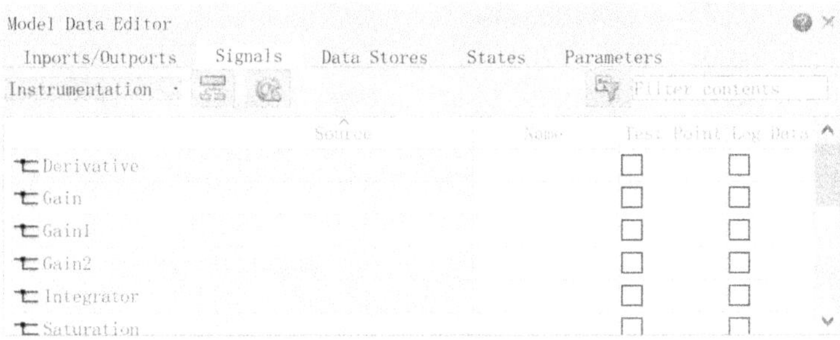

Figure 3.37: The table listing all the signal lines.

(4) **SIMULATE** portion. The users are allowed to specify **Stop Time** and use **Run** and **Stop** buttons to start and stop the simulation process. Single step execution of Simulink model is also allowed, and with **Step Back** button, the simulation process is rolling back one step. The **Fast Restart** button is a very important one. Normally, when parameters are changed in a model, the model must be recompiled before the simulation process begins. If the button is pressed, the recompiling process is bypassed, and simulation may begin. The use of the button will be demonstrated in Chapter 8.

(5) **REVIEW RESULTS** portion. The icons of **Data Inspector** or **Logic Analyzer** can be used to observe results. Of course, scopes or output ports can be used to observe simulation results in a more straightforward manner.

3.4.2 Model debugging

If the pane **DEBUG** is chosen, the left portion of the toolbar is shown in Figure 3.38, while the right portion is the same as others. The toolbar is used to assess simulation algorithms.

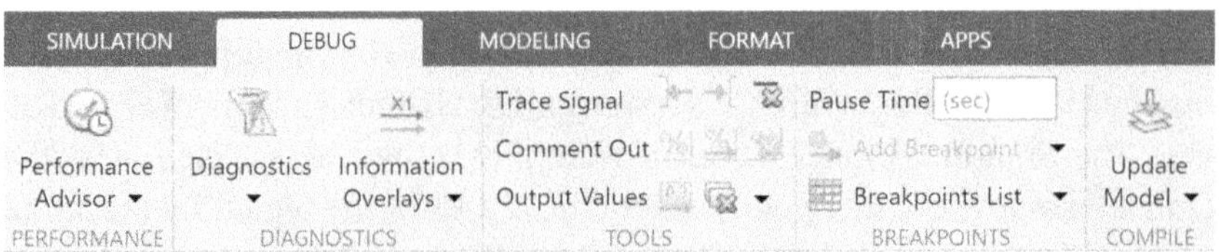

Figure 3.38: The left portion of debug pane toolbar.

If the **Performance Advisor** button is pressed, the corresponding menu is shown in Figure 3.39. For instance, if the **Solver Profiler** program is called, the profiler window is opened. Under the interface, click button to start the simulation process, and the algorithm profiler interface is opened, as shown in Figure 3.40. It can be seen that the algorithm selected is ode45, with maximum step-size of 0.2. In 87.27 % of the time, maximum step-size is used in simulation. Therefore the assessment result is satisfactory. In the modeling and simulation of large-scale systems, the statistical

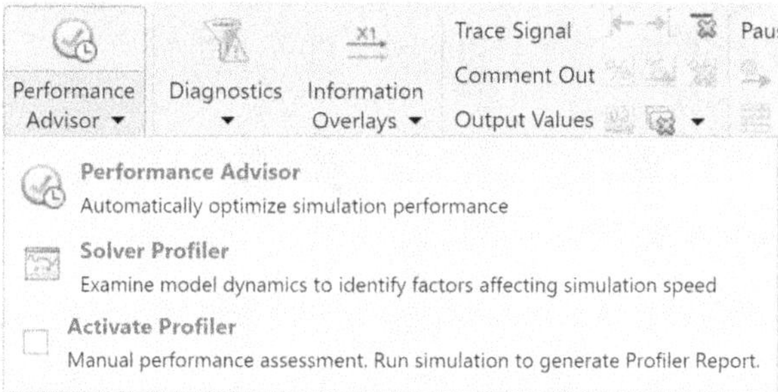

Figure 3.39: Performance adviser interface.

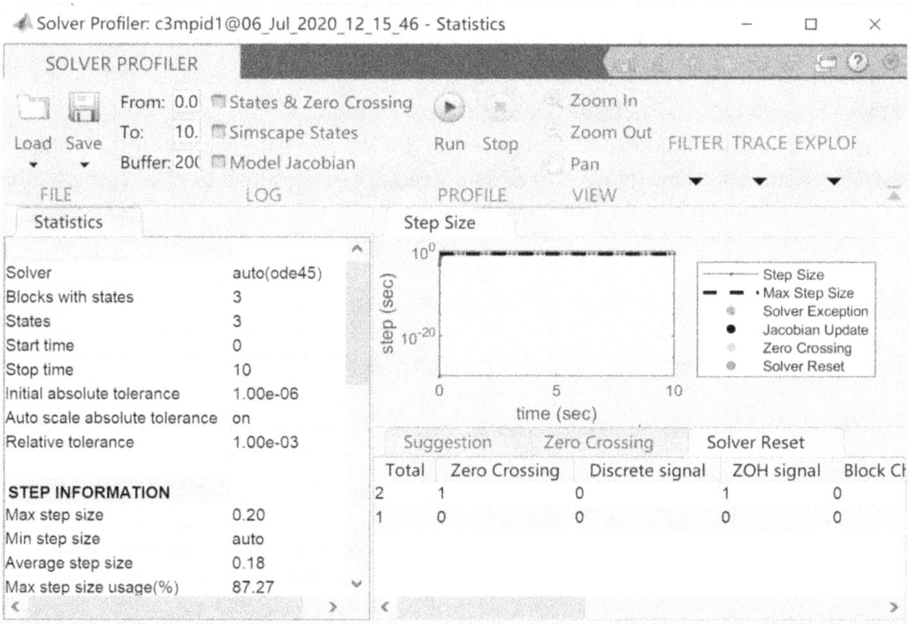

Figure 3.40: Solver profiler analysis result.

analysis results provided in this tool is meaningful. It can be used to guide the users to improve the simulation model, and promote the simulation efficiency.

3.4.3 Model creation and parameter setting

If the **MODELING** pane is selected, the left portion of the toolbar is shown in Figure 3.41, while the right is the same. It can be seen that this pane is used to specify the model and simulation control parameters. It can be seen from the toolbar that the tools such as **Model Advisor** can be called directly. It can be seen that **Model Data Editor** program can be opened, and the interface is similar to the **Signal Table** one.

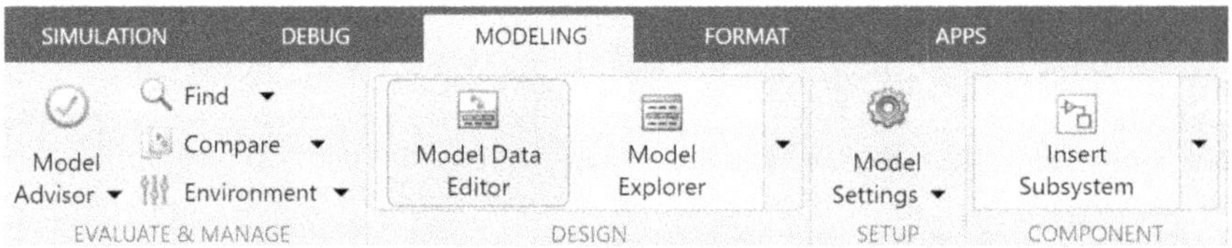

Figure 3.41: The left portion of the modeling pane toolbar.

Besides, the toolbar also allows to search and compare models, where the **Compare Models** menu is shown in Figure 3.42(a). If the **Model Settings** pop-up menu is opened, its contents are shown in Figure 3.42(b). If the **Model Settings** option is selected, the simulation parameters setting dialog box is shown, which is the same as the one in the previous section, as in Figure 3.30. With such a menu and the opened dialog box, various control parameters can be assessed. If the **Model Properties** option is chosen, the dialog box shown in Figure 3.19 is displayed. These settings are the same as the ones presented in the previous section.

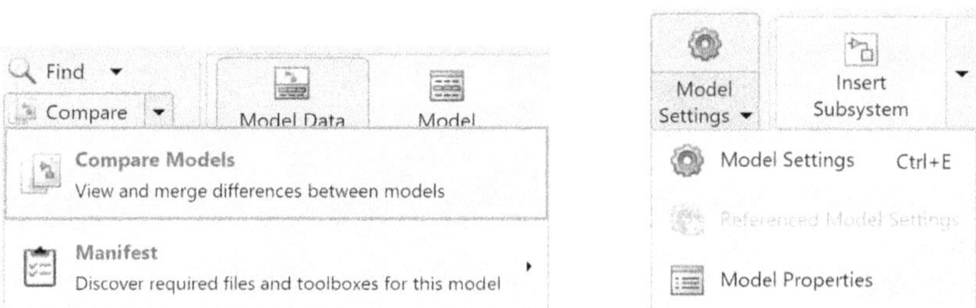

(a) model compare (b) model setting

Figure 3.42: Model Settings menu.

3.4.4 Format setting

The toolbar for the FORMAT pane is shown in Figure 3.43. For instance, the Screenshot option can be selected, and the pop-up menu is shown in Figure 3.44. The options Send Bitmap to Clipboard or Send Windows Metafile to Clipboard can be selected to copy the block model into clipboard. The two screenshot methods were discussed earlier. Besides, if the Windows screenshot facilities are used, use Win+Shift+S keys to initialize the screenshot action, so that the expected model can be copied into the Windows clipboard in bitmap form.

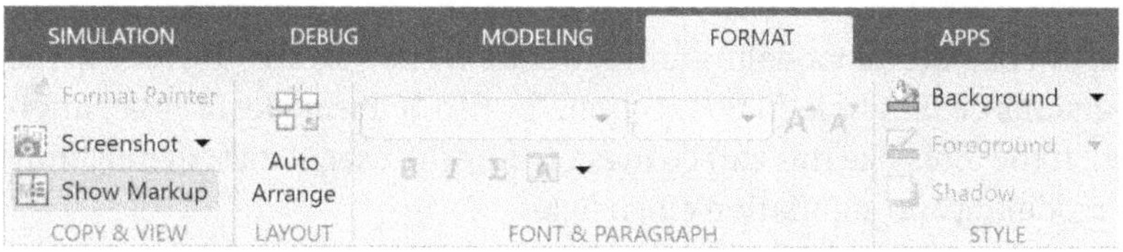

Figure 3.43: Format setting toolbar.

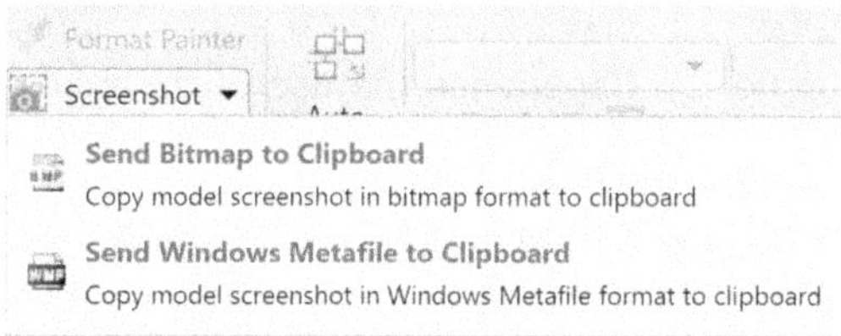

Figure 3.44: Screenshot menu.

The middle part of the toolbar enables the user to set font information, while on the right, the block decorations such as foreground and background color setting are allowed. These facilities are similar to other Microsoft Windows programs.

3.4.5 Calling APPs

Several relevant APPs can be visited directly from Simulink toolbar, as shown in Figure 3.45. It can be seen that the APPs such as Linearization Manager, Model Linearizer, Control System Designer and Parameter Estimator and so on in the toolbar can be called directly.

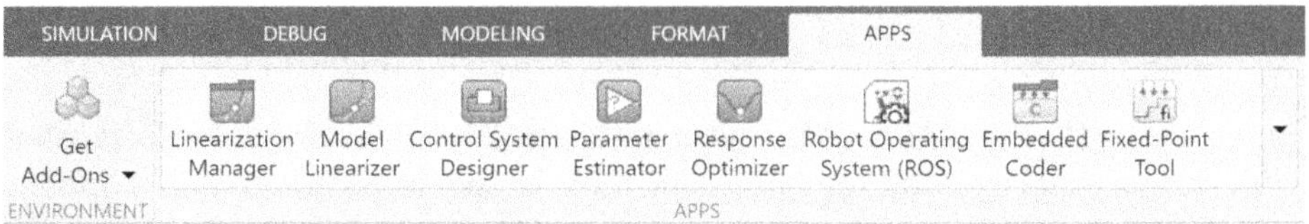

Figure 3.45: APP toolbar.

3.5 Manipulation interface of model browser

A significant amount of tools for modeling and simulation analysis are provided in the relatively new versions of Simulink. In this section, the use of the model browser interface is presented. If the tools like this can be mastered by the user, it may significantly boost the users modeling and simulation efficiencies.

The model browser can be visited directly from many Simulink menu items. For instance, in the old versions, the menu View → Model Explorer can be used to open the browser. In the new versions, the browser can be called from the MODELING pane. In this section, a simple model is used as an example to demonstrate the use of model browser.

Example 3.14. Recall the model in Example 3.4. Experience the use of the model browser.

Solutions. In the model, there are many blocks. If these blocks are visited individually, it is rather complicated. The File → Model Explorer menu item can be used directly, and the model browser interface is opened as shown in Figure 3.46. The browser allows the user to visit any block in the model, and also observe and modify the parameters in the blocks.

The model browser window is shown in left, middle and right portions. On the left, some currently opened Simulink blocks are listed, and `c3mpid2a.slx` is the Simulink model of the example. If other models are opened, they are listed as well in the listbox. To close a Simulink model, right click the model from the listbox, and from shortcut menu, select Close item, and the model can be closed directly.

Assume that there are many models currently opened. Here, `c3mpid2a.slx` model is selected. All of the information of the blocks in the selected model are displayed in the middle portion of the browser, including Name, BlockType and so on.

On the right of the browser, the model information is displayed. In fact, this portion is the same as the one in the Model Properties menu. The model information can be browsed, and some callback facilities such as Preload Fcn can be called.

Now consider the middle portion of the model browser. If the user selects a block name, the information in the parameter dialog box is shown on the right portion of the browser. For instance, if the Saturation block is chosen, the browser is shown

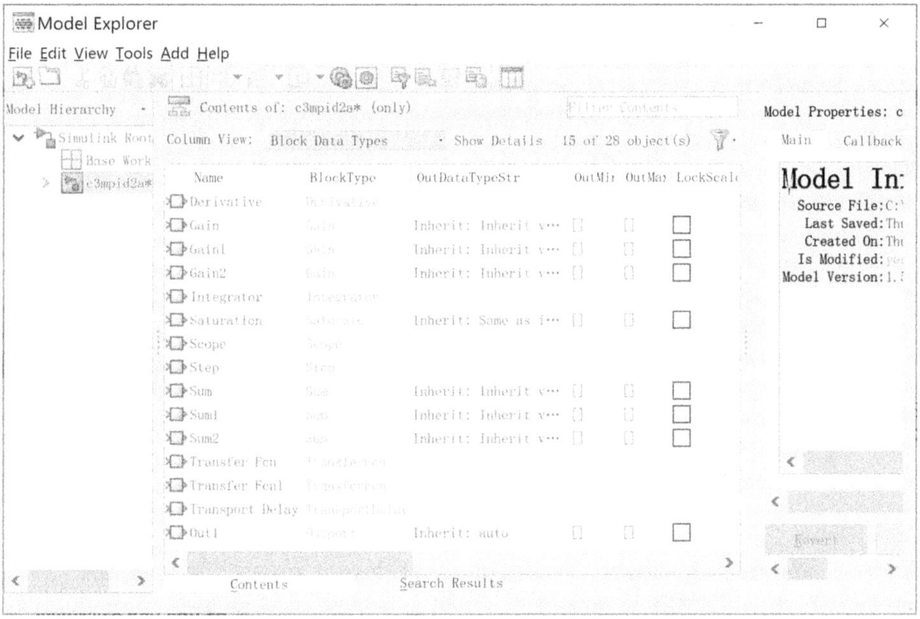

Figure 3.46: Model browser interface.

in Figure 3.47. It can be seen that the right portion of the browser is the same as the Saturation block parameter dialog box. Therefore, to modify the blocks in a model, it is not necessary to open the parameter dialog box of the block, one by one. The model browser can be used to modify them in the same interface.

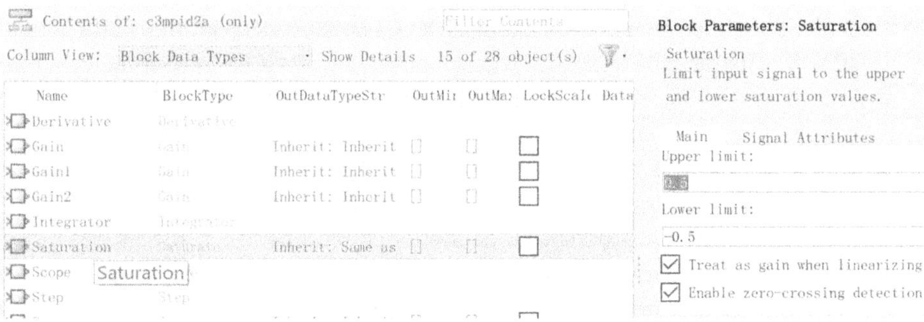

Figure 3.47: Model information display and modification.

Select a block, and right click the block to get the shortcut menu. Select the **Properties** menu item, and the information in the original parameter dialog box can be displayed in the browser. It can be used to modify block parameters.

Apart from the Simulink models, the left portion of the browser also has a Base Workspace entry, displaying the Name, Value, DataType and Dimensions of the variables in MATLAB workspace (Figure 3.48). Each variable is accompanied by an edit box, allowing the user to modify the values of the variable. If a variable name is selected, it can also be modified in the right portion of the browser. Apart from variable values, other properties of the variables cannot be modified.

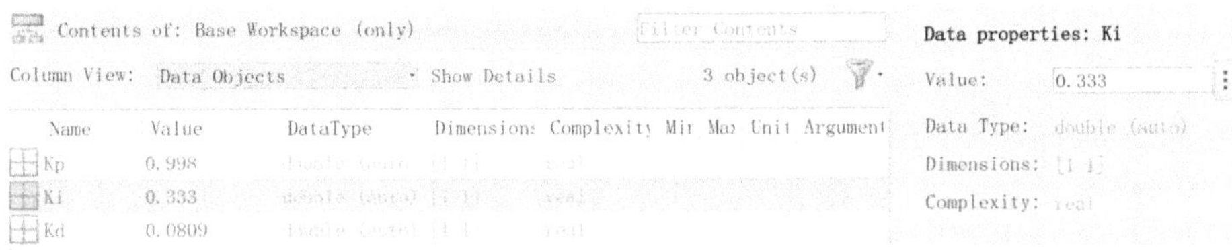

Figure 3.48: MATLAB workspace display.

It can be seen from the above introduction that there is no need to open the blocks one by one to modify the parameters, if such a browser can be used properly. With the browser framework, these actions can be manipulated properly, so as to boost the modeling efficiency.

3.6 Exercises

3.1 Consider the Simulink model in Figure 3.5, where the connection to the dead-zone nonlinearity is not satisfactory, and the feedback loop is not good. Select the block and flip the block directly from the shortcut menu. Observe the flipping results.

3.2 Consider the Simulink model in Example 3.11. Use the tools provided in the model interface, to display the data structures of all the signals. Change the vector signal lines into wide ones, so as to see whether the model has vector signals.

3.3 Select the fixed-step simulation method in Example 2.2 for the staircase signal generator, and set the step-size to h. Set $h = 10^{-4}$ in the pre-load callback function so that each time the model is opened, it can be assigned automatically.

3.4 Use the model browser to visit all of the block parameters in the model in Example 3.1, and compare the conveniences in using the browser, compared with traditional individual block manipulating methods.

3.5 Open a Simulink model and copy it into a Microsoft Word file.

3.6 Open the model browser, and observe which of the models are currently opened. Observe also the variables' names existing in the current workspace. See what are the current values of these variables might be.

4 Modeling and simulation of ordinary differential equations

Ordinary differential equations are the mathematical foundation in describing continuous dynamical systems. In Volume V of the series, analytical solutions as well as numerical solutions are presented systematically, also some attempts are made in introducing numerical solutions with Simulink. In this chapter, ordinary differential equations are used as the background, and introduce Simulink modeling and simulation in a systematic manner. If there is not ambiguity, the term "differential equations" in this book refers specifically ordinary differential equations. Partial differential equations cannot be solved directly with Simulink.

In Section 4.1, some common rules and procedures in differential equation modeling are illustrated. Then concentrations are made on the design of simulation templates, so as to provide convenient setting for later modeling tasks. A simple example is given to demonstrate the modeling process. In Section 4.2, Simulink modeling methods for first-order explicit differential equations are presented. Low-level modeling as well as vectorized modeling methods are both presented. Universal frameworks for time invariant and time-varying differential equations are proposed. In Section 4.3, high-order explicit differential equations are illustrated. Integrator chains are used to define key signals, and around with the signals, Simulink modeling of differential equations are presented. Vectorized signal construction and applications are also introduced. Besides, attempts are made to explore the solutions of implicit differential equations. In Section 4.4, Simulink modeling of high-order differential equations are presented. Explicit differential equations and delay differential equations are both explored. Modeling and solution of special differential equations are discussed, such as switched differential equations, and differential equations with stochastic inputs. In Section 4.5, an introduction is made for fractional-order differential equations modeling and simulation. As an extension to this topic, Section 4.6 discussed the modeling and solutions of nonlinear difference equations with examples.

4.1 Template design in Simulink

As it was said by Confucius, "The craftsman who wishes to work well has first to sharpen his implements." In order to solve properly the modeling and simulation problems of complicated systems, Simulink can be selected, but that is not sufficient. Based on Simulink, the optimizations in setting simulation environment is equally important, since the "implement" can be further sharpened, so as to solve the modeling and simulation problems well.

As it was talked earlier, a blank model window can be opened in Simulink with menu or toolbar buttons. The model is based on the default template in MATLAB. In

https://doi.org/10.1515/9783110734959-004

real applications, some control parameters in the default template is not ideal, or even, misleading results may be obtained. Therefore, model template design may be very important in Simulink modeling and simulation. In this section, a potential problem in the default template is demonstrated, then modification methods on the template is suggested, followed by other improved setting in the template.

4.1.1 General rules in ODE modeling

There are some useful rules in Simulink modeling of differential equations. In practice, the following manipulations can be made on drawing block diagrams:

(1) To define two signals $x(t)$ and $x'(t)$, an integrator block can be used. The output signal can be regarded as signal $x(t)$, then the input signal is naturally $x'(t)$ signal. In this way, the key signals in differential equation modeling are defined.

(2) If two signals are equal, they can be joined directly by signal lines. In simulation modeling, this method is often used to close up the simulation loops.

(3) If signal $u(t)$ is known, it can be followed by a Gain block, and double click the block to set the value of the gain to k. The output of the block is $ku(t)$.

(4) If nonlinear algebraic computing on a signal $v(t)$ is expected, for instance, to compute $(v(t) + v^3(t)/3)\sin v(t)$, signal $v(t)$ can be fed into the Fcn block. Double click the block, in the parameter dialog box, the string `(u+u^3/3)*sin(u)` can be specified. The output signal of the block is the expected nonlinear function. Note that no matter what is the mathematical notation of the input signal, it can only be denoted as `u` in the Fcn parameter dialog box.

(5) Many signals can be fed into the Mux block, so that a vector signal can be composed; if a vector signal is fed into a Demux block, it can be decomposed into scalar or smaller vector signals.

(6) If signal $u(t)$ is known, its delay signal $u(t-d)$ is expected, signal $u(t)$ can be fed into a Transport Delay block. Double click the block, parameter d can be specified in the parameter dialog box. The output of the block is the expected delay signal.

(7) To get the information of a certain signal, it can be connected to a Scope block, or linked to an Out block, so as to handle or display the signal.

4.1.2 Block diagram model of a simple ODE

Differential equations are mathematical foundations of dynamical systems. The modeling methods of differential equations are very important in dynamical system simulation. In this section, a very important differential equation is used as an example to build up its Simulink model, and demonstrate limitations in default simulation parameters setting.

Example 4.1. In physics, an equation of a dropping object is governed by $v(t) = v_0 + gt$, where t is the time, $v(t)$ is the instant velocity of the object, $v_0 = 1\,\text{m/s}$ is the initial velocity and $g = 9.81\,\text{m/s}^2$ is the gravity accelerator. Study and simulate the relationship between time t and displacement $x(t)$.

Solutions. It seems that $v(t) = v_0 + gt$ is not a differential equation, but an algebraic equation of velocity $v(t)$. In fact, the relationship between velocity $v(t)$ and displacement $x(t)$ is $v(t) = x'(t)$. Therefore, the original equation can be converted to a differential equation $x'(t) = v_0 + gt$. Introducing an integrator block, and assume that its output signal is the displacement $x(t)$, its input is naturally $x'(t)$. These two signals are referred to as key signals here.

With the key signals, the Simulink model shown in Figure 4.1 can be constructed, where the initial velocity $v_0 = 1$ is provided by a **Constant** block, and time t is provided by the **Clock** block, through the **Gain** block, with a gain value of 9.81, the signal gt can be obtained. Adding up the two signal with a **Sum** block, the expression on the right side of the differential equation is composed. Since the left side of the equation is $x'(t)$, and it is the same as the key signal in the input of the integrator block, the right side signal can be fed directly into the input of the integrator, such that the Simulink model for the differential equation can be established.

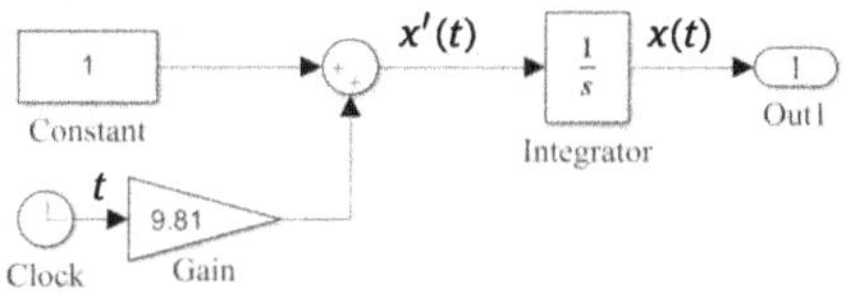

Figure 4.1: Simulink model of the dropping object (file: `c4mmove1.slx`).

Executing the simulation model, the simulation results can be obtained. Since output port block is used, under default setting, a structure variable `out` is returned automatically into MATLAB workspace. All the simulation results are provided in the structured variable `out`. In order to draw the simulation results, the following commands are needed, and the result is displayed in Figure 4.2.

```
>> plot(out.tout, out.yout{1}.Values.Data)
```

It can be seen from the commands that time vector ***t*** is a member in the structure variable `out`. The information of the output signal is also returned, which is member `yout` in the same structured variable. Unfortunately, it cannot be used directly in plotting, since `out.yout` is itself another structured variable of data type `Dataset`. In the command, `{1}` indicates it is a cell array, meaning the first output signal, and `Values` member is another structured variable, composed of many sublevel members. The actual data is in fact stored in its member `Data`.

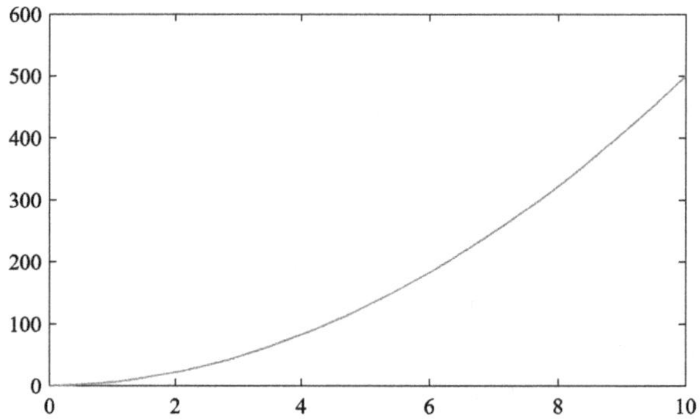

Figure 4.2: Simulation results.

Of course, the command used here is too complicated, and not suitable for real use. To retrieve simulation results in a simpler form, it is necessary to modify the default Simulink template.

Of course, if the default setting is used, the following MATLAB function can be written to draw the simulation results, or extract them. The syntax of the function is `[t,y]=sl_plot(out,v)`, where `out` is the simulation results in structured variable form. The argument v is a vector, indicating which of the signals needed. If v is not given, all the output signals are handled. If no argument is returned, the simulation results are drawn automatically, and if arguments are returned, no plot is drawn.

```
function [t0,yy0]=sl_plot(out,varargin)
t=out.tout; y=out.yout;
no=1:length(y); t=out.tout; yy=[];
if nargin==2, no=varargin{1}; end
for i=1:no, yy=[yy, out.yout{i}.Values.Data]; end
if nargout>0, t0=t; yy0=yy; else, plot(t,yy), end
```

4.1.3 Template design for the new model

Normally, the default template in Simulink is used directly. If a new model is created by the dialog box shown in Figure 2.2, the default template is also used. Therefore, the phenomenon show above may also appear, and the simulation results are returned in such a lengthy name in the structure variable, they are complicated to be visited. Therefore, it is necessary to customize the user's own template.

An example is given next to demonstrate the template customizing methods to convert the setting of output variable data types.

Example 4.2. Design a new template such that the default output data type are made simple and straightforward to use.

Solutions. It can be seen from the default setting in Figure 2.2 that **My Template** is empty. Customized template for Simulink can be made.

Several procedures are needed to create a new template:

(1) Open a blank model window. Click the ▾ sign beside the **Save** button, the file saving listbox is shown in Figure 4.3(a). Select **Template** option, a dialog box shown in Figure 4.3(b) is given. Relevant information in this dialog box can be filled in, so as to create a new template. Click the **Export** button, the new template can be saved. The file name `my_new_template.sltx` can be chosen. Note that the suffix is `slxt`.

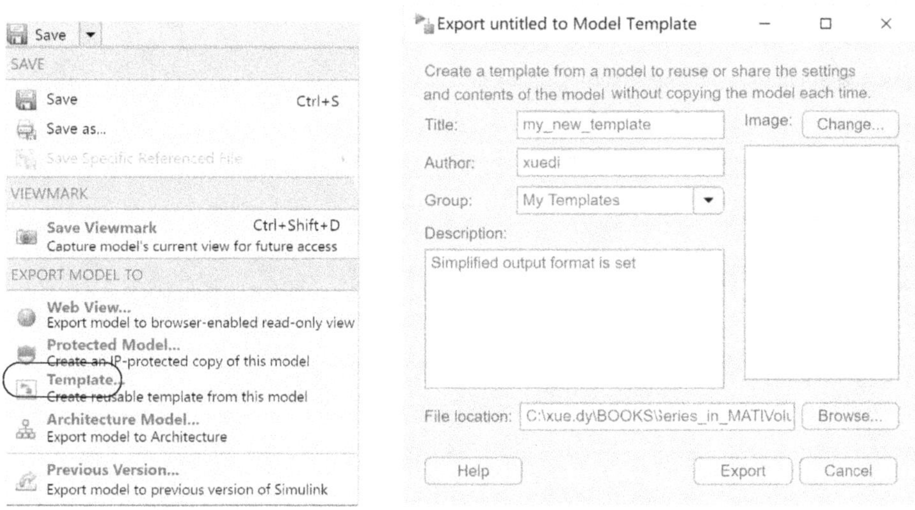

(a) Save menu (b) new template information

Figure 4.3: Creation of a new template.

(2) Of course, the new template thus saved is exactly the same as the default one, since no modifications are made yet. If the user wants to introduce his own modifications, for instance, use menu item **Simulation → Model Configuration Parameters** in the old versions, or click **Model Setting** button or ⚙ icon in the **MODELING** pane, in the new versions of Simulink, the control parameter setting dialog box shown in Figure 4.4 can be opened.

If the default output format is to be changed, the **Data Import/Export** item from the listbox on the left can be selected, and the dialog box is changed into the one in Figure 4.5, where the checkbox at the bottom, **Single simulation output**, is by default checked. Therefore, the structured variable `out` in Example 4.1 is returned. If it is unchecked, according to the setting above, `tout` and `yout` are returned, where `tout` is the time vector, `yout` for the output is still a structured variable. Its default data type is **Dataset**, and it is hard to manipulate. Therefore, it is suggested to change the data type to **Array**. The returned variable `yout` is then a matrix composed of many columns,

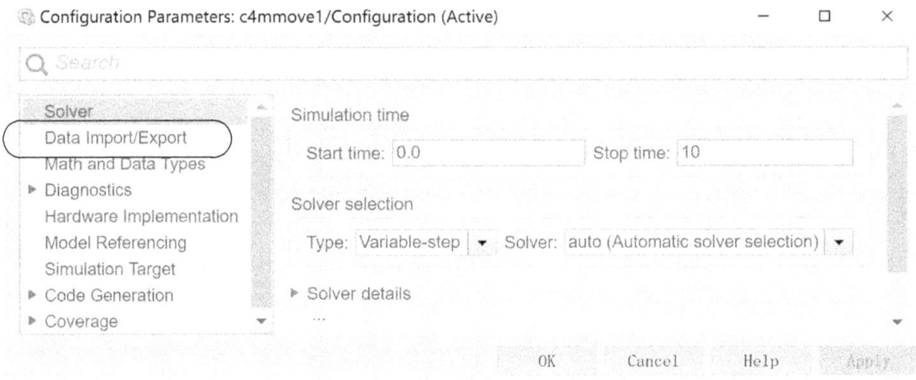

Figure 4.4: Simulation parameter setting dialog box.

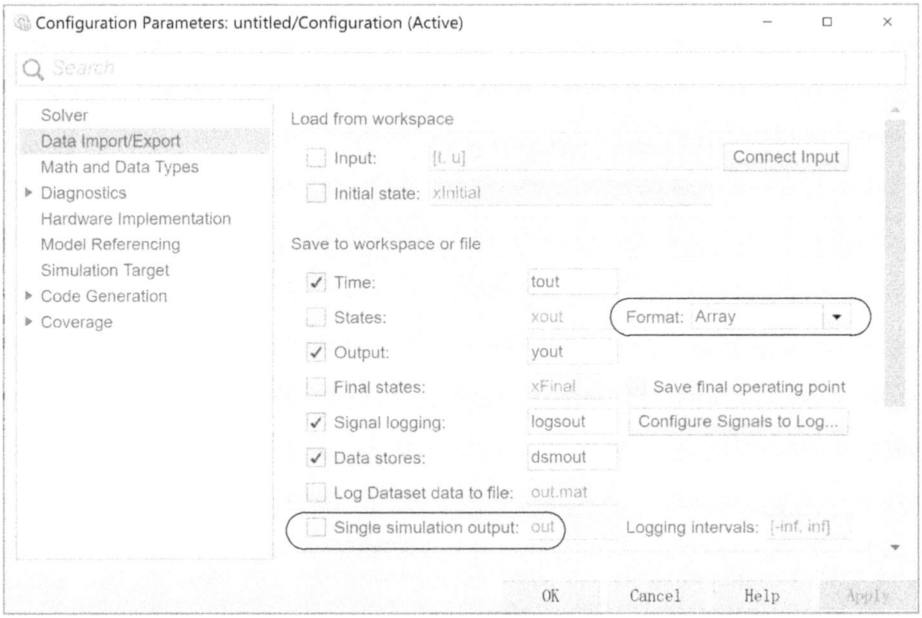

Figure 4.5: Input and output format setting.

and the number of columns equals to the number of output ports. For the model in Example 4.1, since there is only one output port, the automatically returned `yout` is a column vector. Therefore, the plotting command is simplified as

```
>> plot(tout,yout)
```

It can be seen that the output signal format is significantly simplified, such that their handling are made much easier.

(3) After the model template is modified, the procedure in (1) is used to click **Export** button the modified template can be saved. Therefore, if Simulink is invoked, to create a blank model, the new template can be opened directly. The options **Blank Model** should not be used; otherwise, the default template is still used.

It is recommended that the readers can set up Simulink model template for his own convenience, and improve gradually the template, so as to provide better environment for his modeling and simulation tasks.

4.1.4 Terminating conditions in simulation

Normally, stop time for simulation can be specified in the simulation model, by filling in a value in the **Stop time** edit box in the toolbar. The default value is 10 seconds. If the user wants the simulation process to stop automatically at a preset condition, the **Stop Simulation** block in the **Sinks** group can be used to control the simulation progress.

Example 4.3. Consider again the dropping problem in Example 4.1. If the point to the drop object is 15 meters above the ground, how long is needed for the object to reach the ground?

Solutions. Since the stop time is not known, the terminate conditions cannot be specified in the **Stop time** edit box in the toolbar. A very large value, for instance 1000 or even `inf` can be specified instead. The real terminate condition is $x(t) = 15$, and for convenience in numerical implementation, it can be changed to the inequality $x(t) \geqslant 15$. The inequality relationship can be implemented in the **Compare To Constant** block, from the **Logic & Bit Operations** group. If the condition is satisfied, a logic "1" is generated. The signal can be used to drive the **Stop Simulation** block, so as to terminate the simulation process. A Simulink model is created as shown in Figure 4.6.

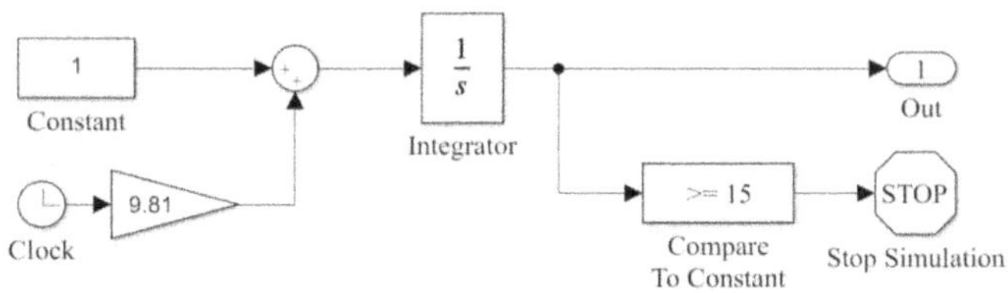

Figure 4.6: New simulation model (file: `c4mmove2.slx`).

Various of inequality relationships are supported in **Compare To Constant** block. Double click the block icon, the parameter dialog box shown in Figure 4.7 is opened. The **Operator** listbox provides different inequality relationships, and the data type of the output signal is "logical." This block can be used to compare the relationship between a signal and a constant. To compare two signals, the **Relationship Operator** block in the same group can be used instead.

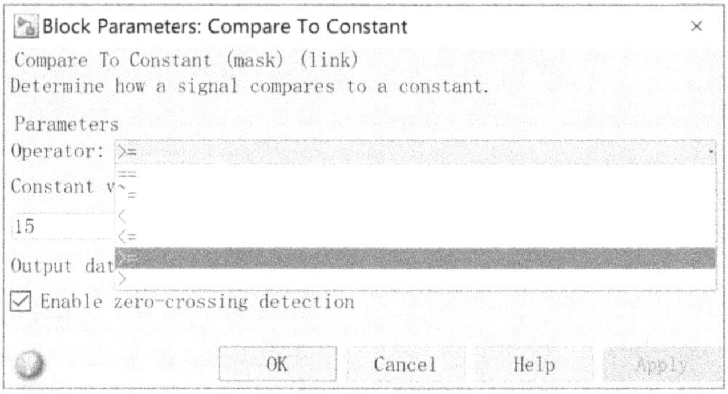

Figure 4.7: Inequality parameter dialog box.

With command `tout(end)`, the returned value is 1.6498 seconds. Of course, symbolic computation can be used to solve the equation, and consistent result of t_1 = 1.6497752452346536061617496370942 can be obtained.

```
>> syms t; t1=vpasolve(t+9.81*t^2/2==15)
```

4.1.5 Zero-crossing detection

There is an **Enable zero-crossing detection** checkbox in the dialog box shown in Figure 4.7. If it is checked, the zero-crossing detection facilities are enabled. Zero-crossing is a very important concept in simulation as well as in many fields. Consider a signal $u(t)$. If at the kth time instance t_k, we have $u(t_k) > 0$, while in the next time instance t_{k+1}, we have $u(t_{k+1}) < 0$, then in the interval (t_k, t_{k+1}), signal $u(t)$ must travel across zero value at least once. The exact point t_ξ when $u(t)$ equals zero is referred to as a zero-crossing point. If exact simulation is expected, the precise value of zero-crossing point t_ξ must be found. Otherwise, unavoidable simulation errors may be deduced. An example demonstrating zero-crossing behavior is given next, and the necessities and costs in finding zero-crossing points are addressed.

Example 4.4. Draw the curve for the function $y = |\sin t^2|$ when $t \in (0, \pi)$.

Solutions. Select a step-size of $T = 0.02$ seconds, the following MATLAB commands can be issued to draw the curve for the given function, as shown in Figure 4.8.

```
>> t=0:0.02:pi; y=abs(sin(t.^2)); plot(t,y)
```

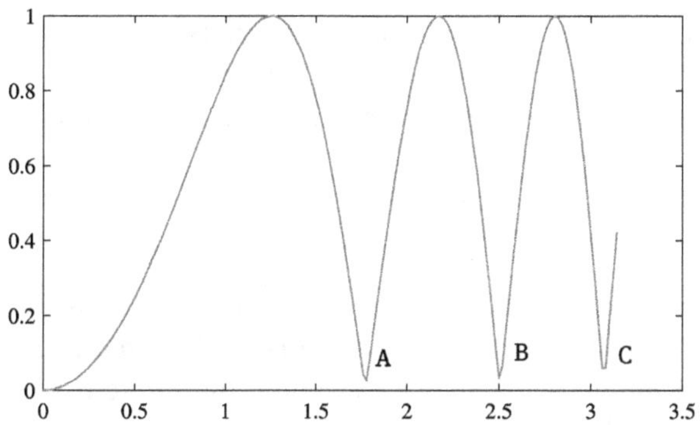

Figure 4.8: Zero-crossing demonstration of the curve $y = |\sin t^2|$.

There are obvious errors at some particular points on the curve. For instance, the behaviors of the curves at points A, B and C, when the function value is close to zeros, the points are suspicious. At point A, the curve should be decreasing to zero then rising, rather than stay at a suspended point. If the exact point when function value is 0 can be found, by, for instance, equation solution methods, and add the point to the $\boldsymbol{t}$ vector, the problem can be solved successfully. The ways of finding exact point at A is referred to as a zero-crossing detection method.

If there is a mechanism to find the zero-crossing point, for instance, with the equation solver provided in Volume IV of the series, the zero-crossing points can be found, and then embedded into the time vector $\boldsymbol{t}$, then the curve in Figure 4.9 can be drawn. It can be seen that the zero-crossing problem have been successfully solved. Therefore, the curve thus obtained is correct.

```
>> f=@(x)abs(sin(x^2)); more_sols(f,zeros(1,1,0),[0,pi])
   t0=X(:); t0=t0(t0>0 & t0<pi); % find the zero-crossing points
   t=sort([t,t0']); y=abs(sin(t.^2)); plot(t,y) % draw the curve again
```

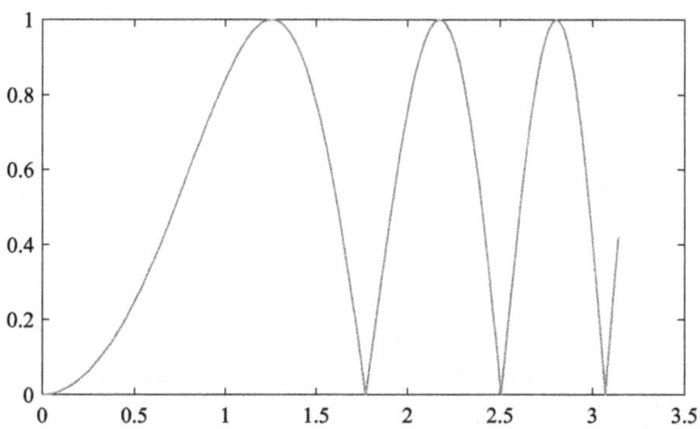

Figure 4.9: Correct curve of $y = |\sin t^2|$.

In variable-step simulation algorithms, if the error tolerance is selected small enough, the zero-crossing points may be handled well in effect, although the precision of them cannot be guaranteed. In fixed-step simulation algorithms, the zero-crossing points may not be found. Therefore, the phenomenon shown in Figure 4.8 is unavoidable. Therefore, variable-step simulation algorithms and small error tolerances are necessary conditions for accurate numerical simulation.

By default, most of the Simulink blocks allow zero-crossing detection automatically. Accurate solutions to zero-crossing points are positive conditions to ensure accurate simulation. From the simulation performance viewpoint, the zero-crossing options are not recommended to be altered. The disadvantages of activating zero-crossing detection facilities are that the speed of simulation is reduced. Since at the points near the zero-crossing ones, algebraic equations are solved automatically to find the zero-crossing points, which may reduce the speed in the entire system simulation process.

4.2 Modeling of first-order explicit ODEs

Command-line driven numerical solution methods for first-order explicit differential equations are extensively studied in Volume V of the series. To understand better MATLAB-based solution methods, the readers are recommended to read the book. In this book, Simulink based solutions of various differential equations are explored. The mathematical form of first-order explicit differential equations is presented first, followed by explorations of Simulink modeling and solutions. Universal Simulink framework is proposed.

Definition 4.1. The mathematical form of the first-order explicit differential equation is given by

$$\boldsymbol{x}'(t) = \boldsymbol{f}(t, \boldsymbol{x}(t)), \tag{4.2.1}$$

where, $\boldsymbol{x}(t) = [x_1, x_2, \dots, x_n]^{\mathrm{T}}$ is referred to as the state vector

$$\boldsymbol{f}(\cdot) = [f_1(\cdot), f_2(\cdot), \dots, f_n(\cdot)]^{\mathrm{T}} \tag{4.2.2}$$

is of any nonlinear functions, t is the time. Besides, initial state vector $\boldsymbol{x}(t_0)$ is known.

The solution method presented in Volume V is that, the right side of the first-order explicit differential equations is described by anonymous functions or MATLAB functions first, then solvers such as `ode45()` can be used to find the numerical solutions. In this section, Simulink based modeling and solution are studied.

4.2.1 Defining and generating key signals

Key signal selection is the basic step in Simulink modeling. For a given first-order explicit differential equation, it is recommended to select $x_i(t)$ and $x_i'(t)$ as key signals. For the equations with n states, n integrators can be selected, and assume that the output of the integrators as $x_i(t)$, while the input signals are $x_i'(t)$. With the key signals, the Simulink model can be constructed for the first-order differential equations based on these key signals. In this section, examples are used to demonstrate Simulink modeling methods for first-order explicit differential equations.

Example 4.5. Solve the following differential equation:[19]

$$\begin{cases} y'(x) = -2xy(x)\ln z(x), \\ z'(x) = 2xz(x)\ln y(x), \end{cases}$$

with known initial conditions $y(0) = \mathrm{e}$ and $z(0) = 1$. It is known that the analytical solutions of the differential equations are $y(x) = \mathrm{e}^{\cos x^2}$ and $z(x) = \mathrm{e}^{\sin x^2}$. Find the numerical solutions.

Solutions. It is seen from the given differential equation that the independent variable is x, while Simulink model is driven by time, the default independent variable should be t. It is better to substitute the variable manually to t. Therefore, the original differential equation can be modified as

$$\begin{cases} y'(t) = -2ty(t)\ln z(t), \\ z'(t) = 2tz(t)\ln y(t), \end{cases}$$

and the analytical solutions are $y(t) = \mathrm{e}^{\cos t^2}$ and $z(t) = \mathrm{e}^{\sin t^2}$.

If Simulink model is needed for the differential equation, key signals $y(t)$, $z(t)$, $y'(t)$ and $z'(t)$ should be assigned. According to the ideas given earlier, two integrators should be introduced to define the key signals, as shown in Figure 4.10(a). The initial values of the differential equation can be filled in the corresponding integrator parameter dialog boxes.

It can be seen from the first equation that the right side of the equation is $-2ty(t)\ln z(t)$. Key signal $z(t)$ can be fed into a Math Function block and assign the function to log. Therefore, the output of the block is $\ln z(t)$. Specifying the time t, key signal $y(t)$ and the obtained $\ln z(t)$ signals as the input to a Product block, and let the output connected to a gain block, with gain value of -2, the right side expression can be established. Since the right size is $y'(t)$ and $y'(t)$ is also a key signal defined earlier, the signal can be fed directly into the integrator, such that a simulation closed-loop can be expressed. Therefore, the simulation model of the first equation can be set up, as shown in Figure 4.10(b).

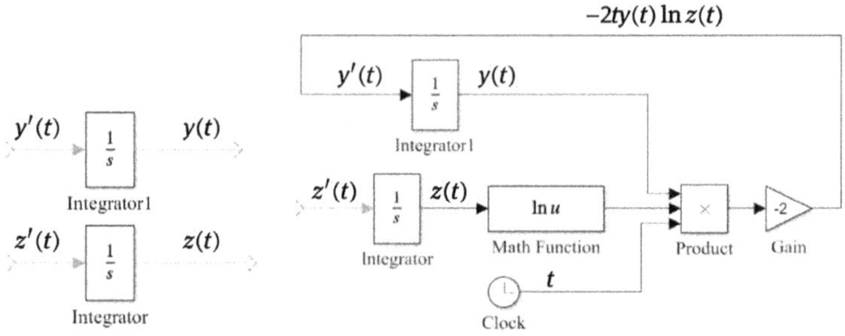

(b) first equation (file: c4mkey1b.slx)

Figure 4.10: Simulink modeling process.

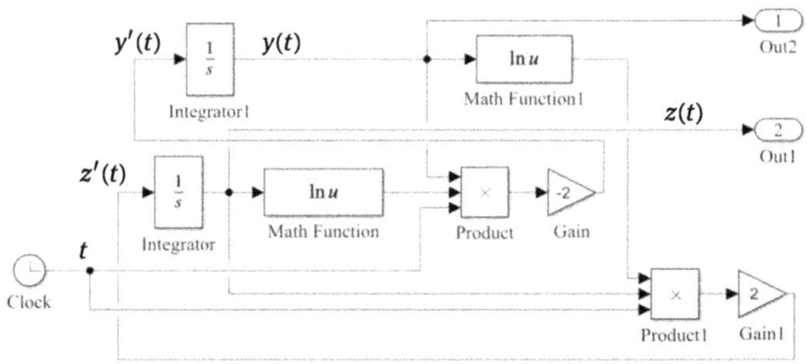

Figure 4.11: Simulink model of the equation (file: c4mkey2.slx).

With the above idea, the second differential equation can also be modeled. Therefore, the Simulink model for the entire differential equations can be established, as shown in Figure 4.11.

With the **Run** button on the Simulink window, the simulation process can be started. The simulation results can be returned into MATLAB workspace in variables `tout` and `yout`. Since there are two output ports in the model, `yout` is a matrix with two columns, representing respectively state variables $y(t)$ and $z(t)$. Function `plot()` can be used to draw the two states, as shown in Figure 4.12. It can be seen from the results that the curves of the numerical solution is rather rough, since the step-size of t at some points are too large. Densely distributed vector $\boldsymbol{t}_0$ vector can be generated to superimpose the theoretical curves on the results, with dash lines. It can be seen that although the values at `tout` samples are close, the curve itself has very large errors, since the values at two adjacent samples are joined by straight lines, not in genuine curves.

```
>> t0=0:0.001:10;
   y1=exp(cos(t0.^2)); y2=exp(sin(t0.^2));
   plot(tout,yout, t0,y1,'--',t0,y2,'--')
```

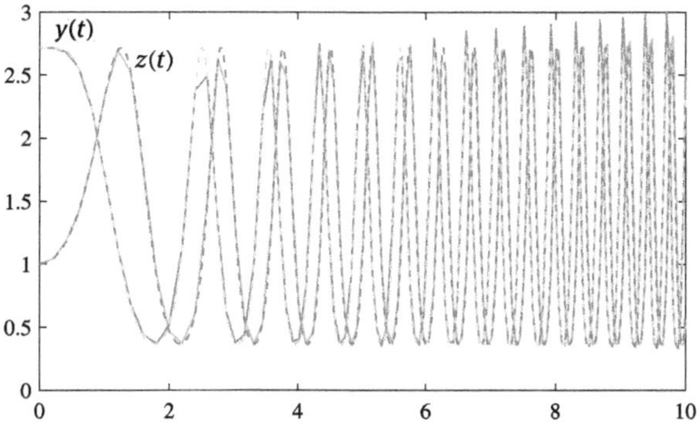

Figure 4.12: Comparisons of numerical and analytical solutions.

4.2.2 Numerical solution validation and control parameter setting

Several steps are recommended in Volume V for finding numerical ODE solutions. Among them, the most important one is solution validation. Under the default variable-step solution algorithms, the relative and absolute error tolerances can be set to smaller values such as 3×10^{-14}, which ensure the most accurate results under double precision data type.

Simulink modeling and simulation should also undergo such validation processes. Similar to the ODE solvers such as `ode45()`, variable-step algorithms are also supported in Simulink. The default value of the relative error tolerance is set to 10^{-3}, corresponding to one-thousandth of relative error. This default value is far too much for ordinary applications. For most differential equations, accurate simulation results cannot be found. The default absolute error tolerance in Simulink is set to 10^{-6}, which is also very large. Therefore, in real solution process, these parameters should be assigned to the smallest possible value of 3×10^{-14} (the real smallest allowed one is 2.8422×10^{-14}). Similar to the output setting discussed earlier, the default template should also be updated.

As before, the user-defined template can be opened again, and then click the button; the simulation parameters setting dialog box shown in Figure 4.13 can be opened. This dialog box can only be used to assign simple simulation parameters. Open the Solver listbox, the desired simulation algorithm can be selected from the list. The default algorithm is Auto(Automatic solver selection), meaning Simulink may select the most suitable algorithm for the model.

Under the Solver listbox, suitable algorithms should be selected for different differential equations. The characteristics of specific algorithms may refer to Volume V. If a simulation comes with no continuous state variable, the algorithm labeled discrete (no continuous states) is recommended. If there are continuous state variables, ode45 (Dormand–Prince) algorithm is recommended, since the accuracy of the algorithm is rather high. Unfortunately, the algorithm has its own limitations, for instance, it is not

Figure 4.13: Simulation parameter setting dialog box.

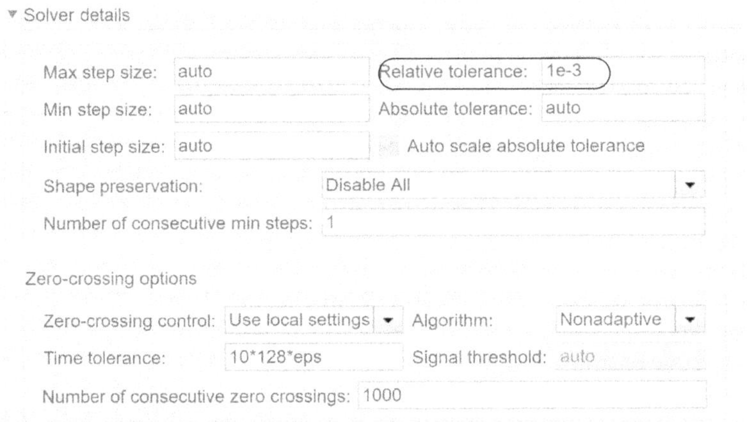

Figure 4.14: Solver details setting.

suitable for dealing with stiff differential equations. If the simulation process takes too long to complete, it is likely a stiff differential equation. The solver ode15s (stiff/NDF) can be tried, which may rapidly complete the simulation work, and the cost is the possible sacrifice of simulation accuracy. Of course, the user may try other algorithms to cross validate the results.

To further specify simulation control parameters, click the Solver details mark, and the parameter dialog box is further expanded as shown in Figure 4.14. It is recommended to set Relative tolerance to 3×10^{-14}, and set Absolute tolerance to the same value, which may ensure to have the most accurate solutions under double precision framework. Of course, other options may be visited, click ... mark, there are other options to set. The options are not discussed here.

After all the control parameters are set, it can be saved to update the template, so that next time Simulink is opened, the modified template can be used, to better complete modeling and simulation tasks.

Example 4.6. Find accurate numerical solutions to the differential equation in Example 4.5. Compare the solution with analytical solution, and assess the precision of the simulation results.

Solutions. Open `c4mkey2.slx` model, set the relative error tolerance to different values, and the maximum errors of the two states can be obtained as shown in Table 4.1. Under default error tolerance, the errors in the solutions are too large. The solution process may be considered fail for this example. If the relative error tolerance is decreased, the solution accuracy becomes higher. If the relative error tolerance is set to 3×10^{-14}, the maximum error may reach the level of 10^{-12}. The solution curves under such error tolerance are shown in Figure 4.15. The errors cannot be witnessed from the curve, and the accuracy is satisfactory.

```
>> t0=tout; y1=exp(cos(t0.^2)); y2=exp(sin(t0.^2));
   plot(tout,yout, t0,y1,'--',t0,y2,'--')
   e1=max(abs(yout(:,1)-y1)), e1=max(abs(yout(:,2)-y2))
```

In fact, the result is the most accurate possible one under double precision data type. Therefore, it is recommended to set the relative error tolerance to 3×10^{-14} in teal simulation.

Table 4.1: Maximum errors under different relative error tolerance.

error tolerance	10^{-3}	10^{-5}	10^{-8}	10^{-10}	10^{-12}	3×10^{-14}
y_1 error	0.3139	0.00304	1.193×10^{-6}	1.184×10^{-8}	1.236×10^{-10}	3.921×10^{-12}
y_2 error	0.3309	0.00314	1.204×10^{-6}	1.191×10^{-8}	1.241×10^{-10}	3.961×10^{-12}

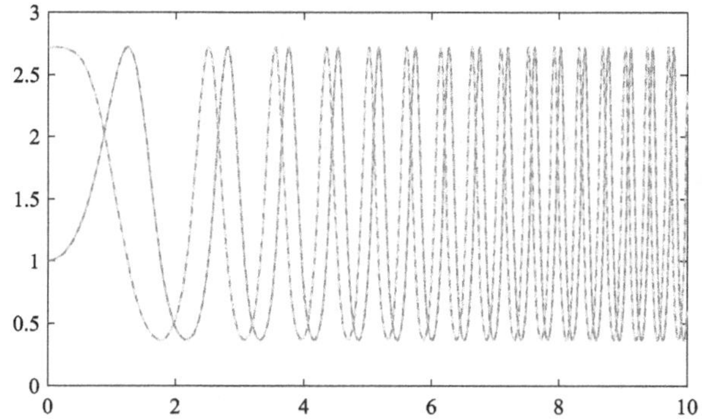

Figure 4.15: Comparing numerical and analytical solutions with modified parameters.

4.2.3 Vectorized modeling methods

Compared with the command-line driven solution method discussed in Volume V, that is, use functions such as `ode45()` for the solutions; the method here is rather complicated. A more concise modeling and simulation method is expected. A simple example is given, and command-line driven methods and vectorized solution ideas are presented.

Example 4.7. Solve the differential equations in Example 4.5 again with the command-line driven method.

Solutions. Select state variables $x_1(t) = y(t)$ and $x_2(t) = z(t)$, and the standard form of the state space model can be obtained as

$$\begin{bmatrix} x_1'(t) \\ x_2'(t) \end{bmatrix} = \begin{bmatrix} -2tx_1(t)\ln x_2(t) \\ 2tx_2(t)\ln x_1(t) \end{bmatrix}.$$

Based on the above first-order explicit differential equation, an anonymous function can be written to describe the ODE, then the numerical solution can be found directly.

```
>> f=@(t,x)[-2*t*x(1)*log(x(2)); 2*t*x(2)*log(x(1))];
   x0=[exp(1); 1]; [t,x]=ode45(f,[0,10],x0);
```

It can be seen from the above first-order explicit differential equation that the command-line driven method is more concise to use. There is naturally a question: "Can Simulink be used in solving such equations in simple ways?" The answer is yes, since vectorized modeling ideas can be introduced.

The so-called vectorized model is that the signals in Simulink models are no longer in scalar forms, but in vector forms. Again an integrator can be used, and assume that the output signal is vector $\boldsymbol{x}(t)$. The input is naturally $\boldsymbol{x}'(t)$. The initial values of the integrator are also given as a vector.

Example 4.8. Establish a Simulink model for the differential equation in Example 4.5 using vectorized methods.

Solutions. Two key points should be noted in building the simulation model. The first one is to use a vector integrator, and the other is to use Fcn block provided in Simulink. Fcn blocks can be used to accept vector input, while unfortunately, it can only provide scalar output signals. Also since time t is explicitly involved, the third input signal $x_3(t)$ in the vector should be introduced, for time t. Therefore, the right side of the equation can be written as

$$\boldsymbol{y}(t) = \begin{bmatrix} -2x_3(t)x_1(t)\ln x_2(t) \\ 2x_3(t)x_2(t)\ln x_1(t) \end{bmatrix}. \tag{4.2.3}$$

From the original vector $\boldsymbol{x}(t)$ and time t, a new vector can be composed, with the **Mux** block in the **Signal Routing** group. Two **Fcn** blocks can be used to describe the signals in $\boldsymbol{x}'(t)$, and with **Mux1** block, a vector signal is composed. Therefore, the Simulink model in Figure 4.16 can be constructed. It can be seen that the simulation results are exactly the same as the model built earlier.

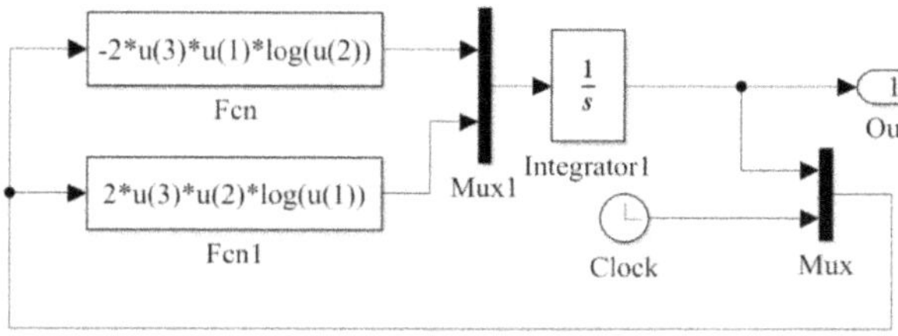

Figure 4.16: Vectorized Simulink model (file: `c4mkey4.slx`).

Note the expressions in **Fcn** and **Fcn1** blocks. In the block, the signals used are `u(i)`, while their mathematical form is x_i. These blocks can only accept `u` as its input signals, not x.

4.2.4 A general framework for first-order explicit ODEs

Block **Fcn** was used earlier to build up right size expressions, There is a fatal weakness in the block, since only scalar signals can be obtained from its output ports, not vector signals. Therefore, the modeling in this method is rather complicated. Blocks **Interpreted MATLAB Fcn** or **MATLAB Function** can be used to build up blocks with vector input and vector output signals. The difference between the two blocks are that in the former block, a `.m` file is used to describe the relationship in functions, while in the latter, the function is embedded in the block. In this section, a more general and concise modeling method is presented.

Example 4.9. Set up a Simulink model for the following Lorenz differential equation:

$$\begin{cases} x_1'(t) = -\beta x_1(t) + x_2(t)x_3(t), \\ x_2'(t) = -\rho x_2(t) + \rho x_3(t), \\ x_3'(t) = -x_1(t)x_2(t) + \sigma x_2(t) - x_3(t), \end{cases}$$

where $\beta = 8/3$, $\rho = 10$, $\sigma = 28$, and the initial values are $x_1(0) = x_2(0) = 0$, $x_3(0) = \epsilon$, with ϵ the small value. For instance, it can be selected as a small positive number $\epsilon = 10^{-10}$.

Solutions. Lorenz differential equations are named after American meteorologist, Edward Norton Lorenz (1919–2008). It seems that **Fcn** block can be used to describe the right side of the equation, and the expression can be filled in the edit box, such that a simulation model can be constructed as shown in Figure 4.17. Executing such a model, a Diagnostic Viewer window is given, as shown in Figure 4.18, indicating that there is a syntax error in the expression. Unfortunately, vector signal output is not allowed in the Fcn block; only scalars are accepted. Therefore, this block cannot be used to describe the equations.

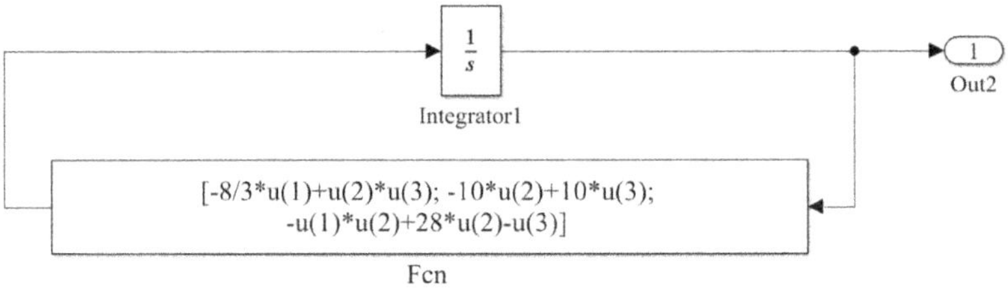

Figure 4.17: Erroneous Lorenz model (file: `c4mlor1.slx`).

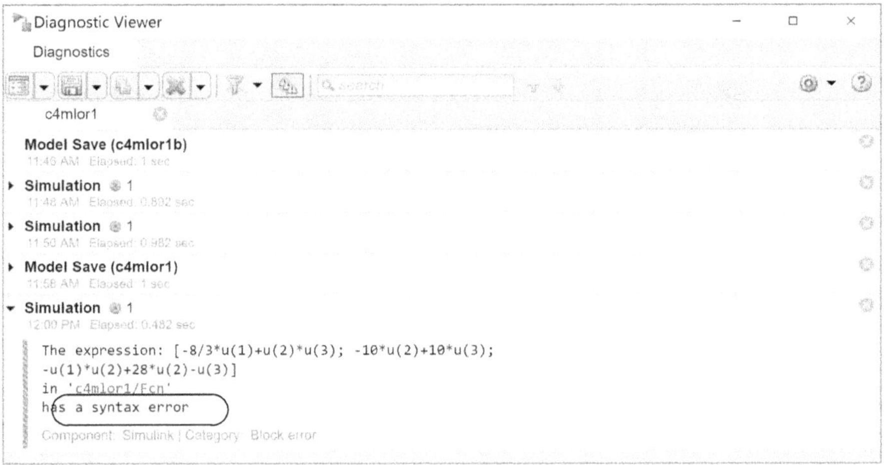

Figure 4.18: Diagnostic viewer window.

Now let us try to use embedded MATLAB function block to describe the right side of the differential equation. The Simulink model can be constructed as shown in Figure 4.19(a). Double click the MATLAB Function block, and the function editor window shown in Figure 4.19(b) is opened. The right side of the equation can be described in the model. Note that the function input argument is u, and it must be the same as the variable name in the function.

Perform simulation to the model, the simulation can be obtained. The following MATLAB commands can be used to draw three-dimensional phase space trajectory, as shown in Figure 4.20:

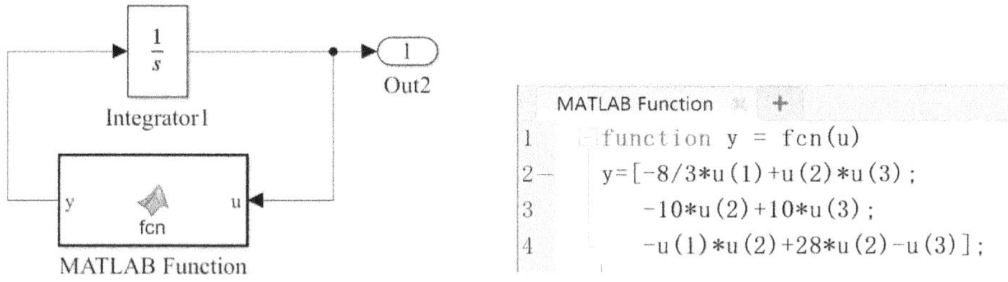

(a) simulation model (file: `c4mlor1a.slx`) (b) function editor

Figure 4.19: Simulink model of Lorenz differential equation.

```
>> plot3(yout(:,1),yout(:,2),yout(:,3)), grid
```

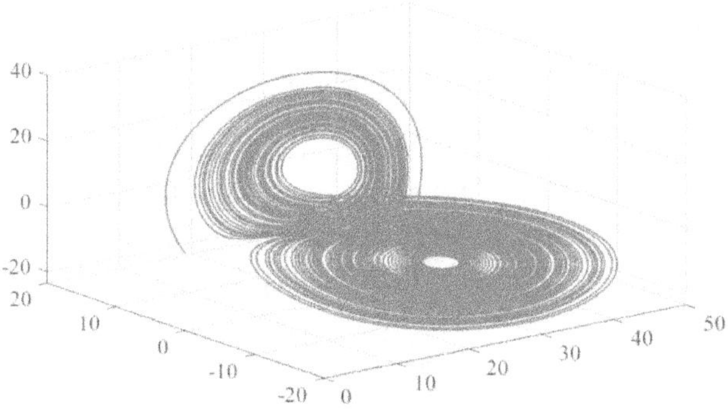

Figure 4.20: Phase space trajectory of Lorenz equations.

Now consider using Interpreted MATLAB Fcn block to describe the functions. The Simulink model shown in Figure 4.21(a) can be established. The following MATLAB function can be written to describe the right side of the equations:

```
function y=c4mlorz(x)
y=[-8/3*x(1)+x(2)*x(3); -10*x(2)+10*x(3); -x(1)*x(2)+28*x(2)-x(3)];
```

Double click the Interpreted MATLAB Fcn block icon, and the parameter dialog box is opened, as shown in Figure 4.21(b). In the MATLAB function edit box, the MATLAB function file name should be specified. The simulation results are exactly the same as the ones obtained earlier.

It can be seen from the example that, if the standard form of the first-order explicit differential equation is known, the method can be used to build the model framework. With the method, the complexity in solving differential equations are similar to the

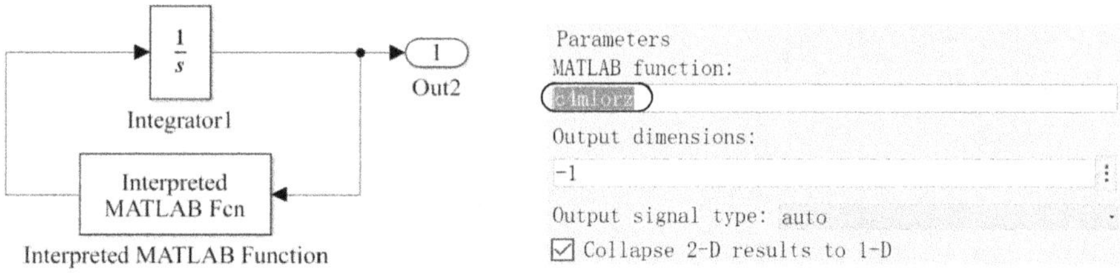

(a) Simulink model (file: `c4mlor1b.slx`) (b) parameter dialog box

Figure 4.21: Interpretive function modeling framework.

`ode45()` function. There is a limitation in the model framework that the time variable t cannot be contained in the expression, that is, it cannot be used in handling the time-varying differential equation. The time-varying differential equation modeling and the solution will be discussed next.

4.2.5 Model framework for first-order time-varying explicit ODEs

The modeling of the time-varying first-order explicit differential equation will be explored in this section. In the Simulink modeling of time-varying differential equations, since the right side contains time t, but Interpreted MATLAB Fcn and MATLAB Function blocks only accept one input signal, a Mux block is needed to augment t as x_{n+1}, such that the state signal vector is augmented, as the input of the block. Therefore, a more general Simulink modeling framework is needed to describe first-order explicit time-varying differential equations. An example is given next to demonstrate the modeling problem of time-varying differential equations.

Example 4.10. Rebuild the Simulink model for the differential equation in Example 4.5, using function blocks.

Solutions. If block Mux is used to generate augmented state vector $\boldsymbol{x}$, the Simulink model shown in Figure 4.22(a) can be established, where the user needs to write the following MATLAB function to describe the relationship in (4.2.3):

```
function y=c4extest1a(x)
y=[-2*x(3)*x(1)*log(x(2)); 2*x(3)*x(2)*log(x(1))];
```

Of course, the following statements can also be used to describe the function:

```
function y=c4extest1b(x)
t=x(3); y=[-2*t*x(1)*log(x(2)); 2*t*x(2)*log(x(1))];
```

The parameter dialog box of Interpreted MATLAB Function block is shown in Figure 4.22(b). Usually, the dimensions of the input and output signals are equal by default. Therefore, in the Output dimensions edit box, the value specified is −1, indicat-

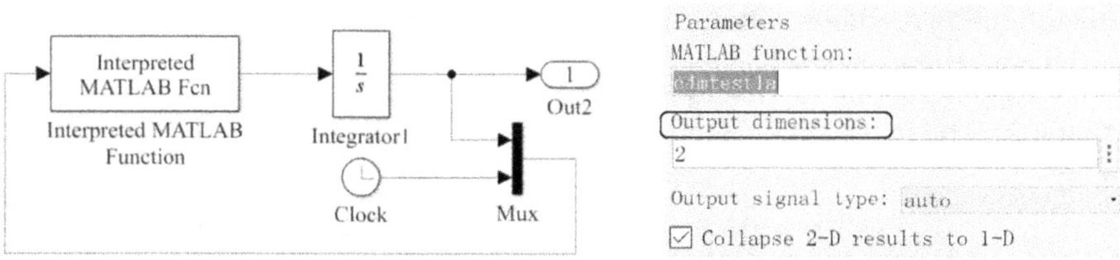

(a) simulation model (file: `c4mmod2a.slx`) (b) parameter dialog box

Figure 4.22: Simulation model framework with interpretive MATLAB function block.

ing the two signal sizes are equal. For this particular framework, the dimensions are different. The dimension of the output signal is 2. Therefore, 2 should be filled in the edit box. Therefore, the model can be used directly in simulation, and the simulation results are the same.

Now consider using MATLAB Function block to describe the functions. The Simulink model in Figure 4.23(a) should be established. Double click the block, and the MATLAB editor interface is opened automatically, as shown in Figure 4.23(b). The user needs only describe the relationship of the output signal $\boldsymbol{y}$ and the input signal $\boldsymbol{u}$. There is no need to specify the dimensions of the output signals. Note that the input variable name must be $\boldsymbol{u}$.

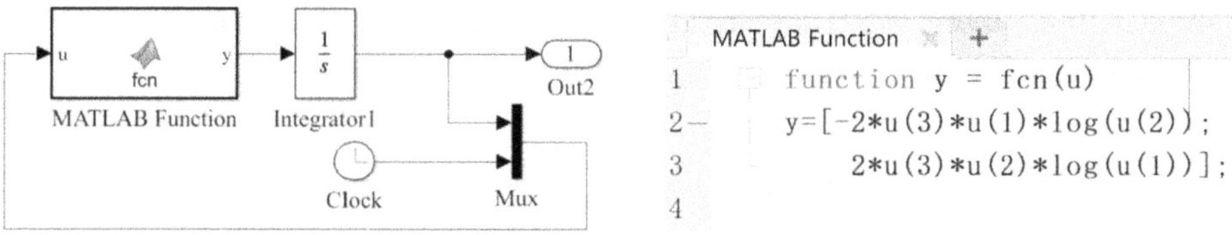

(a) simulation model (file: `c4mmod2b.slx`) (b) function editor

Figure 4.23: Universal framework with embedded function blocks.

Of course, the modeling method here still has restrictions, since the constants must be given values, not MATLAB variable names. If variables in MATLAB are to be used, the S-function based modeling method should be used. The topics such as S-functions will be further presented in Chapter 7.

4.3 Modeling of high-order ODEs

In the previous sections, first-order explicit differential equations are studied, and Simulink modeling methods are presented. In real applications, the differential equations to be studied are not always given in the standard form. High-order differential equations are usually encountered. In this section, the general form of a high-order differential equation is presented, followed by key signal selection methods, and

Simulink modeling methods. Finally, the modeling and solution of high-order implicit differential equations are studied.

Definition 4.2. The mathematical model of an nth order explicit differential equation is given by

$$y^{(n)}(t) = f(t, y(t), y'(t), y''(t), \dots, y^{(n-1)}(t)). \tag{4.3.1}$$

Also, the initial values of the output signal and its high-order derivatives are known, and denoted as $y(t_0), y'(t_0), y''(t_0), \dots, y^{(n-1)}(t_0)$.

4.3.1 Defining key signals

For the high-order differential equation specified in (4.3.1), the key signals should be selected as $y(t), y'(t), y''(t), \dots, y^{(n)}(t)$. Since these signals are associated each other, the selections of the key signals are the foundation in Simulink modeling. The Simulink model for the entire differential equation can be established upon the selected key signals.

Derivative blocks are visited earlier. It seems that the key signals can be defined with the derivative chain shown in Figure 4.24. If derivative chain is used, the input signal to the left most derivative is regarded as $y(t)$, then the other signals defined are in turn the key signals $y'(t), y''(t), \dots, y^{(n)}(t)$. It seems reasonable that the key signals can be defined in this way. However, there are many negative issues in the modeling method presented this way:

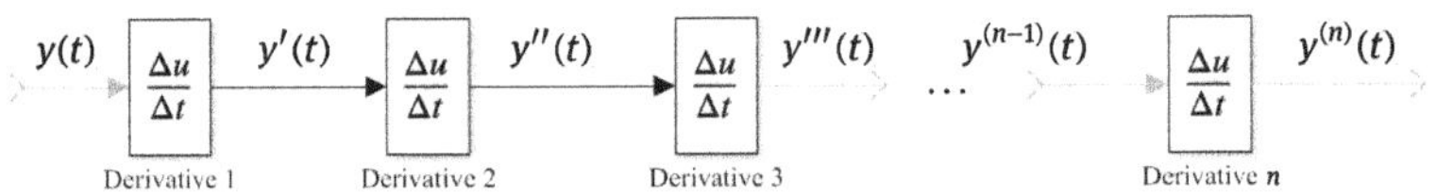

Figure 4.24: Key signals defined by derivative chain.

(1) Numerical differentiation algorithms may always produce errors when the input signal subjects to sudden changes. The errors are inevitable. These kinds of errors may affect the numerical stability of simulation results.

(2) Since the standard form of the differential equations comes with given initial values $\boldsymbol{x}(t_0)$, and the initial value information cannot be expressed in derivative chain based modeling.

(3) Equation (4.3.1) is the explicit expression of $y^{(n)}(t)$, and in the chain shown in Figure 4.24; signal $y^{(n)}(t)$ is generated by the previous block, such that the differential equation model in (4.3.1) cannot be described at all.

Due to the above considerations, the derivative chain is not suitable to be selected to generate key signals, and not applicable in differential equation modeling.

A more practical way to define key signals is to use integrator chains, instead, as shown in Figure 4.25. The output of the right most integrator is defined as signal

$y(t)$, then the input signals in the left integrators are respectively defined as $y'(t)$, $y''(t), \dots, y^{(n)}(t)$. Besides, double click the integrators, and the initial values of the signals can be specified accordingly to the parameter dialog boxes, such that a set of key signals with specified initial values can be defined. This method is used throughout the book to define key signals.

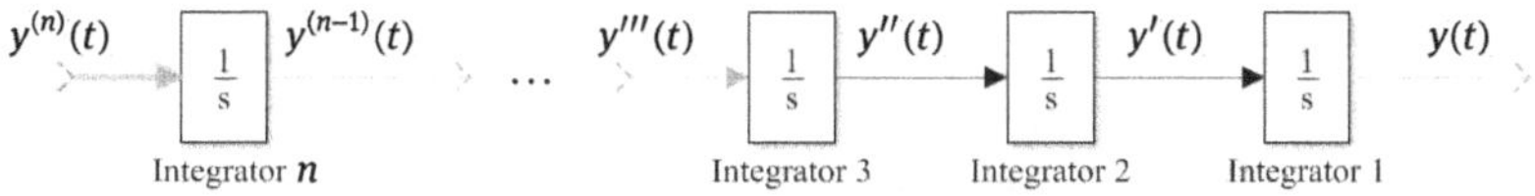

Figure 4.25: Defining key signals with an integrator chain.

With the key signals selected, the Simulink models for the differential equations can be established, together with other blocks, such as sum, product and other function blocks.

4.3.2 Modeling of high-order linear ODEs

In this section, the mathematical form of linear differential equations is presented first. Simulink modeling methods are the explored, and the impact of simulation control parameters on simulation accuracies will be studied.

Definition 4.3. The general form of linear differential equation with constant coefficients is given by

$$\begin{aligned} y^{(n)}(t) + a_1 y^{(n-1)}(t) + \cdots + a_{n-1}y'(t) + a_n y(t) \\ = b_1 u^{(m)}(t) + b_2 u^{(m-1)}(t) + \cdots + b_m u'(t) + b_{m+1}u(t), \end{aligned} \tag{4.3.2}$$

and at time t_0, the initial values are known as $y(t_0), y'(t_0), \dots, y^{(n-1)}(t_0)$. The constant coefficient linear differential equation is also known as linear time invariant (LTI) differential equations.

With the key signals $y(t), y'(t), y''(t), \dots, y^{(n)}(t)$ defined in the integrator chain, the Simulink model for the differential equation can be established around these key signals. The particular method is to build up the right side expression in (4.3.1) based on key signals $y(t), y'(t), y''(t), \dots, y^{(n)}(t)$, as shown in Figure 4.25, and feed the result to the input of the integrator on the left. The closed-loop in the simulation model can be created, such that the model can be used in describing the differential equation. An example is given next to demonstrate the modeling and simulation of high-order differential equations.

Example 4.11. Solve the following linear time invariant differential equation:

$$y^{(4)}(t) + 5y'''(t) + 9y''(t) + 7y'(t) + 2y(t) = u'(t) + 2u(t),$$

where $u(t) = \sin t$, $y(0) = 1$, $y'(0) = y''(0) = y'''(0) = 0$.

Solutions. Generally, the analytical solutions of the linear time invariant differential equations can be found. The following commands can be issued and find the analytical solution of the differential equation, with solver `dsolve()` in Symbolic Math Toolbox:

```
>> syms t y(t); u=sin(t);
   y1=diff(y); y2=diff(y1); y3=diff(y2);
   y0=dsolve(diff(y3)+5*y3+9*y2+7*y1+2*y==2*diff(u)+u,...
              y(0)==1, y1(0)==0, y2(0)==0, y3(0)==0)
   simplify(y0) %simplified analytical solution
```

The following analytical solution can be found:

$$y_0 = \frac{7}{4}e^{-t} - \frac{2}{5}e^{-2t} + te^{-t} + \frac{3}{4}t^2e^{-t} - \frac{\sqrt{2}}{4}\cos\left(t - \arctan\frac{1}{7}\right).$$

Now consider the modeling and simulation methods with Simulink. Since the ODE is of fourth order, the integrator chain with four integrators can be set up, to define the key signals $y(t)$, $y'(t)$, $y''(t)$, $y'''(t)$ and $y^{(4)}(t)$. In fact, with slight modification to the model, the explicit expression of $y^{(4)}(t)$ can be obtained:

$$y^{(4)}(t) = -5y'''(t) - 9y''(t) - 7y'(t) - 2y(t) + 2u'(t) + u(t).$$

With the integrator chain to define the key signals, the weighted sum of the right side signal can be constructed, and fed into the left most integrator, as shown in Figure 4.26, where the block names of the gain blocks are omitted, by setting their status as default **Auto**. Since only $y(0)$ is nonzero, while others are zero, the initial value of the integrator for $y(t)$, that is, the rightmost one, can be set to 1, while others are kept the default value of 0.

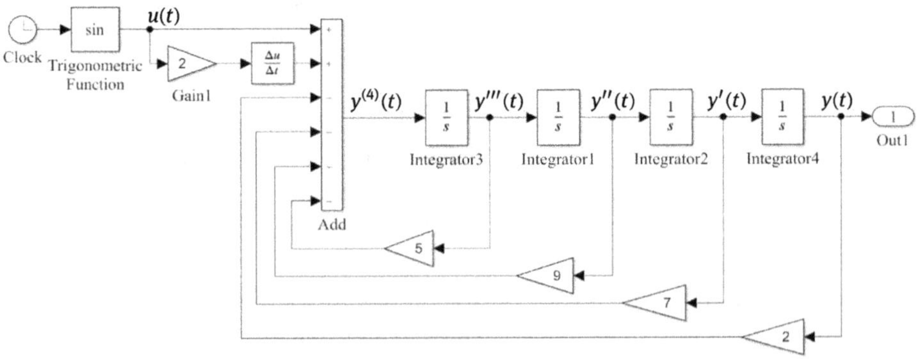

Figure 4.26: Simulink model (file: `c4mlin1.slx`).

The following commands can be used to implement simulation analysis, and find the analytical solutions. The curves of the solutions are as shown in Figure 4.27, where

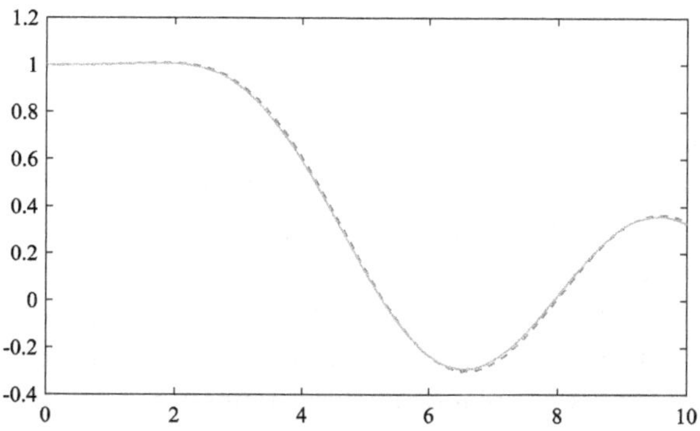

Figure 4.27: Comparisons of simulation results.

the theoretical values are given in a solid line. It can be seen that there are errors in the solutions. The errors are caused by the selected relative error tolerance ϵ. It is necessary to reduce the value of ϵ. The simulation result obtained as exactly the same as the one with `ode45()` algorithm, indicating that the automatically selected algorithm is the `ode45` algorithm. Note that, in order to measure time elapse, the simulation command `sim()` is used to carry out simulation studies. The syntaxes of the function will be presented later in the book:

```
>> tic, [t1,~,y1]=sim('c4mlin1'); toc, length(t1)
   y0=7*exp(-t1)/4-2*exp(-2*t1)/5+t1.*exp(-t1)+...
     3*t1.^2.*exp(-t1)/4-sqrt(2)*cos(t1-atan(1/7))/4;
   max(abs(y0-y1)), plot(t1,y1,'--',t1,y0)
```

Selecting different control parameters and simulation algorithms, the maximum errors and time elapses are compared as shown in Table 4.2. It can be seen that, if $\epsilon = 3 \times 10^{-14}$, the best possible result can be obtained.

Table 4.2: The impact of control parameters and algorithms on simulation results.

compare	default ones	10^{-6},ode45	3×10^{-14},ode45	10^{-6},ode15s	3×10^{-14},ode15s
maximum errors	0.0147	0.0036	2.5012×10^{-7}	0.0014	1.3680×10^{-7}
number of points	76	344	4545766	1453	14147733
time elapse	0.3722	0.55	21.30	0.5425	64.99

Although very tough error tolerance is selected, the simulation result is still not very accurate. The main reason is the use of the derivative $\Delta u/\Delta t$ blocks. Due to the limitations of the numerical differentiation itself, the simulation step-size must be selected extremely small, and time elapse may become very large. Even though, the accuracy cannot be further increased. Better solutions are expected.

Example 4.12. Substitute the derivative block in Figure 4.26, and assess again the simulation accuracies.

Solutions. It is easily seen that since $u(t) = \sin t$, then $u'(t) = \cos t$. Therefore, the signal can be generated directly by the cos block, and the approximation of derivative is bypassed. The new simulation model can be constructed, as shown in Figure 4.28. It can be seen that at tough error tolerance, the number of points is reduced to 3776, with maximum error reduced to 5.8009×10^{-15}. The time elapse is only 0.5697 seconds. It can be seen that by manual processing of the input signals, the solution efficiency is significantly improved. Therefore, in real simulation analysis, numerical derivative block should be avoided.

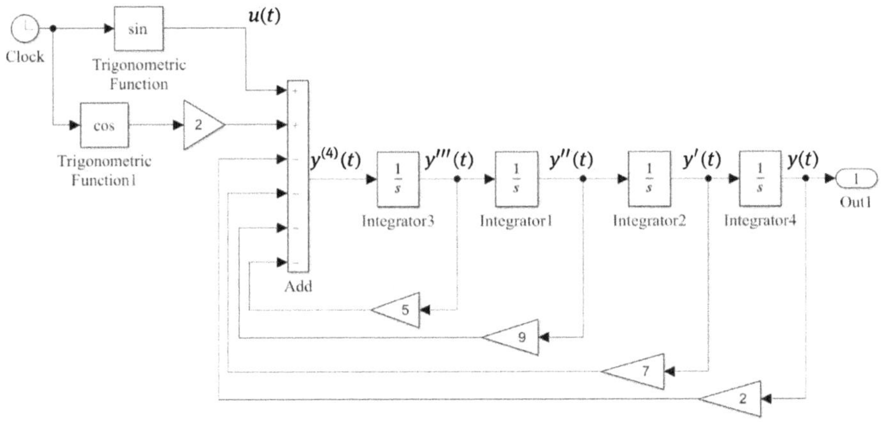

Figure 4.28: Linear system simulation model (file: `c4mlin1a.slx`).

In fact, it can be seen from the simulation diagram that the structure in the Simulink model is not quite tidy. Vectorized modeling procedures can be introduced to construct simple simulation models.

Example 4.13. A concise vectorized form can be tried to establish the Simulink model for the differential equation shown in Example 4.11.

Solutions. It can be seen that if key signals such as $y(t)$, $y'(t)$ and time t are grouped as a vector signal, then the signal is $\boldsymbol{u}(t) = [y(t), y'(t), y''(t), y'''(t), t]$. With Fcn block, the right side of the equation can be a model, and the Simulink model can be set up, as shown in Figure 4.29. Perform simulation to the model, the results are exactly the same.

It is worth mentioning that the simulation model established this way needs the user to provide an analytical form of the input signal derivatives. This is too demanding in real simulation, since for many such systems, the derivative expressions are not

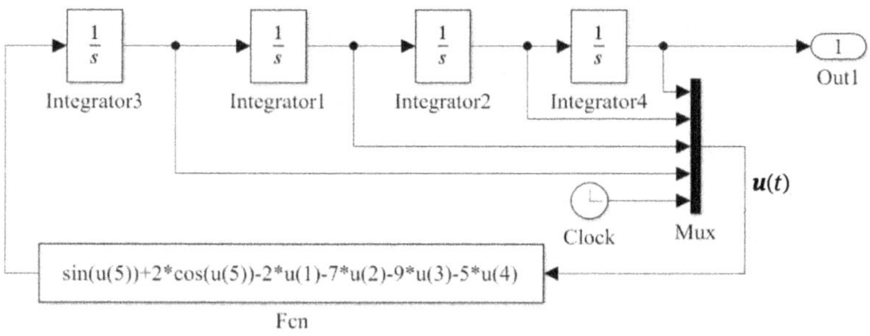

Figure 4.29: Vectorized Simulink model (file: `c4mlin2.slx`).

known. In Chapter 5, discussions for these type problems will be presented, and an algorithm is proposed to increase accuracy, and promote simulation speed.

4.3.3 Transfer function model for constant linear ODEs

Linear time invariant differential equations are the commonly used differential equations in the fields of differential equations and in automatic control systems. If the coefficients of the differential equation are given, Laplace transform can be employed. The concept of transfer function is given next, and transfer function based simulation methods will be explored.

Definition 4.4. If the initial values of the input, output signals and their derivatives of different orders are all zero, the transfer function model in (4.3.2) can be derived:

$$G(s) = \frac{b_1 s^m + b_2 s^{m-1} + b_2 s^{m-2} + \cdots + b_m s + b_{m+1}}{s^n + a_1 s^{n-1} + a_2 s^{n-2} + \cdots + a_{n-1} s + a_n}. \tag{4.3.3}$$

It can be seen that transfer functions are essentially the gain of the system in s domain. That is, the ratio of the output signal to the input signal in s domain.

Transfer function model is a powerful tool in describing linear time invariant differential equations; however, it has inevitable limitations. That is, the initial values of the input, output signals are their derivatives of different orders must be zero. This is rather tough, and the application fields may be narrowed. Besides, in certain circumstances, transfer functions may also cause problems. In this section, examples are used to show the restrictions of transfer function models.

Example 4.14. Find the transfer function model between signals $y(t)$ and $u(t)$ for the differential equation in Example 4.11.

Solutions. From the given differential equation, refer to the definitions in transfer function, and it is not hard to find its transfer function model:

$$G(s) = \frac{s+2}{s^4 + 5s^3 + 9s^2 + 7s + 2}.$$

While, since the initial values $y(t)$ and $u(t)$ signals in the original differential equations are nonzero, transfer functions cannot be used to build the Simulink model for the differential equation. This problem will be explored later in Section 5.1.4.

Although transfer function has certain limitations, it is still a very useful model. Control system theory is established upon transfer functions, and it is rather useful in describing systems whose initial states are at rest. Besides, another example is given to demonstrate such a case, where classical delay differential equation solvers may get wrong results. Simulink modeling and solutions are illustrated.

Example 4.15. A typical feedback control system is shown in Figure 4.30, where the plant $G(s)$ and controller $G_c(s)$ are all given in transfer function model:[81]

$$G(s) = \frac{e^{-s}}{(s+1)(s+2)} = \frac{e^{-s}}{s^2 + 3s + 2}, \quad G_c(s) = 1.66 + \frac{0.91}{s}.$$

Find the step response of the system with zero initial conditions.

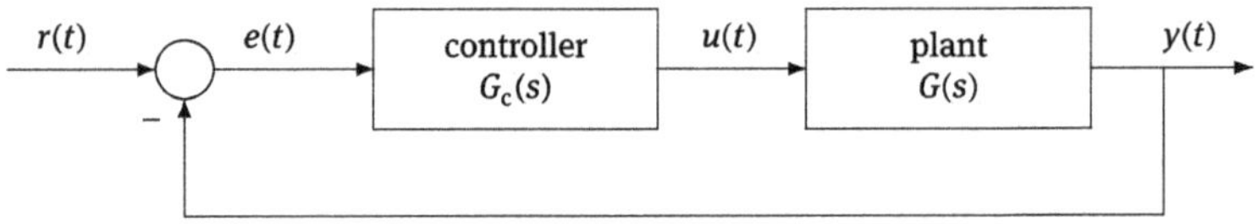

Figure 4.30: Typical feedback control system model.

Solutions. With the properties of Laplace transform, the transfer function model can be restored to the differential equation model. Since the plant model has delay term, the corresponding delay differential equation can be obtained. Observing the relationship between signals $u(t)$ and $y(t)$, the following delay differential equation can be obtained:

$$y''(t) + 3y'(t) + 2y(t) = u(t-1).$$

Now look at the mathematical model of $u(t)$, derived as

$$\begin{aligned} u'(t) &= 1.66(r(t) - y(t))' + 0.91(r(t) - y(t)) \\ &= 1.66\delta(t) - 1.66y'(t) + 0.91 - 0.91y(t), \end{aligned}$$

where $\delta(t)$ is a Dirac function. The Dirac function is named after British theoretic physicist, Paul Adrien Maurice Dirac (1902–1984). It is defined as the derivative of the step function, which can be regarded that the value of the function is infinity at $t = 0$, and the value 0 elsewhere. It is an impulse function.

Selecting the state variables as $x_1(t) = y(t)$, $x_2(t) = y'(t)$ and $x_3(t) = u(t)$, the standard form of the explicit differential equation can be found:

$$\mathbf{x}'(t) = \begin{bmatrix} x_2(t) \\ x_3(t-1) - 3x_2(t) - 2x_1(t) \\ 1.66\delta(t) - 1.66x_2(t) + 0.91 - 0.91x_1(t) \end{bmatrix}.$$

Although MATLAB function `dirac()` is provided, the value of the function at $t = 0$ is set to `Inf`. It cannot be used in effective simulation. Approximations can be introduced, for instance, using an approximate Dirac function of width 10^{-4} and magnitude 10^4 to replace $\delta(t)$, which is represented in MATLAB as `1e4-1e4*heaviside(t-1e-4)`. An anonymous function can be written to describe the delay differential equation, and the find the numerical solutions. The output signal of the equation can be obtained as shown in Figure 4.31. Note that in the transformation from transfer function to differential equations, do not forget the $\delta(t)$ function term; otherwise, wrong results will be obtained:

```
>> f=@(t,x,Z)[x(2); Z(3)-3*x(2)-2*x(1);
      1.66*(1e4-1e4*heaviside(t-1e-4))-1.66*x(2)+0.91-0.91*x(1)];
   ff=ddeset; ee=1e-10; ff.RelTol=ee; ff.AbsTol=ee;
   tau=1; x0=zeros(3,1);                    % delay constants and initial values
   tic, tx=dde23(f,tau,x0,[0,20],ff); toc %solve the equation
   plot(tx.x,tx.y(1,:))                     % draw the output curves
```

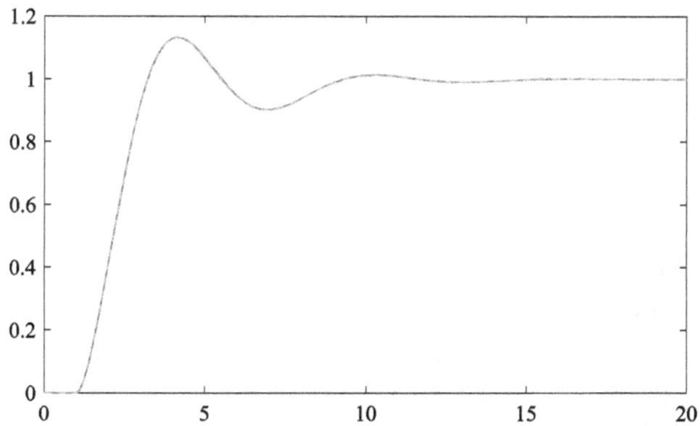

Figure 4.31: Output of the delay system.

In fact, this example is rather suitable for Simulink modeling and solutions. Simulink can be used to set up the model directly in Figure 4.32 and perform simulation. The

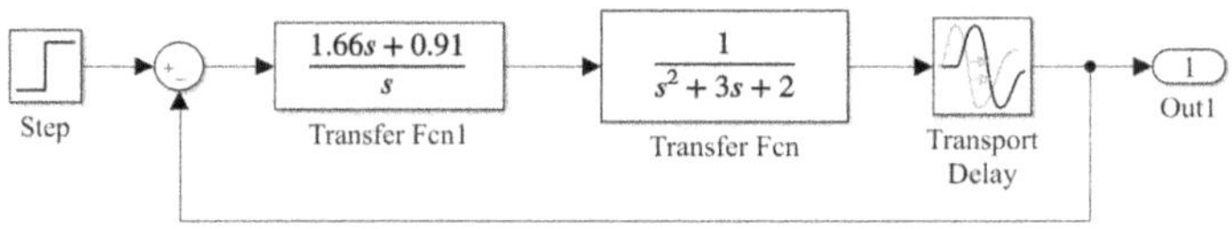

Figure 4.32: Simulink model (file: c4mdde.slx).

same results can be obtained. Regular methods are used in the entire modeling and simulation process. There is no need to introduce Dirac functions and its approximations.

4.3.4 Modeling of high-order nonlinear ODEs

In fact, with key signals defined, the above mentioned methods can also be used in modeling nonlinear differential equations, so as to find numerical solutions. Examples are used to demonstrate differential equation modeling methods.

Example 4.16. Consider the following high-order nonlinear differential equation:[54]

$$y(t)y^{(6)}(t) + 6y'(t)y^{(5)}(t) + 15y''(t)y^{(4)}(t) + 10(y'''(t))^2 = a\sin^m \lambda t.$$

If it is known that $a = 2$, $m = 3$, $\lambda = 2$ and $y(0) = 1$, $y^{(i)}(0) = 0$, $i = 1, 2, \ldots, 5$. Establish a Simulink model, and draw the state variable curves.

Solutions. It can be seen from the differential equation that $y(t) = 0$ is not a singularity of the original differential equation. Since the analytical equation of $y(t)$ is not known, it is not possible to find the singularities of the differential equations theoretically. Numerical methods are the only ways in studying the properties of the differential equations.

The highest order derivative term in the original equation is $y^{(6)}(t)$, and the explicit expression can be derived directly that

$$y^{(6)}(t) = \frac{1}{y(t)}\left[-6y'(t)y^{(5)}(t) - 15y''(t)y^{(4)}(t) - 10(y'''(t))^2 + a\sin^m \lambda t\right],$$

where it is temporarily assumed that $y(t) \neq 0$. Since the differential equation is a sixth-order one, 6 integrators are used to define the key signals. Vectorized modeling can be adopted, and with the key signals and time t, the vector signal $\boldsymbol{u}(t)$ is composed. The Fcn block can be used to model the expression on the right side. Therefore, the signal can be fed into the input port of the leftmost integrator, to close the simulation loop, as shown in Figure 4.33. Set the initial value of the rightmost integrator to 1, indicating $y(0) = 1$, the model can then be simulated. The result obtained is exactly the same, as the one obtained in Volume V, using command-line driven methods.

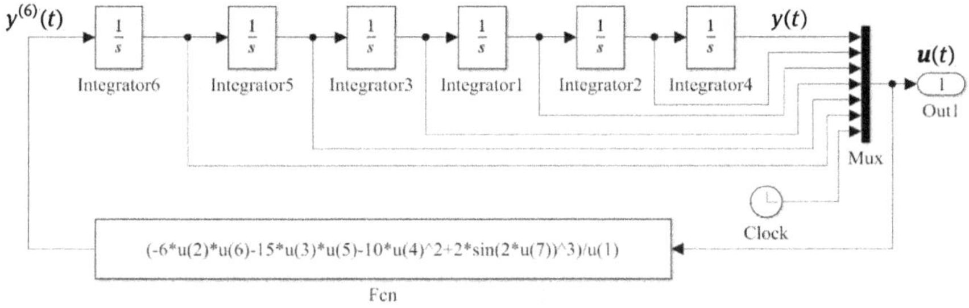

Figure 4.33: Simulink model of a nonlinear ODE (file: `c4mnlin.slx`).

4.3.5 Modeling of implicit ODEs

The differential equations studied so far are all explicit ones. That is, the term with highest-order derivative is given as an explicit formula. If in a certain differential equation, the term with highest order cannot be written as an explicit formula, the differential equation is referred to as implicit differential equation. See the next example.

Example 4.17. Solve the following implicit differential equation:

$$(y''(t))^3 + 3y''(t)\sin y(t) + 3y'(t)\sin y''(t) = \mathrm{e}^{-3t}, \quad y(0) = 1, \quad y'(0) = -1.$$

Solutions. In the earlier examples, the explicit form of $y''(t)$ are given, while in this example, the explicit formula of $y''(t)$ cannot be obtained, thus the input signal to the left most integrator in the chain is not known, and the closed-loop cannot be built up directly. This may bring difficulties to Simulink modeling of differential equations.

The crucial point in solving the implicit differential equation is how to find the value of $y''(t)$ from other key signals. For this particular example, if other key signals are known, the algebraic equation regarding $y''(t)$ should be solved once, in each simulation step. In the Math Operations group, Algebraic Constraint block is provided to solve algebraic equation $f(x) = 0$. The input signal of the block is the expression of the left side of the algebraic equation $f(x)$, and the output is the solution signal x. Examples are used to explore modeling and solutions of implicit differential equations.

Example 4.18. Solve the implicit differential equation in Example 4.17 with Simulink. To check the accuracy of the solutions, compare the numerical solutions obtained with the given analytical solution of $y(t) = \mathrm{e}^{-t}$.

Solutions. To solve the differential equation, the chain of two integrators are needed to define the key signals $y(t)$, $y'(t)$ and $y''(t)$. Denote $p(t) = y''(t)$, an algebraic equation of $p(t)$ can be established:

$$p^3(t) + 3p(t)\sin y(t) + 3y'(t)\sin p(t) - \mathrm{e}^{-3t} = 0.$$

Feed the equation to the input port of Algebraic Constraint block, its output signal can be fed directly to the left most integrator, so as to close the simulation loop. The Simulink model thus established is as shown in Figure 4.34.

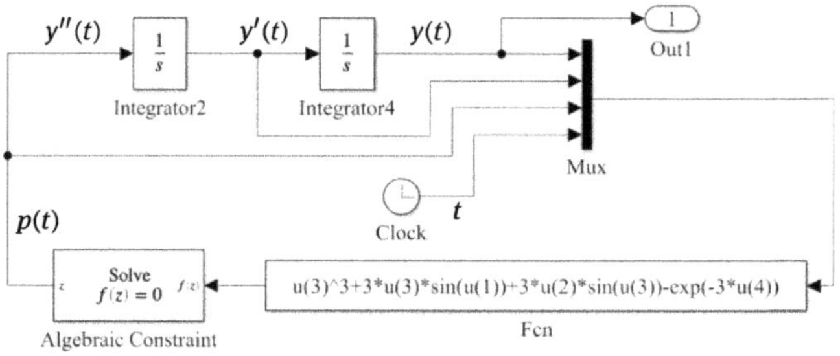

Figure 4.34: Simulink model for implicit differential equation (file: `c4mimp.slx`).

Error message shown in Figure 4.35 is given if the model is simulated directly, indicating that there is an algebraic loop, and the default trust region algorithm is not suitable for this model. Line search algorithm is recommended instead. In the error dialog box, the recommended MATLAB command is provided to modify control parameters. If the parameters are modified, the solution can be found directly, with a maximum error of 6.7856e^{-13}. The handling of algebraic loops will be demonstrated later.

```
>> set_param('c4mimp','AlgebraicLoopSolver','LineSearch')
   [t x y]=sim('c4mimp',5); err=max(abs(y-exp(-t)))
```

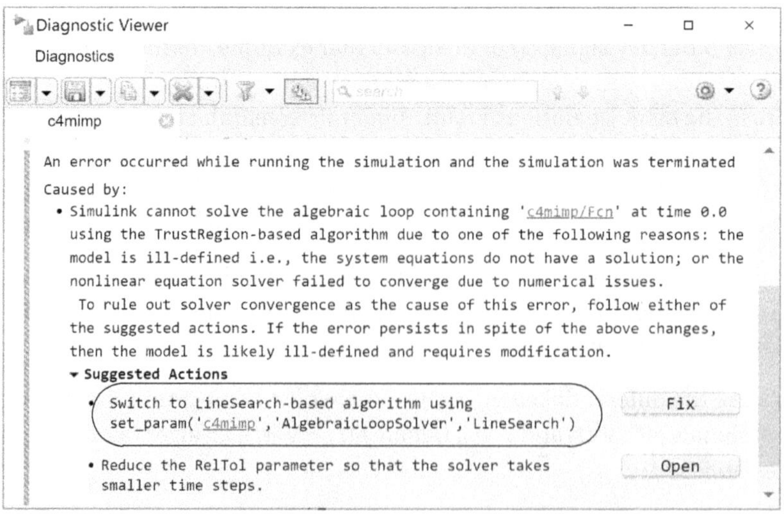

Figure 4.35: Error message dialog box.

Example 4.19. The Algebraic Constraint block has its own adjustable parameters. Select different values for the adjustable parameters and observe their impact on the simulation results for the problem in Example 4.17.

Solutions. Double click the **Algebraic Constraint** block in the `c4mimp.slx` model, a parameter dialog box in Figure 4.36 is displayed, where the **Solver** listbox provides three options: the default **auto**, and the other two, **trust-region** and **line search**. Apart from algorithm, the **Tolerance** edit box is provided.

Figure 4.36: Parameter dialog box of Algebraic Constraint block.

In Table 4.3, the solution results for different parameter values are recorded. It can be seen that for this particular example, the automatic algorithm and trust-region algorithm both failed to find the solutions. If line search algorithm is selected, different error tolerance may yield different accuracies. Under the 10^{-15} error tolerance, the best solution can be found, without increasing the computer burden. Further reducing the tolerance may not improve the results:

```
>> tic, [t x y]=sim('c4mimp',5); toc
   err=max(abs(y-exp(-t)))
```

Table 4.3: Comparisons on simulations results for different parameters.

	Auto	**trust region**	**line search**	**line search**	**line search**	**line search**
error tolerance	–	10^{-10}	10^{-5}	10^{-10}	10^{-14}	10^{-15}
maximum error	none	none	5.57×10^{-10}	1.597×10^{-11}	5.676×10^{-13}	5.713×10^{-13}
time elapse	–	–	0.7893	0.3806	0.2971	0.2888

There are inevitable limitations in the manipulating methods:

(1) **Algebraic Constraint block.** It can only be used to handle scalar algebraic equations. Therefore, it cannot be used in solving implicit differential equation sets.

(2) Since algebraic loops are present in the simulation model, in each simulation step, the algebraic equation is solved once. Therefore, the speed of simulation may be rather slow. Algebraic loops should be avoided, if necessary.

(3) Algebraic equations may have more than one solution. If they are ones with multiple solutions, the simulation algorithm presented here is not suitable, since the solutions obtained may be incorrect.

Even though, if Simulink is used to solve implicit differential equations, this method is unique. There are no other better solution strategies.

4.3.6 Modeling of discontinuous ODEs

Discontinuous differential equations are those whose signals have discontinuities at certain points, or jumping phenomena occur. In normal cases, in order to find numerical solutions to discontinuous differential equations, no special action is needed. The error tolerance can be assigned to larger values, such as 10^{-5}. Besides, stiff solvers should not be adopted. An example is given next to show the solution process of discontinuous differential equations.

Example 4.20. Consider the following discontinuous differential equation:[29]

$$y''(t) + 2Dy'(t) + \mu\,\mathrm{sign}(y'(t)) + y(t) = A\cos\omega t,$$

where $D = 0.1$, $\mu = 4$, $A = 2$, $\omega = \pi$. The initial values are $y(0) = 3$ and $y'(0) = 4$.

Solutions. The term $\mu\,\mathrm{sign}(y'(t))$ on the left is Coulomb friction. For the equation, when the velocity $y'(t)$ is altering the sign, sudden jumps may occur in the Coulomb friction, between $[-4, 4]$. The size of the jumping signal is rather large, so that discontinuous effect exists in the model. These jumping behaviors cannot be correctly handled in many of the numerical algorithms.

With the Simulink method discussed earlier, the simulation model shown in Figure 4.37 can easily be constructed. Note that around the output port of block Fcn2, there is an  icon, which is the virtual scope assigned with the shortcut menu item in Figure 2.19. After simulation process is completed, double the icon, and the signal can be displayed. In fact, the signal is $2Dy'(t) + \mu\,\mathrm{sign}(y'(t)) + y(t)$.

When the simulation process is executed, the results can be found as shown in Figure 4.38. Meanwhile, double click the virtual scope block, and the signal $2Dy'(t) + \mu\,\mathrm{sign}(y'(t)) + y(t)$ can be drawn, as shown in Figure 4.39. It can be seen that the discontinuous Coulomb friction is switching all the time in a specific region, which may slow down the entire simulation process.

For this specific problem, the error tolerance should not be assigned to extremely small values; otherwise, the simulation speed will be significantly reduced, or the solutions cannot be found. In this example, the relative error tolerance is 10^{-7}, which

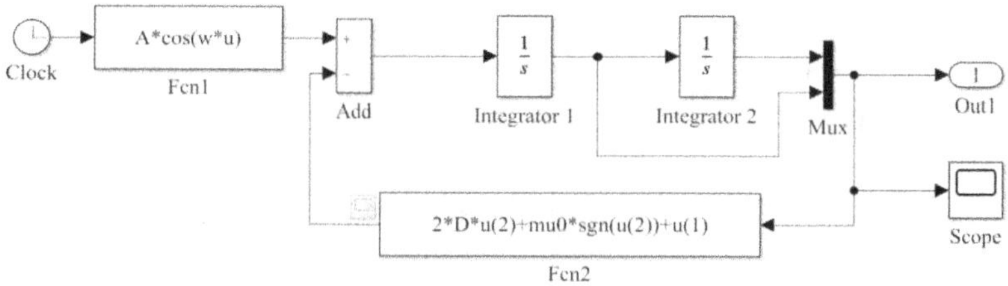

Figure 4.37: Simulink model of discontinuous ODE (file: `c4mdisc.slx`).

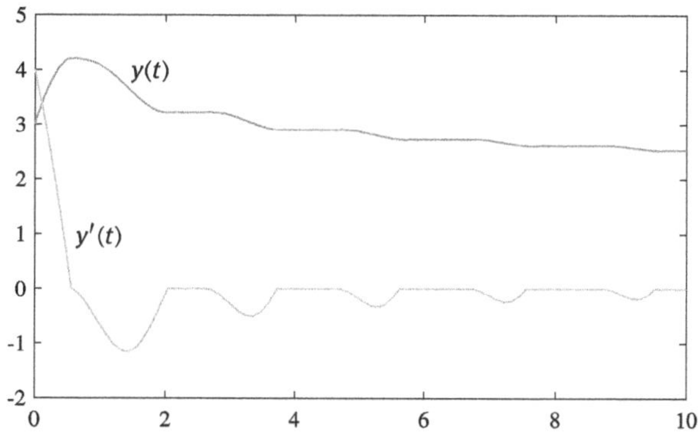

Figure 4.38: Solution of the discontinuous ODE.

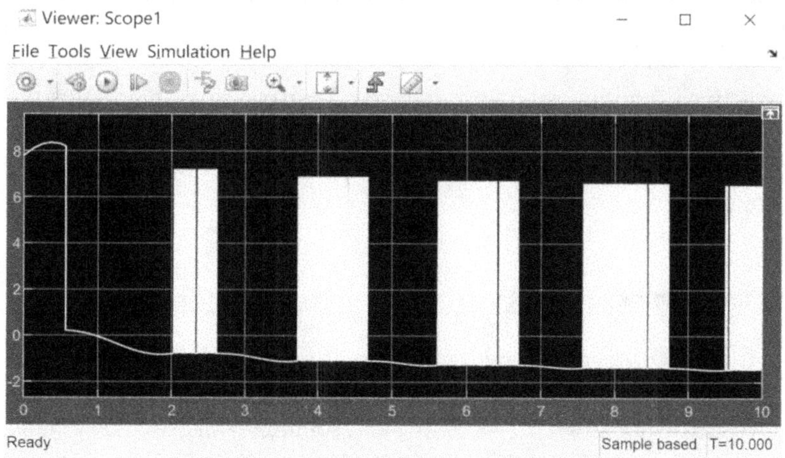

Figure 4.39: Jumping signal $2Dy'(t) + \mu\,\text{sign}(y'(t)) + y(t)$.

is lower than the one in Volume V, where the command-line driven solver was used, so that more accurate solutions may be found.[81]

It should be pointed out that stiff solvers may not be suitable for discontinuous differential equations. For instance, if the error tolerance is assigned to 10^{-8}, erroneous results may be found, as shown in Figure 4.40, where the solid line is obtained with ode15s algorithm, while the dashed line is from the ode23tb solver. In the two algorithms, the result in the first 2 seconds are fine. Later, due to the effect of Coulomb friction, the signal keeps jumping, and to avoid such effect, low-order approximation is taken automatically, thus they may lead to wrong results. If the ode45 solver is used, small step-size is automatically chosen. Although the speed is sacrificed in ode45 solver, accuracy in the solution is ensured.

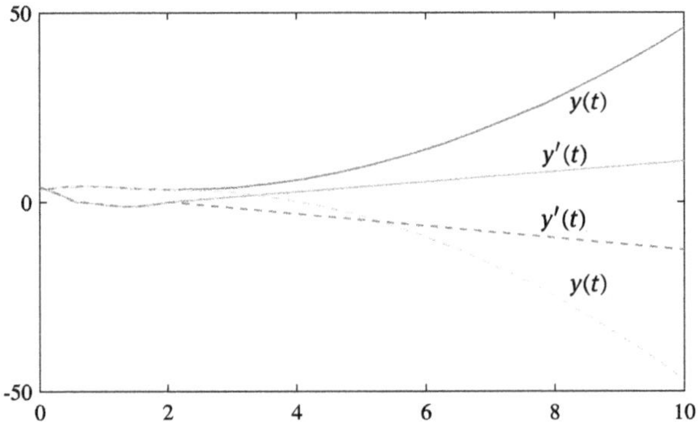

Figure 4.40: The erroneous solution with stiff solvers.

4.4 Modeling of high-order ODE sets

In the previous sections, Simulink modeling and simulation of a single high-order differential are studied. In real applications, simultaneous high-order differential equation sets may also be encountered. In this section, modeling and solution of high-order ODE sets are explored.

4.4.1 Modeling methodologies of ODEs

Consider the high-order ODE sets with two equations. The general mathematical form is given first, then concentrations are made on the Simulink modeling of the ODEs. Examples are given to demonstrate the modeling and solution of more ODEs. Finally, the modeling and solution of delay differential equations are discussed.

Definition 4.5. Explicit ODE sets with two equations can be written as

$$\begin{cases} x^{(m)}(t) = f(t, x(t), x'(t), \dots, x^{(m-1)}(t), y(t), y'(t), \dots, y^{(n-1)}(t)), \\ y^{(n)}(t) = g(t, x(t), x'(t), \dots, x^{(m-1)}(t), y(t), y'(t), \dots, y^{(n-1)}(t)), \end{cases} \tag{4.4.1}$$

with the initial conditions given as $x(t_0)$, $x'(t_0), \dots, x^{(m-1)}(t_0)$, $y(t_0)$, $y'(t_0), \dots, y^{n-1}(t_0)$.

Of course, two sets of integrator chains can be used. The former one is used to describe $x(t)$ and its derivatives, and the latter for $y(t)$ and its derivatives. Therefore, around these key signals, the right sides of the two equations can be expressed, and the signals can be fed into the two leftmost integrators, to close the simulation loops. Therefore, the Simulink model can be established, and examples are used in this section to demonstrate the modeling and solution of the ODEs.

Example 4.21. The position (x, y) of the Apollo satellite is described with the following nonlinear differential equations:[21]

$$\begin{cases} x''(t) = 2y'(t) + x(t) - \mu^*(x(t) + \mu)/r_1^3(t) - \mu(x(t) - \mu^*)/r_2^3(t), \\ y''(t) = -2x'(t) + y(t) - \mu^* y(t)/r_1^3(t) - \mu y/r_2^3(t), \end{cases}$$

where $\mu = 1/82.45$, $\mu^* = 1 - \mu$, and

$$r_1(t) = \sqrt{(x(t) + \mu)^2 + y^2(t)}, \quad r_2(t) = \sqrt{(x(t) - \mu^*)^2 + y^2(t)}.$$

Establish a Simulink model. If the initial values are $x(0) = 1.2$, $x'(0) = 0$, $y(0) = 0$, $y'(0) = -1.04935751$, solve the differential equation, and draw the trajectory (x, y) of the Apollo satellite.

Solutions. Two pairs of integrators can be used to define the key signals $x(t)$, $x'(t)$, $x''(t)$, $y(t)$, $y'(t)$ and $y''(t)$. Then with these key signals, the intermediate signals $r_1(t)$ and $r_2(t)$ can be generated. Then the right sides of the two equations can be modeled, and fed into the corresponding integrators, to close the two simulation loops. The Simulink model can the be established, as shown in Figure 4.41.

For the convenience of modeling, two blocks, Mux and Mux1, can be used, and the former one merges the four key signals into a vector signal, the latter combines also the two new signals $r_1(t)$ and $r_2(t)$, such that a vector signal with 6 scalar signals can be formed. The code in blocks Fcn1 and Fcn2 can be described as

`sqrt((u(1)+mu)^2+u(3)^2)` and `sqrt((u(1)-mu1)^2+u(3)^2)`

In this example, the Selector block is deliberately used, and the parameter can be set to `[1,3]`, meaning that the signals $x(t)$ and $y(t)$ can be extracted from the obtained vector. After simulation, the following commands can be used to draw the trajectory of the Apollo satellite, as shown in Figure 4.42:

```
>> plot(yout(:,1),yout(:,2))
```

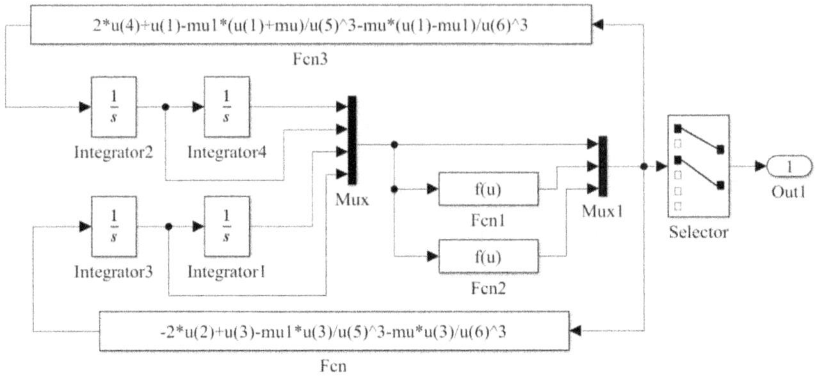

Figure 4.41: Simulink model of Apollo satellite (file: `c4mnlin2.slx`).

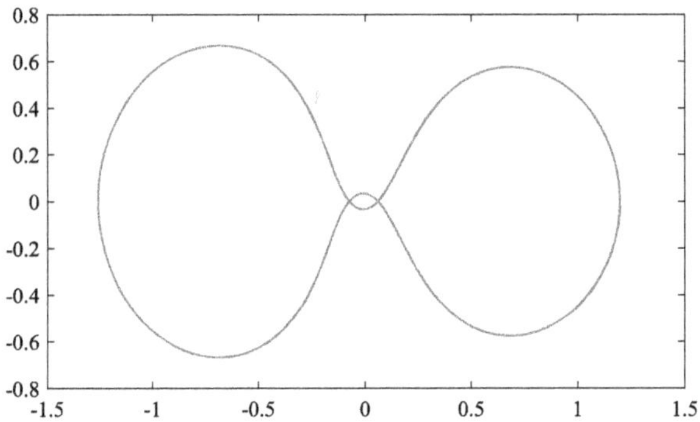

Figure 4.42: Trajectory of the Apollo satellite.

Of course, the trajectory is obtained with tough error tolerance. If the default error tolerance is used, wrong results may be obtained.

4.4.2 Solutions of ODEs

The solution methods for the ODE set are presented earlier. In fact, ODEs with more equations can be handled in a similar way. If there are m equations, then m integrator chains can be constructed to define the key signals. It is suggested to compose the key signals into a vector signal, then Fcn blocks can be used and the outputs can be fed into the integrators to close the simulation loops. With the methods thus presented, the Simulink model can be established. An example is given to demonstrate the modeling and simulation methods.

Example 4.22. Build a Simulink model for the following differential equations:

$$\begin{cases} x''(t) - x(t) + y(t) + z(t) = 0, \\ x(t) + y''(t) - y(t) + z(t) = 0, \\ x(t) + y(t) + z''(t) - z(t) = 0, \end{cases}$$

where $x(0) = 1$, $y(0) = z(0) = x'(0) = y'(0) = z'(0) = 0$.

Solutions. The explicit form of the equation can be written directly:

$$\begin{cases} x''(t) = x(t) - y(t) - z(t), \\ y''(t) = -x(t) + y(t) - z(t), \\ z''(t) = -x(t) - y(t) + z(t). \end{cases}$$

The key signals in the three integrator chains can be used to build a vector signal, such that the Simulink model in Figure 4.43 can be constructed. Since first-order derivatives are not used in the right side of the equations, they can be omitted when composing the signal vector. If the initial values in the integrators are assigned properly, numerical solutions can be obtained for the differential equations.

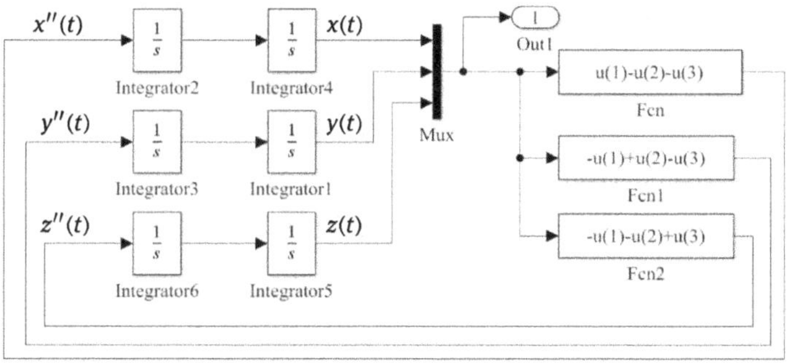

Figure 4.43: ODE sets with three equations (file: `c4m3eq.slx`).

4.4.3 Solutions of delay differential equations

In the previous differential equations, all of the signals involved happen at the same time instance t. If the equations contain not only the current signal values, but also the signal values in the past, the equations are referred to as delay differential equations.

Definition 4.6. The general form of delay differential equations with constant delays can be expressed as

$$\boldsymbol{x}'(t) = \boldsymbol{f}(t, \boldsymbol{x}(t), \boldsymbol{x}(t-\tau_1), \boldsymbol{x}(t-\tau_2), \dots, \boldsymbol{x}(t-\tau_m)), \tag{4.4.2}$$

where $\tau_i \geqslant 0$ are the delay constants for state variables $\boldsymbol{x}(t)$.

In the mathematical expressions, $\boldsymbol{x}(t)$ is the state vector at current time t, and $\boldsymbol{x}(t-\tau_i)$ can be regarded as the values of the states τ_i seconds ago. Assume that there are m delay terms, $\boldsymbol{\tau} = [\tau_1, \tau_2, \dots, \tau_m]$, the delay terms can be a specific constant of functions of time. Based on the information, the first-order derivative $\boldsymbol{x}'(t)$ of the state vector at time t can be defined.

Command-line driven solution methods are provided in MATLAB for delay differential equations. Besides, the solvers can also be used in solving delay differential equations with state dependent delays. These kinds of delay differential equations are difficult for Simulink. In this section, the modeling and solution of various delay differential equations are illustrated.

Example 4.23. For the given delay differential equations with constant delays

$$\begin{cases} x'(t) = 1 - 3x(t) - y(t-1) - 0.2x^3(t-0.5) - x(t-0.5), \\ y''(t) + 3y'(t) + 2y(t) = 4x(t), \end{cases}$$

where when $t \leqslant 0$, $x(t) = y(t) = y'(t) = 0$. Find the numerical solution of the delay differential equation.

Solutions. Two integrator chains can be used to define the key signals. Alternatively, it can be seen from the second equation that $x(t)$ signal can be fed into a transfer function model $4/(s^2+3s+2)$, and the signal $y(t)$ can be generated. It is not difficult to create a Simulink model, as shown in Figure 4.44. In the model, signal $x(t-0.5)$ can be obtained by feeding the key signal $x(t)$ through the Transport Delay block. From the simulation model, the output signal $y(t)$ can be obtained as shown in Figure 4.45.

Example 4.24. Assume that the history functions of the state variables are all zero. Solve the following delay differential equations with variable delays:

$$\begin{cases} x_1'(t) = -2x_2(t) - 3x_1(t - 0.2|\sin t|), \\ x_2'(t) = -0.05x_1(t)x_3(t) - 2x_2(t-0.8) + 2, \\ x_3'(t) = 0.3x_1(t)x_2(t)x_3(t) + \cos(x_1(t)x_2(t)) + 2\sin 0.1t^2. \end{cases}$$

Solutions. The difficulties in the modeling is the description of the term $x_1(t - 0.2|\sin t|)$, where the delay time $0.2|\sin t|$ is a function of time t. Then the Variable Transport Delay block in the Continuous group can be used. The first input signal

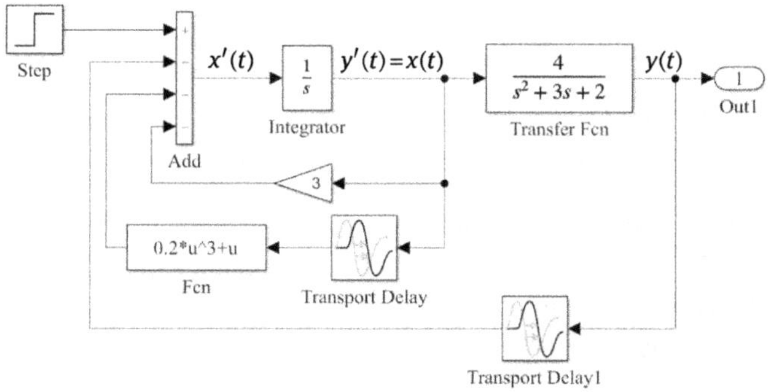

Figure 4.44: Simulink model of the delay differential equation (file: `c4mdde1.slx`).

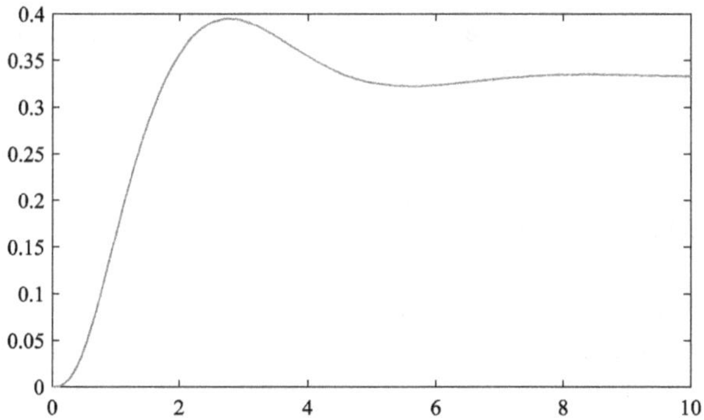

Figure 4.45: Output signal curve.

is connected to signal $x_1(t)$, and the second one is connected to the delay function $0.2|\sin t|$. The output signal of the block is then $x_1(t - 0.2|\sin t|)$. Based on the considerations, the Simulink model shown in Figure 4.46 can be established. In the Interpreted MATLAB Function block, the corresponding function is

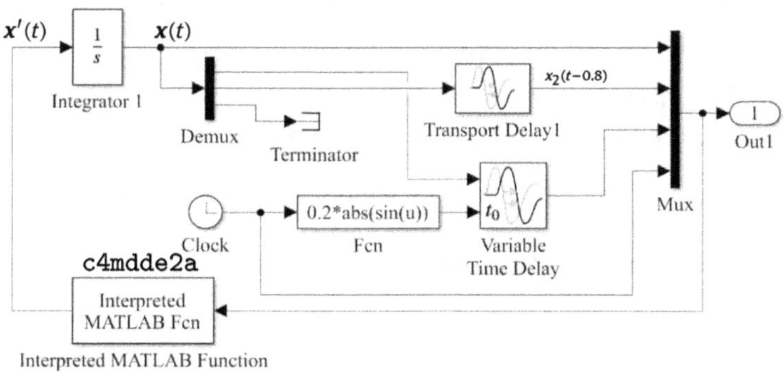

Figure 4.46: Simulink model with variable delays (file: `c4mdde2.slx`).

```
function dx=c4mdde2a(x)
dx=[-2*x(2)-3*x(5);
    -0.05*x(1)*x(3)-2*x(4)+2;
    0.3*x(1)*x(2)*x(3)+cos(x(1)*x(2))+2*sin(0.1*x(6)^2)];
```

In practical modeling, the **Demux** block can be used to divide $\boldsymbol{x}(t)$ vector into three scalar signals, where $x_1(t)$ and $x_2(t)$ are connected to other blocks, while $x_3(t)$ is an unconnected one. To avoid warning messages, it should be connected to a genuine block. For instance, connect the signal to the **Terminator** block in the **Sinks** group.

Compared with the command-line driven solvers provided in MATLAB, Simulink modeling is simpler. This method also has limitations. For instance, if the delay term τ_i is a function of state, or αt, with $\alpha < 1$, it is hard to construct a Simulink model. Command-line driven methods can be used to solve these types of problems. For details, please refer to Volume V.

4.4.4 Neutral-type delay differential equations

Apart from the simple delay differential equations, in some applications, neutral-type delay differential equations are also involved. The mathematical form is provided. Modeling and simulation methods of neutral-type delay differential equations are then explored. In particular, the modeling and simulation of delay differential equations with nonzero history functions are also discussed.

Definition 4.7. The mathematical form of a neutral-type delay differential equation is given by

$$\boldsymbol{x}'(t) = \boldsymbol{f}(t, \boldsymbol{x}(t), \boldsymbol{x}(t-\tau_{p_1}), \dots, \boldsymbol{x}(t-\tau_{p_m}), \boldsymbol{x}'(t-\tau_{q_1}), \dots, \boldsymbol{x}'(t-\tau_{q_k})), \tag{4.4.3}$$

where in the equation, not only the delay signals in the state variables can be included, the delay of the derivative of states is also contained. They are expressed by the two vectors $\boldsymbol{\tau}_1 = [\tau_{p_1}, \tau_{p_2}, \dots, \tau_{p_m}]$ and $\boldsymbol{\tau}_2 = [\tau_{q_1}, \tau_{q_2}, \dots, \tau_{q_k}]$.

In the above definition, τ_p and τ_q can be constant vectors to represent delay constants. They can also be $\tau_p(t)$ and $\tau_q(t)$, the functions of time.

Example 4.25. Solve the following simple neutral-type delay differential equation:[9]

$$y'(t) = y(t) + y'(t-1),$$

where when $t \leqslant 0$, $y(t) = 1$ and $t \in (0, 3)$. It is known that the analytical solution of the equations is

$$y(t) = \begin{cases} e^t, & 0 \leqslant t \leqslant 1, \\ e^t + (t-1)e^{t-1}, & 1 < t < 2, \\ e^t + e^{t-1} + (t-2)(t+2e)e^{t-2}/2, & 2 \leqslant t \leqslant 3. \end{cases}$$

Solutions. There are difficulties in Simulink modeling of the delay differential equation. When $t \leqslant 0$, the initial value of the state variable $y(t)$ is nonzero. The Switch block can be used to describe the history function of the signal when $t \leqslant 0$. The neutral-type delay differential equations can be modeled as shown in Figure 4.47.

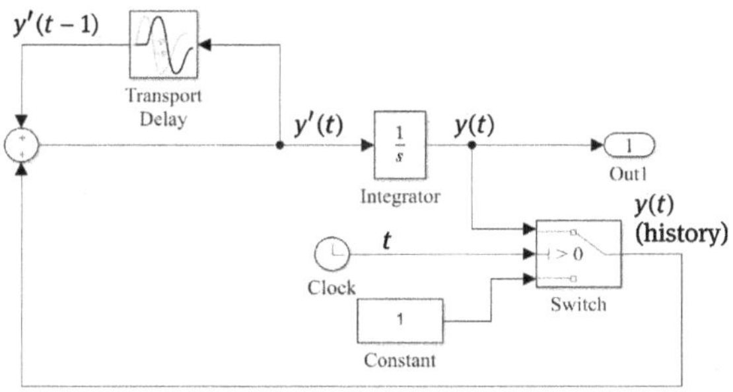

Figure 4.47: Neutral-type delay differential equations (file: `c4mdde3.slx`).

In fact, although the sum block can be used to extract signal from $y'(t)$, and the closed-loop is composed, the loop is not an algebraic loop, since the $y'(\cdot)$ signals on the two sides are not happening at the same time. After simulation, the solution can be compared with the theoretical one, and it is found that the maximum error is 4.9360×10^{-5}, and the accuracy is higher than the command-line driven solver:[81]

```
>> t=tout;
   z=exp(t)+(t-1).*exp(t-1).*(t>=1 & t<2)+...
     (exp(t-1)+(t-2).*(t+2*exp(1)).*exp(t-2)/2).*(t>=2);
   max(abs(z-yout))
```

The Simulink modeling and solution of several types of commonly used differential equations are presented earlier. Other types of differential equations such as switched differential equations, differential equations subject to stochastic input and fractional-order differential equations are shown.

4.4.5 Switched ODEs

The switched system is composed of a series continuous or discrete systems, known as subsystems. The subsystems can be switched from one to another under certain switching laws. Various switched system examples can be found in [61, 86, 87]. In Volume V, command-line driven solution methods are presented. In this section, concentrations are made on presenting Simulink modeling and solution methods.

Definition 4.8. The mathematical model of the switch differential equation is

$$\boldsymbol{x}'(t) = \boldsymbol{f}_{\sigma(t,\boldsymbol{x}(t))}(t, \boldsymbol{x}(t), \boldsymbol{u}(t)), \tag{4.4.4}$$

where $\boldsymbol{x}(t_0)$ are given, and $\boldsymbol{u}(t)$ is the external input signal. The value $\sigma(t, \boldsymbol{x}(t)) = \{1, 2, \dots, m\}$ is referred to as the switching law.

Under a certain switching law, the whole system is switched among different subsystems. With switched control theory, controllers can be designed such that the whole system can be made stable, even if the subsystems $\boldsymbol{f}_{\sigma(t,\boldsymbol{x}(t))}(\cdot)$ are unstable.

The physical meaning of switching law is that if the ith preassigned condition is satisfied, the system can be switched to the ith subsystem. It can be seen that if the switching law can be described clearly, the system can be described by anonymous functions or MATLAB functions, so that the whole switched system can be described. The command-line driven method can then be used to find the numerical solutions of the switched differential equations.

If Simulink is used in the modeling and simulation, the Switch block or Multiport switch block can be used. The switching law can be applied to the control terminal, and the subsystems can be connected to other input ports of the switch block individually, so that a Simulink model can be established. Examples are used to demonstrate the modeling and simulation of switched differential equations.

Example 4.26. Assume that the system is given by $\boldsymbol{x}'(t) = \boldsymbol{A}_i\boldsymbol{x}(t)$, where

$$\boldsymbol{A}_1 = \begin{bmatrix} 0.1 & -1 \\ 2 & 0.1 \end{bmatrix}, \quad \boldsymbol{A}_2 = \begin{bmatrix} 0.1 & -2 \\ 1 & 0.1 \end{bmatrix}.$$

If $x_1(t)x_2(t) < 0$, that is, the states are located in the II and IV quadrants, the system is switched to $\boldsymbol{A}_1$; if $x_1(t)x_2(t) \geqslant 0$, that is, the states are in the I and III quadrants, the system is switched to $\boldsymbol{A}_2$. If the initial states are $x_1(0) = x_2(0) = 5$, draw the state variables in $t \in (0, 20)$.

Solutions. The two matrices can be entered in MATLAB, and the eigenvalues of them can be found:

```
>> A1=[0.1,-1; 2,0.1]; A2=[0.1,-2; 1,0.1]; eig(A1), eig(A2)
```

It can be seen that there are $\mathrm{Re}(\lambda_j) > 0$ eigenvalues in the two matrices, indicating that the two subsystems are all unstable. If switch block is used in describing the switching law, the Simulink model in Figure 4.48(a) can be established. The parameter dialog box of the switch block is shown in Figure 4.48(b). To implement the switching laws, the Threshold of the block is set to 0, and set Criterion for passing first input to the

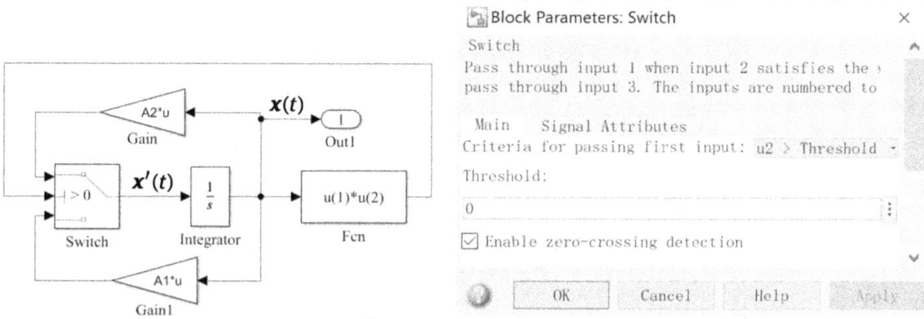

(a) Simulink model (file: `c4mswi1.slx`) (b) parameter dialog box

Figure 4.48: Simulation model of switched differential equation.

default u2 > Threshold. Besides, in order to get accurate results, Enable zero-crossing detection should be checked.

After the simulation process, the results are returned to the MATLAB workspace. With the following commands, the state variables are shown in Figure 4.49. To better view the status of the state variables, the phase plane trajectory can be drawn as shown in Figure 4.50. It can be seen that under the given switching law, the whole system is stable.

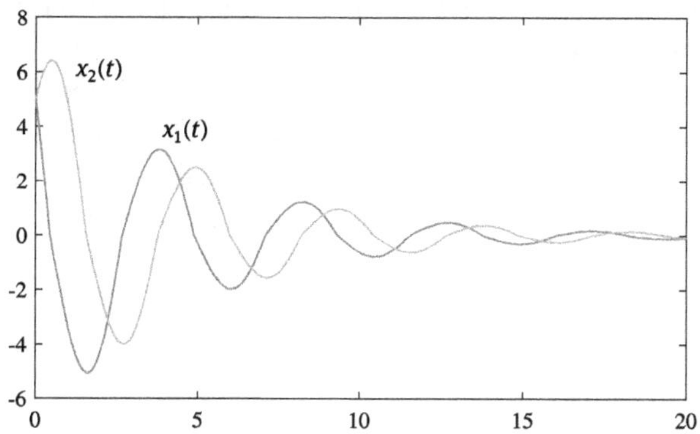

Figure 4.49: Solutions of switched differential equation.

4.4.6 ODEs with stochastic inputs

If the input signal to the differential equation is a stochastic one, the solutions of such equation are considered in this section. For stochastic differential equations, simple simulation is useless. A series of simulations should be carried, and statistical analysis should be made to the simulation results.

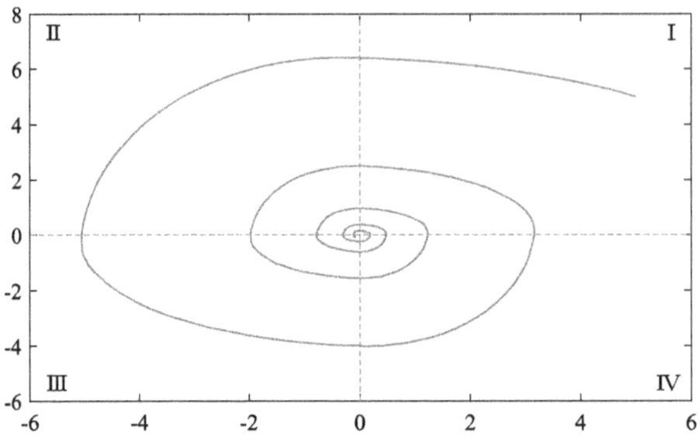

Figure 4.50: Phase plane trajectory of switch differential equation.

Assume that the signal is a Gaussian white noise, and consider using the block provided in Simulink to approximate white noise excitation. Several random signal blocks are provided in the **Sources** group in Simulink. Which one of them should be used? It has been studied in [81], and the conclusion is that the signal intensity of the **Uniform Random Number** and **Random Number** blocks is not sufficient, and should not be used to excite continuous systems; otherwise, the simulation results are wrong.

The correct stochastic process for the excitation is the **Band-Limited White Noise** block, whose parameter dialog box is shown in Figure 4.51, where **Noise power** is the variance of the stochastic process. The **Sample time** should be specified, and fixed-step simulation algorithm is recommended. The parameter **Seed** is the seed needed to generate pseudorandom numbers. Normally, different seeds may yield different simulation results, but the statistical conclusion may not be affected. If repeated experiment is to be carried out, the same value of seed should be selected each time.

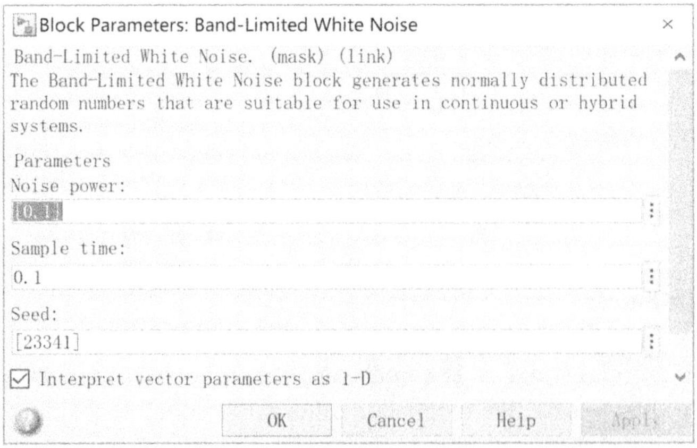

Figure 4.51: Parameter dialog box of Band-Limited White Noise block.

Example 4.27. Consider the nonlinear feedback control model in Figure 4.52. The transfer function model $G(s)$ is given by

$$G(s) = \frac{s^3 + 7s^2 + 24s + 24}{s^4 + 10s^3 + 35s^2 + 50s + 24},$$

and saturation nonlinearity is described as

$$\mathcal{N}(e) = \begin{cases} 2e, & |e| \leqslant 1, \\ 2\,\mathrm{sign}(e), & |e| > 1. \end{cases}$$

If the deterministic input signal is $r(t) = 0$. The variance of the Gaussian white noise signal $\gamma(t)$ is $\sigma^2 = 4$. Find the probability density function of $e(t)$ signal. In [76], using the approximate Fokker–Planck equation, the analytical solution of the probability density function of signal $e(t)$ is given below. Compare the numerical and quasi-analytical solutions:

$$p(e(t)) = \begin{cases} 0.2950\mathrm{e}^{-0.3496e^2(t)}, & |e(t)| \leqslant 1, \\ 0.3629\mathrm{e}^{-0.1114e^2(t)-0.4454e(t)\mathrm{sign}(e(t))}, & |e(t)| > 1. \end{cases}$$

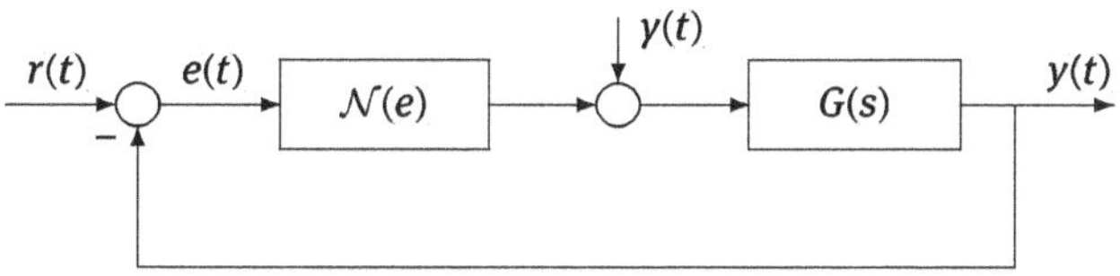

Figure 4.52: Nonlinear feedback control system with stochastic input.

Solutions. With the Band-Limited White Noise block, the simulation model is constructed, as shown in Figure 4.53. Select a fixed step-size of $T = 0.08$ and carry out simulation, and the simulation results can be obtained from vector `yout`.

With the simulation results, the histogram can be drawn for its probability density function, superimposed by analytical solution, shown in Figure 4.54. It can be seen that the simulation results are rather accurate. It is worth mentioning that if simulation analysis is needed, the step-size T should not be set to extremely small values:

```
>> ee=-5:0.1:5; % theoretical values
   p=0.2950*exp(-0.3496*ee.^2).*(abs(ee)<1)+...
     0.3629*exp(-0.1114*ee.^2-...
     0.4454*ee.*sign(ee)).*(abs(ee)>=1);
   c=linspace(-5,5,40); y1=hist(yout,c);
   bar(c,y1/(length(tout)*(c(2)-c(1)))), line(ee,p)
```

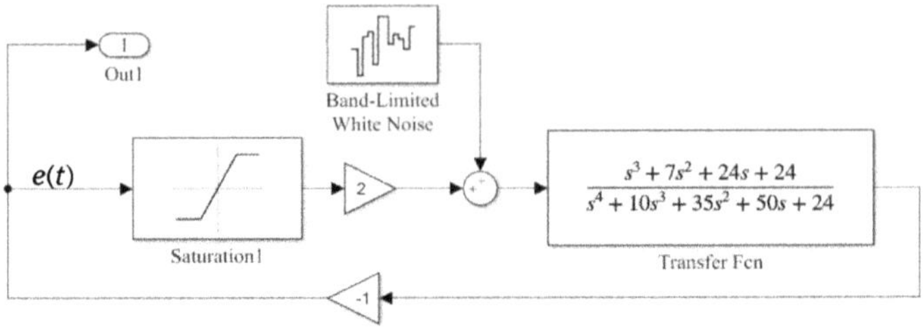

Figure 4.53: Simulink model (file: `c4mbandc.slx`).

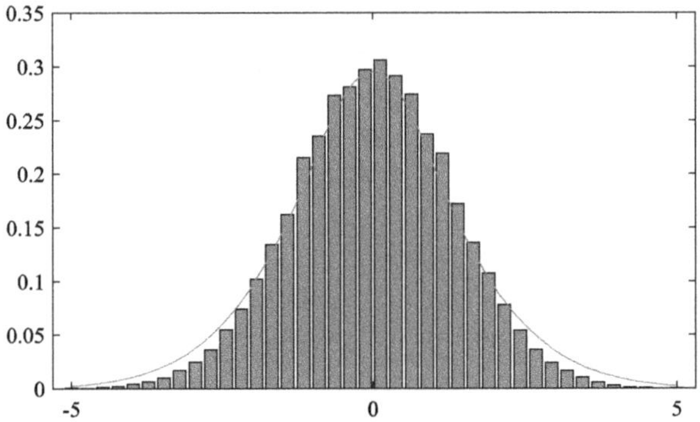

Figure 4.54: Probability density function and histograms.

4.5 Modeling and solving fractional-order ODEs

4.5.1 An introduction to fractional calculus

Fractional calculus is a relatively new application field in calculus. The so-called fractional means that the order of differentiation or integral is no longer integers, and it can be any fractional number or even irrational ones. In [79, 80], numerical methods are addressed, and in Volumes II and V, numerical methods for fractional calculus and fractional-order differential equations are presented.

The commonly used definitions in fractional calculus are the Grünwald–Letnikov definition, Riemann–Liouville definition and Caputo definition.[79, 80] For ordinary functions, the former two definitions and equivalent. The definitions and properties are presented first, and the modeling and simulation of fractional-order differential equations are explored.

Definition 4.9. The Riemann–Liouville fractional-order integral is defined as

$$ {}_{t_0}^{\mathrm{RL}}\mathscr{D}_t^{-\alpha} f(t) = \frac{1}{\Gamma(\alpha)} \int_{t_0}^{t} \frac{f(\tau)}{(t-\tau)^{1-\alpha}} \mathrm{d}\tau, \tag{4.5.1} $$

where $\Gamma(\cdot)$ is the Gamma function, with $\alpha > 0$, and t_0 is the initial time. If $t_0 = 0$, the differentiation notation is simplified as ${}^{\mathrm{RL}}\mathscr{D}_t^{-\alpha}f(t)$. If there is no conflict, the superscript RL can be omitted. The Riemann–Liouville definition is a commonly used definition in fractional calculus. In particular, the subscripts on the left and right of $\mathscr{D}$ are respectively lower and upper bounds on the integral.[36]

Riemann–Liouville fractional-order integral definition are named after German mathematician, Georg Friedrich Bernhard Riemann (1826–1866) and French mathematician, Joseph Liouville (1809–1882). They proposed relevant definitions respectively in 1847 and 1834.

From the integral, the fractional-order derivative can also be defined. For rounded integer $m = \lceil\alpha\rceil$, the fractional-order derivative is defined as

$$\begin{aligned} {}^{\mathrm{RL}}_{t_0}\mathscr{D}_t^{\alpha}f(t) &= \frac{\mathrm{d}^m}{\mathrm{d}t^m}\left[{}^{\mathrm{RL}}_{t_0}\mathscr{D}_t^{-(m-\alpha)}f(t)\right] \\ &= \frac{1}{\Gamma(m-\alpha)}\frac{\mathrm{d}^m}{\mathrm{d}t^m}\left[\int_{t_0}^{t}\frac{f(\tau)}{(t-\tau)^{1+\alpha-m}}\mathrm{d}\tau\right]. \end{aligned} \tag{4.5.2}$$

Definition 4.10. Caputo fractional-order derivative is defined as

$${}^{\mathrm{C}}_{t_0}\mathscr{D}_t^{\alpha}f(t) = \frac{1}{\Gamma(m-\alpha)}\int_{t_0}^{t}\frac{f^{(m)}(\tau)}{(t-\tau)^{1+\alpha-m}}\mathrm{d}\tau, \tag{4.5.3}$$

where $m = \lceil\alpha\rceil$ in an integer.

Caputo definition was proposed by an Italian mathematician and geophysicist, Michele Caputo in 1971, aiming at solving nonzero initial value problems in fractional calculus.[14]

The Caputo fractional-order integral is exactly the same, as the one in the Riemann–Liouville definition.

4.5.2 Approximating fractional-order operators

To effectively establish the Simulink model for fractional-order differential equations, the Laplace transform techniques can be adopted. The Laplace transform formulas under different definitions are given next.

Theorem 4.1. *Under the Riemann–Liouville definition, Laplace transform of the fractional-order derivative can be written as*

$$\mathscr{L}\left[{}^{\mathrm{RL}}_{t_0}\mathscr{D}_t^{\alpha}f(t)\right] = s^{\alpha}\mathscr{L}[f(t)] - \sum_{k=1}^{n-1}s^k\,{}^{\mathrm{RL}}_{t_0}\mathscr{D}_t^{\alpha-k-1}f(t)\Big|_{t=t_0}. \tag{4.5.4}$$

The Laplace transform of fractional-order integrals under Caputo and Riemann–Liouville definitions are the same, and not given here.

Theorem 4.2. *The Laplace transform for Caputo fractional-order derivative satisfies*

$$\mathscr{L}\left[{}_{t_0}^{\mathrm{C}}\mathscr{D}_t^{\gamma} f(t)\right] = s^{\gamma} F(s) - \sum_{k=0}^{n-1} s^{\gamma-k-1} f^{(k)}(t_0). \tag{4.5.5}$$

In particular, if the initial values of $f(t)$ and its derivatives of different orders are all zeros, $\mathscr{L}[{}_{t_0}\mathscr{D}_t^{\alpha} f(t)] = s^{\alpha}\mathscr{L}[f(t)]$.

It can be seen from the Laplace transform properties that only initial values of integer-order derivatives are involved in the Caputo definition, while in the Riemann–Liouville definition, initial values of fractional-order derivatives are involved. Therefore, the Caputo definition is more suitable for describing fractional-order dynamical systems with nonzero initial conditions.

To have effective computation for fractional-order derivatives and integrals, the approximation of s^{α} is crucial. Various approximation methods can be used for s^{α}, among them the filter introduced in [52] is the most effective one, and it is referred to as Oustaloup filter in this book.

Theorem 4.3. *Assume that the interested frequency range is* $(\omega_{\mathrm{b}}, \omega_{\mathrm{h}})$, *the integer-order transfer function of an* N*th order filter is*

$$G_{\mathrm{f}}(s) = K \prod_{k=1}^{N} \frac{s + \omega_k'}{s + \omega_k}, \tag{4.5.6}$$

where the zeros, poles and gain are

$$\omega_k' = \omega_{\mathrm{b}} \omega_{\mathrm{u}}^{(2k-1-\gamma)/N}, \quad \omega_k = \omega_{\mathrm{b}} \omega_{\mathrm{u}}^{(2k-1+\gamma)/N}, \quad K = \omega_{\mathrm{h}}^{\gamma}, \tag{4.5.7}$$

with $\omega_{\mathrm{u}} = \sqrt{\omega_{\mathrm{h}}/\omega_{\mathrm{b}}}$, *known as base frequency.*

Based on the above algorithm, the following MATLAB function can be written to design the continuous filter. If signal $f(t)$ is fed into the filter, the output signal is the approximation of ${}^{\mathrm{RL}}\mathscr{D}_t^{\gamma} f(t)$:

```
function G=ousta_fod(gam,N,wb,wh)
if round(gam)==gam, G=tf('s')^gam; %if already integer-order
else, k=1:N; wu=sqrt(wh/wb);          %find base frequency
  wkp=wb*wu.^((2*k-1-gam)/N); wk=wb*wu.^((2*k-1+gam)/N);
                                                  %zeros and poles
  G=zpk(-wkp,-wk,wh^gam); G=tf(G); %construct the Oustaloup filter
end
```

The syntax of the function is G_1=`ousta_fod(`γ`, `N`, `ω_{b}`, `ω_{h}`)`, where γ is the fractional order, and it can either be positive or negative. N is the filter order. The frequencies ω_b and ω_h are the lower and upper bounds of the interested frequency range. The approximation within the specified frequency range is usually very good, while for those outside the range, the fitting is poor. Since powerful and high efficiency facilities are provided in MATLAB, the interested frequency range that can be selected is very large. For instance, $(10^{-5}, 10^5)$, and a high order can be selected, such as $N = 25$. In this case, fractional-order derivatives can be approximated very accurately.

A fractional-order operator approximation block was implemented in Simulink, which can be used to design various filters to approximate fractional-order operators, including Oustaloup filter. The block is provided in the FOTF Toolbox,[80, 79] and can be used directly. If the FOTF Toolbox is installed, with the `fotflib` command, the FOTF blockset is opened, as shown in Figure 4.55, where the **Fractional operator** block implements the Oustaloup filter, and can be used in Simulink modeling. Assume that the order of the block is γ, and the input signal to the block is $u(t)$, then the output of the block is the γth order Riemann–Liouville derivative. Of course, if $\gamma < 0$, the output is the $-\gamma$th order Riemann–Liouville integral of the input signal $u(t)$.

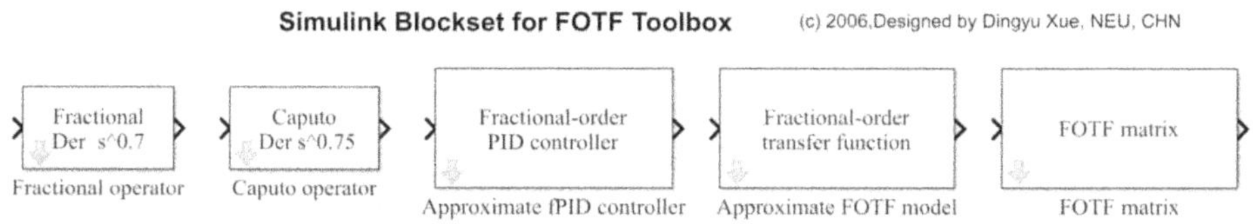

Figure 4.55: FOTF blockset.

Double click the block, and the parameter dialog box shown in Figure 4.56 is opened, allowing the users to specify parameters, such as the **Derivative order**, **Frequency range** and **Approximation order**.

The block cannot be used in evaluating Caputo derivatives or integrals directly. The following alternative methods can be used to approximate Caputo derivatives.

Theorem 4.4. *The Caputo fractional-order derivative can be generated from Oustaloup filter under the following method:*

$$ {}_{t_0}^{\mathrm{C}}\mathscr{D}_t^{\gamma} y(t) = {}_{t_0}^{\mathrm{RL}}\mathscr{D}_t^{-(\lceil\gamma\rceil-\gamma)} \left[y^{(\lceil\gamma\rceil)}(t) \right]. \tag{4.5.8} $$

The physical explanation of the theorem is that the γth order Caputo derivative of $y(t)$ can be obtained, if the integer-order derivative $y^{(\lceil\gamma\rceil)}(t)$ is fed into the Oustaloup block for $(\lceil\gamma\rceil - \gamma)$th order integral.

It can be seen that by using the integrator chain, the integer-order derivatives of $y(t)$ can be constructed. If the 2.4th-order Caputo derivative is expected, the signal $y'''(t)$ can be fed into the 0.6th-order Oustaloup integral filter. The expected output is the expected ${}^{\mathrm{C}}\mathscr{D}^{2.4}y(t)$ signal.

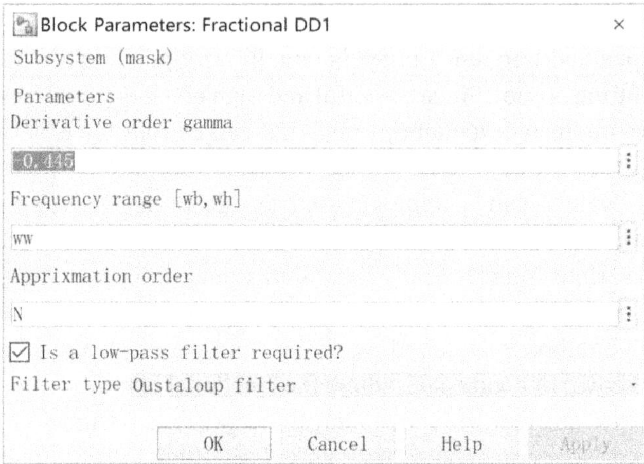

Figure 4.56: Parameter dialog box of Oustaloup filter.

Theorem 4.5. *If* $(\lceil\gamma\rceil - \gamma)$*th order Riemann–Liouville derivative is taken to both sides of* (4.5.8), *the following relationship can be found:*

$$ {}^{\mathrm{RL}}_{t_0}\mathscr{D}_t^{\lceil\gamma\rceil-\gamma}\left[{}^{\mathrm{C}}_{t_0}\mathscr{D}_t^{\gamma}y(t)\right] = y^{(\lceil\gamma\rceil)}(t). \tag{4.5.9} $$

The physical explanation of the theorem is that, for a γth order Caputo derivative ${}^{\mathrm{C}}_{t_0}\mathscr{D}^{\gamma}y(t)$, to find $(\lceil\gamma\rceil - \gamma)$th order Riemann–Liouville derivative, the integer-order derivative $y^{(\lceil\gamma\rceil)}(t)$ can be found. That is when Caputo derivatives are fed into the Oustaloup filter, and integer-order derivatives can be found.

Based on the above two theorems, Oustaloup filters can be used to take suitable actions on signals, and create Caputo derivatives for unknown signals, so as to build up simulation models for Caputo differential equations.

4.5.3 Solving fractional-order ODEs

With the above theoretical knowledge, a Simulink block diagram for the Caputo equations can be constructed, so as to find the solutions. In this section, two examples are used to demonstrate modeling and solutions of fractional-order differential equations.

Example 4.28. Use Simulink to solve the following Caputo differential equation:[81]

$$ {}^{\mathrm{C}}_{0}\mathscr{D}_t^{1.455}y(t) = -t^{0.1}\frac{\mathrm{E}_{1,1.545}(-t)}{\mathrm{E}_{1,1.445}(-t)}\mathrm{e}^{t}y(t)\,{}^{\mathrm{C}}_{0}\mathscr{D}_t^{0.555}y(t) + \mathrm{e}^{-2t} - \left[y'(t)\right]^2, $$

where $y(0) = 1$, $y'(0) = -1$. It is known that the analytical solution of the equation is $y(t) = \mathrm{e}^{-t}$.

Solutions. Since the highest order of the unknown signal $y(t)$ is 1.455, then $q = 2$, that is, a chain with two integrators can be used to define the key signals $y(t)$, $y'(t)$ and $y''(t)$, filling in the two initial values into the corresponding integrators. From (4.5.8), it is known that the signal ${}_0^{\mathrm{C}}\mathscr{D}_t^{0.555}y(t)$ can then be generated from $y'(t)$, and feed it into a 0.445th-order Oustaloup integrator, and the output of the filter is the expected signal ${}_0^{\mathrm{C}}\mathscr{D}_t^{0.555}y(t)$. With these signals, low-level modeling can be used to establish the right side of the equation, that is, ${}_0^{\mathrm{C}}\mathscr{D}_t^{1.455}y(t)$. To close the simulation loop, it is known from (4.5.9) that the signal is fed into a 0.445th-order Oustaloup filter block, and the signal $y''(t)$ can be generated. This signal is the starting signal of the integrator chain. Therefore, the output of the Oustaloup filter can be connected with $y''(t)$ to close the simulation loop, as shown in Figure 4.57. For simplicity, the time t based nonlinear operation can be expressed with a Interpreted MATLAB Fcn block, whose function is

```
function y=c4mmlfs(u) %MATLAB function for the ODEs
y=u^0.1*exp(u)*ml_func([1,1.545],-u)./ml_func([1,1.445],-u);
```

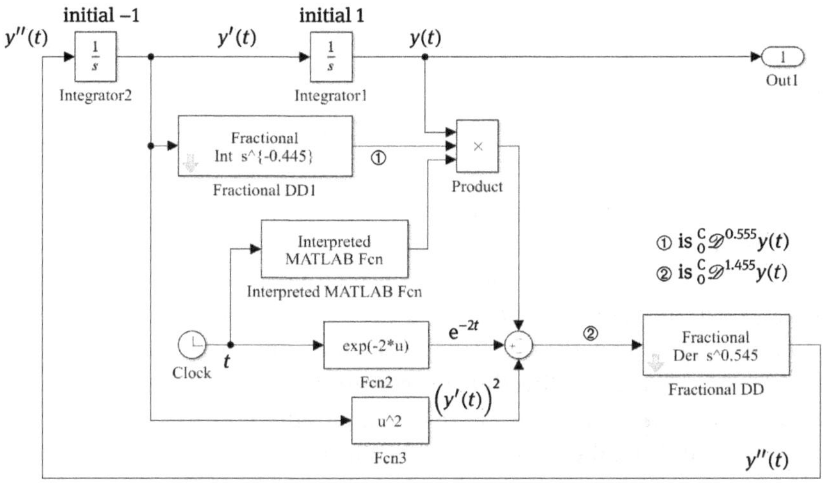

Figure 4.57: Simulink model for Caputo equation (file: `c4mfode.slx`).

To select filter parameters for the Oustaloup filter, the maximum error and time elapse can be measured, as shown in Figure 4.4. It can be seen that, if the interested frequency range is selected large enough, and the filter order is high enough, accurate solutions can be found:

```
>> N=30; ww=[1e-6 1e6]; %select Oustaloup filter parameters
   tic, [t,x,y]=sim('c4mfode'); toc
   max(abs(y-exp(-t))) %solve and validate results
```

Table 4.4: Accuracy and speed under different filter parameters.

frequency range	$(10^{-3}, 10^3)$	$(10^{-5}, 10^5)$	$(10^{-6}, 10^6)$	$(10^{-6}, 10^6)$	$(10^{-6}, 10^6)$	$(10^{-6}, 10^6)$
order	5	15	15	20	25	30
maximum error	0.0014	1.07×10^{-5}	3.82×10^{-5}	7.53×10^{-6}	1.14×10^{-6}	5.52×10^{-7}
time elapse	0.4066	0.3088	0.2876	0.4238	0.3734	0.5141

Example 4.29. Solve the following implicit differential equation:

$$\begin{aligned}&{}_0^{\mathrm{C}}\mathscr{D}_t^{0.2}y(t)\,{}_0^{\mathrm{C}}\mathscr{D}_t^{1.8}y(t)+{}_0^{\mathrm{C}}\mathscr{D}_t^{0.3}y(t)\,{}_0^{\mathrm{C}}\mathscr{D}_t^{1.7}y(t)\\&\quad=-\frac{t}{8}\left[\mathrm{E}_{1,1.8}\left(-\frac{t}{2}\right)\mathrm{E}_{1,1.2}\left(-\frac{t}{2}\right)+\mathrm{E}_{1,1.7}\left(-\frac{t}{2}\right)\mathrm{E}_{1,1.3}\left(-\frac{t}{2}\right)\right],\end{aligned}$$

where $y(0) = 1$, $y'(0) = -1/2$. It is known that the analytical solution of the equation is $y(t) = \mathrm{e}^{-t/2}$.

Solutions. Convert first the implicit Caputo differential equation into the standard form:

$$\begin{aligned}&{}_0^{\mathrm{C}}\mathscr{D}_t^{0.2}y(t)\,{}_0^{\mathrm{C}}\mathscr{D}_t^{1.8}y(t)+{}_0^{\mathrm{C}}\mathscr{D}_t^{0.3}y(t)\,{}_0^{\mathrm{C}}\mathscr{D}_t^{1.7}y(t)\\&\quad+\frac{t}{8}\left[\mathrm{E}_{1,1.8}\left(-\frac{t}{2}\right)\mathrm{E}_{1,1.2}\left(-\frac{t}{2}\right)+\mathrm{E}_{1,1.7}\left(-\frac{t}{2}\right)\mathrm{E}_{1,1.3}\left(-\frac{t}{2}\right)\right]=0.\end{aligned}$$

Based on the above modeling method, the key signals $y(t)$, $y'(t)$ and $y''(t)$ can be defined first. The Caputo fractional-order derivatives $\mathscr{D}^{0.2}y(t)$, $\mathscr{D}^{0.3}y(t)$, $\mathscr{D}^{1.7}y(t)$ and $\mathscr{D}^{1.8}y(t)$ can be generated. With the key signals, the left side of the above equation can be constructed, and it can be fed into the Algebraic Constraint block. The output signal is $\mathscr{D}^{1.8}y(t)$. Taking 0.2th-order derivative to the signal, the signal $y''(t)$ can be obtained. The signal $y''(t)$ can be connected directly to the starting signal of the integrator chain, and the complete Simulink model for the implicit Caputo equation can be established, as shown in Figure 4.58. Since the initial values are already specified in the integrators, the standard Oustaloup filters can be used to describe the required derivatives. The function for the Interpreted MATLAB Fcn block is

```
function y=c4mimpfs(u) % describing the implicit function
y=1/8*u*(ml_func([1,1.8],-u/2)*ml_func([1,1.2],-u/2)+...
   ml_func([1,1.7],-u/2)*ml_func([1,1.3],-u/2)); % left side function
```

If the following parameters for the Oustaloup filter are chosen, the numerical solution of the implicit differential equation can be found, with a maximum error of 3.8182×10^{-5}, and the time elapse of 151.2 seconds. Compared with other models, the time consumption of this solution process is rather large, since there exists an algebraic loop in the systems. In each simulation step, the algebraic equation is solved

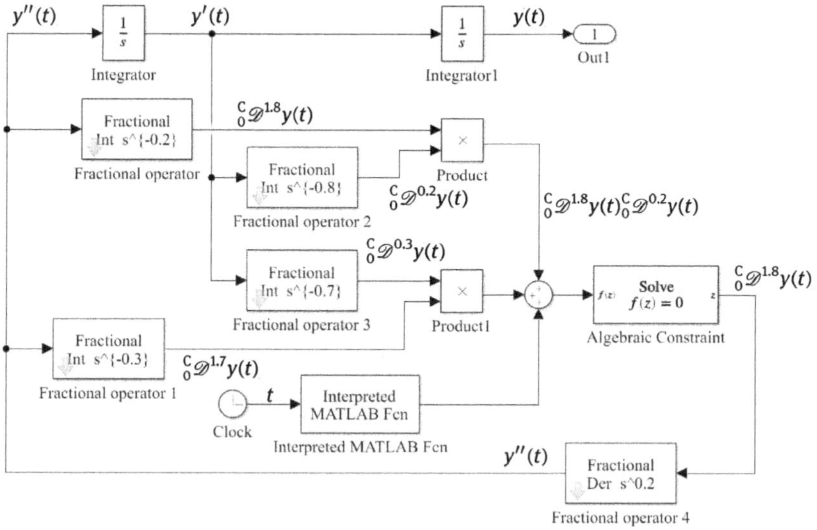

Figure 4.58: Simulink model for the implicit differential equation (file: `c4mimps.slx`).

once. In the solution process, a warning message "Block diagram contains 1 algebraic loop" is displayed. This is inevitable, and it can be ignored:

```
>> ww=[1e-5 1e5]; n=30; tic, [t,x,y]=sim('c4mimps'); toc
   max(abs(y-exp(-t/2)))
```

4.6 Model of difference equations

Difference equations are another category of mathematical models for describing dynamical systems. The difference between differential equations and difference equation is that in differential equations, the relationship in continuous signals are described, while in difference equations, the relationships among discrete signals are described. The so-called discrete signal is that the signal is not continuous, but the function values at specific time instances are involved. Since the modeling and low-level processing methods of difference equations are similar to differential equations, the modeling and solution of difference equations with Simulink are presented in this section.

Definition 4.11. The explicit form of ordinary difference equation is

$$y((k+n)T) = f(k, y(kT), y((k+1)T), \dots, y((k+n-1)T)), \tag{4.6.1}$$

where T is the sample time, and n is referred to as the order of the difference equation.

In the literature, difference equation are also simply expressed as

$$y(t+n)=f\big(t,y(t),y(t+1),\dots,y(t+n-1)\big). \tag{4.6.2}$$

In describing the key signals in differential equations, the integrator chain can be used to define the signals. For difference equations, the delay chain shown in Figure 4.59 can be used to define the key signals. With the **Unit Delay** blocks, the key signals $y(t), y(t+1), y(t+2), \dots, y(t+n)$ can be defined. Based on these signals, the Simulink models for difference equations can be established. The specific method is that using the key signals $y(t), y(t+1), y(t+2), \dots, y(t+n-1)$ defined in Figure 4.59 to construct the right side of (4.6.2), then the signal can be fed into the leftmost delay block to close the simulation system loop, so as to create the Simulink model.

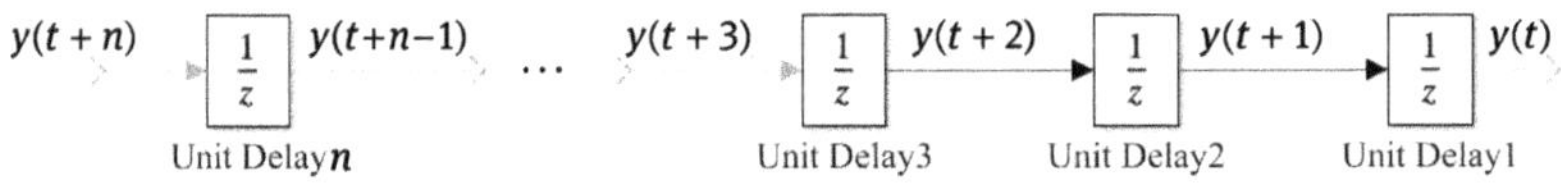

Figure 4.59: Definition of the key signals.

Example 4.30. Build the Simulink model for the following nonlinear difference equation:

$$y(t+3)=\frac{y^2(t+2)+1.1y(t+1)}{1+y^2(t+2)+0.2y(t+1)+0.4y(t)}+0.1u(t),$$

where the input signal is $u(t)=\sin t$, and sample time is $T=0.05$ seconds.

Solutions. Since the order of the difference equation is $n=3$, three unit delay blocks can be used to build up the chain. Thus the key signals $y(t)$, $y(t+1)$, $y(t+2)$ and $y(t+3)$ can be defined. Therefore, the right side of the difference equation can be built up with these signals, and fed into the left most unit delay block, to form the closed loop, as shown in Figure 4.60.

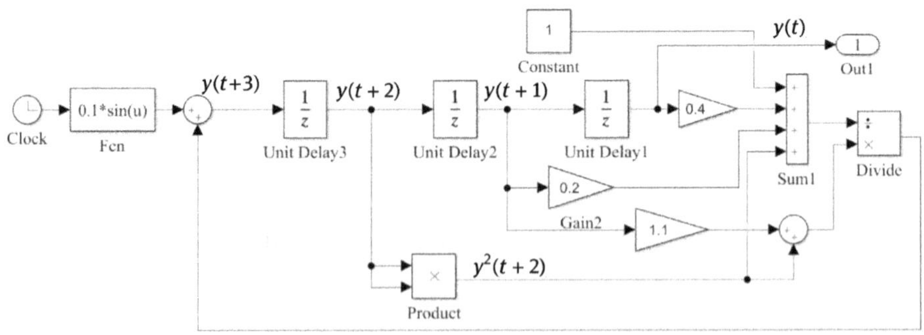

Figure 4.60: Simulink model for the difference equation (file: `c4mdis1.slx`).

An important question is given here: how to express $T = 0.05$ seconds in the model? Double click the leftmost unit delay block, the sample time can be assigned. The sample time in other delay blocks are unchanged, by using the default −1, indicating it inherits the sample time of the previous signal. Therefore, there is no need to modify the sample time values for the blocks individually. The simulation model thus constructed is easy to maintain. Besides, since the model here is discrete, the fixed-step simulation algorithm is adopted automatically. To start the simulation process, the simulation results shown in Figure 4.61 can be obtained.

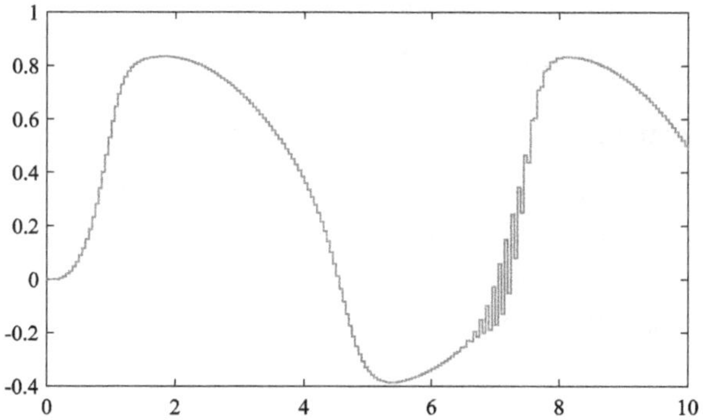

Figure 4.61: Simulation result of difference equations.

In fact, it can be seen from the model that the low-level modeling is rather complicated. Therefore, the **Mux** block can be used to compose the vector key signal. Therefore, a more concise model in Figure 4.62 can be constructed. Start simulation to the model, and the simulation results can be obtained, which is the same as the one in Figure 4.61.

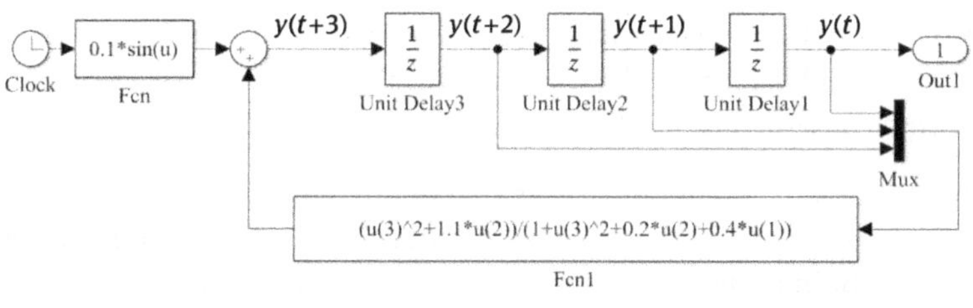

Figure 4.62: A more concise difference equation model (file: `c4mdis2.slx`).

Example 4.31. The discrete implementation of tracking and differentiator is:[32]

$$\begin{cases} x_1(k+1) = x_1(k) + Tx_2(k), \\ x_2(k+1) = x_2(k) + T\mathrm{fst}(x_1(k), x_2(k), u(k), r, h), \end{cases} \tag{4.6.3}$$

where T is the sample time, $u(k)$ is the input signal at the kth time instance. The parameter r determines the tracking speed, and h is the parameter for filtering results, when the signal is corrupted. The value of the fst(·) function can be evaluated directly with the following formulas:

$$\delta = rh, \quad \delta_0 = \delta h, \quad y_0 = x_1(k) - u + hx_2(k), \quad a_0 = \sqrt{\delta^2 + 8r|y_0|}, \tag{4.6.4}$$

$$a = \begin{cases} x_2 + y_0/h, & |y_0| \leqslant \delta_0, \\ x_2 + 0.5(a_0 - \delta)\,\text{sign}(y_0), & |y_0| > \delta_0, \end{cases} \tag{4.6.5}$$

$$\text{fst} = \begin{cases} -ra/\delta, & |a| \leqslant \delta, \\ -r\,\text{sign}(a), & |a| > \delta. \end{cases} \tag{4.6.6}$$

Build the low-level Simulink model using switches for the tracking and differentiator.

Solutions. In the mathematical expression of the tracking and differentiator, the model itself is a nonlinear discrete transfer function, where $u(k)$ is the input signal, $x_1(k)$ and $x_2(k)$ are the output signals. Parameters r, h and T are given constants.

If the fst(·) function is regarded as the input signal $v(t)$, the above discrete state space model can be regarded as a linear discrete state space model block:

$$\begin{bmatrix} x_1(k+1) \\ x_2(k+1) \end{bmatrix} = \begin{bmatrix} 1 & T \\ 0 & 1 \end{bmatrix} \begin{bmatrix} x_1(k) \\ x_2(k) \end{bmatrix} + \begin{bmatrix} 0 \\ T \end{bmatrix} v(k). \tag{4.6.7}$$

Observing (4.6.3) and (4.6.7), it can be seen that $v(k) = \text{fst}(\cdot)$. It can be seen from the expressions that to find the value of fst, the variable a must be known. If one wants to find a, the values such as δ_0 must be given. In real computing order, the value of δ_0 should be evaluated first, then find a, and finally find the value of fst, so as to find $v(k)$, where a and δ_0 must be computed with switches.

Based on the above mathematical model, the Simulink model in Figure 4.63 is established. Discrete linear state space model is expressed with Discrete State-Space block. These conditions can be expressed individually with Switch blocks. The function in Fcn2 block can be assigned as `u(2)+0.5*(u(3)-u(4))*sign(u(5))`.

It can be seen from the block diagram that the modeling method discussed here is rather complicated, and easy to induce errors, and not easy to validate. In fact, the correctness of the model built earlier is not ensured, since the model cannot be used correctly. This model is not recommended for real use. The intention of introducing such an example is to show the complicities in building the models. Concise and reliable modeling methods will be introduced later. For instance, S-function modeling methods will be given in Chapter 6.

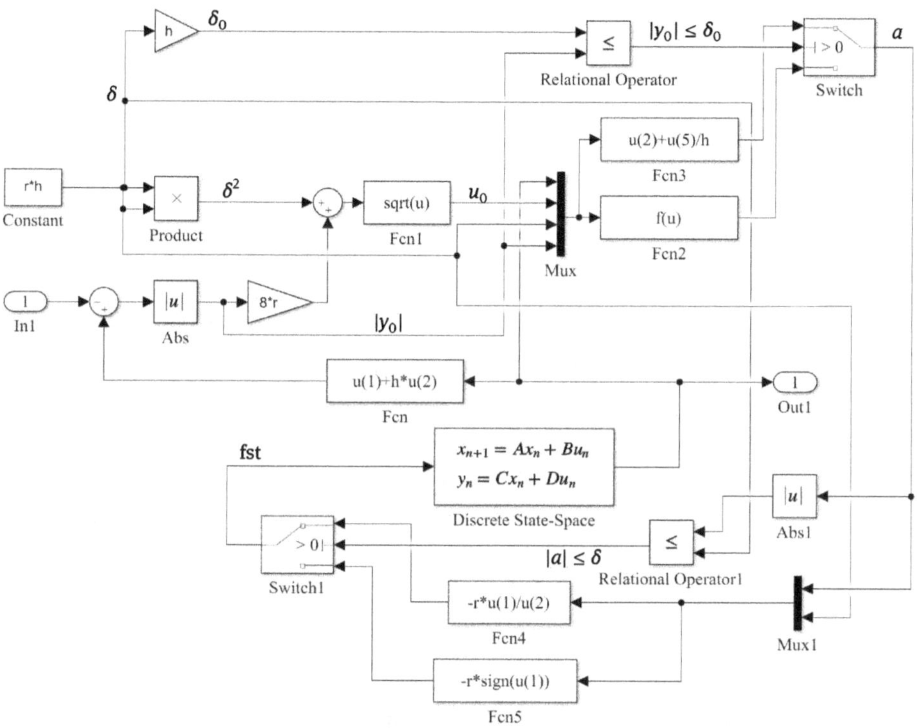

Figure 4.63: Simulink model of tracking and differentiator (file: `c4mhan.slx`).

4.7 Exercises

4.1 Judge the differences in Figures 4.64(a) and (b). What are the meanings in the two output signals? Are there any differences?

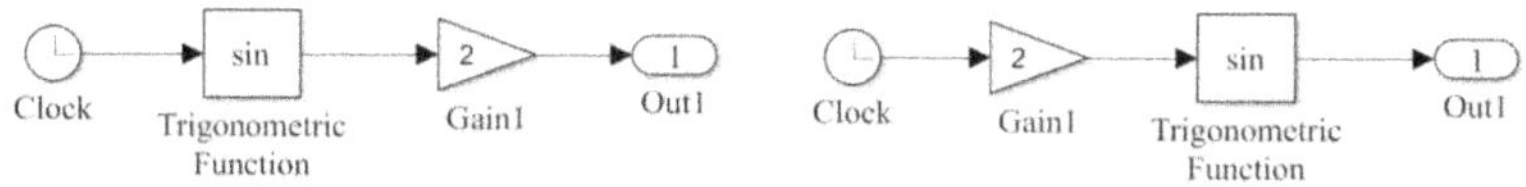

(a) model 1 (b) model 2

Figure 4.64: The output signals in to models.

4.2 Modify the default template, such that fixed-step algorithm is used, with a step-size of 0.001. Note that, although such setting is demonstrated here, it is not recommended to use the setting. It is not used unless otherwise used.

4.3 Solve the following linear differential equation:

$$y^{(5)} + 13y^{(4)} + 64y'''(t) + 152y''(t) + 176y'(t) + 80y(t) = \mathrm{e}^{-2t}\left[\sin\left(2t + \frac{\pi}{3}\right) + \cos 3t\right],$$

where $y(0) = 1$, $y'(0) = 3$, $y''(0) = 0$, $y'''(0) = 0$ and $y^{(4)}(0) = -1$. Use Simulink to solve the differential equation.

4.4 Consider the following Rössler differential equation:

$$\begin{cases} x'(t) = -y(t) - z(t), \\ y'(t) = x(t) + ay(t), \\ z'(t) = b + (x(t) - c)z(t). \end{cases}$$

Select $a = b = 0.2$, $c = 5.7$, and $x(0) = y(0) = z(0)$. Establish the Simulink model, and draw the three-dimensional phase space trajectory of the simulation results, and also draw its projection on the xy plane. If $a = 0.2$, $b = 0.5$, $c = 10$, draw the plots again.

4.5 Consider the following differential equation:[57]

$$\begin{cases} y'(t) = \tan\phi(t), \\ v'(t) = -\dfrac{g\sin\phi(t)\gamma v^2(t)}{v(t)\cos\phi(t)}, \\ \phi'(t) = -g/v^2(t), \end{cases}$$

where $g = 0.032$, $\gamma = 0.02$, and initial values are $y(0) = 0$, $v(0) = 0.5$. The parameter $\phi(0)$ can be assigned to 0.3782 and 9.7456 separately. Use Simulink to find the ordinary differential equation.

4.6 The mathematical model of a chemical reaction process is[30]

$$\begin{cases} y_1'(t) = -Ay_1(t) - By_1(t)y_3(t), \\ y_2'(t) = Ay_1(t) - MCy_2(t)y_3(t), \\ y_3'(t) = Ay_1(t) - By_1(t)y_3(t) - MCy_2(t)y_3(t) + Cy_4(t), \\ y_4'(t) = By_1(t)y_3(t) - Cy_4(t), \end{cases}$$

where $A = 7.89 \times 10^{-10}$, $B = 1.1 \times 10^7$, $C = 1.13 \times 10^3$, $M = 10^6$, $y_1(0) = 1.76 \times 10^{-3}$, $y_2(0) = y_3(0) = y_4(0) = 0$, $t \in (0, 1000)$. Established a Simulink model for the stiff differential equation, and find the numerical solution. To further increase the interval to $t \in (0, 10^{13})$, solve the equation with the Simulink model.

4.7 Chua circuit can be described by the following dimensionless differential equation:[22]

$$\boldsymbol{x}'(t) = \begin{bmatrix} \alpha(x_2(t) - h(x_1(t))) \\ x_1(t) - x_2(t) + x_3(t) \\ -\beta x_2(t) \end{bmatrix},$$

where the nonlinear equation is

$$h(a) = m_1 a + \frac{1}{2}(m_0 - m_1)(|a+1| - |a-1|).$$

If the control parameters are chosen as $\alpha = 9$, $\beta = 14.2886$, $m_0 = -1/7$, $m_1 = 2/7$, draw the simulation model for the Chua circuit with Simulink, and draw the phase plane trajectory of $x_1(t)$ and $x_2(t)$.

4.8 Describing the following differential equation with Simulink, $y(0) = 2$, $y'(0) = y''(0) = 0$:

$$y'''(t) + ty(t)y''(t) + t^2 y'(t)y^2(t) = \mathrm{e}^{-ty(t)}.$$

4.9 Solve the polynomial equation $t^4 + 10t^3 + 31t^2 + 46t + 24 = 0$ with the Algebraic Constraint block, and validate the results, with symbolic expressions.

4.10 Use Simulink to build up the seven-body model, and draw the trajectories. The positions of the seven bodies at $(x_i(t), y_i(t))$ satisfying[29]

$$x_i''(t) = \sum_{j \neq i} \frac{m_j}{r_{ij}(t)}\left(x_j(t) - x_i(t)\right), \quad y_i''(t) = \sum_{j \neq i} \frac{m_j}{r_{ij}(t)}\left(y_j(t) - y_i(t)\right),$$

where $m_i = i$, $i = 1, 2, \ldots, 7$ and

$$r_{ij}(t) = \left(\left(x_i(t) - x_j(t)\right)^2 + \left(y_i(t) - y_j(t)\right)^2\right)^{3/2}.$$

If the initial position and speed of the objects are $x_1(0) = 3$, $x_2(0) = 3$, $x_3(0) = -1$, $x_4(0) = -3$, $x_5(0) = 2$, $x_6(0) = -2$, $x_7(0) = 2$, $y_1(0) = 3$, $y_2(0) = -3$, $y_3(0) = 2$, $y_4(0) = 0$, $y_5(0) = 0$, $y_6(0) = -4$, $y_7(0) = 4$, $x_6'(0) = 1.75$, $x_7'(0) = -1.5$, $y_4'(0) = -1.25$, $y_5'(0) = 1$ and the remaining $x_i'(0) = y_i'(0) = 0$. The interval is $t \in (0, 3)$.

4.11 Consider the differential equation in Example 4.24 with variable delays. If when $t \leqslant 0$, $x_1(t) = \sin(t+1)$, $x_2(t) = \cos t$, $x_3(t) = \mathrm{e}^{3t}$, use Simulink to establish the simulation model and solve the equation.

4.12 For the following time-varying differential algebraic equation,[3]

$$\begin{cases} x_1'(t) = (\alpha - 1/(2-t))x_1(t) + (2-t)\alpha z(t) + (3-t)/(2-t), \\ x_2'(t) = (1-\alpha)x_1(t)/(t-2) - x_2(t) + (\alpha - 1)z(t) + 2\mathrm{e}^{-t}, \\ (t+2)x_1(t) + (t^2 - 4)x_2(t) - (t^2 + t - 2)\mathrm{e}^t = 0, \end{cases}$$

if $x_1(0) = x_2(0) = 1$, and α is a constant. The analytical solution is $x_1(t) = x_2(t) = \mathrm{e}^t$, $z(t) = -\mathrm{e}^t/(2-t)$, and use Simulink to solve the differential algebraic equation.

4.13 Use Simulink to solve the delay differential equation,[57] where when $t \leqslant 0$, $y_1(t) = y_4(t) = y_5(t) = \mathrm{e}^{t+1}$, $y_2(t) = \mathrm{e}^{t+0.5}$, $y_3(t) = \sin(t+1)$, and the solution interval is

$t \in [0,1]$:

$$\begin{cases} y_1'(t) = y_5(t-1) + y_3(t-1), \\ y_2'(t) = y_1(t-1) + y_2(t-0.5), \\ y_3'(t) = y_3(t-1) + y_1(t-0.5), \\ y_4'(t) = y_5(t-1)y_4(t-1), \\ y_5'(t) = y_1(t-1). \end{cases}$$

4.14 Use Simulink to solve the following neutral-type delay differential equation:

$$\boldsymbol{x}'(t) = \boldsymbol{A}_1\boldsymbol{x}(t-0.15) + \boldsymbol{A}_2\boldsymbol{x}'(t-0.5) + \boldsymbol{B}u(t),$$

where the input signal is $u(t) \equiv 1$, and the matrices are

$$\boldsymbol{A}_1 = \begin{bmatrix} -13 & 3 & -3 \\ 106 & -116 & 62 \\ 207 & -207 & 113 \end{bmatrix}, \quad \boldsymbol{A}_2 = \begin{bmatrix} 0.02 & 0 & 0 \\ 0 & 0.03 & 0 \\ 0 & 0 & 0.04 \end{bmatrix}, \quad \boldsymbol{B} = \begin{bmatrix} 0 \\ 1 \\ 2 \end{bmatrix}.$$

4.15 Solve the state dependent delay differential equation:[33]

$$x'(t) = -x(t - |x(t)|),$$

where the history function at $t \leqslant 0$ is a piecewise one given below, and the analytical solution is known $y(t) = t + 1$:

$$y(t) = \begin{cases} -1, & t \leqslant -1, \\ 3(t+1)^{1/3}/2 - 1, & -1 < t \leqslant -7/8, \\ 10t/7 + 1, & -7/8 < t \leqslant 0. \end{cases}$$

4.16 Use Simulink to solve the following switched linear differential equation:

$$\begin{cases} x_1'(t) = f(x_1(t)) + x_2(t), \\ x_2'(t) = -x_1(t), \end{cases}$$

where $x_1(0) = x_2(0) = 5$, and $f(x_1(t))$ is a piecewise function described by

$$f(x_1(t)) = \begin{cases} -4x_1(t), & x_1(0) > 0, \\ 2x_1(t), & -1 \leqslant x_1(0) \leqslant 0, \\ -x_1(t) - 3, & x_1(0) < -1. \end{cases}$$

4.17 Consider the linear switched differential equation and a state feedback control structure:[42]

$$x'(t) = A_\sigma x(t) + B_\sigma u(t), \quad u(t) = k_\sigma x(t),$$

where $\sigma = \{1, 2\}$. The two systems are respectively

$$A_1 = \begin{bmatrix} 1 & 0 \\ 1 & 1 \end{bmatrix}, \quad A_2 = \begin{bmatrix} 1 & 1 \\ 0 & 1 \end{bmatrix}, \quad B_1 = \begin{bmatrix} 1 \\ 0 \end{bmatrix}, \quad B_2 = \begin{bmatrix} 0 \\ 1 \end{bmatrix},$$

and the two state feedback vectors are respectively $k_1 = [6, 9]$ and $k_2 = [9, 6]$. If the switching law to switch from subsystem 1 to subsystem 2 is $|x_1(t)| = 0.5|x_2(t)|$, and the one to switch from subsystem 2 into 1 is $|x_1(t)| = 2|x_2(t)|$. If the initial state vector is $x_0 = [100, 100]^T$, use Simulink to draw the phase plane trajectory of the system.

4.18 Solve the following discontinuous differential equation[29] with Simulink, where the initial value is $y(0) = 0.3$:

$$y'(t) = \begin{cases} t^2 + 2y^2(t), & (t + 0.05)^2 + [y(t) + 0.15]^2 \leqslant 1, \\ 2t^2 + 3y^2(t) - 2, & (t + 0.05)^2 + [y(t) + 0.15]^2 > 1. \end{cases}$$

4.19 Consider the linear feedback control system in Figure 4.65, with unity negative feedback, where the plant and controllers are respectively

$$G(s) = \frac{s^3 + 7s^2 + 24s + 24}{s^4 + 10s^3 + 35s^2 + 50s + 24}, \quad G_c(s) = \frac{s + 0.1}{0.1s + 1}.$$

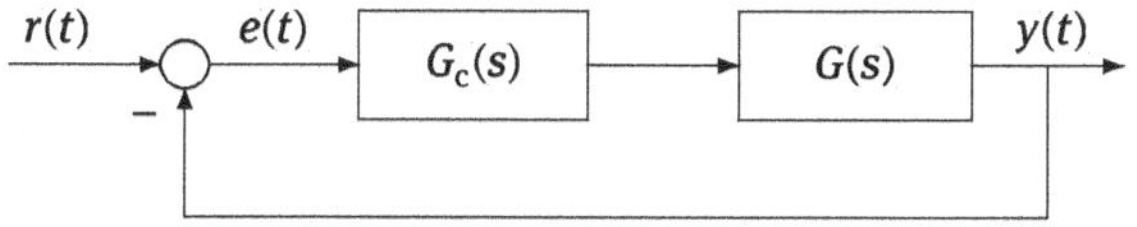

Figure 4.65: Typical feedback control system.

If the Gaussian white noise signal $u(t)$ is with variance of 1 and zero mean, use the simulation method to find the error signal $e(t)$ and draw the probability density function, and find the variance. Note that in theory that the equivalent transfer function from $r(t)$ to $e(t)$ can be written as $\widetilde{G}(s) = 1/[1 + G(s)G_c(s)]$.

4.20 For the nonlinear fractional-order differential equation described in Figure 4.66, write out the mathematical expression of the differential equation and draw the output signal $y(t)$.

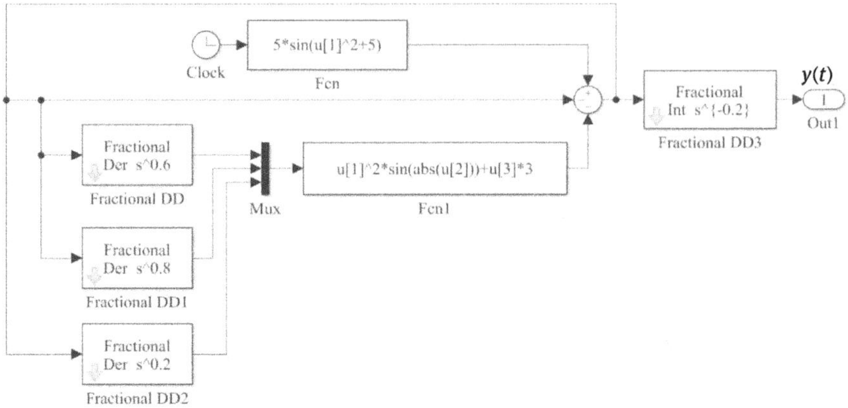

Figure 4.66: Simulink model for nonlinear fractional-order differential equation (file: `c4mfode4.slx`).

4.21 For the Bagley–Torwik equation[17] $Ay''(t) + B\mathscr{D}^{3/2}y(t) + Cy(t) = C(t+1)$, $y(0) = y'(0) = 1$, prove that the solutions of the equations are independent of the constants A, B and C.

4.22 In [79], five benchmark problems for testing fractional-order differential equations are proposed. Two of them have been studied earlier. Use Simulink to solve directly the remaining three, and compare the solutions with analytical solutions:

① ${}_0^{\rm C}\mathscr{D}_t^{1.6}y(t) = t^{0.4}{\rm E}_{1,1.4}(-t)$, $0 \leqslant t \leqslant 100$, $x(0) = 1$, $x'(0) = -1$, with analytical solution of $y(t) = {\rm e}^{-t}$.

② The linear fractional-order differential equation is given by

$$y'''(t) + {}_0^{\rm C}\mathscr{D}_t^{2.5}y(t) + y''(t) + 4y'(t) + {}_0^{\rm C}\mathscr{D}_t^{0.5}y(t) + 4y(t) = 6\cos t,$$

and the known initial values are $y(0) = 1$, $y'(0) = 1$, $y''(0) = -1$, and $0 \leqslant t \leqslant 10$. The analytical solution is $y(t) = \sqrt{2}\sin(t + \pi/4)$.

③ The fractional-order state space model

$$\begin{cases} {}_0^{\rm C}\mathscr{D}_t^{0.5}x(t) = \dfrac{1}{2\Gamma(1.5)}\left(\left[(y(t)-2)(z(t)-3)\right]^{1/6} + \sqrt{t}\right), \\ {}_0^{\rm C}\mathscr{D}_t^{0.2}y(t) = \Gamma(2.2)\left[x(t)-1\right], \\ {}_0^{\rm C}\mathscr{D}_t^{0.6}z(t) = \dfrac{\Gamma(2.8)}{\Gamma(2.2)}\left[y(t)-2\right], \end{cases}$$

where $x(0) = 1$, $y(0) = 2$, $z(0) = 3$. The analytical solution of the fractional-order state space equation is $x(t) = t + 1$, $y(t) = t^{1.2} + 2$, $z(t) = t^{1.8} + 3$, for $0 \leqslant t \leqslant 10$.

4.23 Use Simulink to solve the following difference equation:

$$72y(t)+102y(t-1)+53y(t-2)+12y(t-3)+y(t-4)=12u(t)+7u(t-1),$$

where $u(t)$ is the step signal, and $y(-3)=1$, $y(-2)=-1$, $y(-1)=y(0)=0$.

4.24 Use Simulink to solve the following difference equation, where $u(t)=\mathrm{e}^{-0.1t}$, and the initial value of $y(t)$ is 0:

$$y(t)-0.6y(t-1)+0.12y(t-2)+0.008y(t-3)=u(t).$$

4.25 Use Simulink to solve the following nonlinear difference equation. When $t\leqslant 0$, $y(t)=0$, and signal $u(t)=\mathrm{e}^{-0.2t}$,

$$y(t)=u(t)+y(t-2)+3y^2(t-1)+\frac{y(t-2)+4y(t-1)+2u(t)}{1+y^2(t-2)+y^2(t-1)}.$$

4.26 If the discrete state space model is described below, and $\boldsymbol{x}^{\mathrm{T}}(0)=[1,-1]$, find the step response of the system with Simulink. Find also the analytical solution and compare the accuracy:

① $\boldsymbol{x}(t+1)=\begin{bmatrix}0 & 1\\ -0.16 & -1\end{bmatrix}\boldsymbol{x}(t)+\begin{bmatrix}1\\ 1\end{bmatrix}u(t)$,

② $\boldsymbol{x}(t+1)=\begin{bmatrix}11/6 & -1/4 & 25/24 & -2\\ 1 & 1 & -1 & -1\\ 0 & 1 & -1 & 0\\ 0 & 1 & -3/4 & 0\end{bmatrix}\boldsymbol{x}(t)+\begin{bmatrix}2\\ 1/2\\ -3/8\\ 1/4\end{bmatrix}u(t)$.

4.27 It is indicated that there exist problems in the model built in Example 4.31, and it cannot be used correctly. Now let us select $T=h=0.001$, $r=30$, and the input signal is a sinusoidal signal, and see whether the system returns the input signal and its derivative. The interested readers may check the model and see where there are problems.

5 Simulink modeling of control systems

The earliest target of Simulink was to solve the control system simulation problems. The research on control system simulation also motivated a lot of the development and popularization of Simulink. Before the debut of Simulink, most control system simulation languages are in programming forms. The graphical and block oriented form provided in Simulink makes the research of control system simulation much easier and straightforward. Therefore, Simulink becomes gradually the top selected computer tool in control system simulation research. Significant amount of blocks are provided for control system simulation, however, there are still many problems to be solved. Some of the blocks need to be further extended, such that more suitable blocks for control system simulation can be developed. In this chapter, the modeling and simulation of control systems using Simulink are systematically presented.

In Section 5.1, continuous linear system modeling problems are illustrated. The low-level blocks in Simulink modeling is first presented, and then singular system modeling and the modeling of transfer functions with nonzero initial conditions are studied. Finally, the LTI System block in the Control System Toolbox is introduced to show the unified way in complicated linear system modeling. In Section 5.2, the modeling and simulation of various discrete linear systems are discussed, and the handling in complicated discrete models are presented. In Section 5.3, lookup table based modeling methods for single-valued and double-valued nonlinearities are explored, aiming at finding a universal way for describing arbitrary static nonlinearities. Demonstrations are made to show the use of two-dimensional lookup table blocks. In Section 5.4, automatic sorting rules in Simulink are presented, such that the readers can better understand the modeling and simulation mechanism in Simulink. In Section 5.5, approximate simulation methods for control systems are introduced, for instance, linearization of nonlinear systems and Padé approximation of delay models. The algebraic loop problems in simulation are also demonstrated, and a universal method to eliminate algebraic loops is illustrated. Of course, in theory, Simulink can be used for accurate simulation for systems of any complexity. Unless absolutely necessary, there is actually no need to perform approximate analysis to control systems.

5.1 Simulink modeling of linear continuous systems

In conventional control courses and control engineering, linear systems are classified into two types of models, that is, transfer function models and state space models. For multivariable systems, transfer function models can further be extended to transfer function matrices. In standard MATLAB and Simulink, three groups of linear continuous blocks are provided: Continuous group and Additional Linear group in Simulink Extras, and the blockset provided in the Control System Toolbox. In this section, the

https://doi.org/10.1515/9783110734959-005

Simulink representation of various linear continuous blocks are demonstrated. In particular, transfer functions with nonzero initial conditions are studied, and this problem is hard to be handled with the current versions in Simulink.

5.1.1 Transfer function models

In linear control system theory, transfer function is the most important mathematical model for control systems. Transfer function models are established upon the linear differential equations with constant coefficients. The general form of a transfer function is given in (2.3.1), whose mathematical expression is

$$G(s)=\frac{b_1s^m+b_2s^{m-1}+\cdots+b_ms+b_{m+1}}{a_1s^n+a_2s^{n-1}+\cdots+a_ns+a_{n+1}},\tag{5.1.1}$$

where a_i, $i=1,2,\ldots,n+1$ and b_i, $i=1,2,\ldots,m+1$ are constants, and n is referred to as the order of the model. If $m<n$, the system is strictly proper, while if $m=n$, the system is proper. If $n<m$, the system is not physically realizable.

Transfer function is the ratio of the Laplace transform of the output signal over the Laplace transform of the input signal. It can be regarded as the gain of the system in s domain. It is worth mentioning that the concept of the transfer function was created upon the assumption that the initial values of input, output and their derivatives are all zeros.

Two methods are supported in the MATLAB Control System Toolbox to input transfer function models in MATLAB. The first method is to create numerator and denominator coefficients vectors, `num` and `den`, in the descending order of s. Then with command `G=tf(num,den)`, the transfer function object G can be created. The variable G can be filled into the **LTI System** object so that the transfer function model can be created in Simulink directly. An alternative method is to use `s=tf('s')` command to define operator s in MATLAB, then simple commands can be used to set up the transfer function model. If there is a delay term, with delay constant of T, three methods can be used to input the model:

```
G=tf(num,den,'ioDelay',T);
G=tf(num,den); G.ioDelay=T;
s=tf('s'); G=tf(num,den)*exp(-T*s);
```

Example 5.1. The transfer function models are given below. Input the transfer function models with MATLAB commands:

$$G_1(s)=\frac{s^3+7s^2+24s+24}{s^4+10s^3+35s^2+50s+24},\quad G_2(s)=\frac{s^2+4s+3}{s^2(s+4)((s+1)^2+3)}.$$

Solutions. It can be seen that the numerator and denominator polynomials are available in $G_1(s)$, function `tf()` can be used directly to input the transfer function model.

In model $G_2(s)$, the coefficients in the denominator are not provided, regular `tf()` function call can only be made when the factors are expanded. Therefore, the s operator should be declared first, and mathematical expression can be used to describe the transfer function model. The following commands can be used to input the two models:

```
>> G1=tf([1 7 24 24],[1 10 35 50 24])
   s=tf('s'); G2=(s^2+4*s+3)/s^2/(s+4)/((s+1)^2+3)
```

If the transfer function is known, the Transfer Fcn block in the Continuous group can be used to describe it directly. However, if the transfer function is not physically realizable, or if the transfer function has nonzero initial conditions, the block cannot be used. If transfer function object G has been created in Example 5.1, the command `[n,d]=tfdata(G,'v')` can be used to extract the numerator and denominator coefficient vectors, and return them in variables $\boldsymbol{n}$ and $\boldsymbol{d}$. If the model has time delay, it can be extracted with `T=G.ioDelay`. The information can be filled into the Transfer Fcn block, such that the transfer function can be represented directly in Simulink.

5.1.2 State space models

An alternative commonly used mathematical model for linear systems is the state space model. For physically realizable systems, it can also be represented as state space models. The mathematical form is presented first in this section, then limitations are pointed out, and the state space model of linear singular systems is introduced.

Definition 5.1. The mathematical model of linear constant system is

$$\begin{cases} \boldsymbol{x}'(t) = \boldsymbol{A}\boldsymbol{x}(t) + \boldsymbol{B}\boldsymbol{u}(t), \\ \boldsymbol{y}(t) = \boldsymbol{C}\boldsymbol{x}(t) + \boldsymbol{D}\boldsymbol{u}(t), \end{cases} \tag{5.1.2}$$

where $\boldsymbol{x}(t) \in \mathscr{R}^{n\times 1}$ is referred to as the state vector. The constant matrices $\boldsymbol{A} \in \mathscr{R}^{n\times n}$, $\boldsymbol{B} \in \mathscr{R}^{n\times q}$, $\boldsymbol{C} \in \mathscr{R}^{p\times n}$ and $\boldsymbol{D} \in \mathscr{R}^{p\times q}$ are known. The initial values of the state $\boldsymbol{x}(t_0)$ at time $t = t_0$ are also known, with a default of zero vector. To avoid conflict with other state space models, if necessary, this type of state space model is referred to as conventional state space model in this book.

State space models can be specified in MATLAB with the Control System Toolbox function `ss()`, with syntax `G=ss(A,B,C,D)`. State space models can be used to describe directly multivariable systems.

Example 5.2. The multivariable model can be described by the method, and no other special manipulation is necessary. Input the following model with two input, two output into MATLAB workspace:

$$\begin{cases} \boldsymbol{x}'(t) = \begin{bmatrix} -12 & -17.2 & -16.8 & -11.9 \\ 6 & 8.6 & 8.4 & 6 \\ 6 & 8.7 & 8.4 & 6 \\ -5.9 & -8.6 & -8.3 & -6 \end{bmatrix} \boldsymbol{x}(t) + \begin{bmatrix} 1.5 & 0.2 \\ 1 & 0.3 \\ 2 & 1 \\ 0 & 0.5 \end{bmatrix} \boldsymbol{u}(t), \\ \boldsymbol{y}(t) = \begin{bmatrix} 2 & 0.5 & 0 & 0.8 \\ 0.3 & 0.3 & 0.2 & 1 \end{bmatrix} \boldsymbol{x}(t). \end{cases}$$

Solutions. The $\boldsymbol{A}$, $\boldsymbol{B}$, $\boldsymbol{C}$ and $\boldsymbol{D}$ matrices can be recognized from the given model. Therefore, they can be specified in MATLAB, and with function `ss()`, the state space model can be described in MATLAB directly:

```
>> A=[-12,-17.2,-16.8,-11.9; 6,8.6,8.4,6; 6,8.7,8.4,6;
      -5.9,-8.6,-8.3,-6];
   B=[1.5,0.2; 1,0.3; 2,1; 0,0.5];
   C=[2,0.5,0,0.8; 0.3,0.3,0.2,1]; D=zeros(2,2); %input the matrices
   G=ss(A,B,C,D) %input and dispace state space modek
```

If the transfer function object G is known, command G_1=`ss(`G`)` can be used to convert it to the equivalent state space model. After conversion, command `[`$\boldsymbol{A}$`,`$\boldsymbol{B}$`,`$\boldsymbol{C}$`,`$\boldsymbol{D}$`]=ssdata(`G_1`)` can be used to extract the matrices. If the state space object is G_1, the command G=`tf(`G_1`)` can be used to find the transfer function model object G.j

State space models can be represented directly with **State Space** block in the Continuous group. The user needs to specify $\boldsymbol{A}$, $\boldsymbol{B}$, $\boldsymbol{C}$ and $\boldsymbol{D}$ matrices in the parameter dialog box. If the model has nonzero initial conditions $\boldsymbol{x}(t_0)$, it can be specified in the **Initial States** edit box.

More generally, state space models with internal delays[85] can be handled in the Control System Toolbox. Since the description of such a form is complicated, the mathematical model and block diagram is not presented here. Interested readers may refer to [85].

5.1.3 Linear singular state space models

It can be seen from the mathematical expression that conventional state space models can only be used to describe proper transfer function models, that is, the order of numerator is not higher than that of the denominator. In some situations, conventional state space models are not adequate. For instance, for a strictly proper transfer function, its inverse system is not physically realizable. Conventional state space models cannot be used to describe such inverse systems. Therefore, the conventional state space models should be extended. In this section, state space models for linear singular systems are introduced.

Definition 5.2. The mathematical model for linear singular system is

$$\begin{cases} \boldsymbol{E}\boldsymbol{x}'(t) = \boldsymbol{A}\boldsymbol{x}(t) + \boldsymbol{B}\boldsymbol{u}(t), \\ \boldsymbol{y}(t) = \boldsymbol{C}\boldsymbol{x}(t) + \boldsymbol{D}\boldsymbol{u}(t), \end{cases} \tag{5.1.3}$$

where $\boldsymbol{E} \in \mathscr{R}^{n\times n}$ is referred to as a descriptor matrix.

If $\boldsymbol{E}$ is an identity matrix, the system is a conventional state space model in Definition 5.1. If $\boldsymbol{E}$ is a singular matrix, the system is singular. If $\boldsymbol{E}^{-1}$ exists, a new state variable $\boldsymbol{z}(t) = \boldsymbol{E}\boldsymbol{x}(t)$ can be used such that conventional state space model of states $\boldsymbol{z}(t)$ can be found.

Singular system input is supported in the MATLAB Control System Toolbox. If the state space model is known, G=`dss(`$\boldsymbol{A}$`,`$\boldsymbol{B}$`,`$\boldsymbol{C}$`,`$\boldsymbol{D}$`,`$\boldsymbol{E}$`)` command can be used to input the model. If the five matrices are input into the **Descriptor State-Space** block, singular systems can also be expressed in Simulink.

Example 5.3. Consider the strictly proper transfer function $G_1(s)$ in Example 5.1. Write out the state space model of its inverse system.

Solutions. Since $G_1(s)$ model is strictly proper, its inverse $G_2(s) = 1/G_1(s)$ is not physically realizable. Its conventional state space model does not exist. The singular system model should be used to describe its state space model. With the following statements, the state space model can be found:

```
>> G1=tf([1 7 24 24],[1 10 35 50 24]); G2=1/G1;
   G3=ss(G2), step(G3) % find state space model and step response
```

such that the singular state space model can be found

$$\begin{bmatrix} 0 & 0.25 & 0 & 0 & 0 \\ 0 & 0 & 0.25 & 0 & 0 \\ 0 & 0 & 1 & 0 & 0 \\ 0 & 0 & 0 & 1 & 0 \\ 0 & 0 & 0 & 0 & 1 \end{bmatrix} \boldsymbol{x}'(t) = \begin{bmatrix} 1 & 0 & 0 & 0 & 0 \\ 0 & 1 & 0 & 0 & 0 \\ 0 & 0 & -7 & -6 & -3 \\ 0 & 0 & 4 & 0 & 0 \\ 0 & 0 & 0 & 2 & 0 \end{bmatrix} \boldsymbol{x}(t) + \begin{bmatrix} 0 \\ 0 \\ 8 \\ 0 \\ 0 \end{bmatrix} u(t),$$

and the output equation is $y(t) = [2, 5, 4.375, 1.5625, 0.375]\boldsymbol{x}(t)$.

Of course, since this system is not physically realizable, command `step(`G_3`)` may yield error message "Cannot simulate the time response of improper (non-causal) models."

Example 5.4. In the modeling of certain systems, if the state variables are not selected properly, or if the redundant states are selected, it is likely to find state space models in singular form. Consider the simple resistor, compactor and inductor circuit in Figure 5.1. If the states are selected as $i(t)$, $u_l(t)$, $u_c(t)$ and $u_r(t)$[15], write out the state space model of the circuit.

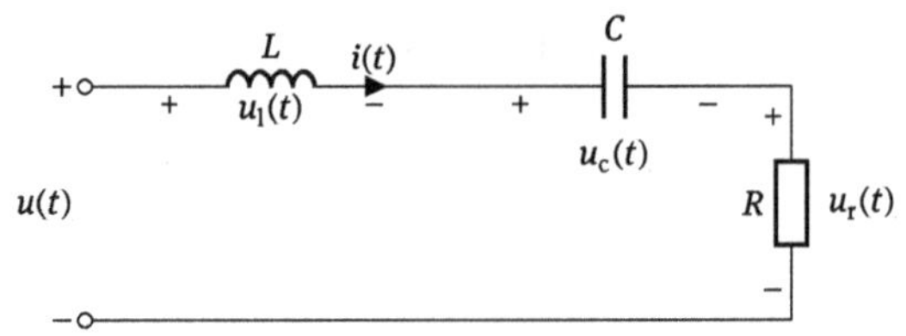

Figure 5.1: Series resistor, compactor and inductor circuit.

Solutions. Under normal cases,[81] the state variables are selected as $x_1(t) = u_c(t)$, $x_2(t) = i(t)$, and conventional state space model can be constructed. Unfortunately, if the state variables are chosen in the examples, four differential equations (some are algebraic equations) can be established:

$$\begin{cases} i'(t) = u_l(t), \\ u_c'(t) = i(t)/R, \\ u_r(t) = Ru(t), \\ u(t) = u_l(t) + u_c(t) + u_r(t). \end{cases}$$

It is not hard to rewrite them in matrix form, such that the following singular state space model can be found:

$$\begin{bmatrix} L & 0 & 0 & 0 \\ 0 & 0 & 1 & 0 \\ 0 & 0 & 0 & 0 \\ 0 & 0 & 0 & 0 \end{bmatrix} \begin{bmatrix} i'(t) \\ u_l'(t) \\ u_c'(t) \\ u_r'(t) \end{bmatrix} = \begin{bmatrix} 0 & 1 & 0 & 0 \\ 1/C & 0 & 0 & 0 \\ -R & 0 & 0 & 1 \\ 0 & 1 & 1 & 1 \end{bmatrix} \begin{bmatrix} i(t) \\ u_l(t) \\ u_c(t) \\ u_r(t) \end{bmatrix} + \begin{bmatrix} 0 \\ 0 \\ 0 \\ -1 \end{bmatrix} u(t),$$

and the output equation is $y(t) = u_c(t)$.

Example 5.5. Consider the singular state space model in Example 5.4. If $R = L = C = 1$, draw its step response curve.

Solutions. If $R = L = C = 1$ are substituted into the model, the following singular state space model can be obtained:

$$\begin{bmatrix} 1 & 0 & 0 & 0 \\ 0 & 0 & 1 & 0 \\ 0 & 0 & 0 & 0 \\ 0 & 0 & 0 & 0 \end{bmatrix} \boldsymbol{x}'(t) = \begin{bmatrix} 0 & 1 & 0 & 0 \\ 1 & 0 & 0 & 0 \\ -1 & 0 & 0 & 1 \\ 0 & 1 & 1 & 1 \end{bmatrix} \boldsymbol{x}(t) + \begin{bmatrix} 0 \\ 0 \\ 0 \\ -1 \end{bmatrix} u(t), \quad y(t) = x_3(t).$$

The following commands can be used to input the singular model, and `step()` function can be called to draw the step response, as shown in Figure 5.2. In fact, if the model is converted into zero-pole form, it is found that $G_1(s) = 1/(s^2+s+1)$, indicating that the model is a second-order one, with redundant states removed automatically:

```
>> E=[1,0,0,0; 0,0,1,0; 0,0,0,0; 0,0,0,0];
   A=[0,1,0,0; 1,0,0,0; -1,0,0,1; 0,1,1,1];
   B=[0; 0; 0; -1]; C=[0,0,1,0]; %input matrices
   G=dss(A,B,C,0,E); step(G);     %draw each states
   G1=zpk(G)                      %find equivalent model
```

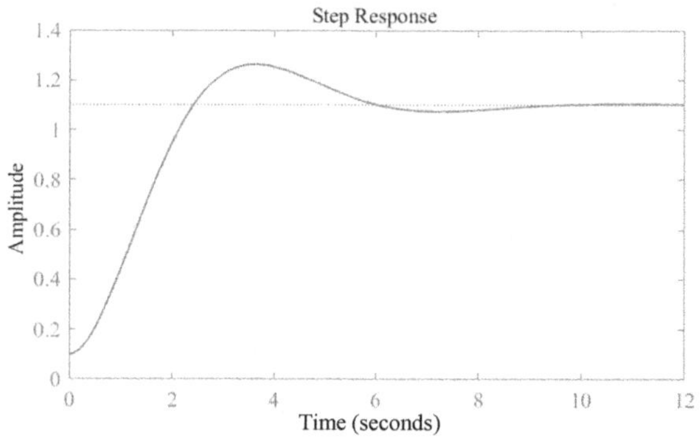

Figure 5.2: Step response of the singular system.

If Simulink is used, the **Descriptor State-Space** block in the **Continuous** group can be used to construct the model, as shown in Figure 5.3. Note that stiff solvers such as **ode15s** should be adopted, and the error tolerance ought to be set to relatively larger values such as 10^{-8}. The simulation result is the same as the one in Figure 5.2.

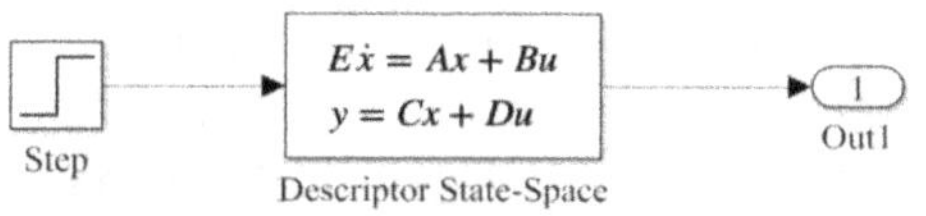

Figure 5.3: Simulink model of a singular system (file: `c5mdss.slx`).

5.1.4 Transfer functions with nonzero initial conditions

The conventional transfer function model has a fatal restriction. That is, it is defined on the assumption that the input and output signals have zero initial conditions. If nonzero initial conditions are involved, it is not easy to solve the problems.

The **Additional Linear** subgroup in **Simulink Extras** blockset is provided, and some of the initial value problems are provided, as shown in Figure 5.4, while it seems that those blocks can only be used to solve problems with given $y(t_0)$ values. The given information such as $y'(t_0)$, $y''(t_0)$ cannot be handled. A transformation algorithm is needed to convert the transfer function with nonzero initial conditions to a block describable in Simulink. In this section, the conversion method is presented. In Chapter 7, the equivalent block can further be masked, so that a reusable block can be constructed.

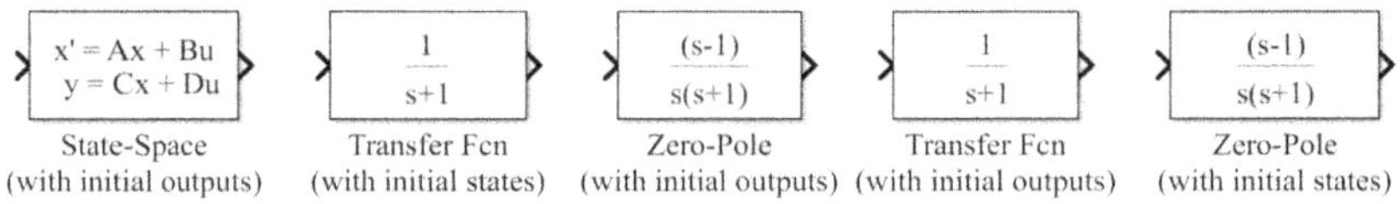

Figure 5.4: Additional Linear group.

Definition 5.3. A linear high-order differential equation with constant coefficients is given by

$$\begin{aligned} y^{(n)}(t) + a_1 y^{(n-1)}(t) + a_2 y^{(n-2)}(t) + \cdots + a_{n-1} y'(t) + a_n y(t) \\ = b_1 u^{(n-1)}(t) + b_2 u^{(n-2)}(t) + \cdots + b_{n-1} u'(t) + b_n u(t), \end{aligned} \tag{5.1.4}$$

where the input signal $u(t)$ is given, and the initial values $y(t_0), y'(t_0), \dots, y^{(n-1)}(t_0)$ are known.

Note that in the expression the order of $u(t)$ is deliberately set to 1 less than that of $y(t)$; if the order of $u(t)$ is the same as $y(t)$, it can be handled individually; if the order of $u(t)$ is higher than $y(t)$, the differential equation is not physically realizable, and not considered in the book.

Such a differential equation model cannot be expressed directly with the existing blocks in Simulink. A new block should be designed. In fact, the Simulink modeling methods for the differential equation have been studied. For instance, in Figure 4.26, derivative block was used, and the accuracy is rather low. In Figure 4.28, the mathematical form of $u'(t)$ must be provided, and this request is too rigorous, since the mathematical expression of the input signal is not usually known. Therefore, a better solution strategy for Simulink modeling is needed.

Theorem 5.1. *Selecting a set of states $x_1(t), x_2(t), \dots, x_n(t)$, the following state space model can be constructed:*[44]

$$\begin{cases} x_1'(t) = x_2(t), \\ x_2'(t) = x_3(t), \\ \quad \vdots \\ x_{n-1}'(t) = x_n(t), \\ x_n'(t) = -a_n x_1(t) - a_{n-1} x_2(t) - \cdots - a_1 x_n(t) + u(t), \\ y(t) = b_n x_1(t) + b_{n-1} x_2(t) + \cdots + b_2 x_{n-1}(t) + b_1 x_n(t). \end{cases} \tag{5.1.5}$$

From the state space model, the block diagram can be constructed, as shown in Figure 5.5.[44] The simulation model of the differential equation can also be expressed with MATLAB commands.

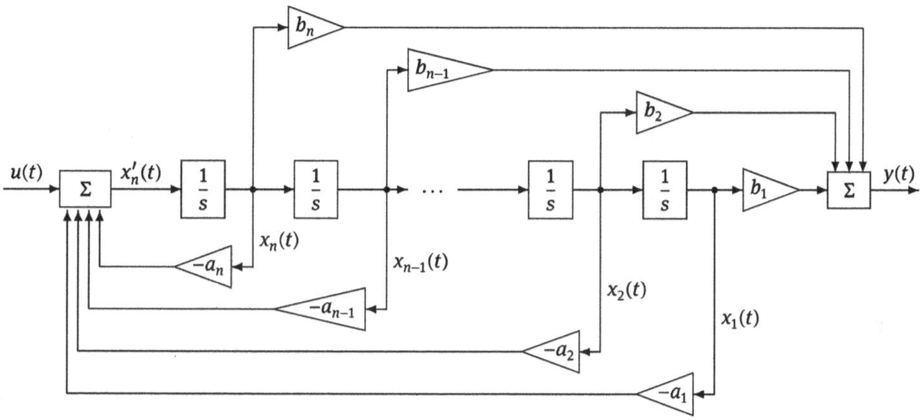

Figure 5.5: Block diagram representation of linear differential equation.

Since the initial values of $y(t)$ and its derivatives are known, and the initial state vector is not, the conventional state system blocks cannot be used to describe the transfer function model directly. Low-level similarity transform must be carried out such that the equivalent initial state vector can be found. Then state space model or block diagram can be drawn for the Simulink model, so as to solve the differential equation. It is rather complicated to draw the block diagram model. An alternative method is to find the equivalent state space model, and then use the **State-Space** block to describe the differential equation model in Simulink.

Theorem 5.2. *The matrix form of the state space model in Theorem* 5.1 *is*

$$\boldsymbol{A} = \begin{bmatrix} 0 & 1 & 0 & \cdots & 0 \\ 0 & 0 & 1 & \cdots & 0 \\ \vdots & \vdots & \vdots & \ddots & \vdots \\ -a_n & -a_{n-1} & -a_{n-2} & \cdots & -a_1 \end{bmatrix}, \quad \boldsymbol{B} = \begin{bmatrix} 0 \\ 0 \\ \vdots \\ 1 \end{bmatrix}, \quad \boldsymbol{C}^{\mathrm{T}} = \begin{bmatrix} b_n \\ b_{n-1} \\ \vdots \\ b_1 \end{bmatrix}. \tag{5.1.6}$$

It can be seen from the state space model that the matrices $\boldsymbol{A}$, $\boldsymbol{B}$, $\boldsymbol{C}$ and $\boldsymbol{D}$ can be composed from the given transfer function model. Unfortunately, the initial state vector $\boldsymbol{x}(t_0)$ cannot be written directly. The given information $y(t_0)$, $y'(t_0)$ can be used to find the equivalent initial state vector.

Denote $\beta_i^1 = b_i$, $i = 1, 2, \ldots, n$, the output equation can be written as

$$y(t_0) = \beta_n^1 x_1(t_0) + \beta_{n-1}^1 x_2(t_0) + \cdots + \beta_2^1 x_{n-1}(t_0) + \beta_1^1 x_n(t_0). \tag{5.1.7}$$

Taking derivative to both sides, and substitute in the state space equation, it can be found that

$$\begin{aligned} y'(t_0) &= \beta_n^1 x_1'(t_0) + \beta_{n-1}^1 x_2'(t_0) + \cdots + \beta_2^1 x_{n-1}'(t_0) + \beta_1^1 x_n'(t_0) \\ &= \beta_n^1 x_2(t_0) + \beta_{n-1}^1 x_3(t_0) + \cdots + \beta_2^1 x_n(t_0) \\ &\quad + \beta_1^1(-a_n x_1(t_0) - a_{n-1}x_2(t_0) - \cdots - a_1 x_n(t_0) + u(t_0)) \\ &= \beta_n^2 x_1(t_0) + \beta_{n-1}^2 x_2(t_0) + \cdots + \beta_1^2 x_n(t_0) + \beta_1^2 u(t_0), \end{aligned} \tag{5.1.8}$$

where $\beta_n^2 = -\beta_1^1 a_n$, $\beta_i^2 = \beta_{i+1}^1 - \beta_1^1 a_i$, $i = 1, 2, \ldots, n-1$.

Find further the derivative, it is found that

$$\begin{aligned} y''(t_0) &= -\beta_n^2 x_1'(t_0) + \beta_{n-1}^2 x_2'(t_0) + \cdots + \beta_1^2 x_n'(t_0) + \beta_1^1 u'(t_0) \\ &= \beta_{3,n} x_1(t_0) + \beta_{3,n-1} x_2(t_0) + \cdots \\ &\quad + \beta_{3,1} x_n(t_0) + \beta_{2,1} u(t_0) + \beta_{1,1} u'(t_0), \end{aligned} \tag{5.1.9}$$

where $\beta_n^3 = -\beta_1^2 a_n$, $\beta_i^3 = \beta_{i+1}^2 - \beta_1^2 a_i$, $i = 1, 2, \ldots, n-1$.

It can be seen that

$$\begin{aligned} y^{(k-1)}(t_0) &= \beta_n^k x_1(t_0) + \beta_{n-1}^k x_2(t_0) + \cdots + \beta_1^k x_n(t_0) \\ &\quad + \beta_1^{k-2} u^{(k-2)}(t_0) + \cdots + \beta_1^2 u'(t_0) + \beta_1^1 u(t_0), \end{aligned} \tag{5.1.10}$$

where $\beta_n^k = -\beta_1^{k-1} a_n$, $\beta_i^k = \beta_{i+1}^{k-1} - \beta_1^{k-1} a_i$, $i = 1, 2, \ldots, n-1$, $k = 2, 3, \ldots, n$.

Theorem 5.3. *If the initial values of input signal, output signal and their derivatives are known, the equivalent initial state vector can be found from*

$$\begin{bmatrix} \beta_n^1 & \beta_{n-1}^1 & \cdots & \beta_1^1 \\ \beta_n^2 & \beta_{n-1}^2 & \cdots & \beta_1^2 \\ \vdots & \vdots & \ddots & \vdots \\ \beta_n^n & \beta_{n-1}^n & \cdots & \beta_1^n \end{bmatrix} \begin{bmatrix} x_1(t_0) \\ x_2(t_0) \\ \vdots \\ x_n(t_0) \end{bmatrix} = \begin{bmatrix} y(t_0) \\ y'(t_0) - \beta_1^1 u(t_0) \\ \vdots \\ y^{(n-1)}(t_0) - \sum\limits_{i=1}^{n-1} \beta_1^i u^{(i-1)}(t_0) \end{bmatrix}. \tag{5.1.11}$$

If the coefficient matrix is a full-rank one, the equation in (5.1.11) *has a unique solution, such that the equivalent initial state* $\boldsymbol{x}(t_0)$ *can be found.*

Theorem 5.4. *It can be shown that the coefficient matrix obtained is the observability test matrix*

$$\begin{bmatrix} \beta_n^1 & \beta_{n-1}^1 & \cdots & \beta_1^1 \\ \beta_n^2 & \beta_{n-1}^2 & \cdots & \beta_1^2 \\ \vdots & \vdots & \ddots & \vdots \\ \beta_n^n & \beta_{n-1}^n & \cdots & \beta_1^n \end{bmatrix} = \begin{bmatrix} \boldsymbol{C} \\ \boldsymbol{CA} \\ \vdots \\ \boldsymbol{CA}^{n-1} \end{bmatrix}. \tag{5.1.12}$$

Therefore, if the system is observable, and the equivalent initial state vector can be uniquely determined.

With the above algorithm, it is not hard to write a MATLAB function, to find out the equivalent state space model. Note that to express β_i^k in a matrix, a straightforward method is $\boldsymbol{b}(k,i) = \beta_i^k$.

```
function [x0,G1,T,Y]=find_x0(G,y0,u0)
[num,den]=tfdata(G,'v'); n=length(den)-1;
d0=den(1); den=den/d0; num=num/d0; Y=y0(1); D=num(1);
y0=[y0(:); zeros(n-length(y0),1)];
u0=[u0(:); zeros(n-length(u0)-1,1)];
num=num-D*den; num=num(2:end); b=num; a=den(2:end);
A=diag(ones(1,n-1),1); A(end,:)=-den(end:-1:2);
B=[zeros(n-1,1); 1]; C=num(n:-1:1); G1=ss(A,B,C,0);
for k=2:n
   b(k,n)=-b(k-1,1)*a(n);
   for i=1:n-1, b(k,i)=b(k-1,i+1)-b(k-1,1)*a(i); end
   Y=[Y; y0(k)-b(1:k-1,1).'*u0(1:k-1)];
end
T=fliplr(b);  x0=inv(T)*Y;
```

The syntax of the function is $[\boldsymbol{x}_0, G_1, \boldsymbol{T}, \hat{\boldsymbol{y}}_0]$ = find_x0$(G, \boldsymbol{y}_0, \boldsymbol{u}_0)$, where G is the transfer function model, $\boldsymbol{y}_0 = [y(t_0), y'(t_0), \dots, y^{(n-1)}(t_0)]$ and $\boldsymbol{u}_0 = [u(t_0), u'(t_0), \dots, u^{(n-2)}(t_0)]$ are respectively the initial values of the output and input signals. The returned $\boldsymbol{x}_0$ is the equivalent initial state vector, and G_1 is the equivalent state space model. Matrix $\boldsymbol{T}$ is the transformation matrix, and also the observability test matrix. Vector $\hat{\boldsymbol{y}}_0$ returns the actual initial values.

Example 5.6. Build a Simulink model for the problem in Example 4.11, and assess the accuracy and speed in the solution.

Solutions. In Example 4.11, derivative blocks were used, such that the simulation accuracy is very low and rather time consuming. Although in Example 4.12, an improved method was introduced, but the method is too rigorous since the mathematical expression of the derivative of the input signal is needed. Here, the equivalent state space model is used, and a Simulink model can be established to solve the differential equation.

It is not hard to write out the equivalent transfer function model

$$G(s) = \frac{2s+1}{s^4+5s^3+9s^2+7s+2}.$$

Input the transfer function into MATLAB workspace, and also the initial value vectors of the input and output signals can also be entered. The new function `find_x0()` can be called to find the equivalent state space model, and the initial states:

```
>> num=[2,1]; den=[1,5,9,7,2]; G=tf(num,den);
   u0=[0;1;0]; y0=[1;0;0;0];
   [x0,G1,T]=find_x0(G,y0,u0) %find the equivalent model
```

The equivalent state space model obtained is as follows, and the equivalent initial state vector is $\boldsymbol{x}_0 = [-29, 16, -8, 4]^{\mathrm{T}}/3$:

$$\boldsymbol{x}'(t) = \begin{bmatrix} 0 & 1 & 0 & 0 \\ 0 & 0 & 1 & 0 \\ 0 & 0 & 0 & 1 \\ -2 & -7 & -9 & -5 \end{bmatrix} \boldsymbol{x}(t) + \begin{bmatrix} 0 \\ 0 \\ 0 \\ 1 \end{bmatrix} u(t), \quad y(t) = [1, 2, 0, 0]\boldsymbol{x}(t).$$

These information can be specified in the **State Space** block, and the Simulink model can be constructed, as shown in Figure 5.6. The model can be simulated, and the results can be compared to assess the accuracy:

```
>> [A,B,C,D]=ssdata(G1);
   tic, [t1,~,y1]=sim('c5mlin1'); toc, length(t1)
   y0=7*exp(-t1)/4-2*exp(-2*t1)/5+t1.*exp(-t1)+...
      3*t1.^2.*exp(-t1)/4-sqrt(2)*cos(t1-atan(1/7))/4;
   max(abs(y0-y1)), plot(t1,y1,'--',t1,y0)
```

It can be seen that the maximum error found is 1.4211×10^{-14}, and the number of point is 2142. The total time elapse is 0.2219 seconds. It can be seen that the simulation method here is easy to use, and it is more accurate.

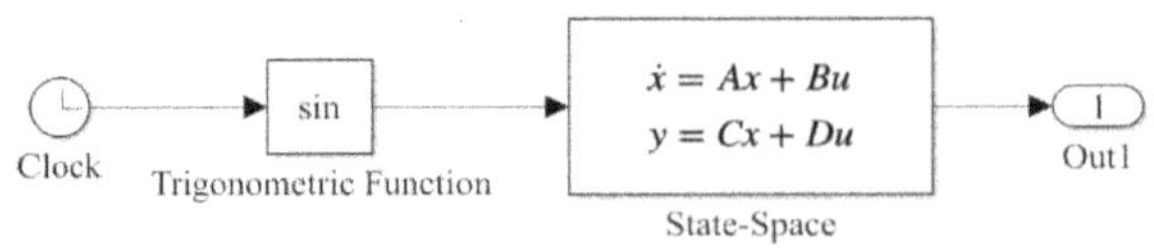

Figure 5.6: Simulink model of transfer function with nonzero initial conditions (file: `c5mlin1.slx`).

In Chapter 7, the transfer function can be masked, such that a reusable block for the transfer function with nonzero initial conditions can be constructed.

5.1.5 Simulink modeling of transfer function matrices

Transfer function matrix is a commonly used mathematical model for multivariable linear systems. In this section, the definition of transfer function matrix is presented, then Simulink modeling of the transfer function matrix is explored, and a concise and universal modeling method is proposed.

Definition 5.4. The general form of a continuous transfer function matrix of a multivariable system is given by

$$G(s)=\begin{bmatrix} g_{11}(s) & g_{12}(s) & \cdots & g_{1p}(s) \\ g_{21}(s) & g_{22}(s) & \cdots & g_{2p}(s) \\ \vdots & \vdots & \ddots & \vdots \\ g_{q1}(s) & g_{q2}(s) & \cdots & g_{qp}(s) \end{bmatrix}, \tag{5.1.13}$$

where $g_{ij}(s)$ is the radio of the ith output signal to the jth input signal, referred to as the (i,j)th subtransfer function. The matrix is a $q \times p$ one, indicating that there are q output signals, and p input signals.

Substitute the input and output vectors into the system, and it can be found that

$$\begin{bmatrix} y_1(s) \\ y_2(s) \\ \vdots \\ y_q(s) \end{bmatrix} = \begin{bmatrix} g_{11}(s) & g_{12}(s) & \cdots & g_{1p}(s) \\ g_{21}(s) & g_{22}(s) & \cdots & g_{2p}(s) \\ \vdots & \vdots & \ddots & \vdots \\ g_{q1}(s) & g_{q2}(s) & \cdots & g_{qp}(s) \end{bmatrix} \begin{bmatrix} u_1(s) \\ u_2(s) \\ \vdots \\ u_p(s) \end{bmatrix}. \tag{5.1.14}$$

It can be seen that the low-level Simulink description of transfer function matrices are very complicated. A 2×2 transfer function matrix is used as an example to demonstrate low-level modeling.

Example 5.7. Use Simulink to describe the following multivariable transfer function matrix with time delays:[51]

$$G(s)=\begin{bmatrix} \dfrac{0.1134\mathrm{e}^{-0.72s}}{1.78s^2+4.48s+1} & \dfrac{0.924}{2.07s+1} \\ \dfrac{0.3378\mathrm{e}^{-0.3s}}{0.361s^2+1.09s+1} & \dfrac{-0.318\mathrm{e}^{-1.29s}}{2.93s+1} \end{bmatrix}.$$

Solutions. Transfer function and transport delay blocks can be used in Simulink; however, they can only be used to describe directly single variable systems, not multivariable ones. If a 2×2 transfer function matrix is to be described by low-level blocks, the following explicit formula can be obtained from (5.1.14):

$$y_1(s) = g_{11}(s)u_1(s) + g_{12}(s)u_2(s),$$
$$y_2(s) = g_{21}(s)u_1(s) + g_{22}(s)u_2(s).$$

If each of the subtransfer function $g_{ij}(s)$ is modeled with low-level blocks such as Transfer Fcn and Transport Delay, a Simulink model can be constructed, as shown in Figure 5.7. It can be seen that the low-level modeling method is rather complicated, and may lead to errors. Therefore, this method is not good in multivariable system modeling.

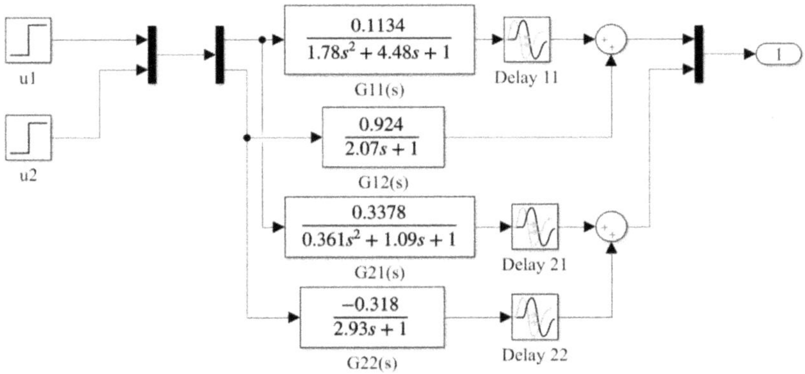

Figure 5.7: Low-level modeling of multivariable system (file: `c5mmimo1.slx`).

For transfer function matrices or other linear models, the linear time invariant (LTI) object in MATLAB Control System Toolbox can be used to describe it directly. Functions `tf()` and `ss()` in the Toolbox can be used to express the transfer function and state space models, including descriptor ones.

The Control System Blockset is also provided for Simulink, as shown in Figure 5.8, where the LTI System block can be used to describe LTI object in Simulink. The LTI object can be specified with the above functions, and the name of the LTI object can be specified on the block. Examples are given next to show the use of LTI System block in Simulink modeling.

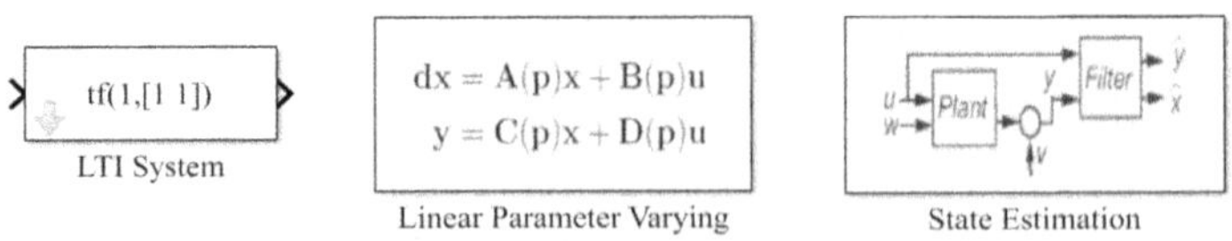

Figure 5.8: Control System Toolbox Blockset.

Example 5.8. Build a Simulink model with LTI System block for the multivariable transfer function in Example 5.7.

Solutions. To input the transfer function matrix in Example 5.7, with delay terms, a direct method is to input the four subtransfer functions. The matrix input command can be used to compose the transfer function matrix model G, and then filled G to the parameter dialog box of LTI System block. Therefore, the complicated transfer function matrix model can be expressed in Simulink. There is no need to build the low-level model in Figure 5.7:

```
>> g11=tf(0.1134,[1.78 4.48 1],'ioDelay',0.72);
   g12=tf(0.924,[2.07 1]);
   g21=tf(0.3378,[0.361 1.09 1],'ioDelay',0.3);
   g22=tf(-0.318,[2.93 1],'ioDelay',1.29);
   G=[g11, g12; g21, g22]; %input the transfer function matrix as a matrix
```

This method can also be used to represent complicated models, including those composite transfer functions, which cannot be described by transfer function blocks. State space models with internal delays are supported in the Control System Toolbox,[85] the models with variable parameters are also supported. Examples are used to demonstrate the model representation of complicated linear time invariant systems.

Example 5.9. Consider the composite transfer function model in Example 2.7. The low-level Simulink model has been created, and that was rather complicated and error-prone. Build a Simulink model again with simpler method.

Solutions. If manual conversion is used, the model cannot be expressed to the standard form of the transfer function model shown in (5.1.1). With the model processing method provided in the Control System Toolbox, it can be converted automatically to the state space model with internal delays. Therefore, the LTI System block can be used to describe the composite transfer function model directly. The model should be expressed in the MATLAB workspace in variable G, then G can be specified to the edit box of the parameter dialog box of LTI System block. Therefore, this model can be expressed directly in Simulink:

```
>> s=tf('s');                      % define s operator
   G=(1+3*exp(-s)/(s+1))/(s+2);    % input the composite transfer function
```

5.1.6 Simulink models with varying parameters

Definition 5.5. The mathematical form of a linear state space model with varying parameters is described as

$$\begin{cases} \boldsymbol{x}'(t) = \boldsymbol{A}(\boldsymbol{p})\boldsymbol{x}(t) + \boldsymbol{B}(\boldsymbol{p})\boldsymbol{u}(t), \\ \boldsymbol{y}(t) = \boldsymbol{C}(\boldsymbol{p})\boldsymbol{x}(t) + \boldsymbol{D}(\boldsymbol{p})\boldsymbol{u}(t), \end{cases} \tag{5.1.15}$$

where, $\boldsymbol{p} = [p_1, p_2, \ldots, p_m]$ are variable parameters.

In the MATLAB Control System Toolbox, function `realp()` can be used to describe varying parameters.

In Simulink, these kinds of models can be expressed directly with the blocks in the Linear Parameter Varying group shown in 5.8. Double click the group icon, and the blocks and subgroups in the group are shown in Figure 5.9.

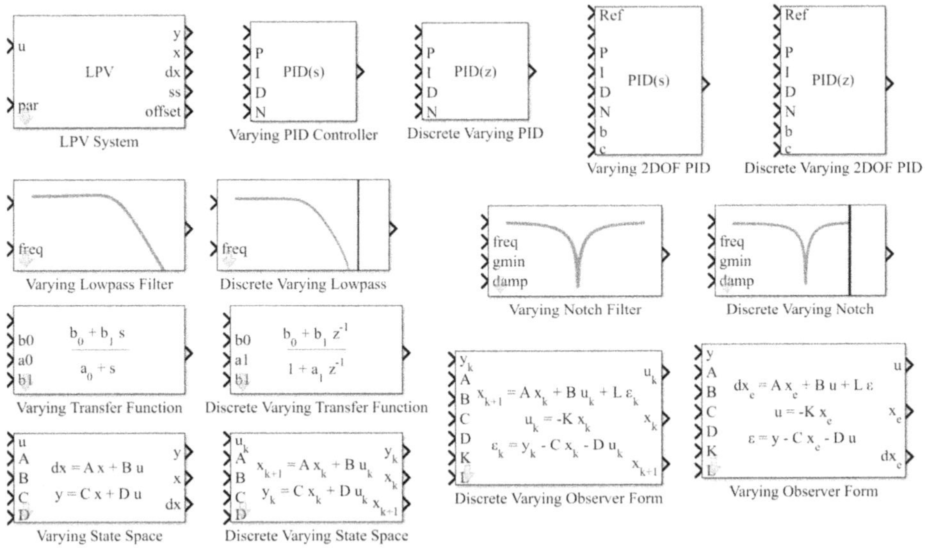

Figure 5.9: Blocks and subgroups for models with varying parameters.

5.2 Modeling and simulation of linear discrete systems

Theoretical and model representation of continuous system are presented earlier. The so-called continuous refers that the signals in the systems are continuous functions. If the signals are not continuous, but described by a series samples, there are discrete components. Special methods should be introduced to describe them. In this section, various mathematical models are presented, and illustrations are given to show how to input the discrete models in MATLAB. Illustrations are also made on how to convert the model between continuous and discrete formats, and how to represent discrete models in Simulink. Finally, examples are used to demonstrate the impact of control parameters and other factors on simulation results.

5.2.1 Mathematical models of linear discrete systems

Compared with continuous systems, difference equations are essential for discrete linear systems. There are also transfer functions and state space representations. In this section, the mathematical models in discrete systems are presented, and then model input methods based on Control System Toolbox are illustrated.

Definition 5.6. The mathematical form of a difference equation is

$$\begin{aligned} & a_1 y((k+n)T) + a_2 y((k+n-1)T) + \cdots + a_{n+1} y(kT) \\ & \quad = b_1 u((k+n)T) + b_2 u((k+n-1)T) + \cdots + b_{n+1} u(kT), \end{aligned} \tag{5.2.1}$$

where T is the sample time of the discrete system. More simply, denote $t+n = (k+n)T$, the symbol T can be removed in describing difference equations:

$$a_1 y(t+n) + a_2 y(t+n-1) + \cdots + a_{n+1} y(t) = b_1 u(t+n) + b_2 u(t+n-1) + \cdots + b_{n+1} u(t), \tag{5.2.2}$$

or more concisely

$$a_1 y_{t+n} + a_2 y_{t+n-1} + \cdots + a_{n+1} y_t = b_1 u_{t+n} + b_2 u_{t+n-1} + \cdots + b_{n+1} u_t. \tag{5.2.3}$$

Definition 5.7. The linear discrete transfer function is given by

$$G(z) = \frac{b_1 z^n + b_2 z^{n-1} + \cdots + b_n z + b_{n+1}}{a_1 z^n + a_2 z^{n-1} + \cdots + a_n z + a_{n+1}}. \tag{5.2.4}$$

Definition 5.8. The mathematical representation of linear discrete state space model is given by

$$\begin{cases} \boldsymbol{x}_{t+1} = \boldsymbol{F}\boldsymbol{x}_t + \boldsymbol{G}\boldsymbol{u}_t, \\ \boldsymbol{y}_t = \boldsymbol{C}\boldsymbol{x}_t + \boldsymbol{D}\boldsymbol{u}_t. \end{cases} \tag{5.2.5}$$

Functions `tf()` and `ss()` in the MATLAB Control System Toolbox can be used to input discrete transfer functions and state space models as well. The functions can be called in a similar way as in continuous systems. The difference is that the sample time should be specified:

```
G=tf(num,den,'Ts',T);
z=tf('z',T);
G=ss(A,B,C,D,'Ts',T);
```

Example 5.10. If the sample time is $T = 0.1$ seconds, input the following two discrete transfer functions:

$$G_1(z) = \frac{2z+1}{z^4+2.6z^3+2.52z^2+1.08z+0.1728}, \quad G_2(z) = \frac{2z+1}{(z+0.6)^3(z+0.8)}.$$

Solutions. It can be seen from the transfer functions that in the former one, the coefficients of the numerator and denominator are all given, therefore, the first input method can be used. In the latter, the second input method can be used, that is, to define z operator, and then the mathematical expression can be used to describe the transfer function model:

```
>> G1=tf([2 1],[1 2.6 2.52 1.08 0.1728],'Ts',0.1)
   z=tf('z',0.1); G2=(2*z+1)/(z+0.6)^3/(z+0.8)
```

5.2.2 Conversions between continuous and discrete models

Conversions between continuous and discrete models can be completed with the functions `c2d()` and `d2c()` provided in the Control System Toolbox. For a continuous LTI object G, with given sample time of T, command G_1=`c2d(`G,T`)` can be used to find directly the equivalent discrete model G_1. If G is a transfer function model, G_1 after conversion is a discrete transfer function model. If G is a state space model, the model G_1 after discretization is a discrete state space model.

If discrete LTI object G_1 is known, command G=`d2c(`G_1`)` can be used to find the equivalent continuous model G. In the function call, sample time T is not used, since the information is already contained in model G_1.

Example 5.11. Consider the continuous transfer function in Example 5.6. If the sample time is selected respectively as $T = 0.1$ seconds and $T = 0.01$ seconds, discretize the system, and compare the step responses of the models.

Solutions. The continuous transfer function should be input first, then under different sample time, discretization is carried out. The step responses of the continuous and discrete systems can be obtained, as shown in Figure 5.10. It can be seen that the signals in discrete systems are expressed by staircase forms. Within each sample time, the signals are kept constant. In effect, it is equivalent to the signal appended by a zero order hold:

```
>> num=[2,1]; den=[1,5,9,7,2]; G=tf(num,den);
   G1=c2d(G,0.1), G2=c2d(G,0.01), step(G,G1,'--',G2,':')
```

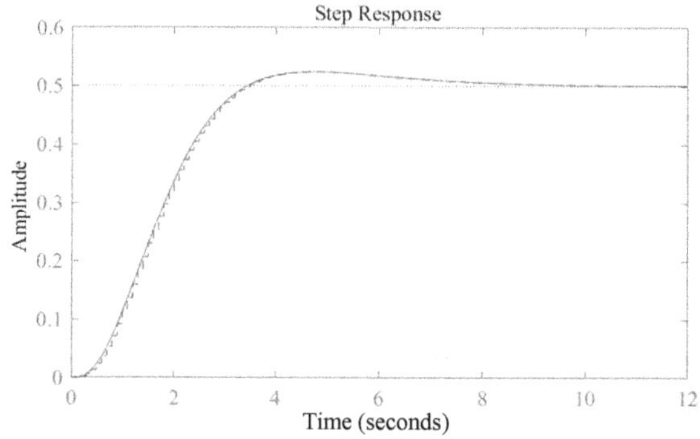

Figure 5.10: Step responses after discretization.

The discrete transfer function models obtained are respectively

$$G_1(z) = \frac{0.000298z^3 + 0.0007821z^2 - 0.0007757z - 0.0002263}{z^4 - 3.533z^3 + 4.679z^2 - 2.752z + 0.6065},$$

$$G_2(z) = \frac{3.296 \times 10^{-7}z^3 + 9.757 \times 10^{-7}z^2 - 9.749 \times 10^{-7}z - 3.207 \times 10^{-7}}{z^4 - 3.95z^3 + 5.852z^2 - 3.853z + 0.9512}.$$

The following commands can be used to find the equivalent continuous model:

```
>> G3=d2c(G1)
```

and the result is shown below. It can be seen that the original continuous transfer function model is restored. In some coefficients, very small errors are introduced:

$$G_3(s) = \frac{1.291 \times 10^{-17}s^3 - 3.248 \times 10^{-16}s^2 + 2s + 1}{s^4 + 5s^3 + 9s^2 + 7s + 2}.$$

5.2.3 Simulink modeling of discrete systems

In the Discrete group studied earlier, discrete transfer functions, discrete state space model blocks are provided. Also, blocks like delays and zero-order hold are provided. These blocks can be used to build discrete models in Simulink.

Due to the birth defect in transfer function models, the transfer function blocks can only be used to deal with transfer function models with zero initial conditions. If a transfer function block capable of handling nonzero initial conditions is needed, the Additional Discrete group in Simulink Extras blockset shown in Figure 5.11 can be used. Of course, to better solve the problem, the algorithm similar to the one for continuous model can be studied, and MATLAB implementation can be made.

Apart from these two groups, the blockset in the Control System Toolbox can be used, where the LTI System block can be used to handle the discrete LTI object directly.

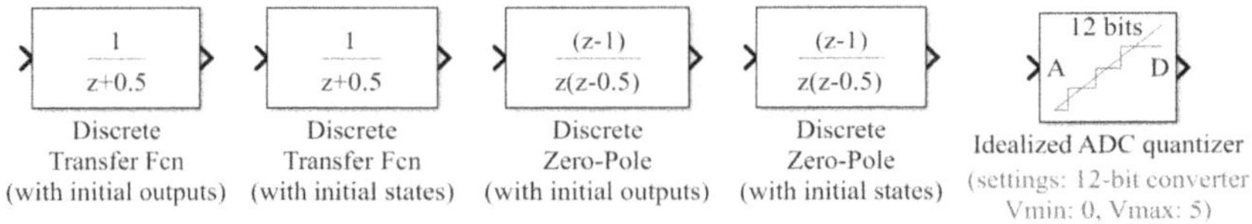

Figure 5.11: Additional Discrete group.

There is no need to convert the model into continuous form. Assign the discrete LTI object to the **LTI System** block, and it can be directly used in modeling and simulation.

Example 5.12. Consider the discretization problem in Example 5.11. Use Simulink to model and simulation the continuous and discrete systems, and find the step responses.

Solutions. Two methods can be used to describe the discrete models, with the corresponding Simulink models shown in Figure 5.12, where **Discrete Transfer Function** and **LTI System** blocks can be used respectively for the discrete model. For ease in modeling and simulation, the following information should be included in the **Model Properties**:

```
num=[2,1]; den=[1,5,9,7,2]; G=tf(num,den);
G1=c2d(G,0.1); [numd,dend]=tfdata(G1,'v');
```

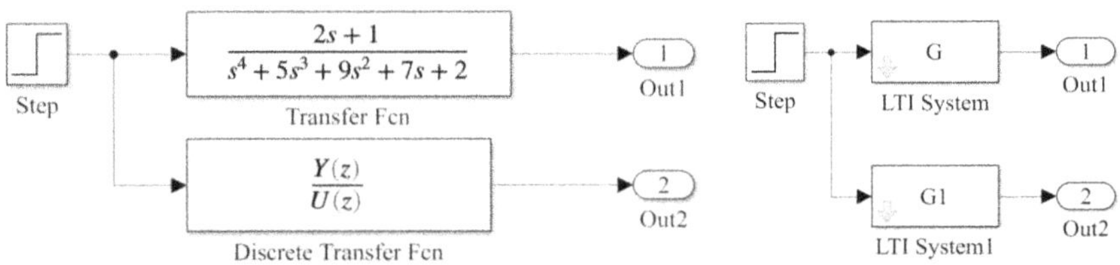

(a) method (file: `c5mc2d1.slx`) (b) method 2 (`c5mc2d2.slx`)

Figure 5.12: Discretization and simulation.

In the **Discrete Transfer Fcn** block, the variable names `numd` and `dend` can be filled into the numerator and denominator edit boxes. The sample time of 0.1 can also be specified. Since the parameters of the discrete transfer function are already saved in the variables, the information is not displayed in the icon. In the **LTI System** block, the model edit box should specify G_1, and there is no need to specify the sample time values, since it has been contained in G_1 object.

Start simulation process for the two models, the results are exactly the same, as the one shown in Figure 5.10. Note that when drawing the simulation results, the curves of continuous systems can be drawn. However, for discrete systems, discrete signals are given in staircase signals.

In constructing the Simulink model, if discrete blocks are used, the sample time should be specified in the parameter dialog box. The sample time can be specified as the actual value, or specify −1, indicating the sample time of the input signal is inherited in this block. In this way, if the sample time of several blocks are the same, there is no need to assign sample time individually. They can all be set to −1, except the first one. The other blocks inherit this sample time automatically. The Simulink modeling is made simpler in this way.

5.2.4 Impact of simulation control parameters

Generally, if there are only discrete blocks, with no continuous ones, and the sample time in all the blocks are the same, fixed-step simulation algorithms can be accepted for the simulation process. If there are continuous blocks, discrete algorithms are not recommended. The variable-step algorithm should be used; otherwise, simulation accuracies cannot not ensured. Here, an example is used to demonstrate the impact to simulation accuracies by the control parameters, in variable-step simulation algorithms.

Example 5.13. Assume that the continuous system is described in Example 5.11. Selecting sample time of $T = 0.1$ seconds, a discretized model can be found. Analyze the impact of control parameters to simulation results.

Solutions. Consider the simulation model in Example 5.11. Since there are both continuous and discrete blocks, fixed-step algorithms cannot be used. Variable-step algorithms should be adopted instead. If default control parameters are used, the simulation results in Figure 5.13 can be found.

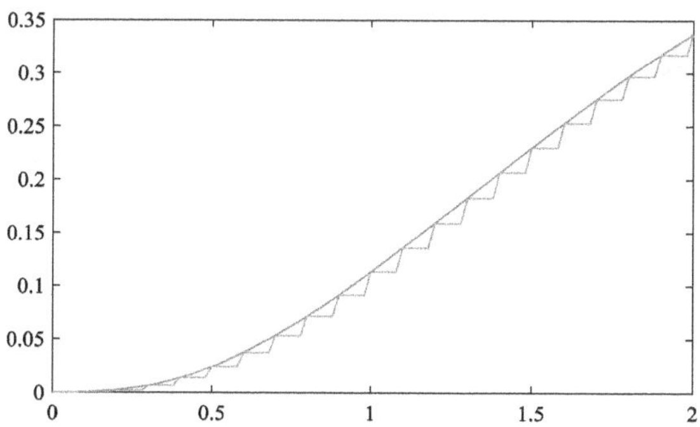

Figure 5.13: Simulation results under default parameters.

In order to observe better the simulation results, the stop time is deliberately set to 2 seconds. It can be seen that although the simulation results on the discrete samples are correct, this model is not the same as the expected discrete systems, since in normal cases, the outputs of the discrete systems are in staircase form. To solve such a problem, the default error tolerance in Simulink should be set to very small values, such as 3×10^{-14}. If the relative and absolute error tolerances are all set like this, the correct staircase curves in Figure 5.10 can be obtained. It can be seen that the control parameters may significantly impact the simulation results for discrete systems. If correct simulation results are expected, if possible, the error tolerance should be set to the smallest possible value.

5.2.5 Impact of discrete parameter precisions

If the poles of discrete transfer functions are close to unit circles, the slight variations on the denominator coefficient vectors may significantly impact the simulation results of the entire systems. An example is used here to demonstrate the impact to the simulation results by the denominator parameters in discrete systems.

Example 5.14. Assume that the continuous model is given in Example 5.11. Selecting a sample time of $T = 0.1$ seconds, the discretized models can be found. Analyze the number of digits reserved in the denominator coefficients to the simulation results.

Solutions. The continuous transfer function model can be entered first, and discretization can be found. Function `tfdata()` can be used to extract the numerator and denomination coefficients. If 2 digits after the decimal points are to be reserved, the value can be multiplied by 100 and find the integers. The result can be multiplied by 0.01 such that the remaining digits are omitted. The converted denominator is set to G_{1a}. With such methods, the denominator with different numbers of reserved digits can be found, and the step responses can be obtained, as shown in Figure 5.14. Here, only the ones with 4 and 6 reserved digits are shown, since the discretized models G_{1a} and G_{1b} are unstable. It can be seen that the number of digits reserved may affect significantly on the simulation results. Too less reserve digits may make the systems unstable. If 4 digits are reserved, the errors are significant. There are also differences on the steady-state values. If 6 digits are reserved, there are also observable differences from the discretized model:

```
>> num=[2,1]; den=[1,5,9,7,2]; G=tf(num,den);
   G1=c2d(G,0.1); [n,d]=tfdata(G1,'v');
   G1a=G1; G1b=G1; G1c=G1; G1d=G1;
   d0=0.01*round(d*100); G1a.den{1}=d0;
   d0=0.001*round(d*1000); G1b.den{1}=d0;
```

```
d0=0.0001*round(d*10000); G1c.den{1}=d0;
d0=1e-6*round(d*1e6); G1d.den{1}=d0;
step(G,G1c,G1d)
```

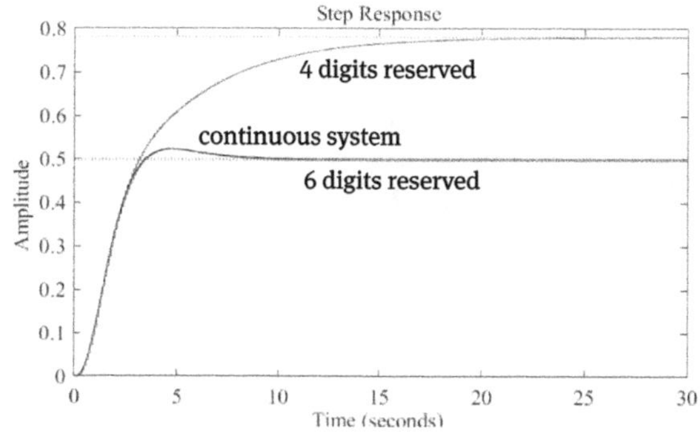

Figure 5.14: Comparisons in discrete simulation results.

Now let us analyze the discretized model G_1. It is found that there are triple poles at 0.9048, rather close to the unit circle, and another one located at 0.8781. If two digits are reserve in the discretized model, the poles of G_{1a} are located at 1.0931 ± 0.2490j and 0.6719 ± 0.1841j. The magnitude of the first pair is 1.1211, outside the unit circle. Therefore, the discretized model is unstable. If 4 digits are reserved, the poles of G_{1c} are located at 0.9759, 0.8858 ± 0.0796j and 0.7857. The first pole is located close to the unit circle. It can be seen that the poles are different from the discretized model G_1. There are significant difference in the simulation results:

```
>> eig(G1), d1=max(abs(ans))
   eig(G1a), d2=max(abs(ans)), eig(G1c), d3=max(abs(ans))
```

5.3 Simulink modeling of nonlinear components

It can be seen from the Discontinuities group that the current blocks are quite limited for nonlinearities. The complicated nonlinearities, which can be decomposed into simpler ones, such as the one in Example 2.8, the number is also very restricted. A more general way in describing nonlinearities is expected in simulation. In this section, single-valued static nonlinearities of any shape can be established with lookup table blocks. Further, double-valued static nonlinearities can be constructed with lookup table and switch blocks.

5.3.1 Lookup table blocks

Double click the **1-D Lookup Table** block icon in the **Lookup Table** group shown in Figure 2.36, the parameter dialog box is shown in Figure 5.15, where the **Number of table dimensions** given here is one-dimensional, and it can be specified as larger integers.

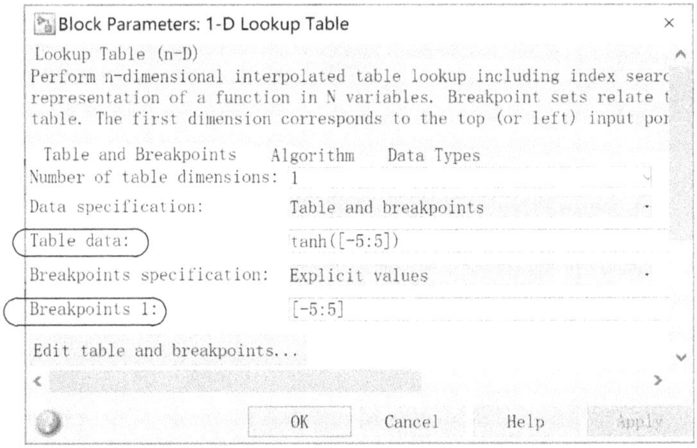

Figure 5.15: Parameter dialog box of 1D lookup table block.

The block has two important vectors, **Breakpoints 1** and **Table data**. The values of them are respectively the horizontal vector $\boldsymbol{u} = [u_1, u_2, \dots, u_m]$ and vertical vector $\boldsymbol{y} = [y_1, y_2, \dots, y_m]$. The vectors should be specified by the users. With the two vectors, the output signals of the block can be uniquely determined from the current values of the input signal. In real applications, these two vectors should be specified first.

In normal cases, if the input signal in the block satisfied $u_k \leqslant u(t) < u_{k+1}$, the output signal $y(t)$ can be evaluated from

$$y(t) = \frac{y_{k+1} - y_k}{u_{k+1} - u_k}(u(t) - u_k). \tag{5.3.1}$$

In fact, this algorithm is referred to as linear interpolation. If the user wants to compute output signals with other algorithms, the **Algorithms** pane in Figure 5.15 can be selected. The **Interpolation** or **Extrapolation** algorithms in the new dialog box can be selected.

To edit the values of the turning points, click the **Edit table and breakpoints** button, and the table editor interface shown in Figure 5.16 is provided. If the first column elements are selected, horizontal values can be modified, while if the second column elements are selected, the vertical values can be modified. In this way, the turning points can be modified easily.

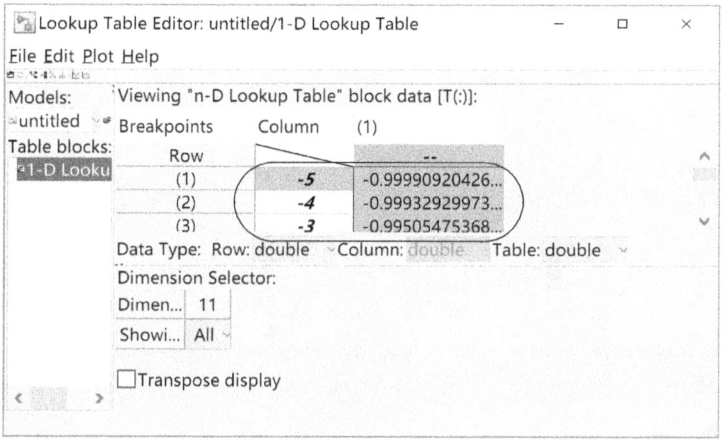

Figure 5.16: Table entry editor.

When the parameter setting is completed, if the lengths of the two vectors are the same, the shape of the function can be displayed as the icon of the block. This is automatically drawn by the Simulink mechanism.

5.3.2 Construction of single-valued nonlinearities

For the convenience of presentation, the static nonlinearities of single-valued and double-valued nonlinear functions are presented separately. Then Simulink modeling methods are presented.

Definition 5.9. Single-valued nonlinear function indicates that, the output signal $y(t)$ can be determined uniquely by the current value of the input signal $u(t)$ at any time instance t. The mathematical description is $y(t) = f(u(t))$.

Now let us consider the modeling method for piecewise linear static nonlinearity with Simulink blocks. Consider the piecewise linear nonlinear characteristics shown in Figure 5.17, with the turning points of the nonlinearity are known as $(x_1, y_1), (x_2, y_2), \dots, (x_{n-1}, y_{n-1}), (x_n, y_n)$. If a Simulink lookup table block is to be used to describe the nonlinear function, a point x_0 to the left of x_1 should be selected, such that $x_0 < x_1$. The corresponding value of the function y_0 can be computed. Also another point x_{n+1} should be selected such that $x_{n+1} > x_n$, and the value of y_{n+1} can be obtained. Therefore, two vectors xx and yy can be created, such that

$$\mathtt{xx=[}x_0, x_1, x_2, \dots, x_n, x_{n+1}\mathtt{];}\quad \mathtt{yy=[}y_0, y_1, y_2, \dots, y_n, y_{n+1}\mathtt{];}$$

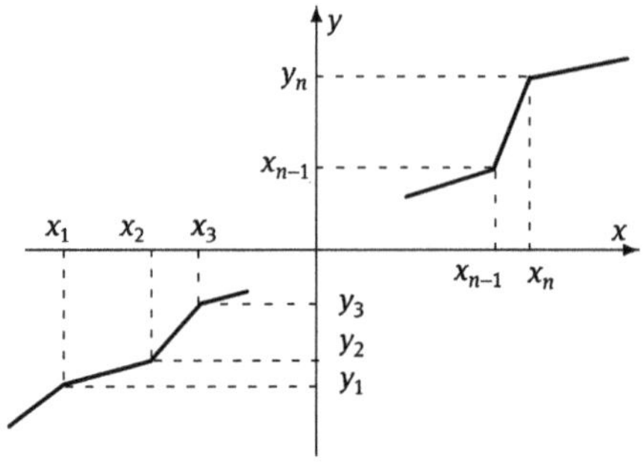

Figure 5.17: General form of piecewise linear singled-valued nonlinearity.

Assign xx to the edit box **Breakpoints 1** in the **Lookup Table** block, and yy to the edit box Table data, and the lookup table can be used to describe the nonlinearity in Simulink models. A simple example is given next to demonstrate the Simulink modeling of single-valued nonlinearities.

Example 5.15. Use Simulink lookup table block to describe the nonlinearity in Example 2.8.

Solutions. From the nonlinearities in Figure 2.30(a), it can be seen that the turning points are $(-1.2, 1)$, $(-0.2, 0)$, $(0.2, 0)$ and $(1.2, 1)$. If the lookup table block is to be used, two extra points should be selected. For instance, two points $(-1.5, 1)$ and $(2, 1)$ can be selected, such that two vectors `xx=[-1.5,-1.2,-0.2,1.2,2]` and `yy=[-1,-1,0,0,1,1]` can be created, The two vectors can be entered into MATLAB workspace. The two variables can be set to the edit boxes in the **Lookup Table** block. The expected nonlinear characteristics can be described by the block. It can be seen that this method is more straightforward than the one in Example 2.8, and it can be extended directly to describe any static single-valued nonlinearities.

5.3.3 Construction of double-valued nonlinearities

Double-valued nonlinear function means that for the same input function $u(t)$, the output signal $y(t)$ may take one of the two different values, under different external conditions, for instance, the trend of $u(t)$. The modeling of double-valued nonlinearities are no longer simple. Here, two examples are used to demonstrate the Simulink modeling of a double-valued nonlinearity, and general modeling rules are presented.

It can be seen that any single-valued nonlinearity can be modeled by the method shown earlier, while if there are loops of double-valued nonlinearities, this method is not valid. Other blocks such as switches should also be used.

Example 5.16. Construct a Simulink model to describe the double-valued nonlinearity shown in Figure 5.18.

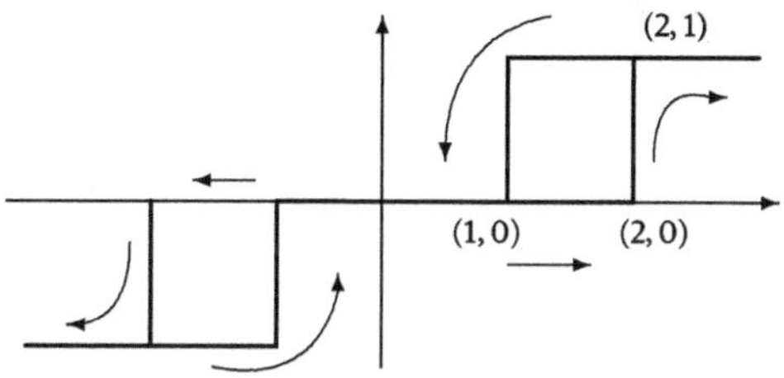

Figure 5.18: Relay nonlinearity with loops.

Solutions. It can be seen that this function is no longer single-valued. When the input signal is increasing, one piecewise linear function is followed, while when it is decreasing, another function is taken. The nonlinear function can be decomposed into two single-valued nonlinearities, shown in Figure 5.19. Of course, the two single-valued functions are conditional. They are determined by whether the input signal is increasing or decreasing.

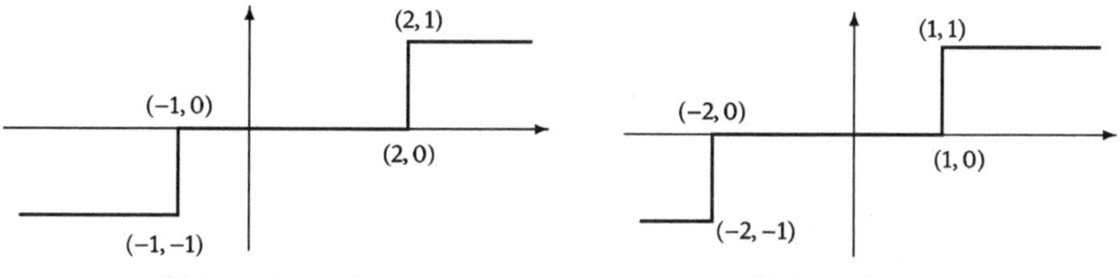

Figure 5.19: Decompose loops into single-valued functions.

The crucial point in double-valued nonlinearity modeling is how to check whether the input signal is increasing or decreasing. If the trend of the input signal can be recognized, a switch block can be used to control the selection of the two branches, such that one of them can be connected. A Memory block in the Continuous group can be used to find the value of the signal at a previous step. Therefore, a Simulink model can be constructed in Figure 5.20. A comparison block is used in the model to compare the sizes of the current signal at its previous value. The output is a logic one, whose value is 1 when the signal is increasing, 0 for decreasing. This signal can be used to control the switch block, and set the Threshold value of the switch to 0.5. Therefore, when the signal is increasing, the upper path can be used to evaluate the output of the block, while when the signal is decreasing, the lower path is used.

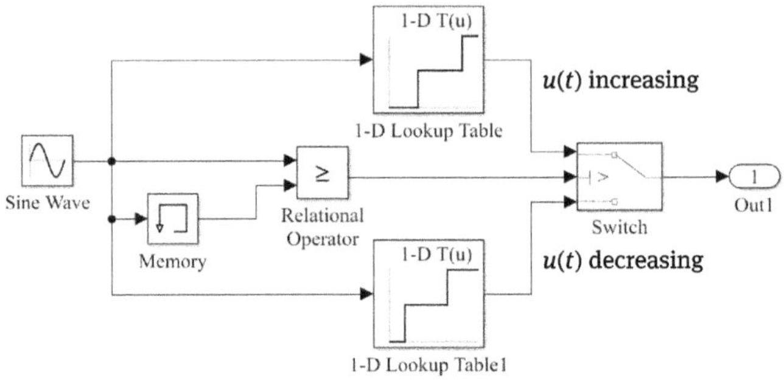

Figure 5.20: Simulink model of the nonlinear element (file: `c5mloop1.slx`).

Two vectors for the turning points can be expressed respectively as

$$x_1 = [-3, -1, -1 + \epsilon, 2, 2 + \epsilon, 3], \quad y_1 = [-1, -1, 0, 0, 1, 1]$$
$$x_2 = [-3, -2, -2 + \epsilon, 1, 1 + \epsilon, 3], \quad y_2 = [-1, -1, 0, 0, 1, 1]$$

where ϵ can be taken as a small value, for instance, 10^{-10}. Note that the value should not be taken too small, such as the MATLAB reserved constant `eps`, otherwise the icon cannot be drawn correctly.

Example 5.17. Set up a Simulink model for the relay nonlinearity with saturation, shown in Figure 5.21.

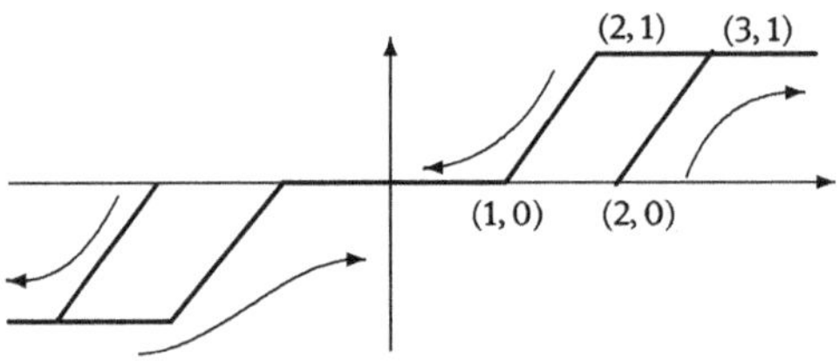

Figure 5.21: Relay with saturation.

Solutions. It can be seen that the nonlinearity is essentially the same as the one studied in the previous example. Loop nonlinearity can be decomposed into two single-valued nonlinearities. The differences are the turning points of the single nonlinear functions. The turning points should be extracted again from the given curves. The same Simulink model can be used, and the two vectors can be modified in the lookup table blocks, as

$$x_1 = [-3, -2, -1, 2, 3, 4], y_1 = [-1, -1, 0, 0, 1, 1]$$
$$x_2 = [-4, -3, -2, 1, 2, 3], y_2 = [-1, -1, 0, 0, 1, 1]$$

Therefore, the new Simulink model in Figure 5.22 can be established.

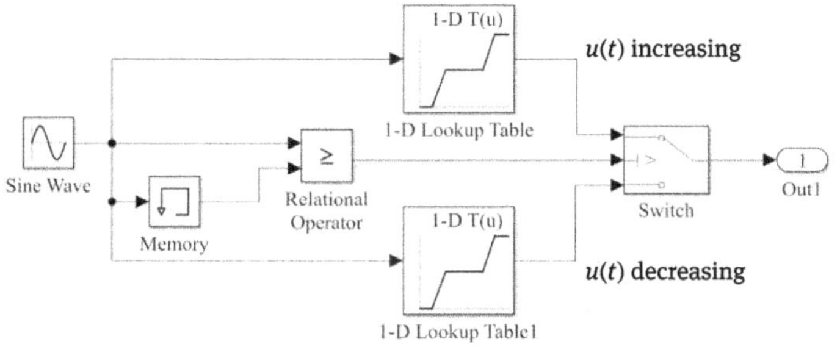

Figure 5.22: Simulink model of another nonlinearity (file: `c5mloop2.slx`).

It can be seen that any static nonlinearities, single- or double-valued, can be represented in Simulink and can be used in simulation directly.

5.3.4 Multidimensional lookup table blocks

Apart from one-dimensional lookup table blocks, two-dimensional and multidimensional lookup table blocks are also provided in Simulink. The output of multidimensional blocks can be evaluated using linear interpolation algorithms. The two-dimensional lookup table block is demonstrated next to show the use of multidimensional lookup table blocks in simulation.

Example 5.18. Consider the two-dimensional function $z = f(x,y) = (x^2 - 2x)e^{-x^2-y^2-xy}$. If sparsely distributed mesh grid points on the x-y plane can be generated, and the function values at these points can be evaluated with

```
>> xx=linspace(-3,3,15); yy=linspace(-2,2,15);
   [x,y]=meshgrid(xx,yy);
   z=(x.^2-2*x).*exp(-x.^2-y.^2-x.*y);
   surf(xx,yy,z);
```

the three-dimensional surface of the samples can be drawn, as shown in Figure 5.23. Establish a Simulink model from these samples.

Solutions. If the data are stored in the variables xx, yy and z, where xx and yy are respectively the scale vectors in x and y axes, and z is a matrix. Two Random Number blocks can be used to generate random numbers, for the input signals x and y. A two-dimensional lookup table block is connected, such that the Simulink block diagram shown in Figure 5.24(a) can be constructed. Note that in the two random generator blocks, the seeds cannot be selected the same; otherwise, the two input signal are identical, and the demonstration here is not carried out correctly. The seed of the

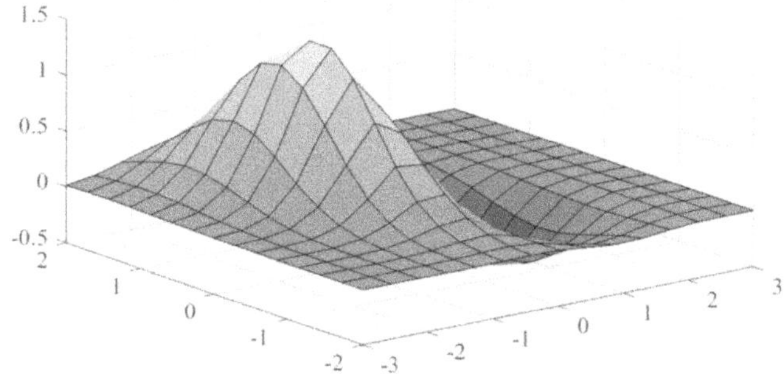

Figure 5.23: The surface plots from the samples.

second random number generator block can be set to 123456, or any other number. Besides, the ranges of the two blocks can be set respectively to (−3, 3) and (−2, 2).

Double click the 2D lookup table block, the parameter dialog box in Figure 5.24(b) van be opened. The three variable names can be specified in the relevant edit boxes, such that the 2D lookup table block can be set up correctly.

Selecting fixed-step algorithm, with a step-size of 0.001 seconds, and stop time of 10. 10 000 points can be generated by simulation, and based on the data, the scattered 3D plot can be obtained as shown in Figure 5.25. It can be seen that the shape of the plots is similar to the given surface plot.

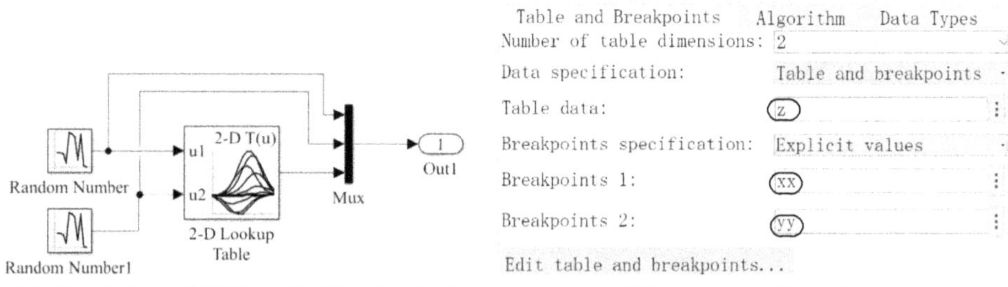

(a) Simulink model (file: c5m2dtab.slx) (b) parameter dialog box

Figure 5.24: Setting and simulation of 2D lookup table block.

```
>> plot3(yout(:,2),yout(:,1),yout(:,3),'.')
```

Under default setting, linear interpolation algorithm is used in the evaluation. If different algorithms are needed, the Algorithm pane in the parameter dialog box of the lookup table blocks can be selected, such that different interpolation algorithms

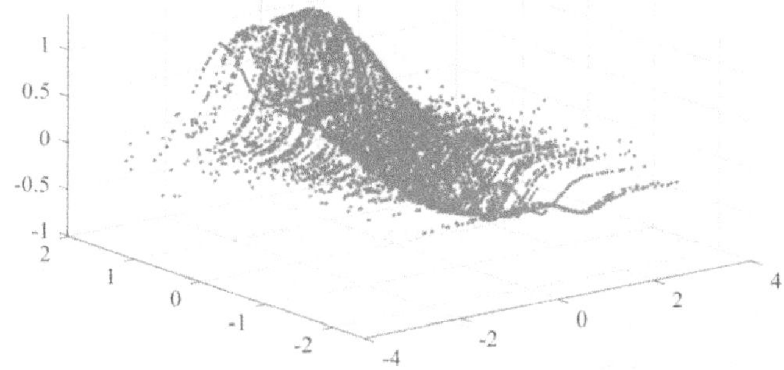

Figure 5.25: Scattered data in simulation results.

are supported. Apart from two-dimensional lookup table blocks, multidimensional lookup table blocks are also provided.

5.4 Automatic sorting of Simulink blocks

Simulink models are composed of blocks. The entire simulation results can be derived by the in-turn execution of the blocks, step by step. The word "in turn" means that the blocks are executed in a certain order. How the orders are decided in the model?

When the Simulink model is completed, its mechanism determines the execution order automatically, by preprocessing the blocks in the model. Generally, the blocks whose current output signal do not depend upon its current input are ranked first. Other blocks are arranged accordingly to the given relationships. The signal source blocks, integrator blocks are ranked on top. Signal source blocks come with no input ports, and its output depends only on its internal time relations, which yield the outputs of the blocks. For integrator blocks, on the other hand, within each simulation step, the output of the block can be evaluated from its initial value only. When the simulation loop is established, the initial values of the integrators can be updated by the input signal afterward, for the output evaluation of the integrator block in the next simulation step.

For automatic sorting of blocks, if the output of the signal source blocks and integrator blocks are available, the computation of other blocks can be computed in turn, according to the signal flow orders in the model, to determine the execution order of all the blocks in the model.

For the blocks such as transfer function and state space models, if they are strictly proper, or $D = 0$ in the state space models, the output of the blocks relies only on the initial values of the integrators in the low-level structure of the block, they can also be ranked top in the sorting process. If the orders of numerator and denominator are equal in transfer function blocks, or $D \neq 0$ in state space models, the output signal of the block is determined also by the input signals, these blocks cannot be ranked in the top positions. If the input signal to the block is known, the output of the block can be evaluated. If the input signal is directly reflected in computing the output signal, the phenomenon is referred to as direct feedthrough blocks, with its `DirectFeedthrough` property set to 1. Block names and properties will be presented in detail later.

In the shortcut menu in a Simulink model window, shown in Figure 5.26, the option Other Displays →Blocks →Sorted Execution Order can be selected to display the execution orders of the blocks. Examples will be given next to show the use of the menu items, and explain the sorting process.

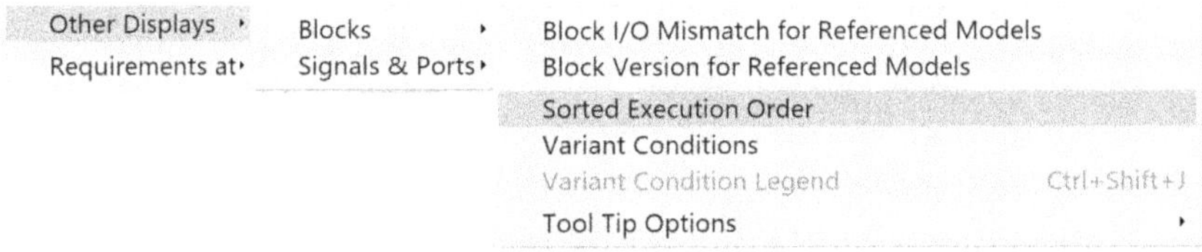

Figure 5.26: Shortcut menu for automatic sorting.

Example 5.19. Display the execution order of the delay differential equation model established in Example 4.23.

Solutions. Open the `c4mdde1.slx` model. In the model window, select the shortcut menu shown in Figure 5.26, the sorted execution order of the model is displayed, as shown in Figure 5.27. Note that the background and foreground colors are set to

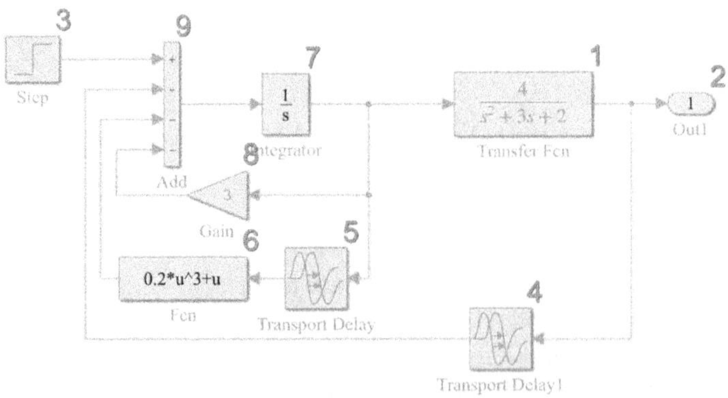

Figure 5.27: The execution order of the blocks.

shallow ones automatically, and some of the foreground colors, as in the cases of the Fcn block, are not changed.

It can be seen from the execution order that, since the Transfer Fcn block is strictly proper, the output of this block is computed directly. Therefore, it is ranked number 1. The signal is fed to Out1 block, therefore, the latter is ranked the second. Besides, Step block is a signal source, and it is automatically ranked third. Since signal 1 is known, the Transport Delay1 block can be computed directly, which is ranked fourth. The Add block should not be considered for the time being, since many of its input signals are not available yet. Now consider the integrator block, with known initial values. The output of the block flows into the Transport Delay and Fcn blocks, and also into the Gain block. Of course, Integrator the execution of the integrator can be carried out directly. Therefore, these blocks can be ranks to 5th to the 8th. Now, all the input signals to the Add block are available, the output of Add can be computed, so that it is ranked the last.

In real computation, the 4th, 5th, 7th and 8th places may not affect the execution results. If the orders of the 3rd and the 5th blocks are swapped, the execution is not affected. Some of the relative orders are important in simulation. For instance, if block 1 is available, block 2 can be computed; if block 5 is available, block 6 can be evaluated; if blocks 3, 4, 6 and 8 are known, block 9 can be worked out. Therefore, these relative orders should be made correctly; otherwise, the simulation results are suspicious. Reliable automatic sorting mechanism are provided in Simulink, and correct results are guaranteed.

Example 5.20. Analyze the automatic sorting the Simulink model for the implicit differential equation in Example 4.17.

Solutions. Open the model file `c4mimp.slx`. Select the above mentioned shortcut menu item discussed earlier, and the automatic sorted result is shown in Figure 5.28.

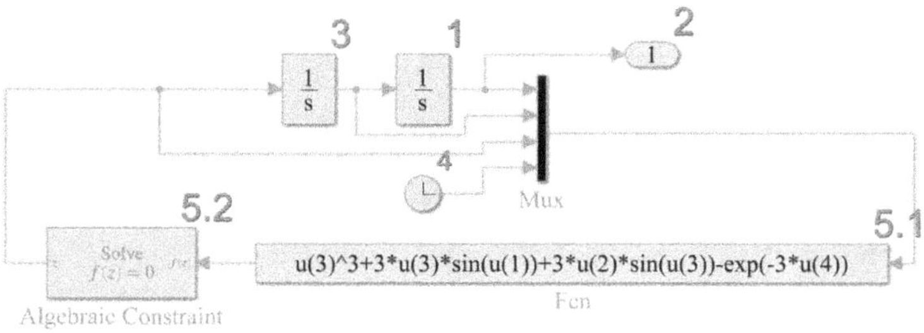

Figure 5.28: Sorting orders in the implicit differential equation model.

It can be seen the sorted result that the integrator blocks are ranked in the front. If the output of integrator is known, it can be used to compute the output port directly. Therefore, the output port block is also ranked top.

Now observe the Fcn block and the Algebraic Constraint block. The relationship of these two blocks are tricky.

(1) To compute first the output of Fcn block, but the input of the Fcn block includes `u(3)` signal, while `u(3)` is the output of the Algebraic Constraint block. Therefore, this order is not either possible.

(2) To compute first the output of the Algebraic Constraint block, but its input is the output of the Fcn block. This order is either possible.

There is a paradox in the computation. Neither of them can be computed first. They must be worked simultaneously. These signals satisfy an algebraic equation, and the phenomenon is referred to as an algebraic loop. More on algebraic loops will be talked about later. To solve the two signals correctly, the algebraic equation should be solved in each step in simulation. Therefore, the two blocks are sorted as 5.1 and 5.2 in Simulink, and the sorting is reasonable.

5.5 Approximate analysis of nonlinear control systems

In traditional nonlinear control theory, since there were no suitable support from simulation tools, the methods for handling nonlinear systems are extremely restricted. Linearization methods are usually adopted, such that the nonlinear systems are approximated in certain small regions as linear systems. Besides, the commonly adopted describing function can also be used for nonlinear elements, so as to find approximate analysis results. If a system has time delay terms, or other irrational terms, Padé approximation techniques can also be adopted, to carry out approximate analysis. In this section, approximate methods for nonlinear systems are discussed, including Padé approximation for delay systems, and linearization methods for ordinary nonlinear systems. The concept of algebraic loops and elimination methods is also studied.

It is worth mentioning that the approximate methods may usually introduce certain errors, and the errors are sometimes very large and may be misleading. The aim of the approximation is to use well-established linear system analysis methods, so as to carry out theoretical analysis. If theoretical analysis is not expected, of course, with powerful tools such as Simulink, accurate analysis of complicated nonlinear systems can be carried out. Also, the accuracies of approximate analysis methods can be assessed with simulation results.

5.5.1 Padé approximation of delay systems

For continuous systems, the pure delay term can be mathematically described as e^{-Ts}, where T is the delay constant. Delay terms cannot be expressed directly using conventional linear terms. In the linearization process, Padé approximation is usually carried out. Padé approximation is named after the French mathematician, Henri Padé (1863–1953), which was originally used to approximate any function with rational expressions. It happens that a rational expression of s is the expected transfer function model.

For a given function $f(s)$, Taylor series expansion can be found such that

$$f(s) = c_1 + c_2 s + c_3 s^2 + \cdots + c_m s^{m-1} + \cdots. \tag{5.5.1}$$

With the Padé approximation method, it is found that

$$G_{r/k}(s) = \frac{\beta_1 + \beta_2 s + \cdots + \beta_r s^{r-1} + \beta_{r+1} s^r}{\alpha_1 + \alpha_2 s + \cdots + \alpha_k s^{k-1} + \alpha_{k+1} s^k}, \tag{5.5.2}$$

where r and k are respectively the orders of the numerator and denominator. If the rational expression reserves the first $r + k + 1$ coefficients of the series, with suitable manipulation, the formulas for finding β_i and α_i can be derived, such that Padé approximation can be found. Function `pade()` is provided in the MATLAB Control System Toolbox, to compute the Padé approximation $G_{r/r}(s)$ of the e^{-Ts} term, with the syntax of G=`pade(`T`,`r`)`. A solver `pade_app()` is written by the author to find Padé approximation of any orders, for any given function. The syntax of the function is G=`pade_app(`$\boldsymbol{c}$`,`r`,`k`)`, where $\boldsymbol{c}$ is the vector containing Taylor series coefficients.[82]

More generally, if G is an LTI object, even though the model contains complicated internal delays, the function `pade()` in the Control System Toolbox can still be used for the approximate model, with the syntax G_1=`pade(`G`,`r`)`, where delay terms are contained in G.

Example 5.21. Consider the transfer function matrix defined in Example 5.7. Assess the fitting quality of second-order Padé approximation.

Solutions. Function `pade()` provided in MATLAB can be used to find the Padé approximation expression not only to pure time delay terms, but also to transfer function matrices. The following commands can be used to draw the step responses of the original transfer function model and its Padé approximation, as shown in Figure 5.29. It can be seen that, apart from $g_{22}(s)$, the approximation quality are satisfactory. The differences cannot be witnessed from the curves. Since the delay of $g_{22}(s)$ term is rather large, there is a small difference at the very beginning, and the total behaviors are acceptable:

```
>> g11=tf(0.1134,[1.78 4.48 1],'ioDelay',0.72);
   g12=tf(0.924,[2.07 1]);
   g21=tf(0.3378,[0.361 1.09 1],'ioDelay',0.3);
   g22=tf(-0.318,[2.93 1],'ioDelay',1.29);
   G=[g11, g12; g21, g22];
   step(G,pade(G,2),'--') %Compare the step responses
```

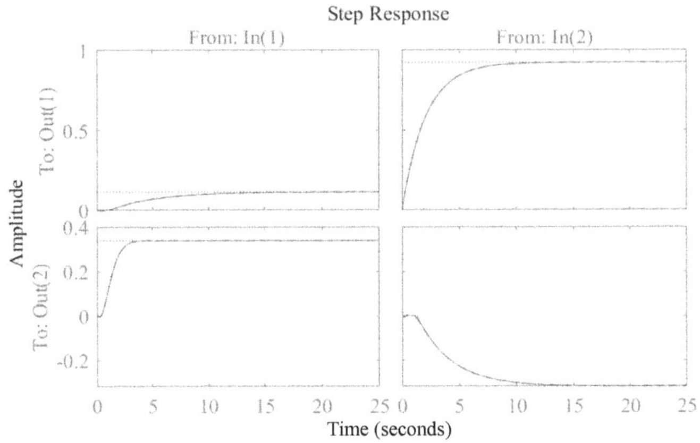

Figure 5.29: Comparisons with Padé approximations.

5.5.2 Linearization of Simulink models

In control theory, the tools for linear system analysis and design are rather complete. Different methods can be used, for instance, time domain and frequency domain methods, to analyze the properties for linear systems. In contrast, apart from simulation methods, only very few methods can be used in the analysis of nonlinear systems. Therefore, linearization approximation at certain points, known as equilibrium points, can be used. Before introducing linearization of nonlinear systems, the concept and solution methods on system equilibrium points are discussed first.

Definition 5.10. If the nonlinear state space model is given by

$$\boldsymbol{x}'(t) = \boldsymbol{f}(\boldsymbol{x}(t)), \tag{5.5.3}$$

and there is no explicit relationship with time t, the system is referred to as an autonomous system.

Definition 5.11. The derivatives $\boldsymbol{x}'(t)$ of the states in the autonomous system in Definition 5.10 can be regarded as the velocities, or understood as the velocity components of the particle in different axes. If all the velocity components are zero, the states of the particle stop. The point when all the state derivatives equal zero is referred to as the equilibrium point.

Theorem 5.5. *The equilibrium point of the autonomous state space model* $\boldsymbol{x}'(t) = \boldsymbol{f}(\boldsymbol{x}(t))$ *can be found directly with the following algebraic equations:*

$$\boldsymbol{f}(\boldsymbol{x}(t)) = \boldsymbol{0}. \tag{5.5.4}$$

The equilibrium point defined here is restricted to autonomous differential equations. For nonautonomous differential equations, the time t is considered, and the number of unknowns is smaller than the number of unknowns. The equilibrium point cannot be found.

For a given Simulink model, if the equilibrium point is expected, an input port and an output port can be added to the system model. With these prepared, the function `trim()` can be called to find directly the equilibrium points:

[$\boldsymbol{x}_0$,$\boldsymbol{u}_0$,$\boldsymbol{y}$,$\Delta\boldsymbol{x}$]=`trim`(model name)
[$\boldsymbol{x}_0$,$\boldsymbol{u}_0$,$\boldsymbol{y}$,$\Delta\boldsymbol{x}$]=`trim`(model name, $\boldsymbol{x}_0$,$\boldsymbol{u}_0$) % specify the initial search points

where the returned variables $\boldsymbol{x}_0$ and $\boldsymbol{u}_0$ are the values of the states and inputs at the equilibrium point. The value $\boldsymbol{y}$ is the output signal value. $\Delta\boldsymbol{x}$ is the increment of the state values at the equilibrium point, which is very close to zero. In the latter syntax, the initial values of the search points for the states and inputs can be specified.

With the equilibrium point $\boldsymbol{x}_0$ and $\boldsymbol{u}_0$, the differential equation can be linearized. In this section, the concept of linearization is introduced, and the solution in MATLAB is then discussed.

Consider a second-order autonomous nonlinear system:

$$\begin{cases} x'(t) = F(x(t), y(t)), \\ y'(t) = G(x(t), y(t)). \end{cases} \tag{5.5.5}$$

The equilibrium point x_0 and y_0 are known. A point in a neighborhood of the equilibrium point is selected as $(x_0+\Delta x, y_0+\Delta y)$, and assume that $|\Delta x| \ll 1$ and $|\Delta y| \ll 1$. It is found that

$$\begin{cases} F(x_0 + \Delta x, y_0 + \Delta y) = \dfrac{\partial}{\partial x}F(x_0, y_0)\Delta x + \dfrac{\partial}{\partial y}F(x_0, y_0)\Delta y + \cdots, \\ G(x_0 + \Delta x, y_0 + \Delta y) = \dfrac{\partial}{\partial x}G(x_0, y_0)\Delta x + \dfrac{\partial}{\partial y}G(x_0, y_0)\Delta y + \cdots, \end{cases} \tag{5.5.6}$$

where the high-order terms of Δx and Δy are omitted, and the constant $F(x_0, y_0)$ and $G(x_0, y_0)$ are also omitted, since at equilibrium points, the values of them are 0. From (5.5.6), it is not hard to find that

$$\begin{cases} \Delta x'(t) \approx \dfrac{\partial}{\partial x}F(x_0, y_0)\Delta x(t) + \dfrac{\partial}{\partial y}F(x_0, y_0)\Delta y(t), \\ \Delta y'(t) \approx \dfrac{\partial}{\partial x}G(x_0, y_0)\Delta x(t) + \dfrac{\partial}{\partial y}G(x_0, y_0)\Delta y(t). \end{cases} \tag{5.5.7}$$

Let $z_1(t) = \Delta x(t)$, $z_2(t) = \Delta y(t)$, the linearized state space model of states $\boldsymbol{z}(t)$ can be obtained

$$\boldsymbol{z}'(t) = \boldsymbol{J}\boldsymbol{z}(t), \tag{5.5.8}$$

where $\boldsymbol{J}$ is the Jacobian matrix, whose mathematical form is

$$\boldsymbol{J} = \begin{bmatrix} \dfrac{\partial}{\partial x}F(x_0, y_0) & \dfrac{\partial}{\partial y}F(x_0, y_0) \\ \dfrac{\partial}{\partial x}G(x_0, y_0) & \dfrac{\partial}{\partial y}G(x_0, y_0) \end{bmatrix}. \tag{5.5.9}$$

Theorem 5.6. *More generally, the linearized model for the ordinary first-order differential equation $\mathbf{x}'(t) = \boldsymbol{f}(\mathbf{x}(t))$ can be linearized as the one in* (5.5.8), *where*

$$\boldsymbol{J} = \begin{bmatrix} \partial f_1/\partial x_1 & \partial f_1/\partial x_2 & \cdots & \partial f_1/\partial x_n \\ \partial f_2/\partial x_1 & \partial f_2/\partial x_2 & \cdots & \partial f_2/\partial x_n \\ \vdots & \vdots & \ddots & \vdots \\ \partial f_n/\partial x_1 & \partial f_n/\partial x_2 & \cdots & \partial f_n/\partial x_n \end{bmatrix}_{(x_1(0),x_2(0),\ldots,x_n(0))}. \tag{5.5.10}$$

Linearization of the differential equation can be evaluated directly with the symbolic computation. Function `jacobian()` in the Symbolic Math Toolbox can be used to find the Jacobian matrix.

If the nonlinear model is described in the Simulink model, and with the input and output ports indicated, function G=`linearize(`model name`)` can be used to directly compute the linearized model G, where G is an LTI object. If no nonlinear elements are contained in the model, simplification to the Simulink model can be carried out, and the LTI object can be found.

Low-level function `linmod2()` can be used to extract the linearization in the Simulink, to find the matrices $\boldsymbol{A}$, $\boldsymbol{B}$, $\boldsymbol{C}$ and $\boldsymbol{D}$ in the linearized model.

`[`$\boldsymbol{A}$`,`$\boldsymbol{B}$`,`$\boldsymbol{C}$`,`$\boldsymbol{D}$`]=linmod2(`model name`,`$\boldsymbol{x}_0$`,`$\boldsymbol{u}_0$`)`

If the system is composed of linear blocks only, the linearized model is in fact the equivalent simplification of the linear model. No model approximation is used in the system. In the linearization function call, the input arguments $\boldsymbol{x}_0$ and $\boldsymbol{u}_0$ can be omitted, since the equivalent model is independent of the model extraction and equilibrium point selection.

For ordinary continuous nonlinear Simulink models, function `linmod2()` is recommended. If the system has delay terms, function `linmod()` is recommended. Padé approximation can be used to fit the delay terms. If there are discrete elements, function `dlinmod()` is recommended, so as to find the discrete state space model.

Example 5.22. Consider the nonlinear model in Example 3.1. Perform linearization to the model and assess the quality of the fitting results.

Solutions. Assume that the width of the saturation nonlinearity is 2, and the width of the dead zone is 0.1, the simulation model in Figure 5.30 can be constructed. This model is different from the one in Figure 3.8, since the input and output ports blocks are added to the model, which is required by linearization process. Therefore, with the following commands, linearization can be made for the original nonlinear system, with the following statements:

```
>> G=linearize('c5mlinr1'), zpk(G)
```

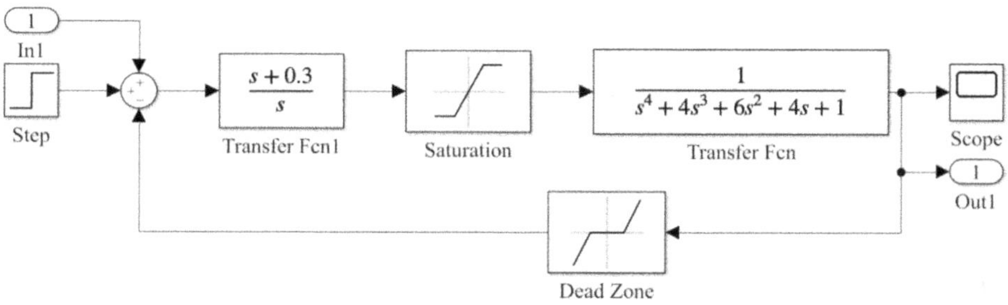

Figure 5.30: Simulation model of the nonlinear system (file: `c5mlinr1.slx`).

The linearized model is given in the state space form, and the mathematical form is too complicated to give here. The converted zero-pole model of the system is written as

$$G(s) = \frac{(s+0.3)}{(s+0.2185)(s^2+0.4498s+0.4221)(s^2+3.332s+3.253)}.$$

To assess the fitting quality, simulation to the original system can be carried out first, then the following commands can be issued. The step responses of the systems are obtained as shown in Figure 5.31. It can be seen that the differences between the original model and the linearized model are rather large:

```
>> G=linearize('c5mlinr1'); G1=zpk(G)
   [y1 t1]=step(G); plot(tout,yout,t1,y1,'--')
```

If the dead-zone nonlinearity become wider, to ±0.3, there is no change in the linearized model. The step responses of the two systems are shown in Figure 5.32. It can be seen that when the dead zone becomes wider, in other words, the degree of nonlinearity becomes higher, the fitting of the linearized models become worse. Therefore, the linearized model cannot be used to describe the properties of the original system well.

It can be seen from the above example that linearization may lead to very large errors. The sizes of the errors depend heavily on the degree of nonlinearity. The higher

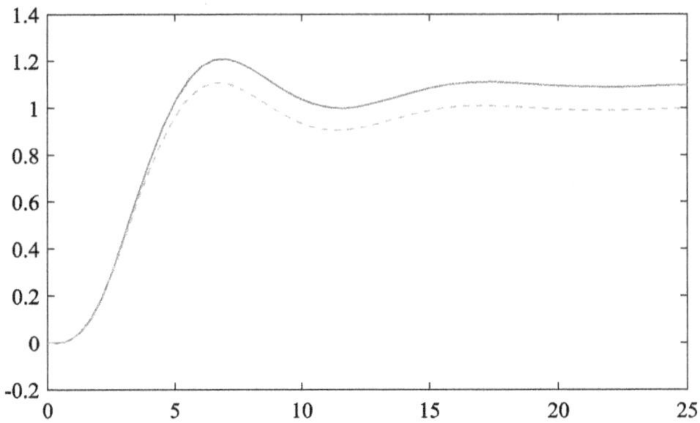

Figure 5.31: Step response fitting of nonlinear systems.

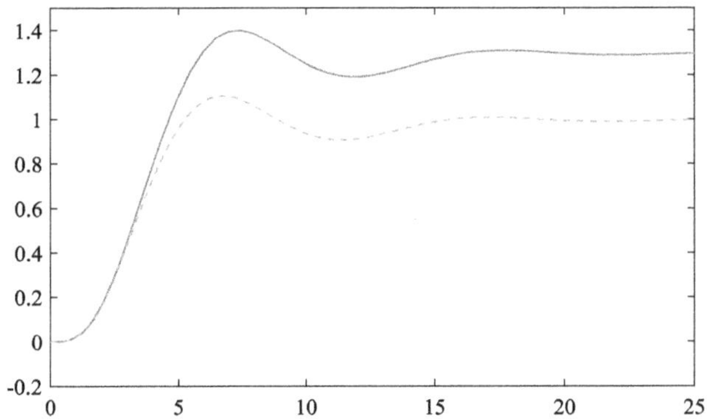

Figure 5.32: Step response approximation when the dead-zone becomes wider.

the degree of nonlinearity, the larger the error. Considering from numerical simulation, care must be taken in the use of linearization, or not using the linearization technique.

Example 5.23. Consider the composite transfer function model in Example 2.7. Use the linearization method to find the approximate linear model, and compare the fitting quality of original and approximate models via step responses.

Solutions. Since the linearization process is needed, an input port and an output port can be added to the model. Since step responses are expected, a step input block can also be added to excite the system. Considering the two factors, the Simulink model in Figure 5.33 can be established.

If the delay model for linearization is expected, double click the delay block, and fill in the Padé order edit box the values such as 2 or 3, and the following commands can be used to find the linearized models. The step responses can be obtained as shown in Figure 5.34, where the original model is given in solid lines, second-order

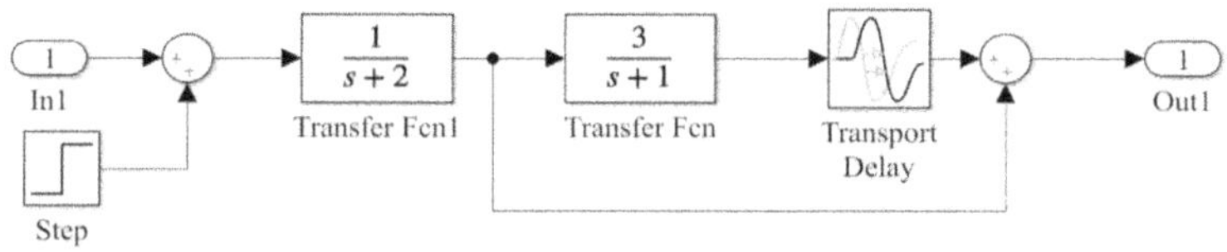

Figure 5.33: Simulation model of the composite model (file: `c5mlinr2.slx`).

approximation model in a dotted line, while third-order model is in dashed lines. It can be seen that the fitting is accurate, and third-order approximation is even more accurate than the second-order one:

```
>> G=linearize('c5mlinr2'); G1=zpk(G)
   [y1 t1]=step(G); plot(tout,yout,t1,y1,'--')
```

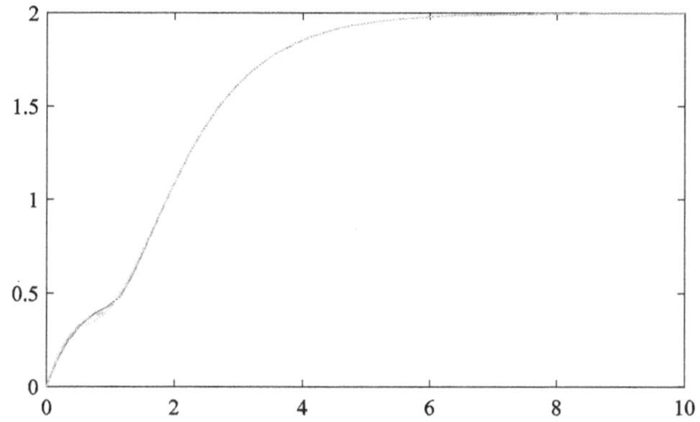

Figure 5.34: Comparison of the models.

The second- and third-order Padé approximation can be found as follows: There is no delay terms is the reduced models:

$$G_2(s) = \frac{(s+10.44)(s^2-0.4404s+4.598)}{(s+2)(s+1)(s^2+6s+12)},$$

$$G_3(s) = \frac{(s^2-0.4289s+4.443)(s^2+10.43s+108)}{(s+4.644)(s+2)(s+1)(s^2+7.356s+25.84)}.$$

In fact, for the linear models, if the linearization methods are used, the same linearized models can be found, and the results are the same:

```
>> s=tf('s'); G=(1+3*exp(-s)/(s+1))/(s+2);
   G1=zpk(pade(G,2)), G2=zpk(pade(G,3))
```

5.5.3 Algebraic loop phenomenon and elimination

The phenomenon of algebraic loops was demonstrated in Example 5.20. An algebraic loop means that in a certain model the input signals of one or many blocks depend on their own output signals. Therefore, in each simulation steps, algebraic equations should be solved once. Since algebraic equation solutions are rather time consuming, the computational load may be increased if there are algebraic loops in the model.

In the control parameter setting dialog box, if the **Diagnostics** item is selected, the first subitem is the **Algebraic loop**, which has three options, **none**, **warning** and **error**. The default one is a **warning** option. If an algebraic loop is detected, the warning message can be given, but the simulation process is not terminated. This is the recommended option. If the **error** option is selected, the simulation process is terminated, if an algebraic loop is detected. An error message is displayed. If the **none** option is chosen, algebraic loops are ignored. Algebraic equations can be solved automatically without giving any messages.

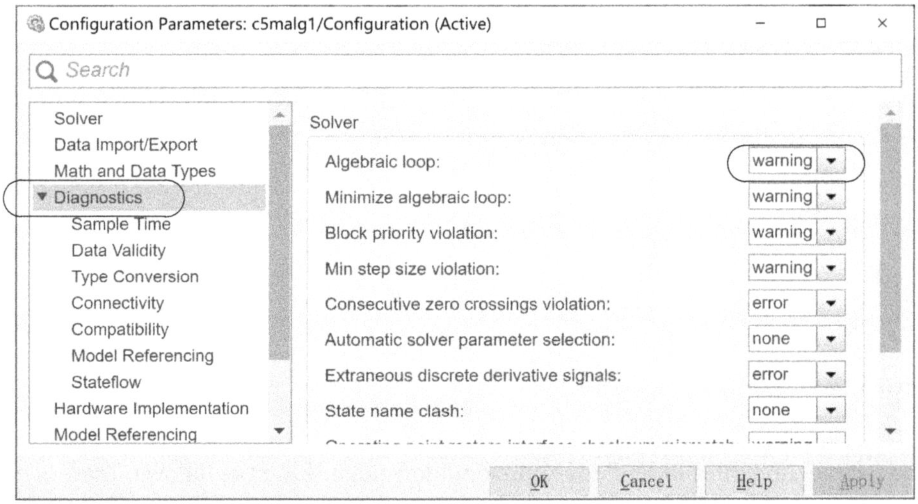

Figure 5.35: Diagnostic dialog box.

An example is given next to show algebraic loop problem, and then explore several methods to eliminate algebraic loops, and introduce an effective solution methods.

Example 5.24. Consider a simple linear feedback system shown in Figure 5.36(a). Set up a Simulink model to observe the algebraic loops.

Solutions. For this particular model, since the orders of the numerator and denominator in the open-loop model are the same, to compute signal output signal $y(t)$, the input signal $u(t)$ must be known. To compute $u(t) = r(t) - y(t)$, signal $y(t)$ is also

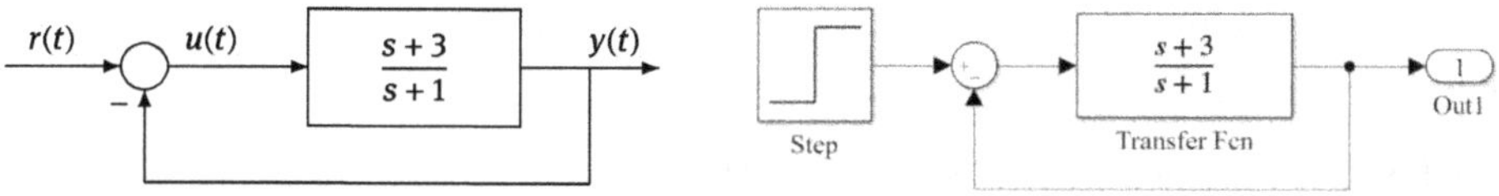

(a) feedback system (b) simulation model (file: `c5malg1.slx`)

Figure 5.36: Simple feedback control system models.

needed. A paradox is demonstrated here, which is referred to as the algebraic loop. In the simulation process, if there is an algebraic loop, an algebraic equation must be solved once in each simulation step, so as to find the signals $u(t)$ and $y(t)$, satisfying the algebraic equation.

A Simulink model for the system can be drawn, as shown in Figure 5.36(b). Perform simulation to such a system, and the following warning message can be displayed in the command window:

Warning: Block diagram '`c5malg1`' contains 1 algebraic loop(s). To see more details about the loops use the command:

```
Simulink.BlockDiagram.getAlgebraicLoops('c5malg1')
```

or the command line Simulink debugger by typing `sldebug('c5malg1')` in the MATLAB command window. To eliminate this message, set the algebraic loop to "none."

The message indicates that there is an algebraic loop. Different commands can be tried to analyze the algebraic loop. If no display is needed to display algebraic loop, the **Algebraic loop** listbox in Figure 5.35 must be set to none.

Example 5.25. Use theoretical method to eliminate the algebraic loop in Example 5.24.

Solutions. In fact, if a negative feedback system formula is used to find the equivalent model, the equivalent transfer function model can be derived directly from

$$G_1(s) = \frac{G(s)}{1+G(s)} = \frac{(s+3)/(s+1)}{1+(s+3)/(s+1)} = \frac{s+3}{2s+4}.$$

Therefore, the **Step** block can be used to drive $G_1(s)$ model, to eliminate completely the algebraic loop.

Although algebraic loops are avoided perfectly, the suitability of this method is not acceptable, since very few problems can be simplified in this way. For instance, if there is nonlinearity in the system, this method cannot be used. Other general purpose methods should be explored.

The reason why algebraic loop exists is that some key input and output signals involved are at the same time. If the two related signals are happening at different times, the algebraic loop is expected to be avoided. One obvious method is to append a delay block, and feed its output signal to the input of another block. If the delay constant is small enough, the algebraic loop is expected to be avoided. An alternative method is that to append a low-pass filter, $1/(Ts+1)$, to avoid feedthrough phenomenon. If T is small enough, algebraic loops may be avoided.

Example 5.26. Introduce a small delay term e^{-Ts} to eliminate the algebraic loop in the system in Example 5.24, and assess the approximation results.

Solutions. Append a delay block in the transfer function block in Figure 5.36(b), the modified Simulink model can be constructed as shown in Figure 5.37.

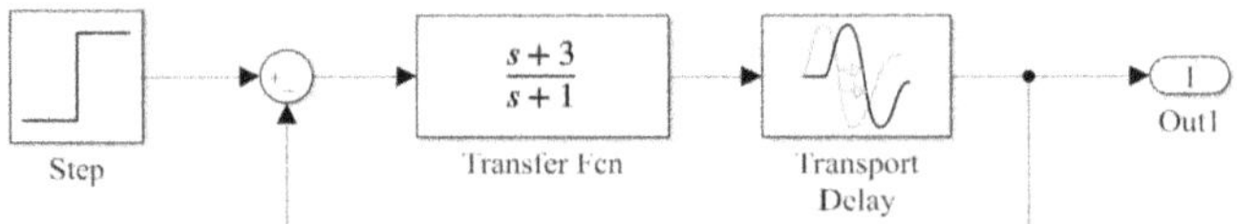

Figure 5.37: With a delay term introduced (file: `c5malg2a.slx`).

If the delay constant is set to 0.01, the algorithm can be set to `ode15s`, and error tolerance is set to 10^{-8}. The simulation result obtained is shown in Figure 5.38. The simulation process is quite time consuming, and extremely large errors are introduced in the simulation results. It is obvious that although the algebraic loop is avoided, new problems are introduced, so that it is not suitable in real simulation.

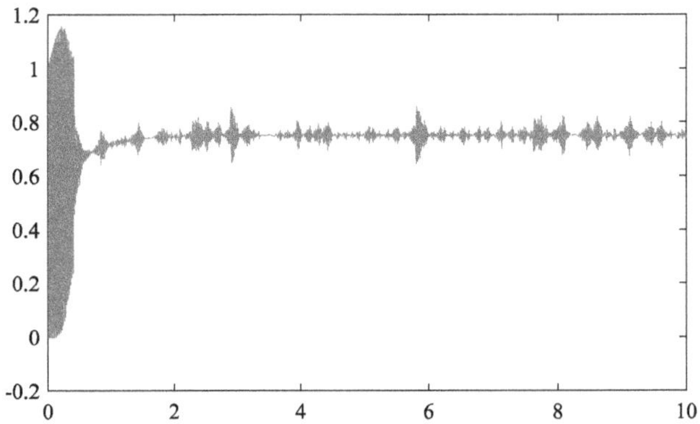

Figure 5.38: Numerical solutions (erroneous results).

Example 5.27. Introduce a low-pass filter to eliminate the algebraic loop in Example 5.24, and assess the approximation results.

Solutions. From the given equivalent transfer function model, the differential equation of the system can be derived:

$$2y'(t) + 4y(t) = u'(t) + 3u(t), \quad y(0) = 0.5,$$

where $u(t)$ is a step function, or Heaviside function. In this case, the following commands can be used to find the analytical solution of the differential equation:

```
>> syms t y(t); u=heaviside(t);
   y0=dsolve(2*diff(y)+4*y==diff(u)+3*u, y(0)==0.5)
   latex(simplify(y0))
```

the analytical solution obtained of the output signal is

$$y_0(t) = \frac{1}{8}e^{-2t} + \frac{3}{8}\,\mathrm{sign}(t) - \frac{3}{8}e^{-2t}\,\mathrm{sign}(t) + \frac{3}{8}.$$

If the transfer function in Figure 5.36(b) is connected, a transfer function block, with numerator of 1, and denominator set as `[T,1]`, the Simulink model shown in Figure 5.39 can be created.

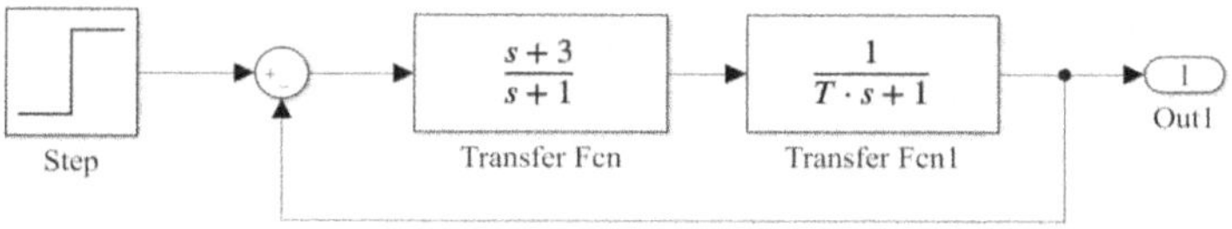

Figure 5.39: With a low-pass filter introduced (file: `c5malg2b.slx`).

Selecting $T = 0.001$, and start simulation, the simulation results can be obtained, together with the theoretical results, as shown in Figure 5.40. No difference can be observed from the displayed curves:

```
>> T=0.001; tic, [t,~,y1]=sim('c5malg2b'); toc
   y0=exp(-2*t)/8+3*sign(t)/8-3*exp(-2*t).*sign(t)/8+3/8;
   plot(t,[y1 y0])
```

In fact, at initial stages, the two models have prominent differences. Since in the original model, there is direct feedthrough phenomenon, the input signal is reflected in the output signal at $t = 0$. The initial value of the output is 0.5. If low-pass filter is used, the direct feedthrough phenomenon is slide over. The output signal is changed from 0 to 0.5 in a very short period of time. The subsequent responses are very close to theoretical ones. The stage is extremely short. If T is small enough, the period can be totally neglected. Apart from the short period, the maximum error is 4.6089×10^{-5}, and the number of points is 6690:

```
>> ii=t>1e-5; max(abs(y(ii)-y1(ii))), length(t)
```

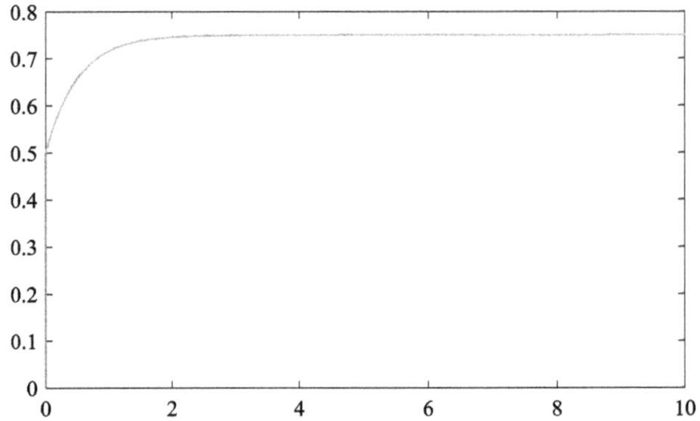

Figure 5.40: Comparisons of time domain responses.

Reducing the value T, the simulations are compared, as shown in Table 5.1. If no low-pass filter is used, that is, the direct feedthrough column, although in presence of an algebraic loop, yields the most accurate one, and the time elapse is larger. In normal cases, the algebraic loops can be neglected, and let Simulink to solve automatically the algebraic equations. In certain cases, when Simulink solutions meet problems, low-pass filters can be introduced to eliminate algebraic loops. It can be seen that, if T is small enough, the approximation is rather satisfactory.

Table 5.1: Simulation results for different values of T.

filter constant T	direct feedthrough	0.001	0.0001	10^{-6}	10^{-10}
maximum error	1.6875×10^{-14}	4.6089×10^{-5}	4.5995×10^{-6}	4.5985×10^{-8}	4.4973×10^{-12}
number of points	823	6690	60675	1353	1128
time elapse	0.4133	0.1722	0.5211	0.1640	0.1961

For a complicated system with algebraic loops, low-pass filters can be introduced to avoid the algebraic loop problems. Besides the introduced approximate methods, the Simulink solution algorithms similar to direct feedthrough can be used. By default, trust region algorithm is adopted in Simulink. If the problem cannot be solved successfully, the algebraic loop problems, line search algorithm can be tried, by setting

```
set_param(model name,'AlgebraicLoopSolver','LineSearch')
```

5.6 Exercises

5.1 A Simulink model `c5mlinr1.slx` for control system is designed for Example 5.22. Judge by yourself the execution orders of the blocks. Use Simulink automatic sorting facilities to validate the sorting orders.

5.2 For the nonlinear control system with backlash nonlinearity shown in Figure 5.41, establish a Simulink model, and draw step responses with different magnitudes.

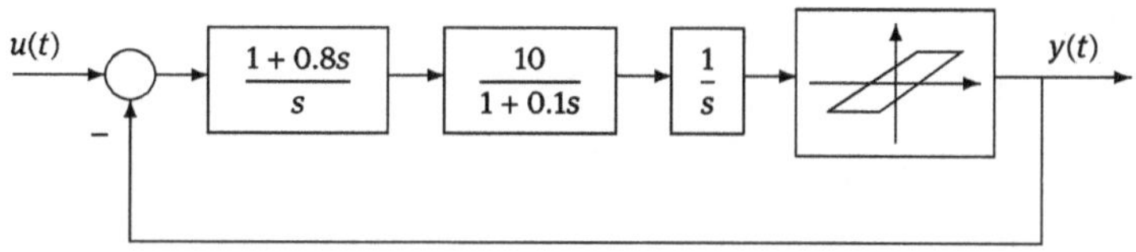

Figure 5.41: Nonlinear system model in exercise 2.

5.3 Draw the Simulink models for the nonlinear feedback control systems in Figures 5.42(a) and 5.42(b).[8] Draw the step response and explore the impact of the nonlinear parameters on the simulation results.

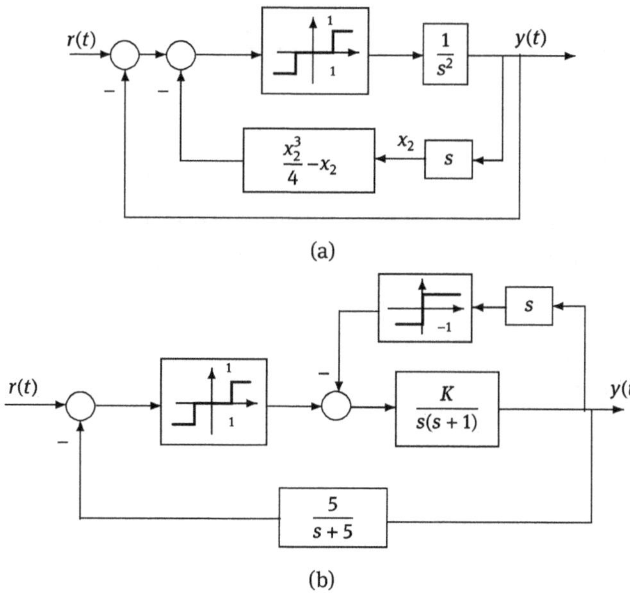

Figure 5.42: Systems for exercise 3.

5.4 Establish the Simulink model for the nonlinear system shown in Figure 5.43.[71] Observe the output and error signals under unit step input signals. Since there are series connection of two nonlinear blocks, explain with simulation method, whether the two nonlinearities can be swapped.

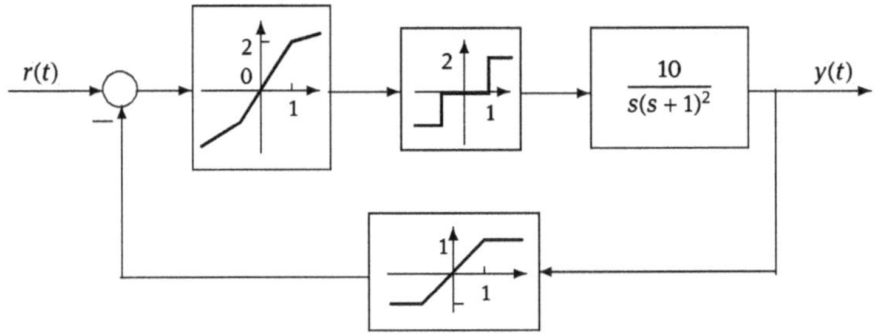

Figure 5.43: System model for exercise 4.

5.5 Establish the Simulink model for the nonlinear system in Figure 5.44. Find its equilibrium point and linearized model. Compare the step responses of the original model and the linearized one.

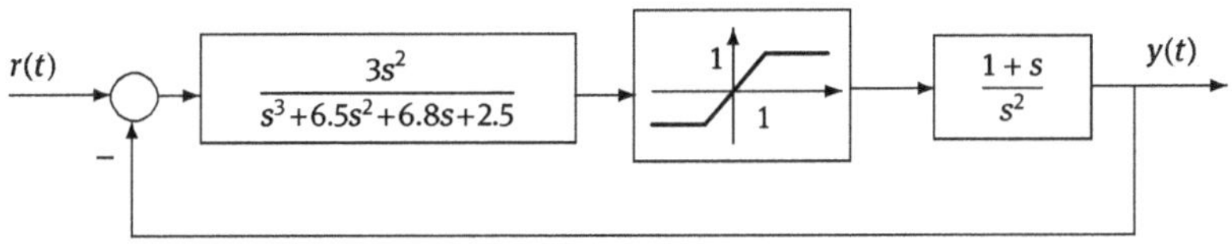

Figure 5.44: Nonlinear system model in exercise 5.

5.6 Establish the Simulink model for the nonlinear system given in Figure 5.45.[45] Observe the output and error signals under unit step input signals.

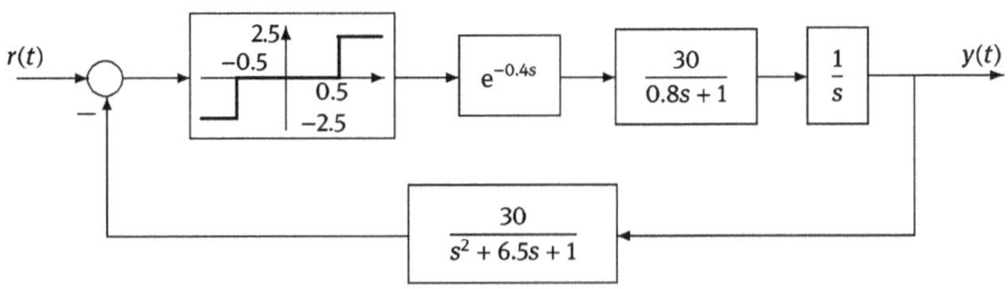

Figure 5.45: Nonlinear system in exercise 6.

5.7 If a Simulink model is given in Figure 5.46, write out its mathematical expression from the given block diagram.

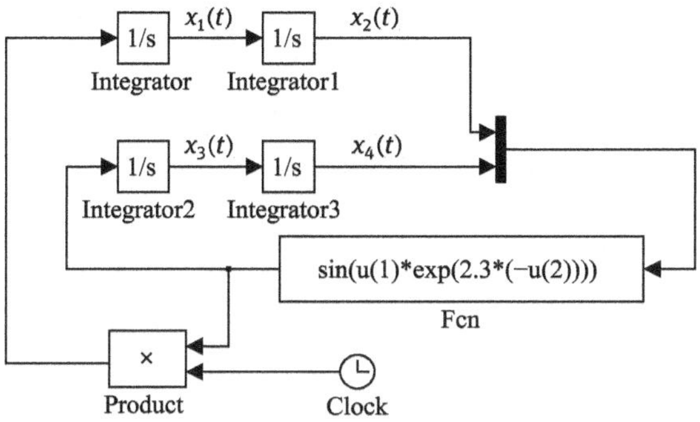

Figure 5.46: Simulink model in exercise 7.

5.8 Assume that the linear differential equation is given by

$$\begin{cases} x_1'(t) = -x_1(t) + x_2(t), \\ x_2'(t) = -x_2(t) - 3x_3(t) + u_1(t), \\ x_3'(t) = -x_1(t) - 5x_2(t) - 3x_3(t) + u_2(t), \end{cases}$$

and $y(t) = -x_2(t) + u_1(t) - 5u_2(t)$, where the two input signals are $u_1(t)$ and $u_2(t)$. Express the model with Simulink using two methods.

5.9 For linear transfer function model $G(s) = \dfrac{6(s+1)}{(s+2)(s+3)(s+4)}$, if the Gaussian white noise signal $y(t)$ is used to excite the model, find the probability density function of the output signal, and compare it with the theoretical results. Also, find the autocorrelation function and spectral function, and compare them with theoretical results.

5.10 Express the transfer function matrix with Simulink:

$$\boldsymbol{G}(s) = \begin{bmatrix} \dfrac{0.806s + 0.264}{s^2 + 1.15s + 0.202} & \dfrac{-15s - 1.42}{s^3 + 12.8s^2 + 13.6s + 2.36} \\ \dfrac{1.95s^2 + 2.12s + 0.49}{s^3 + 9.15s^2 + 9.39s + 1.62} & \dfrac{7.15s^2 + 25.8s + 9.35}{s^4 + 20.8s^3 + 116.4s^2 + 111.6s + 18.8} \end{bmatrix}.$$

5.11 Consider the nonlinear system model in Figure 5.47. If the input stochastic signal $y(t)$ is a Gaussian white noise signal, with 0 mean and variance of 4. Find the probability density function of the error signal $e(t)$ with simulation methods.

5.12 If the plant and controller models are respectively

$$G(s) = \frac{e^{-s}}{s(s+1)^4}, \quad G_c(s) = 0.214 + \frac{0.0049}{s} + 0.793s.$$

Construct a Simulink model, and find its Padé approximation, such that the closed-loop model can be approximated well with Padé approximation.

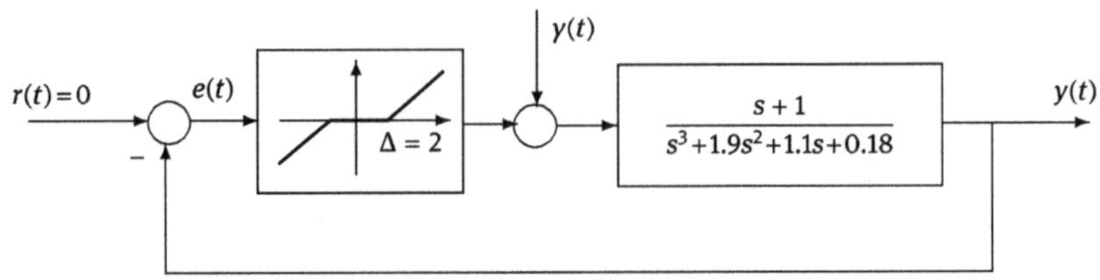

Figure 5.47: Nonlinear system in exercise 11.

5.13 For the given inverse pendulum mathematical model,

$$\begin{cases} y'' = \dfrac{f/m + l\theta^2 \sin\theta - g\sin\theta\cos\theta}{M/m + \sin^2\theta}, \\ \theta'' = \dfrac{-f\cos\theta/m + (M+m)g\sin\theta/m - l\theta^2\sin\theta\cos\theta}{l(M/m + \sin^2\theta)}, \end{cases}$$

where θ is the angle of the pendulum and vertical line, in radians; y is the displacement of the cart, in meters; f is the force to the cart, in Newton; M and m are respectively the masses of the cart and pendulum, in kilograms; l is half of the length of the pendulum, in meters; g is the gravity accelerator of 9.81 m/s^2. Establish a Simulink model for the inverted pendulum. If $m = 0.21$ kg, $M = 0.455$ kg, $l = 0.61/2$ m and f is the input signal to the system. Find the equilibrium point for $y = \theta = 0$, and find out the linearized model. Compare the step responses of the original model and the linearized model.

5.14 In [23], a benchmark problem was presented based on F-14 aircraft model. The problem was used to assess the facilities of control system simulation languages. A Simulink implementation is provided in MATLAB, as shown in Figure 5.48. Find the equivalent linear model from input port 1 to output port 1. Find all the zeros and poles.

5.15 For the nonlinear system model shown in Figure 5.49, create a Simulink model, and find the equilibrium point under step input, and find the linearized model.

5.16 The block diagram of a model reference adaptive control system is given in Figure 5.50,[75] where the parameters of the model and controller are $b_0 = 0.5$, $a_1 = 0.447$, $a_2 = 0.1$, $a_3 = 2$, $d_0 = 1$, $d_1 = 0.5$, $k_1 = 0.03$, $k_2 = 1$, $\hat{b}_0(0) = 0.2$. Two product blocks and related operations are used to adjust the gain $\hat{b}_0(t)$ of the adaptive controller, such that the output of the system follows well the output of the reference model. Based on the give block diagram, establish a Simulink model. Use a square wave to excite the system, draw the response of the system and draw also the curve of the adaptive gain $\hat{b}_0(t)$.

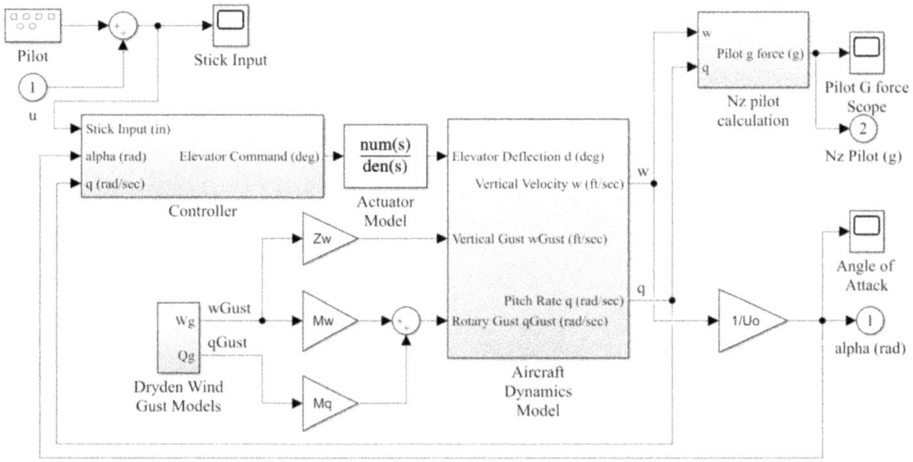

Figure 5.48: Simulink model for F-14 aircraft (file: `f14.slx`).

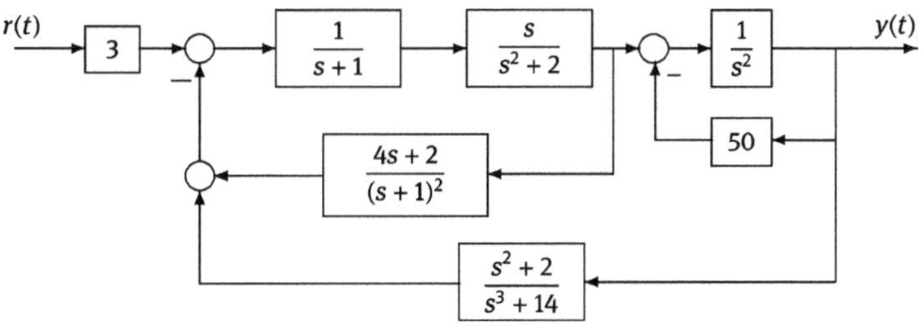

Figure 5.49: Diagram of exercise 15.

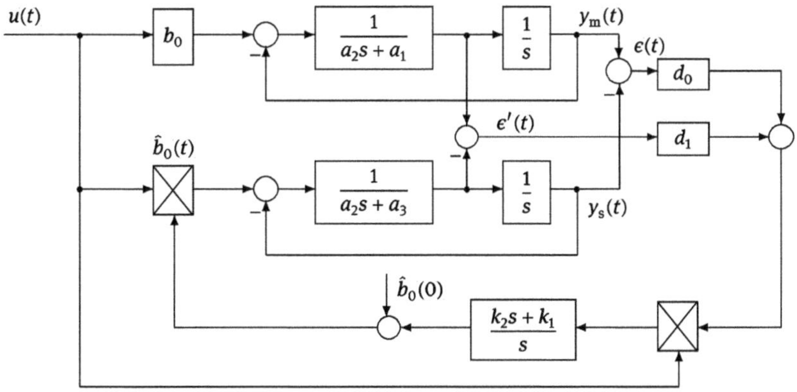

Figure 5.50: Model reference adaptive control system.

6 Subsystems and block masking

Significant amount of blocks are provided in the current versions of Simulink. These blocks can be used directly in constructing simulation block diagrams of complicated systems. In certain research fields, of course, these blocks are far from sufficient. More suitable and practical blocks should be established by the users in their own specialties, so as to make system modeling more simple and straightforward. The concepts and block masking methods are essential for the users to develop their own blocks. In this chapter, the fundamental knowledge on these topics will be systematically presented. The methods of creating Simulink models with MATLAB commands are also presented in this chapter.

In Section 6.1, the fundamental concept and modeling methods of subsystems are presented. The methods of creating and extracting subsystems are illustrated first. The use of simple subsystem models are presented, and the ways to avoid complicated crossover signal lines are also addressed. The method of embedding Simulink files in models are also presented. In Section 6.2, modeling of complicated subsystems are presented, and the applications of enabled and triggered subsystems are demonstrated. Introductions are also given to conditional subsystems, loop subsystems and so on. In Section 6.3, the concepts of block masking are presented, and block masking design methods are illustrated. Icon design of masked blocks is demonstrated first, then the parameter dialog box design and block initialization are presented. In Section 6.4, the Simulink model plotting with MATLAB commands are illustrated. The plotting methods are suitable for complicated system modeling, if certain pattern of the system is regularly organized. In this section, the masking of subsystem blocks with MATLAB commands are also presented. In Section 6.5, the MATLAB commands are presented to control the Simulink process. The acquiring and setting methods for Simulink model parameters with MATLAB commands are illustrated. Parallel simulation methods are presented, and the efficiencies of parallel simulation are assessed with examples.

6.1 Subsystem modeling

In system modeling and simulation, sometimes the system constructions are complicated, and it is difficult to present the whole system block diagram with a single Simulink model. Usually, such block diagrams can be decomposed into several subsystems with independent functions. In Simulink, the subsystem structures are supported. Besides, the users may mask some commonly used subsystems into reusable blocks, and the use of these blocks are the same as standard Simulink blocks. For generality, the users may develop a series of blocks to build their own block groups or blocksets. In this section and the next one, subsystem construction and application are presented. The structures of enabled and triggered subsystems are discussed.

https://doi.org/10.1515/9783110734959-006

6.1.1 Construction and extraction of subsystems

To create a new subsystem, the input and output ports of the subsystem should be assigned. The input port of the subsystem can be expressed by the In block in the Sources group, while the output port can be expressed by the Out block from the Sinks group. These blocks are also provided in the Ports & Subsystems group, and they are exactly the same. In between the input and output ports, the internal structure of the model can be arbitrarily designed. Besides, some blocks or structure can be extracted from an existing complicated model, such that a new subsystem model can be created. In this section, subsystem modeling and extraction will be demonstrated with examples.

PID controller is the most widely used controller type in industry.[5] Powerful and flexible PID controller blocks are provided in the Continuous group in Simulink, in the blocks such as PID Controller and 2DOF PID Controller. Here, a low-level PID controller model is established, to demonstrate the subsystem creation and block masking methods.

Definition 6.1. The mathematical form of a PID controller is

$$G_c(s) = K_p + \frac{K_i}{s} + \frac{K_d s}{K_d s/N + 1}, \tag{6.1.1}$$

where K_p, K_i and K_d are PID controller parameters, while N is the filter constant to approximate PID controllers. Normally, one may chose $N = 10, 100$ or 1000. This type of PID controller is also known as the parallel PID controller.

Example 6.1. Set up a low-level PID controller model, and then create a subsystem block from it.

Solutions. There is one input and one output in a PID controller. The transfer function model $G_c(s)$ is in between the two ports. Therefore, the low-level model for the PID controller can be constructed, as shown in Figure 6.1(a).

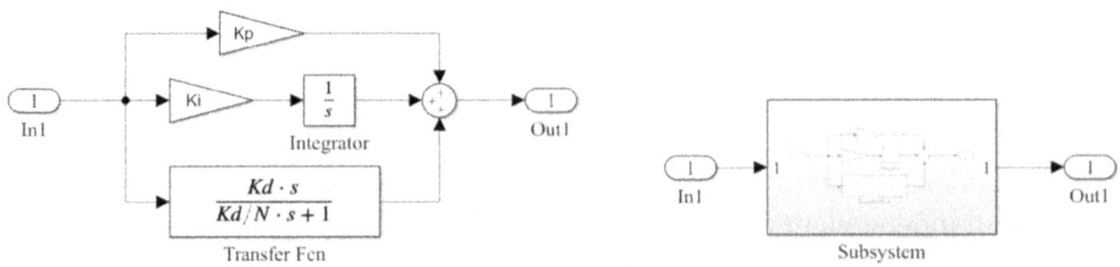

(a) PID controller model (file: c6mpid1.slx) (b) subsystem model (file: c6mpid1a.slx)

Figure 6.1: PID controller and its subsystem models.

Selecting all blocks, or press Ctrl+A keys in the model window, the shortcut menu item Create Subsystem from Selection can be selected, to create a subsystem block, shown in Figure 6.1(b). It can be seen that the sketch of the internal structure of the subsystem block is displayed in shallow color, as the icon. This is helpful in Simulink modeling and maintaining.

Double click the subsystem block, a new model window is presented, and the internal structure of the subsystem in Figure 6.1(a) is displayed. The users are allowed to edit or modify the subsystem model in the model window.

It can be seen from the previous presentation that since Create Subsystem from Selection shortcut menu allows to extract subsystem models from the selected blocks, the facility can be used to extract part of the blocks from the existing Simulink model, to create subsystem block.

Example 6.2. In the control system model `c3mpid1.slx` in Example 3.4, extract the PID controller, and convert it into a subsystem.

Solutions. Open the Simulink model `c3mpid1.slx` first. The mouse drag action can be used to select the blocks in the PID controller, and the saturation nonlinearity, as shown in Figure 6.2(a). It can be seen that the blocks selected are displayed in different colors, to show the differences from the other unselected blocks.

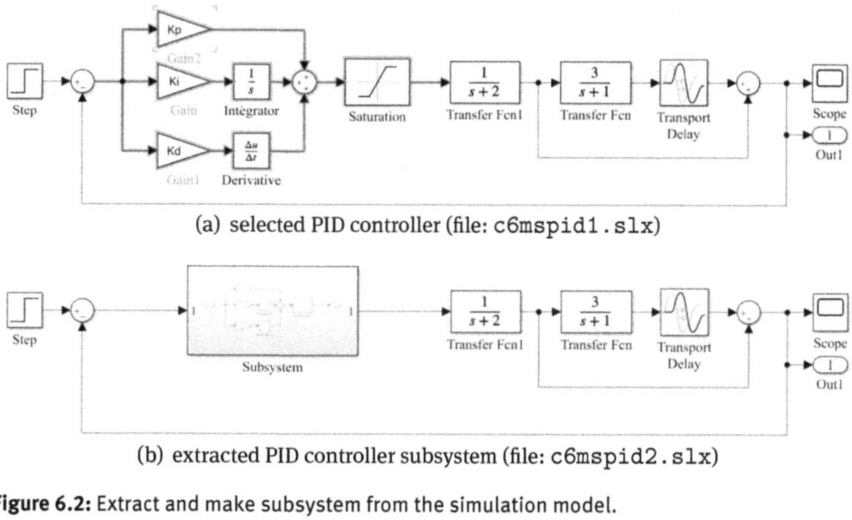

(a) selected PID controller (file: `c6mspid1.slx`)

(b) extracted PID controller subsystem (file: `c6mspid2.slx`)

Figure 6.2: Extract and make subsystem from the simulation model.

Select the related blocks and constructions, and then select Create Subsystem from Selection shortcut menu item, the simulation model is changed to the form shown in Figure 6.2(b). It can be seen that the selected blocks are made into a single subsystem block, whose internal structure is again shown on the icon.

With the subsystem model constructed, select the subsystem block, then select **Subsystem & Model Reference →Expand Subsystem** shortcut menu item, the subsystem block is expanded directly, and the model in Figure 6.2(a) is restored.

With similar method, the plant model can also be made into a subsystem block, shown in Figure 6.3. It can be seen that with subsystem building method, the structures of complicated system can be made concise. To modify a certain subsystem, double click the block, and the maintain and manipulation of the Simulink model are made simpler.

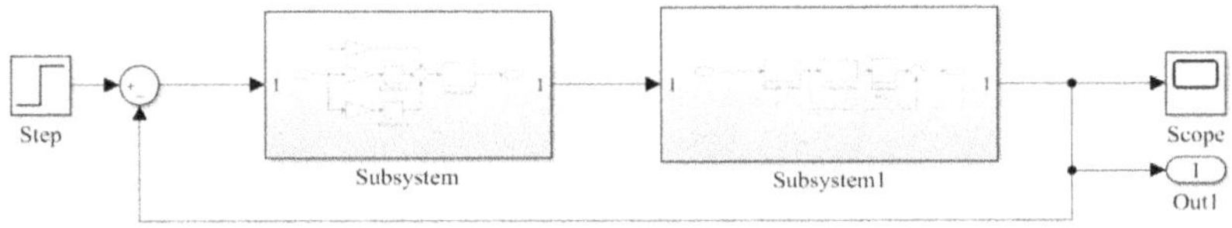

Figure 6.3: The plant is also made a subsystem block (file: `c6mspid3.slx`).

It can be seen that the subsystem with multiple-input and multiple-output can be extracted automatically. The extraction of multivariable subsystem is presented next through an example.

Example 6.3. Consider again the control system model `c3mpid1.slx` in Example 3.4. If the controller and sum blocks are selected simultaneously, extract the subsystem model.

Solutions. Select the blocks in model as shown in Figure 6.4(a), then construct the subsystem. The Simulink model can be constructed as shown in Figure 6.4(b). Note that the subsystem block is constructed automatically with two input ports, In1 and In2, as expected. The first input to the subsystem is the step input signal, while the second one is the output signal of the whole system.

6.1.2 Setting format in crossing lines

Sometimes the signal lines in a complicated Simulink model may be too complicated, and there are too many crossing signal lines, which makes the system look a mess. The understanding and maintaining of the simulation model are complicated. For instance, in large-scale systems, if the output signal at the right most block is to be fed into a block on the left, a long signal line must be drawn, and the signal line may travel across the whole model. If the signal lines in the middle are densely distributed, the lengthy signal line may crossover many signal lines in the middle. To avoid the crossover signal lines, the output line can be connected to the Goto block, while the From block can be used to feed to the left block. Therefore, there is no need to really to draw crossover lines. Of course, the pair of blocks must be matched, to ensure that

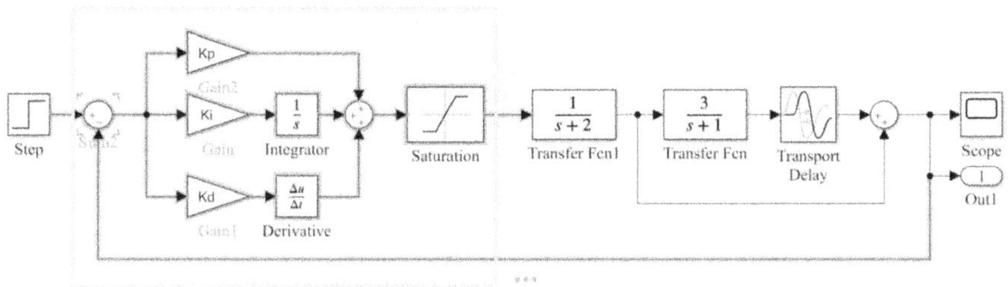

(a) selected PID controller model

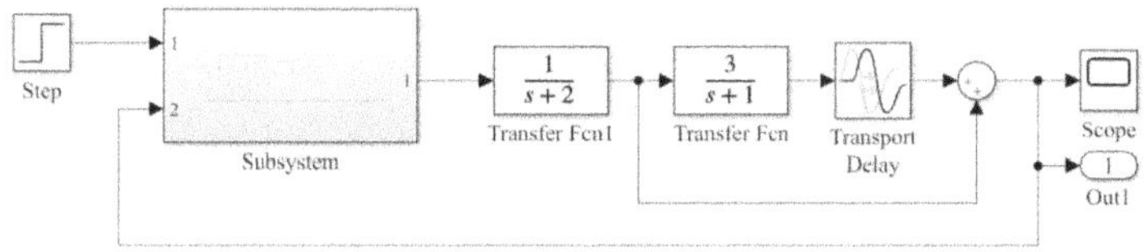

(b) subsystem with two input ports (file: `c6mspid4.slx`)

Figure 6.4: Select blocks again to make a subsystem.

the two blocks can be connected up. Examples are used to demonstrate the linking methods.

Example 6.4. Consider the simulation model in Figure 6.3. If the feedback path is to be linked in this manner, set up a new simulation model.

Solutions. Open the simulation model `c6mspid3.slx` in Figure 6.3. Delete the feedback signal line. Find and copy the **Goto** and **From** blocks in the **Signal Routing** group to the model, and set up a new simulation model, as shown in Figure 6.5.

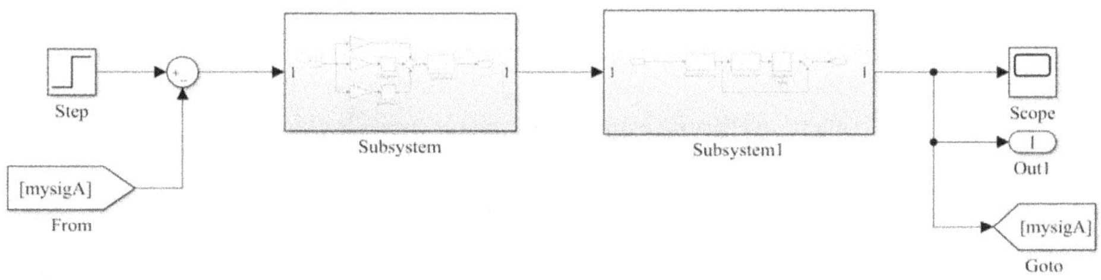

Figure 6.5: Modified model (file: `c6mspid5.slx`).

Double click **Goto** block, a parameter dialog box shown in Figure 6.6 is displayed. The **Goto tag** edit box can be modified. For instance, set it to `mysigA`. In fact, it can be set to any string, while it must be made unique in the same system model. Double click the **From** block, and a similar parameter dialog box can be opened, and the **From tag** edit box must be set to the same tag. Therefore, the two blocks are automatically linked up. Of course, for such a simple problem, there is no need to use the indirect linking method. This method can be used to handle the connections of more complicated models.

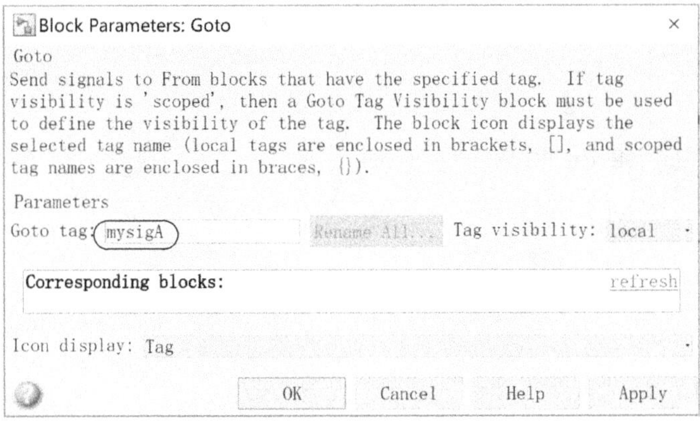

Figure 6.6: Tag setting dialog box.

Example 6.5. Consider the tracking and differentiator model in Example 4.31. Since the low-level block diagram is too complicated, and there are too many crossover signal lines, it is rather difficult to maintain the model. Use Goto and From block pairs to eliminate the unnecessary crossover lines. and simplify the block diagram.

Solutions. Consider setting some signals as tags, such that Goto and From block pairs can be linked up, without the need to line the blocks with signal lines. Therefore, the Simulink model can be established, as shown in Figure 6.7. It can be seen that the structure is much more straightforward than the one in Example 4.31. For instance, on the right of the model, Goto6 block can be used to express the output signal of the switch block to tag a, where when a signal is used in the lower-middle part, and it can be linked to From7 block, without really linking the two blocks. The tag of the signal should be set to the same a. The linking can then be set up, although the link are not displayed directly. The benefit of doing this is to make the block diagram more tidy.

Of course, the low-level modeling method presented here is still very complicated, and error prone. A better modeling method will be introduced through programming implementation.

6.1.3 Embedding Simulink models

Two blocks, Model and Variant Model, are provided in the Ports & Subsystems group. A Simulink model file, with suffixes of `slx` or `mdl`, can be embedded in a Simulink

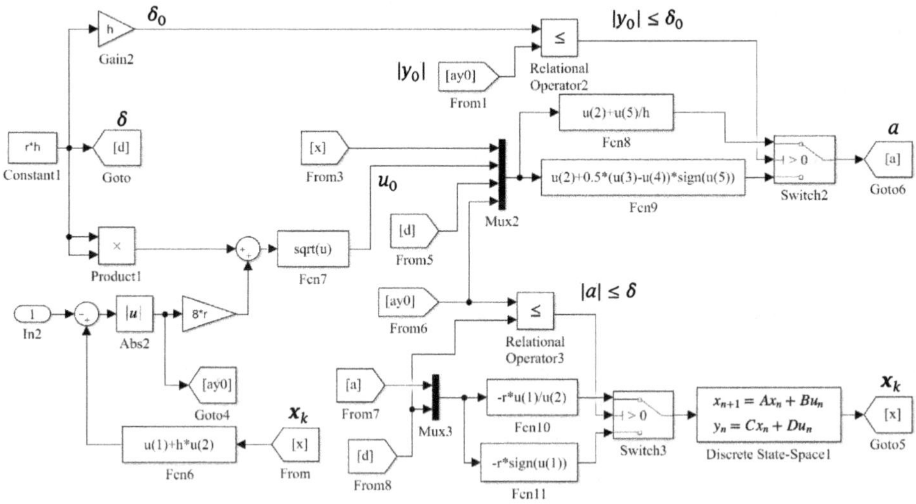

Figure 6.7: Simulink model of tracking and differentiator (file: `c6mhan.slx`).

model. An example is given to show how to embed model files into another Simulink model.

Example 6.6. A Simulink model for the composite transfer function was created in Example 2.7, and saved in the file `c2mtf2.slx`. Embed the file into the PID control system model.

Solutions. In Figure 6.4(c), the model `c6mspid4.slx` created is the PID control system model. Delete the current plant model, and copy the Model block from Subsystems & Ports group to the model window. Note that since this block is not associated with any MATLAB file, and there is no input and output ports in the block. The icon is with display Unspecified Model Name. Double click the block, and a parameter dialog box is opened. In the Model Name edit box, the file name c2mtf2 can be specified. Input and output ports are then added automatically to the block, such that the simulation model in Figure 6.8 can be constructed.

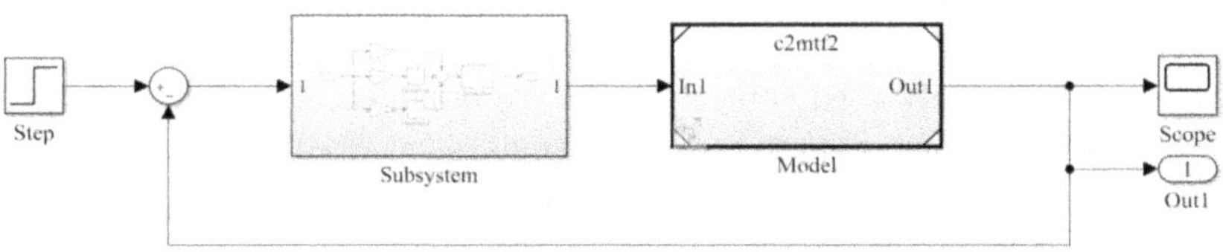

Figure 6.8: Simulink model with an embedded block (file: `c6mspid6.slx`).

6.2 Complicated flow control subsystems

In Volume I of the series, MATLAB programming is presented, where flow control structures in MATLAB are covered. The flow control includes loop structures, conditional structures and switch structures. Similar flow controls can also be implemented in Simulink, since the variations of subsystems also supported these flow controls. In this section, some special subsystem blocks are demonstrated.

6.2.1 Enabled and triggered subsystems

In Simulink, if a subsystem is allowed to execute under certain conditions, the subsystem is referred to as a conditionally executed subsystem. The commonly used conditionally executed subsystems are enabled subsystems and triggered subsystems. In the current version of Simulink, three such subsystem blocks are provided in the Ports & Subsystems group, as shown in Figure 6.9.

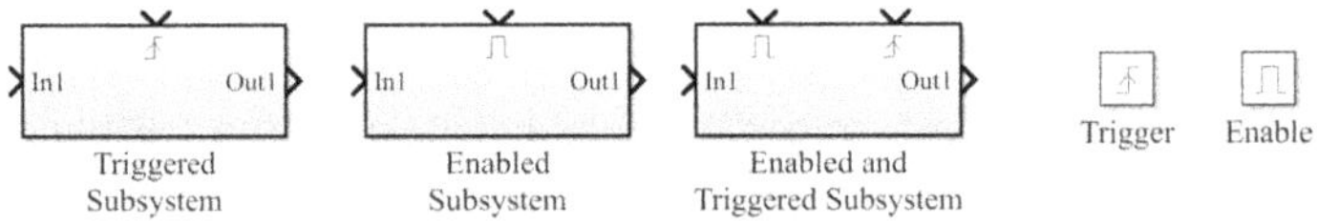

Figure 6.9: Enabled, triggered subsystems and controls.

(1) Enabled Subsystem block. The conditional signal in the block is referred to as control signal. The control signal has two status, as enabled and disabled. The mechanism of such a subsystem is that when the control signal is positive, it is enabled; otherwise, it is disabled. If the subsystem is enabled, the subsystem can be executed, while if it is disabled, to keep the consistency of the block, the user may chose to select maintaining the signal, or reset by forcing zero output of the block.

(2) Triggered Subsystem block. When the control signal satisfies a certain variation, at the instance, the subsystem can be triggered (or activated), then the signal of the subsystem is maintained as a constant, and waits for the next trigger signal. The user may chose rising edge, falling edge, or both, in the control signal to trigger the subsystem.

(3) Enabled and Triggered Subsystem block. When the subsystem is enabled, it may be triggered; otherwise, the subsystem is disabled.

Besides, some commonly used control blocks are provided, such as the Enable block and Trigger block, to generate the control signals, so as to control enabled and triggered subsystems. Of course, these blocks can also be used to control other conditionally execution subsystem blocks.

Example 6.7. Use sinusoidal signal to experience the response of an enabled subsystem block.

Solutions. To experience the actions in the enabled subsystem, the simulation model in Figure 6.10(a) can be created. In the model, the input signal is $u = 8\sin t$. An **Enable Subsystem** block is used, and when the enable signal is active, the output signal is linked directly to the input. If the enable signal is inactive, the output signal remains constant. The enable signal can be represented by the **Pulse Generator** in the **Sources** group, with its parameter dialog box shown in Figure 6.10(b). In the dialog box, **Amplitude** is set to 1, and **Period** is set to 0.4 seconds, **Pulse Width** is set to 20 %. This means that in every 0.4 seconds, the first 20 % of time, the block output is set to 1, while in the remaining time, the output is 0.

To better demonstrate the responses in the enabled subsystem, a fixed-step algorithm is recommended, with a small step-size, such as $h = 0.001$. The following commands can be used and the responses of the system are obtained, as shown in Figure 6.11. It can be seen that when the enabled signal is 1, the output signal reflects the values of the input signal, while its enabled signal is 0, the output signal is kept constant:

```
>> plot(tout,yout,tout,8*sin(tout),'--'), ylim([-8.5, 8.5])
```

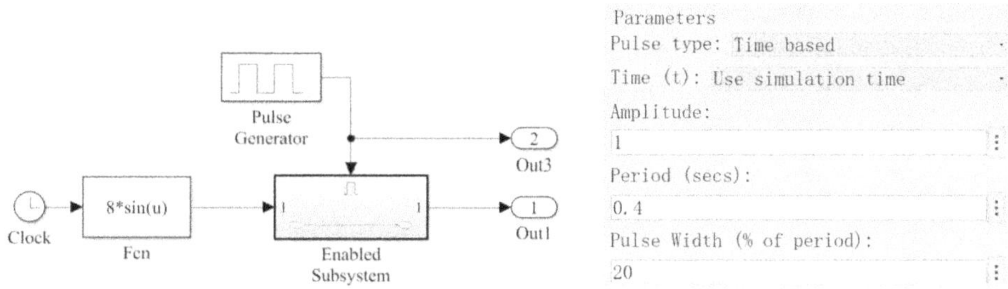

(a) enable subsystem (file: `c6msub1.slx`) (b) parameter dialog box

Figure 6.10: Enable subsystem and enable signal setting.

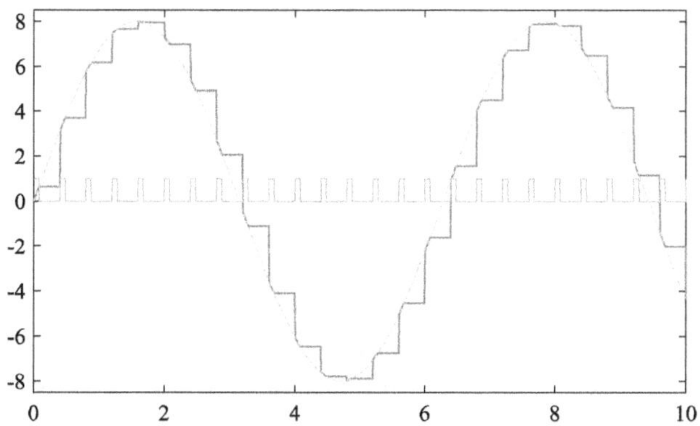

Figure 6.11: Output signal.

Example 6.8. Use sinusoidal signal to demonstrate the responses of a triggered subsystem.

Solutions. If the Enable Subsystem block in the previous example is replaced by a Triggered Subsystem block, the model in Figure 6.12 can be constructed.

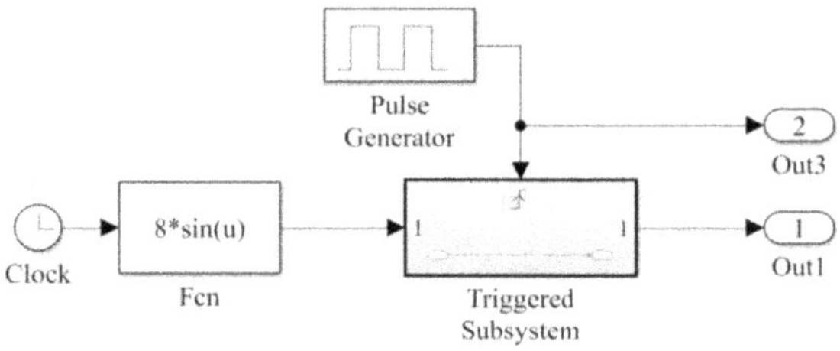

Figure 6.12: Triggered subsystem (file: `c6msub2.slx`).

If the same pulse generator is used as the control signal, the simulation result obtained is as shown in Figure 6.13. It can be seen that when the control signal is changing from 0 to 1, the subsystem is triggered. The instant value of the input is set to the output of the system, and it is kept constant, until the control signal is changed from 0 to 1. The output signal is updated then. Note that at $t = 0$, the value of the control signal is 1, there is no 0 to 1 sudden change, therefore, the subsystem remains with zero output.

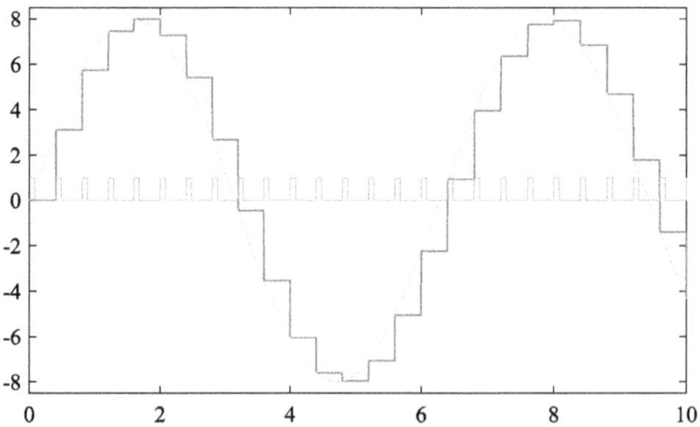

Figure 6.13: Output signal.

The default trigger signal is defined as rising edge trigger, that is, the sudden change from 0 to 1. The sudden change from 1 to 0 is totally neglected. The sudden change like this is referred to as a falling edge trigger. To change the trigger mode, the subsystem block can be double clicked, and the middle portion of its parameter dialog box is shown in Figure 6.14. In the Trigger type listbox, different trigger formats are

Figure 6.14: Parameter dialog box of triggered subsystem.

allowed. If **either** option is selected, the output waveform can be obtained, as shown in Figure 6.15.

It can be seen that no matter whether the trigger signal is increasing or decreasing, the subsystem is triggered, such that the output is set to the current value of the input, and remains constant, until the next trigger signal.

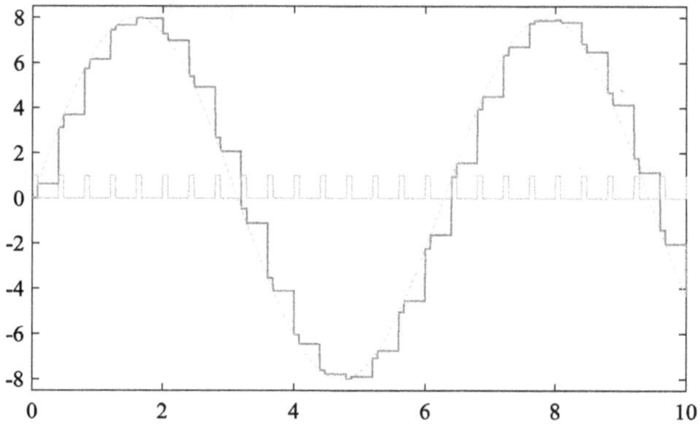

Figure 6.15: Output signal of triggered subsystem.

It appears that the output curves in Figures 6.15 and 6.11 are the same, and in fact, they are different. The curve in Figure 6.11 shows the tracking lines to the input when pulse input signal is at value 1, and they are oblique lines, where in the triggered subsystem, the output signal is staircase one. Besides these two types of subsystems, triggered and enabled subsystems are also supported in Simulink. This is left as an exercise, so that the readers can experience the subsystem themselves.

6.2.2 Conditional subsystems

In MATLAB programming, conditional structures are supported, while with Simulink, conditional subsystems are also provided. That is, when an external control signal is true, the subsystem is activated; otherwise, it is not activated, or other conditions are checked further, to determine the responses. An example is given next to show the modeling using conditional subsystems.

Example 6.9. Consider the mathematical model of a fractal tree. Select arbitrarily an initial point (x_0, y_0) in a two-dimensional plane. Assume that a random number y_i is generated, such that it is uniformly distributed over the $[0,1]$ interval. Based on the size of this value, a new point (x_1, y_1) is generated from the following formula:

$$\begin{cases} x_1 = 0,\ y_1 = y_0/2, & \text{if } y_i < 0.05, \\ x_1 = 0.42(x_0 - y_0),\ y_1 = 0.2 + 0.42(x_0 + y_0), & \text{if } 0.05 \leqslant y_i < 0.45, \\ x_1 = 0.42(x_0 + y_0), y_1 = 0.2 - 0.42(x_0 - y_0), & \text{if } 0.45 \leqslant y_i < 0.85, \\ x_1 = 0.1x_0,\ y_1 = 0.2 + 0.1y_0, & \text{otherwise.} \end{cases}$$

Establish a simulation model with conditional subsystems.

Solutions. Denote $z_1(k) = x$, $z_2(k) = y$. The iterative relationship in the formula is suitable to describe with discrete state space model:

$$\begin{bmatrix} z_1(k+1) \\ z_2(k+1) \end{bmatrix} = \boldsymbol{A} \begin{bmatrix} z_1(k) \\ z_2(k) \end{bmatrix} + \boldsymbol{B}u(k),$$

where $u(k) \equiv 1$, $k = 0, 1, 2, \ldots$. The $\boldsymbol{A}$ and $\boldsymbol{B}$ matrices in the above four iterative expressions can be written as

$$\boldsymbol{A}_1 = \begin{bmatrix} 0 & 0 \\ 0 & 1/2 \end{bmatrix}, \quad \boldsymbol{B}_1 = \begin{bmatrix} 0 \\ 0 \end{bmatrix}, \quad \boldsymbol{A}_2 = 0.42 \begin{bmatrix} 1 & -1 \\ 1 & 1 \end{bmatrix}, \quad \boldsymbol{B}_2 = \begin{bmatrix} 0 \\ 0.2 \end{bmatrix},$$
$$\boldsymbol{A}_3 = 0.42 \begin{bmatrix} 1 & 1 \\ -1 & 1 \end{bmatrix}, \quad \boldsymbol{B}_3 = \begin{bmatrix} 0 \\ 0.2 \end{bmatrix}, \quad \boldsymbol{A}_4 = 0.1 \begin{bmatrix} 1 & 0 \\ 0 & 1 \end{bmatrix}, \quad \boldsymbol{B}_4 = \begin{bmatrix} 0 \\ 0.2 \end{bmatrix}.$$

Based on the idea, the simulation model in Figure 6.16 can be established, where the Uniform Random Number block is used to generate signal y_i, to control the entire simulation process. In the model, an If block is used to describe the conditions. This block has two output ports, corresponding respectively to if and else conditions. In this particular example, since three conditions are given in the formula, the If block can be double clicked, to open the parameter dialog box, shown in Figure 6.17(a). In the If Expression edit box, the first condition u1<0.05 is specified. In the Elseif Expression and edit box, the other logic expressions should be provided. In this example, the conditions u1<0.45 and u1<0.85, separated by a comma, can be specified. Together with Else, four output ports appear automatically in the block.

An If Action Subsystem is designed for each conditions. For instance, the first subsystem is shown in Figure 6.17(b). There is an Action Port port added to the block and used to accept control signals.

Unfortunately, since there are problems in the Simulink model, the fractal tree cannot be drawn correctly. Since one discrete state space model with variable param-

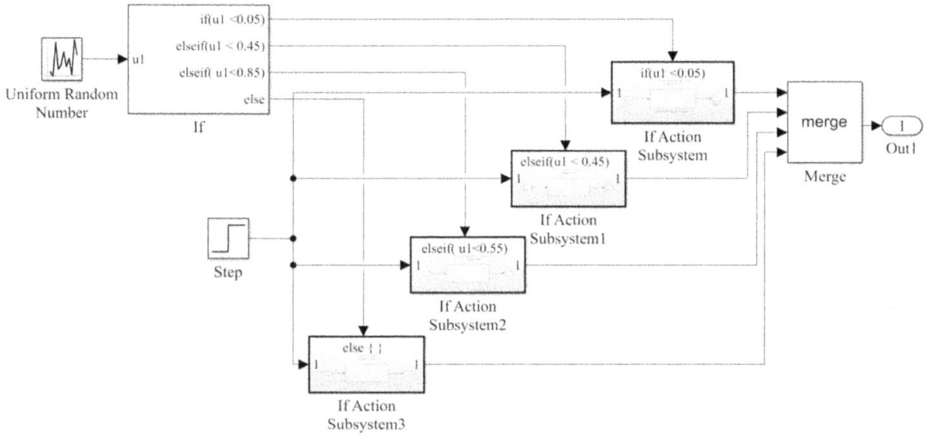

Figure 6.16: The Simulink model (file: `c6mif.slx`).

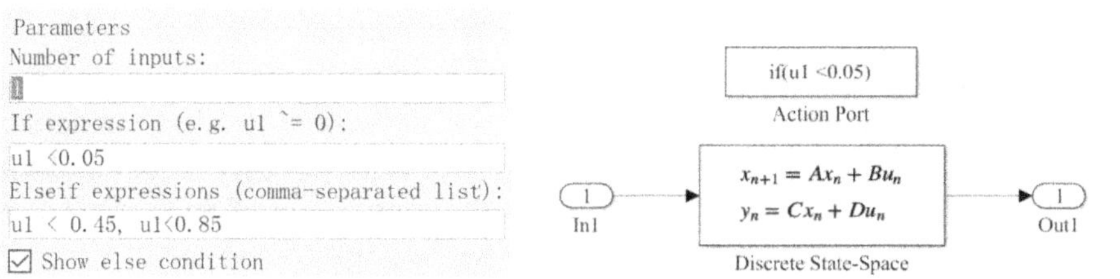

(a) parameter dialog box (b) subsystem model

Figure 6.17: Parameter dialog box and subsystem.

eters is needed, rather than four independent discrete state space blocks. Therefore, only the modeling idea in conditional subsystems are illustrated in the example.

It can be seen from the example that the modeling using this method is rather complicated and there are many trivial things to consider. A better and more concise method by using S-function modeling will be presented later in Chapter 7.

6.2.3 Other flow control subsystems

Other flow control subsystems are also supported in Simulink, such as loop subsystems and switch subsystems. The use of these subsystems are rather complicated, and these structures are more suitable to be described by S-function blocks to be illustrated later. These subsystems are not further presented here. The interested readers are advised to refer relevant materials, or consider using S-function blocks to deal with the related problems.

6.3 Block masking

Subsystem modeling and processing are illustrated in the previous sections. In general cases, the common parts of blocks can be made into subsystems, such that the modeling of large-scale systems can be made more concise, such that the simulation models can be created and maintained easily. The reusable blocks can be made from the subsystem blocks. The necessity of block masking is explained next, and block masking methods, procedures and skills are illustrated. Several examples are given to demonstrate subsystem masking methods.

Example 6.10. The subsystem modeling of PID controllers are presented earlier. It seems that the subsystem model thus created are reusable. Why do we need to mask subsystems? Subsystem blocks are not sufficient in modeling.

Solutions. Consider the PID controller subsystem studied previously. If two PID controllers are needed in a simulation model, the two PID subsystem blocks can be copied into the model window. The readers may then find that no matter how to manipulate the blocks, the parameters of the two PID controller blocks are identical. In real applications, PID controllers with different parameters are usually required in a model, which this is not allowed in subsystem blocks to set independently the parameters. Therefore, a subsystem masking technique is required to build up reusable PID controller blocks, allowing to set the parameters of each controller independently.

The so-called masking means that to hide the subsystem block internal structure, and design its own parameter dialog boxes. The user may specify the parameters of the model with these dialog boxes. In fact, most of the Simulink blocks are masked subsystems, constructed with low-level modeling blocks. For instance, in the transfer function block, the internal structure is invisible. It provides a parameter dialog box to allow the user to specify its numerator and denominator parameters. Also, in the PID controller model, it can also be masked, such that a parameter dialog box can be designed, to accept the four parameters of the controller. Therefore, two or more PID controller blocks can be used in the same model, and the parameters of them can be set independently with the parameter dialog boxes, without affecting the other PID controller block.

In fact, most of the blocks in Simulink library are masked subsystem blocks. Mastering the masking technique may increase the actual capabilities of the reader in building Simulink models. In this section, the concept of block masking handles and icon design are illustrated, then the parameter dialog box design is presented. Finally, masking block programming is demonstrated.

6.3.1 Icon design

In order to mask the users' own model, it should be converted into a subsystem block, using the methods presented earlier. Then select the subsystem icon. Right click the block, and a shortcut menu is opened, and the Mask submenu is shown in Figure 6.18.

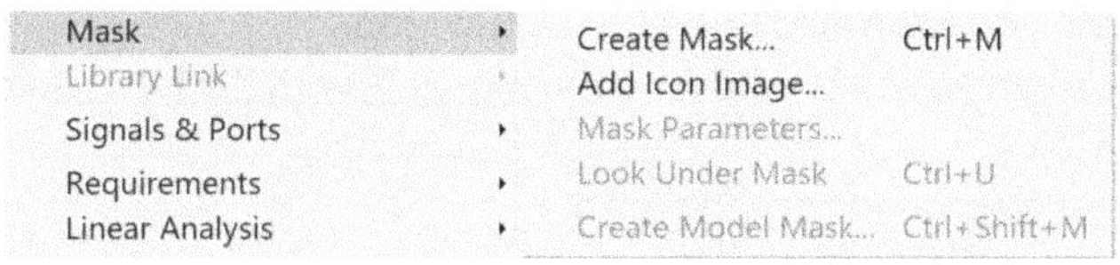

Figure 6.18: Masking shortcut menu.

Selecting the Mask Create item from the shortcut menu, a model mask editor interface is displayed, a shown in Figure 6.19.

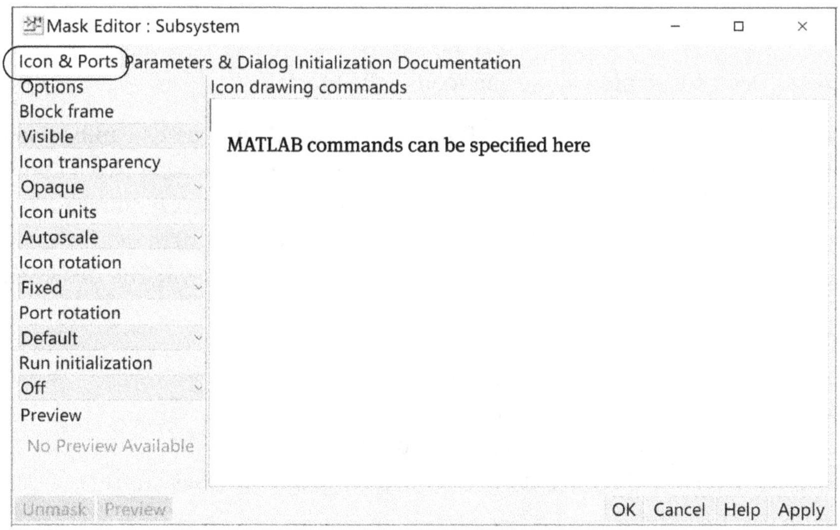

Figure 6.19: Model masking editor interface.

Four panes are provided in the interface, Icon & Port, Parameter & Dialog, Initialization and Documentation. Each pane corresponds to a different page in the interface. In this section, details in each pane is presented, and block masking procedures and skills are shown next.

The default pane is regarding the default icon and port design. On the right, the region labeled **Icon drawing commands** allows the user to write his own MATLAB commands, to design the icon of the block. Normally, three types of commands can be used to draw the icons:

(1) **Draw plots.** Curves or colored patches are allowed to draw in the icon. The `plot()` command can be used to draw the curves. Note that the `plot()` command here is not a genuine MATLAB command. Its syntax is plot(x_1,y_1,x_2,y_2,$\cdots$), and options are not supported here, as in the MATLAB `plot()` function.

Besides, `patch()` function can be used to draw a colored patch, with the following syntax:

patch([x_1,x_2,$\cdots$,x_n,x_1],[y_1,y_2,$\cdots$,y_n,y_1],color)

where the two vectors here describe the coordinates of a closed-path, and the area is filled by the specified color. The option "color" can be set to reserved colors such as `'red'`, or to the color components as $[r,g,b]$ vector, with the range of these three components $[0,1]$.

When drawing curves and displaying text, colors can be set. For instance, if `color`(color) is used, the subsequent drawing commands are made in the assigned color.

Example 6.11. Design a smiling face as an icon of the block.

Solutions. To draw the smiling face, the following MATLAB commands can be used to draw the four curves, where an outer unit circle represents the face; two smaller circles with radius of 0.1, centered respectively at (−0.4,0.2) and (0.4,0.2) represent the eyes; a semioval is drawn to represent the mouth. In the plots, the right eye is drawn with a `patch()` command, such that a red filled circle can be drawn. The designed icon is shown in Figure 6.20(a):

```
t=linspace(0,2*pi,30); t1=linspace(0,pi,30);
plot(cos(t),sin(t),-0.4+0.1*cos(t),0.2+0.1*sin(t))
patch(0.4+0.1*cos(t),0.2+0.1*sin(t),[1,0,0]) %red patch
plot(0.6*cos(t1),-0.1-0.4*sin(t1))           %default black curves
```

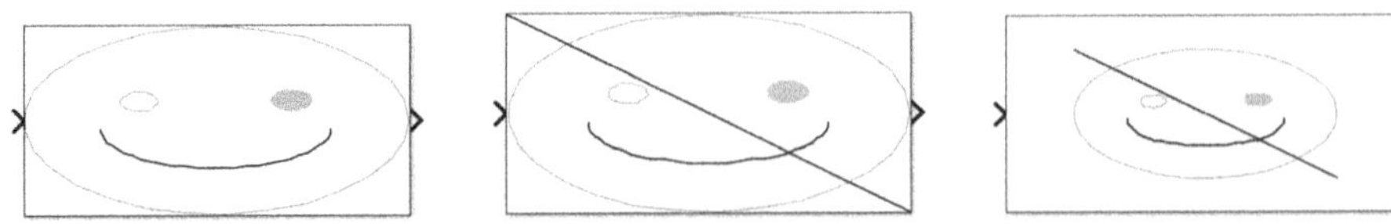

(a) smiling icon (b) with backlash (c) smaller face

Figure 6.20: Smiling face icon.

It should be noted that in older versions of MATLAB, the above commands in the Icon drawing commands edit box cannot be executed, and error messages are given, since the assignment statements cannot be executed in the edit box. It works in new versions of MATLAB.

A diagonal line can be added to the smiling face with the following command:

```
plot([-1,1],[1,-1])
```

The new icon designed is shown in Figure 6.20(b). Note that the commands such as `hold on` and `line()` cannot be used; otherwise, the icon cannot be drawn. Function `plot()` is different from the command `plot()` in MATLAB. New curves can be superimposed on the existing ones directly.

It can be seen from the plot that the drawing will fill automatically the entire icon. If a smaller icon is expected, two extra points can be drawn at respectively lower-left and upper-right corners. For instance, to assign the region to $-1.5 \leqslant x, y \leqslant 1.5$, with the following commands, the new icon is as shown in Figure 6.20(c):

```
plot(-1.5,-1.5), plot(1.5,1.5)
```

With the above commands, the two extra points are set, so that the range of the icon is set automatically to $-1.5 \sim 1.5$, and the two extra points are not actually displayed. The new plots is not stretching over the whole icon. In this way, the size of the plot can be controlled.

(2) Text description. With `disp()` command, text can be superimposed to icons. Multiple lines of text can be displayed. Carriage returns can be expressed with `\n`. In the new versions, `text()` command can also be used, to specify the position of the text, with `text(`x`,`y`,string)`, where (x, y) is the coordinate of the text display, and its coordinates are the same as the one in `plot()` command.

Example 6.12. Add PID Controller text to the icon.

Solutions. If block mask action is initiated, the menu item Mask Create in the shortcut menu is automatically changed to Edit Mask. With such an item, the following commands can be appended to the Figure 6.20(c) setting:

```
disp('PID\nController')
```

Since `disp()` command is used, the text PID Controller can be superimposed to the icon, where `\n` is used, such that Controller is displayed in the next line. Note that the text displayed are center aligned.

To display text near the lower-right corner, the position should be specified. With the `plot()` command, the lower-right corner is at position (1.5, −1.5). Margins can be

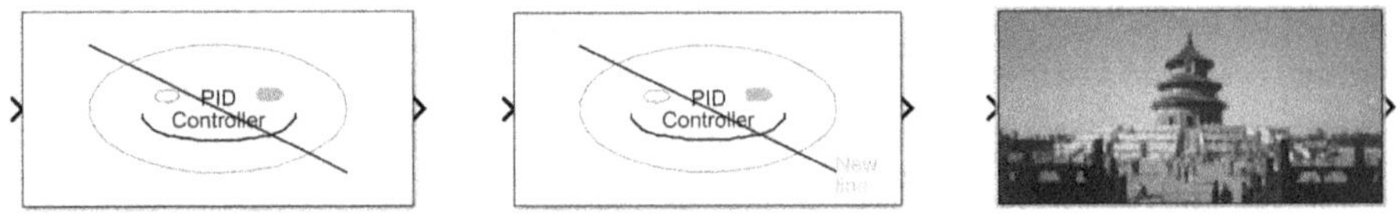

(a) add text (b) specify position (c) add an image

Figure 6.21: Adding text and image on icons.

assigned so that at point (1, −1), text in green can be superimposed, as shown in Figure 6.21(b). Note that the text generated by `text()` is left aligned:

```
color('green'), text(1,-1,'New\nline')
```

(3) **Image file.** MATLAB command `imread(file name)` can be used to load an image file into MATLAB workspace. Then function `image()` can be used to display the image as the icon.

Example 6.13. Assume that there is an image file, `tiantan.jpg`, on the MATLAB search path. Display the image as the icon of the masked block.

Solutions. In the **Icon drawing commands** edit box of the model masking editor interface, the following command can be written:

```
image(imread('tiantan.jpg'))
```

The image is displayed on top of the icon of the block, as shown in Figure 6.21 (c). Note that when the image displayed, all of the plots and text are covered up.

If the **Icon drawing commands** edit box in the model masking editor interface is left blank, the shortcut menu item **Mask →Add Icon Image** can be selected. A dialog box shown in Figure 6.22 is displayed, allowing the user to select an image file.

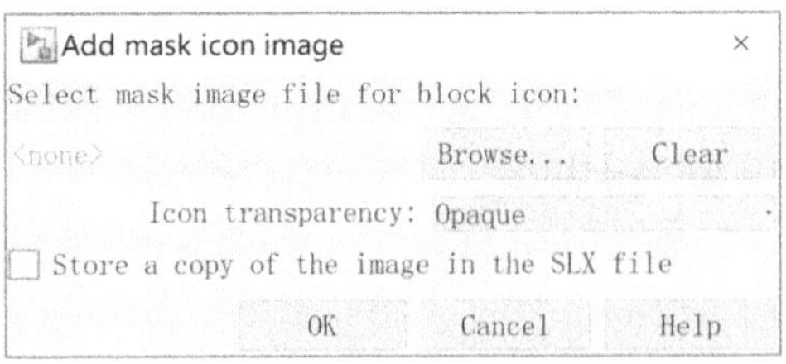

Figure 6.22: File name dialog box.

If the **Look Under Mask** submenu in the **Mask** shortcut menu is selected, the original subsystem model can be opened directly.

6.3.2 Design and decoration of other icon properties

Apart from the above mentioned icon design, on the left of the dialog box in Figure 6.19, some other decoration options for icons and ports are provided.

(1) Block Frame listbox, with two options, Visible and Invisible. The former one is the default one. Most of the existing Simulink blocks have visible frames. If the users do not need the block frame, the Invisible option can be chosen.

(2) Icon Transparency listbox, with three options, Opaque (default one), Transparent and Opaque with Ports. If the default option is chosen, the port information is covered up completely by the icon. Therefore, if image icon is used, and the port name is also needed, the latter two options can be selected.

(3) Icon Rotation listbox, with two options, Fixed (the default one) and Rotates. If the latter one is chosen, when the block is rotated or flipped, the icon is updated as well. In the next example, the icon rotation and flipping are demonstrated.

Example 6.14. Consider again the image icon in Figure 6.21(c). Observe the display results under different manipulations.

Solutions. If Rotates option is selected, the results shown in Figures 6.23(a) and (b) are observed. It seems that when the block is flipped, there is no response. In fact, if it is observed closely, it is found that the image is indeed flipped. If the Fixed option is selected, then the image is not changed, as shown in Figure 6.23(c), if the block is flipped.

(a) rotate 90° (b) block flipping (c) flipping with Fixed option

Figure 6.23: Icon rotation and flipping.

6.3.3 Parameter dialog box design

The most important aim in subsystem masking is that the parameters in the block can be specified independently, without affecting others. Independent parameters are implemented in the parameter dialog box design. For instance, for some blocks such as Transfer Fcn, if it is double clicked, a parameter dialog box is displayed, to prompt the user to specify the numerator and denominator of the transfer function model. The parameter dialog box design is to set up the association between the dialog box variables and internal variables in the model. The parameter dialog box design is the key technique in block masking. In this section, examples are used to demonstrate the design methods.

Select the Parameter & Dialog pane in the mask dialog box in Figure 6.19, and the mask dialog box is changed to the form in Figure 6.24. In the dialog box, it is divided into left, middle and right zones, labeled as Controls and Dialog Box and Property Editor zones.

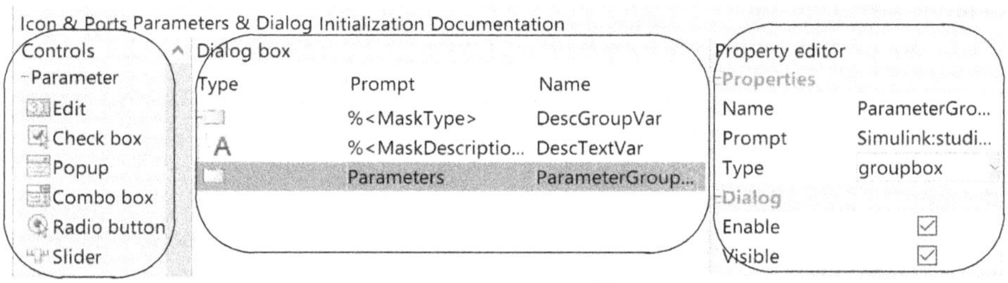

Figure 6.24: Zones in the parameter dialog box design interface.

In the left zone, various controls for parameter dialog box design are provided. Since too many controls are provided, they are given here in four subplots in Figure 6.25. In Figures 6.25(a) and (b), various parameter controls are provided, such as the commonly used ones Edit, Check Box and Popup.

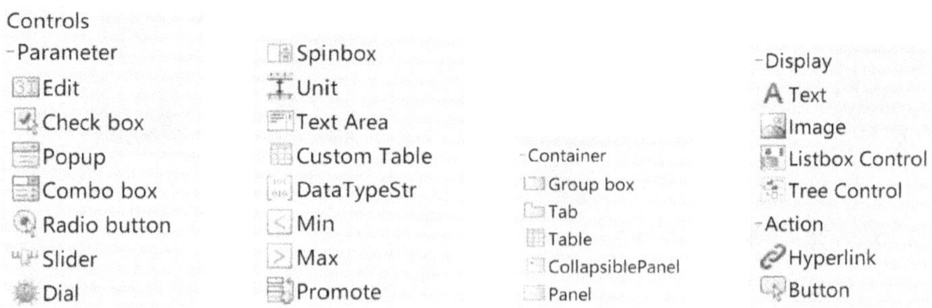

(a) parameter controls I (b) parameter controls II (c) containers (d) display and action

Figure 6.25: The controls provided.

Some Container controls are provided in parameter dialog box design, as shown in Figure 6.25(c), including Tab, Group box, Panel, Collapsible Panel and so on.

In the control zone, the controls such as **Display** and **Action** are also provided, as shown in Figure 6.25(d), including **Text**, **Image**, **Hyper Link**, **Button** and so on.

The middle zone of the dialog box should be designed by the user. Some controls and containers can be created. There are three columns in this zone: in **Type** column, the user needs to specify the type of control. In the **Prompts** column, the prompt for a certain parameter is specified. In the **Name** column, the variable name should be assigned. The variable name should be consistent with the internal variable names in the subsystem, or can be used to compute the internal variables. When this information is specified, the right zone of the dialog box can then be processed. Examples are given next to demonstrate parameter dialog box design.

Example 6.15. Design a parameter dialog box for the masked PID controller block.

Solutions. It can be seen from the Simulink model in Figure 6.1(a) that there are four internal parameters in the model, K_p, K_p, K_p and N. Four controls are needed in the parameter dialog box design, with the former three given by edit boxes, and for the last parameter, N, a listbox is needed.

Select the **Edit** icon in the left zone in Figure 6.24, and drag it to the middle zone, under the **Parameters** column. Release the mouse button, an edit box is added on the parameter list. Drag the **Edit** button twice, to add two more edit boxes, then drag **Popup** icon and release it at the **Parameters** column, the four parameter items are added, as shown in Figure 6.26.

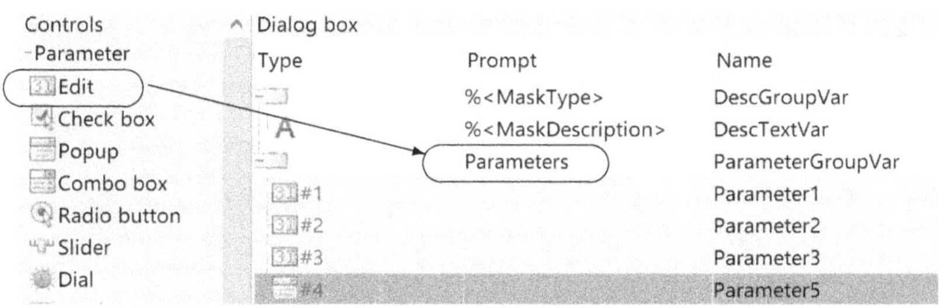

Figure 6.26: Copy the controls.

Now the four controls should be edited. Let us look at the first parameter K_p. A variable name `Kp` should be selected, and a text prompt `Proportional Kp` can be filled in the **Prompt** edit box, and set an default value 1.5 to the **Value** edit box, as shown in Figure 6.27, in the shaded area. The design for the first parameter is then completed.

Similarly, the second and third parameter items can be designed. In the fourth parameter, N, since **Popup** control is selected, the other relevant information can be filled. To set list options such as 10, 100 and 1000, the **Type options** can be clicked. A small edit box appears, allowing the user to fill in the three options. The item 100 can

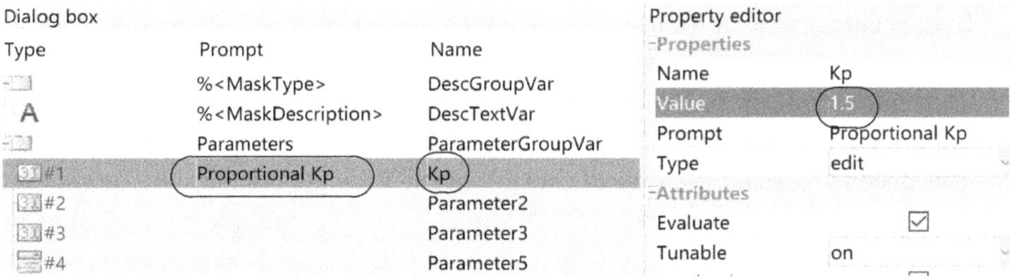

Figure 6.27: Design the first parameter item.

be selected from the **Value** listbox, as the default parameter. The designed four parameters are shown in Figure 6.28. This completes the design of the parameter dialog box.

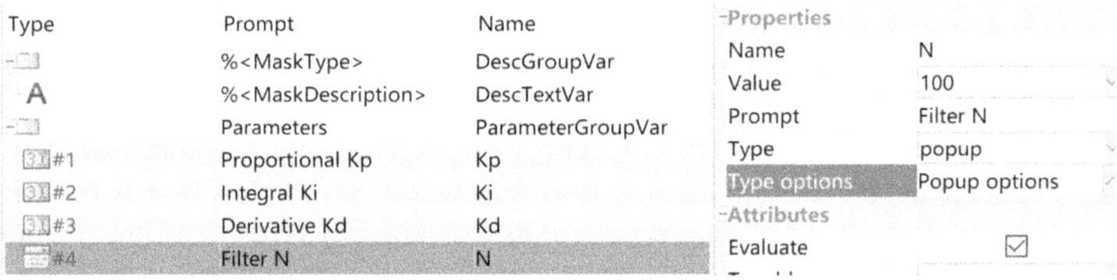

Figure 6.28: Completed the four parameter items.

Double click the designed PID controller block, a parameter dialog box shown in Figure 6.29 is displayed. The PID controller parameters can be specified with such a parameter dialog box.

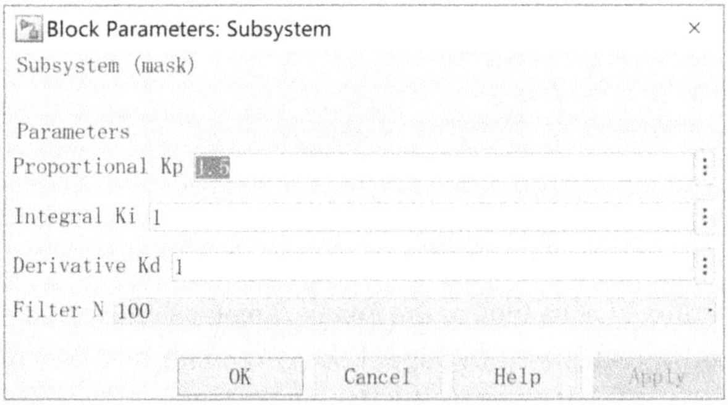

Figure 6.29: The designed parameter dialog box.

The design method for simple parameter dialog boxes are illustrated above. In real applications, in order to enhance the popularity of the dialog box, various other controls such as containers and display controls can be introduced. The controls supported shown in Figure 6.25 can be introduced. For instance, Image control and Tab control can be added. Parameter dialog box with special controls are demonstrated next through examples.

Example 6.16. In real applications, two types of PID controllers are usually used. One is parallel PID controller described in (6.1.1), and the other one is given next:

$$G_c(s) = K_p\left(1 + \frac{1}{T_i s} + \frac{T_d s}{T_d s/N + 1}\right) \tag{6.3.1}$$

This type of PID controller is referred to as standard PID controller. Design a collapsible panel to display the formula images of the two types of PID controllers.

Solutions. Open the mask editor interface again. Select the Parameters & Dialog pane. A popup control with Prompts of PID Controller Type can be added first. The variable name is set to lstPID for later use.

Drag the Collapsible Panel control into the Parameters column, a collapsible panel control is created. The Prompts can be specified PID Controller Mathematical Form, and change the name of the control to conPID for later use. The aim of creating such a panel is to display the relevant the image of the PID controller formula. An Image control can be added, and save the images of the two equations in (6.1.1) and (6.3.1) as image files, `imPID1.jpg` and `imPID2.jpg`. The designed interface is shown in Figure 6.30. Select the imPID control, and in the property editor on the right, set in the File Path edit box the file name `imPID1.jpg`. The image file is automatically associated with the control, and the new model is saved in file `c6mpid5b.slx`.

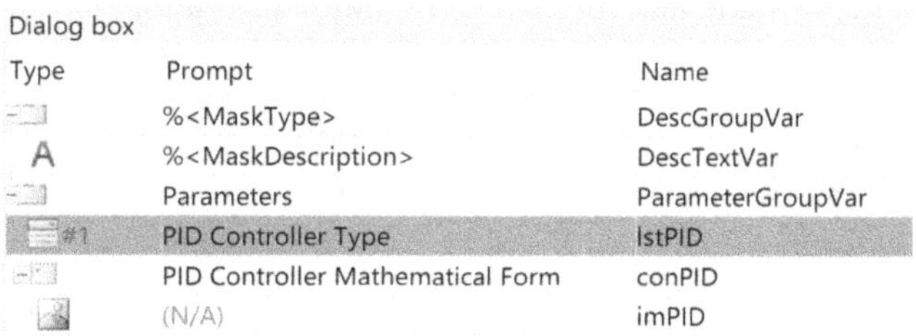

Figure 6.30: Design of Collapsible Panel in the dialog box.

Double click the block, the parameter dialog box is shown in Figure 6.31. In normal cases, the collapsible panel is hidden, and the image is not displayed. To expand the collapsible panel, the ▸ mark to the right should be clicked. The hidden part of the collapsible panel is expanded. Meanwhile, the mark ▸ is automatically changed to ▾. Click the ▾, the panel is hidden. Of course, only dialog box design is presented here, without any programming implementation. Therefore, if the controller type is changed, the images cannot be swapped.

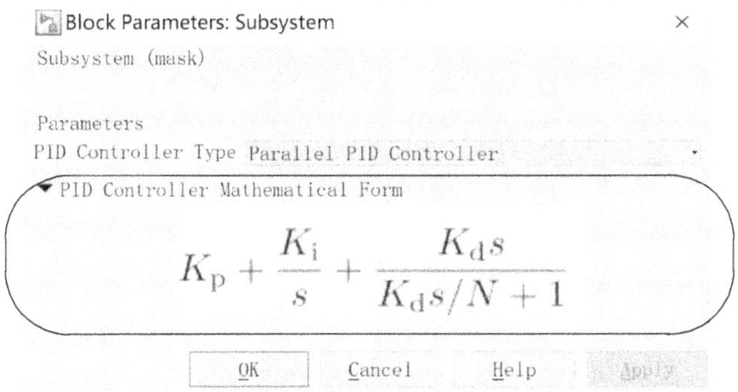

Figure 6.31: The design parameter dialog box.

With the Tab control, panes can also be designed. Following the previous example, the regular PID controller is a linear model, and in real applications, the control signal thus generated may become too large, and the hardware in the system may be damaged. Thus, the controller is companied by a saturation nonlinearity, to restrict the size of the output signal. Therefore, in parameter dialog box design, two panes are needed: one is for the regular PID controller, and the other is for the upper and lower bounds of the saturation component.

Example 6.17. Design two panes in the parameter dialog box: one for ordinary PID controller parameters, with a tag of **Main**, and the other is **Actuator Saturation**, for the actuator saturation parameter input.

Solutions. Open the model `c6mpid5.slx` previous designed. Double click the subsystem model, a switch block and a saturation block can be added, and the subsystem model is modified as Figure 6.32.

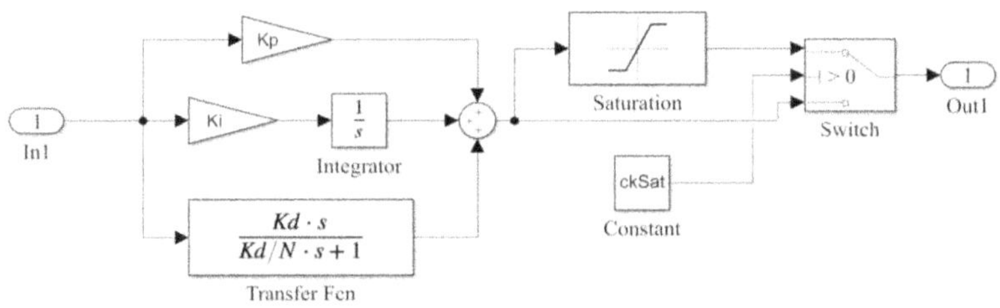

Figure 6.32: Simulink model with actuator saturation (file: `c6mpid5a.slx`).

Return to the subsystem block, select the **Mask** →**Edit Mask** shortcut menu item, and select the **Parameters & Dialog** pane in the editor, and the user can design the parameter dialog box. In the **Image** pane, a **Tag** control can be added. Two extra controls appears automatically, one is **Container**, and the other is **Tag** control, as shown in Figure 6.33. The modified model is saved in file `c6mpid5b.slx`.

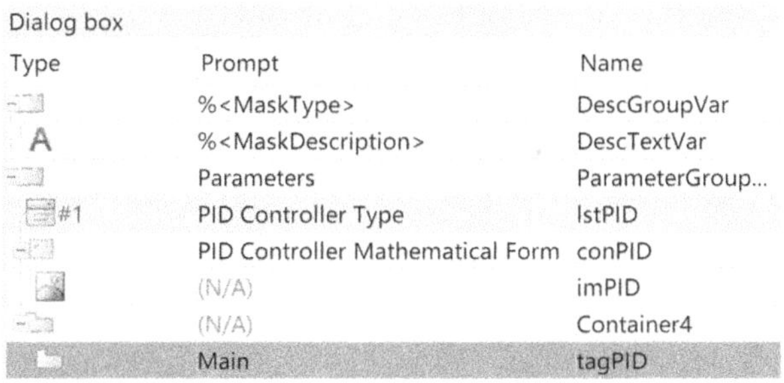

Dialog box

Type	Prompt	Name
	%<MaskType>	DescGroupVar
A	%<MaskDescription>	DescTextVar
	Parameters	ParameterGroup...
#1	PID Controller Type	lstPID
	PID Controller Mathematical Form	conPID
	(N/A)	imPID
	(N/A)	Container4
	Main	tagPID

Figure 6.33: Add a pane.

Modify the **Prompts** for the pane as **Main**, and set the variable name as **tagPID**, the framework of the first pane can be created. Similarly, the controls K_p, K_i, K_d and N can be designed. When the regular PID controller parameter is designed, another **Tag** control can be drawn, with **Prompts** set to **Actuator Saturation**, and the name is set to **tagNL**. A checkbox control named **with Saturation** in the pane can be designed. Two extra edit boxes, **Lower Bound** and **Upper Bound** can also be added. The finalized parameter design is shown in Figure 6.34. The file saved is `c6mpid5c.slx`.

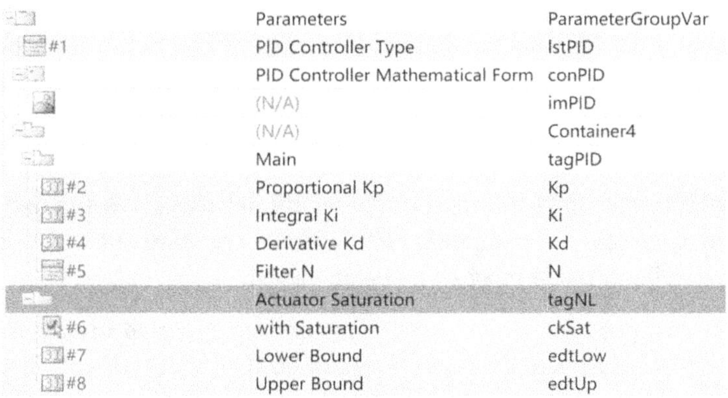

	Parameters	ParameterGroupVar
#1	PID Controller Type	lstPID
	PID Controller Mathematical Form	conPID
	(N/A)	imPID
	(N/A)	Container4
	Main	tagPID
#2	Proportional Kp	Kp
#3	Integral Ki	Ki
#4	Derivative Kd	Kd
#5	Filter N	N
	Actuator Saturation	tagNL
#6	with Saturation	ckSat
#7	Lower Bound	edtLow
#8	Upper Bound	edtUp

Figure 6.34: The parameter dialog box with panes.

When the block masking design process is completed, double click the icon, and the parameter dialog box is opened as shown in Figure 6.35.

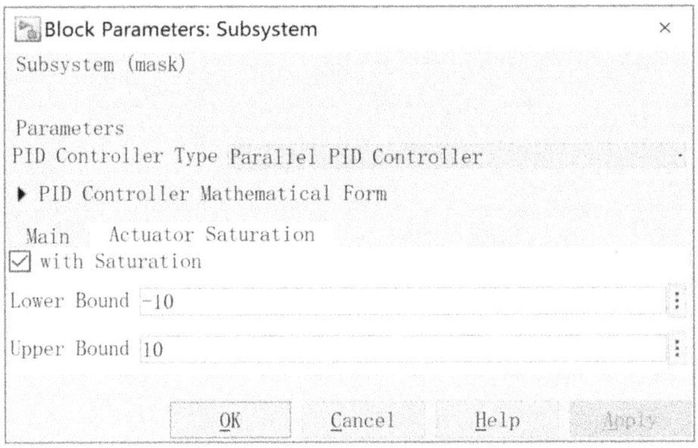

Figure 6.35: Differences in PID controller parameter dialog box.

Of course, this is the designed framework for the parameter dialog box. The parameters in the dialog box cannot be assigned to the parameters in the PID controller model. The assignment should be made through programming methods, and the design method will be presented in the next section.

6.3.4 Callback function for controls

If the properties of some controls are changed, callback functions are needed to set the properties of the controls. For instance, in the PID controller block, two types of PID controllers are supported. If different types are selected, the images of the PID controller formulas can be displayed. An example is used to demonstrate the callback function programming for the controls.

Example 6.18. In the listbox of the PID controller type mentioned earlier. Write a suitable callback function, such that when different PID controller types are selected, the corresponding PID controller formula images can be displayed.

Solutions. Two questions must be answered to implement the expected functions: one is how to find the imPID control, and the second is to identify which PID controller is selected.

To answer the first question, the MATLAB command `Simulink.Mask.get(gcb)` can be used so that the handle of all the controls in the masked block can be extracted. When the statement is called, gcb stands for get current handle, `gcb` the model handle is extracted. Function `getDialogControl()` can be used to find the handle of the

imPID control. In this case, the `FilePath` property can be set to indicate to different image files.

The second question can be solved by calling the function `get_param()` to find the handle of the current handle, from which the type of controller is identified. Based on the above considerations, the following callback function can be written:

```
h=Simulink.Mask.get(gcb); h0=h.getDialogControl('imPID');
h1=get_param(gcb,'lstPID');
if strcmp(h1,'Parallel PID Controller'), h0.FilePath='imPID1.jpg';
else, h0.FilePath='imPID2.jpg'; end
```

Open the mask editor, the **Parameters & Dialog** pane can be selected, select **lstPID** parameter option, as shown in Figure 6.36. Right click the button to the right of the option **Callback**, and the callback function can be specified in the dialog box. The model can be saved in the file `c6mpid5d.slx`.

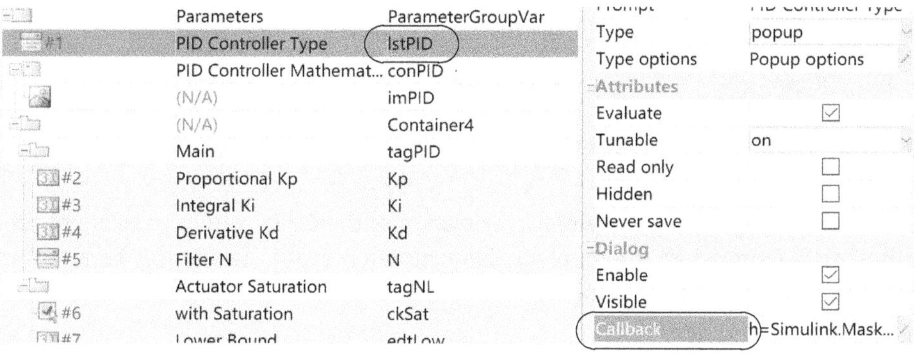

Figure 6.36: Design interface for lstPID control.

6.3.5 Initialization of mask blocks

In the previous description on the parameter dialog box design of the masked block, there are some problem remaining. For instance, the variables K_p, K_i, K_d and N are used in PID controller design. In the standard PID controller pane, the accepted variables are K_{p1}, T_{i1}, T_{d1} and N_1. These variables should be used to convert the variables into the variables used in the model; otherwise, the model cannot be normally executed. Besides, when the collapsible panel is expanded, how can we display the appropriate PID controller formula images? These conversion tasks must be implemented with initialization programming of the block.

Initialization facilities can be implemented in the mask editor, in the **Initialization** pane. Click the **Initialization** pane, and the mask parameter dialog box is displayed as shown in Figure 6.37. It can be seen that there are two zones, left and right ones, in the editor. In the left zone, the parameter list is given, and in the **Parameters & Dialog** pane, the parameters can be designed; the MATLAB command can be filled on the right zone, the blank **Initialization commands** edit box, such that the commands can be executed automatically before the block is run. Click **Preview** button, and the designed parameter dialog box can be displayed.

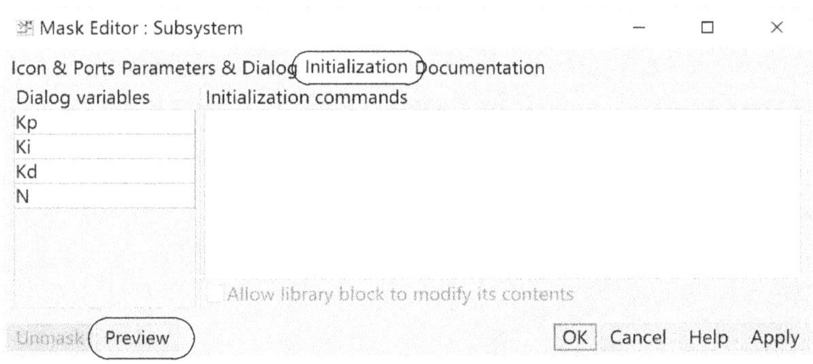

Figure 6.37: Initialization pane design.

In the PID controller mask example, the design of the initialization pane is not presented now. Later when the problem is encountered again, details will be given on initialization pane design methods.

6.3.6 Document and help information

The title and help information can be designed in the **Documentation** pane in the mask edit interface. Click the **Documentation** pane, and the dialog box in Figure 6.38 is displayed.

In the dialog box, there are three edit boxes for title and other help information. The one on the top, **Type** edit box, can be used to accept the title information of the masked block, such as **PID controller**. In the **Description** edit box, useful information is presented in one line to describe the block. For instance, fill in the information **This is a PID controller block** in the edit box. In the third edit box labeled **Help**, help information can be specified. It can be any string.

After the help information is specified, double click the masked block, and the parameter dialog box is shown in Figure 6.39. According to the circled parts, it can be seen that the help information thus designed appear in corresponding parts in the dialog box.

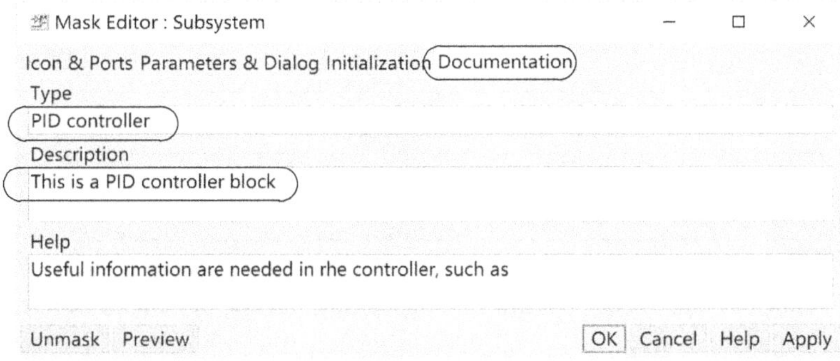

Figure 6.38: Design of help information pane.

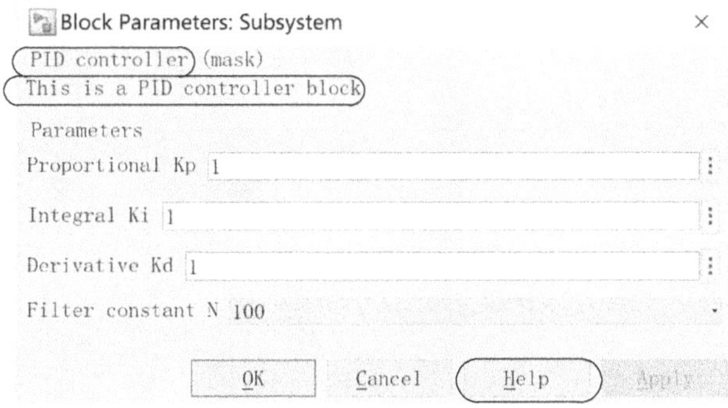

Figure 6.39: Parameter dialog box of the masked block.

In Figure 6.38, the string in Help edit box is not displayed in the parameter dialog box. If Help button in the dialog box is clicked, the help information can be displayed.

6.3.7 Block masking examples

The block mask process of PID controllers are studied earlier. With a series of settings, a powerful reusable PID controller block can be created. Compared with the PID controller block provided in Control System Toolbox, the facilities of the new block is not good enough. In practical modeling, the PID controller in the Control System Toolbox is recommended.

In this section, block masking processes are further demonstrated. Single- and double-valued nonlinearities are used as examples, such that general-purpose block masking methods are introduced.

Example 6.19. The modeling of static piecewise linear nonlinearities are studied in Section 5.3, where single-valued nonlinearity can be implemented directly with lookup table block, while double-valued nonlinearity can be decomposed into two lookup table blocks, with a switch block. Create a unified block for piecewise linear nonlinearities.

Solutions. The following protocols can be proposed. The nonlinear element is represented by two parameters $\boldsymbol{x}$ and $\boldsymbol{y}$. If they are vectors, the nonlinearity is a single-valued one; if they are matrices with two rows, the nonlinearity is a double-valued one. Based on the model `c5mloop2.slx` in Figure 5.22, a unified nonlinearity masked block can be constructed. For ease of presentation, the Simulink model is given again, as shown in Figure 6.40, where an input port is used to replace the sine wave block, and in the **1-D Lookup Table** block, the vectors $\boldsymbol{x}_1$ and $\boldsymbol{y}_1$ are used to represent the turning points, to replace the original vectors. In the **1-D Lookup Table1** block, the parameters are set to vectors $\boldsymbol{x}_2$ and $\boldsymbol{y}_2$.

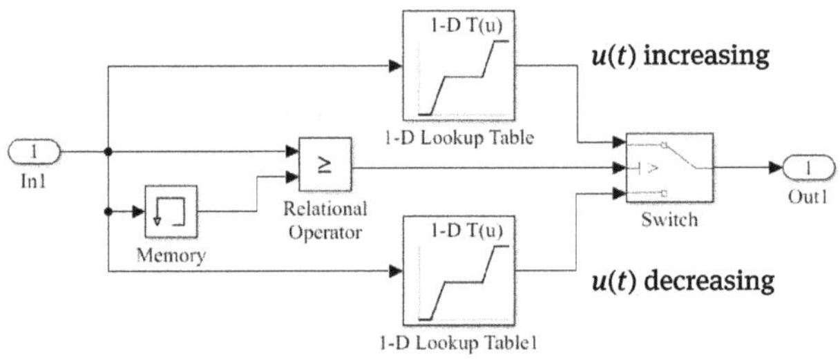

Figure 6.40: Simulink model of the nonlinearity (file: `c6mloop2.slx`)

Select first all the blocks, and a subsystem block can be created. From the subsystem block, the mask editor interface can be launched. In the **Parameters & Dialog** pane, two input variables `xData` and `yData` can be specified in edit boxes, as shown in Figure 6.41.

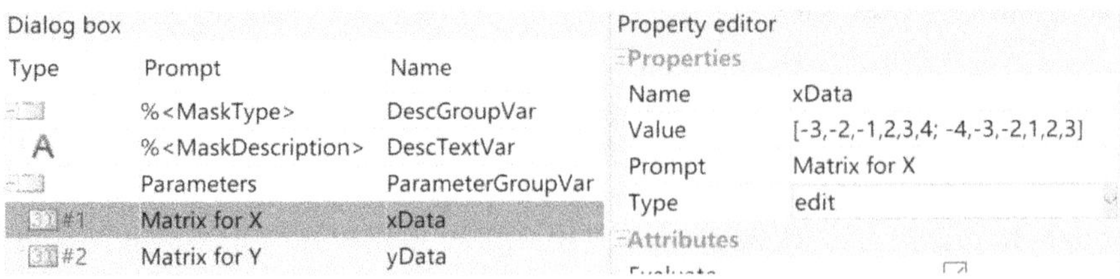

Figure 6.41: Mask editor interface.

An obvious question is to be answered: “How to assign the data in `xData` and `yData` to the needed vectors $\boldsymbol{x}_1$ and $\boldsymbol{y}_1$?” To solve the problem, the initialization program should be written for the masked block. The following commands can be filled into the **Initialization** pane. Then the variable assigned task can be solved. Each time when the model parameters are modified, the automatic assignment can be made:

```
if size(yData,1)==1 %if row vector, then single-valued function
   x1=xData; x2=x1; y1=yData; y2=y1;
else                %for double-valued function
   x1=xData(1,:); x2=xData(2,:); y1=yData(1,:); y2=yData(2,:);
end
```

The second question is "How to draw the icon of the block?" With the vectors x_1 and y_1, it seems that command `plot`(x_1,y_1,x_2,y_2) can be filled in the **Icon & Ports** pane to draw the icon of double-valued nonlinear element. Unfortunately, an error is displayed, indicating that vector x_1 is not defined. Observe the settings in the **Icon & Ports** pane, and it is found that on the lower-left corner, there is a **Run Initialization** listbox, with a default setting of **Off**. It means that the initialization information is not properly passed to the pane. It should be set to **On**. Therefore, the command `plot()` can be used to draw the icon correctly. To better draw the icon of the block, the following commands can be added in the **Icon drawing command** edit box, to add margins in the icon:

```
plot(x1,y1,x2,y2)
dX=max(xData(:))-min(xData(:)); dY=max(yData(:))-min(yData(:));
plot(min(xData(:))-0.08*dX,min(yData(:))-0.08*dY)
plot(max(xData(:))+0.08*dX,max(yData(:))+0.08*dY)
```

The block thus created can be used to process single- and double-valued nonlinearities. It can be seen that the masking facilities are very powerful, so that reusable Simulink blocks can be constructed. Model masking facilities and S-function programming to be presented in Chapter 7 can be combined to enhance the facilities of the masked block.

6.3.8 Canceling mask status

The masking methods for various blocks have been studied earlier. If the masking status of a certain block is to be canceled, the **Mask →Edit Mask** shortcut menu item can be selected, such that the mask editor interface in Figure 6.19 is opened. Click the **Unmask** button on the lower-left corner, and a message box in Figure 6.42 is displayed. Click the **OK** button, and the masking status of the block can be canceled, and changed back to the original subsystem block.

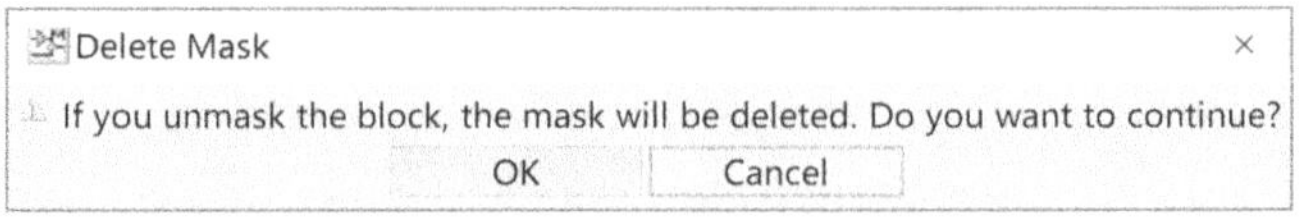

Figure 6.42: Message box to confirm canceling masking.

6.4 Drawing Simulink models with commands

It can be seen from the previous chapters that Simulink models can be created with the blocks provided, and connected in a visual manner. The modeling method is easy and straightforward. In some certain applications, the users may expect to draw the Simulink models with direct use of MATLAB commands.

6.4.1 Create a blank Simulink model

Two steps are used to create a blank model window. A blank logic model should be created, and the model is not displayed. With the logic model, the model can then be opened. The two steps are presented next to demonstrate the blank model window creation methods. Also other modeling manipulation methods are discussed.

(1) Logic model creation. In the MATLAB command window, function `new_system()` can be used to define a blank Simulink model in the MATLAB workspace. The model is not displayed automatically. The model created is referred to as logic model in Simulink. The syntax of `new_system()` function is h=`new_system(`model name`,`options`)`, where the “model” name is given in a string. The logic model thus created is usually the model name. The “options” can be `'Library'` or `'Model'`, while the former one is a blank model library, while the latter is the model window. If the second input argument is not provided, the model created is the logic model. The returned h is the handle of the model. If the model handle h is not returned, the model name can be used instead.

For instance, command `new_system('MyModel')` can be used to create a logic model, named `MyModel`. The model name `'MyModel'` can be used later, as the handle. Besides, if Simulink model is currently opened, command h=`gcs` can be used to get its handle, where `gcs` stands for “get current system.”

(2) Model display. When the logic model is created, command `open_system()` can be used to open the model. The syntax of the function is `open_system(`model name`)`, where “model name” can be a string. It can be a logic model created by `new_system()` function, or an existing Simulink model file. For instance, the logic model created above can be opened directly with `open_system('MyModel')` command. Alternatively, to open created model, the model name can be specified in MATLAB com-

mand window, for instance, with the MyModel command. Of course, open_system() function has other syntaxes to be discussed later.

(3) **Other processing methods.** A series of commands are supported in Simulink to manipulate models. For instance, function find_system() can be used to get the names of all the currently opened models. If more than one model is opened, the model names are returned in a cell array.

(4) **Save the model.** If the model is established, command save_system() can be used to save the model into a file, with suffixes as mdl or slx. The syntax of the function is save_system(model name,options). If the model name is not given, it is automatically saved into the current model file. To save the model into a mdl or slx file, the second input argument should be a string.

As it is presented earlier, function new_system() can be used to setup a blank Simulink model. The program list of the blank model is illustrated later.

Example 6.20. Create a blank model, and find information in the model.

Solutions. The following commands can be used to create and open a new blank model window. It can be saved into a slx file:

```
>> h=new_system('newmodel'); % create a logic model
   open_system(h);           % open the model window
   save_system(h);           % save the model into a file
```

The model can be saved into the newmodel.slx file. The file is not readable. The Simulink models created in different versions of Simulink, the files are not compatible. In the earliest versions of Simulink, the models are saved in m files, and the model is fully readable. Later the model is saved as mdl files, and the source file is still ASCII files. The readability of the model is not quite simple. In the new versions, slx suffix is used, and the model is given as binary files, which are totally unreadable.

To construct a readable Simulink file, the following commands can be used:

```
>> save_system(h,'newmodel1.mdl'); % save as readable file
```

The command edit can be used to open the source code of the Simulink model, and the commands in the model are

```
Model {
  Name                    "newmodel1"
  Version                 10.0
  SavedCharacterEncoding  "GBK"
  GraphicalInterface {
    NumRootInports         0
    NumRootOutports        0
```

```
ParameterArgumentNames     ""
ComputedModelVersion       "1.0"
```

The source code in the blank model is composed of 1000 source lines, and the program is rather lengthy, and it cannot be presented here. In the presented code segment, the model properties and property values are mainly described. All the properties in the Simulink models are defined in the source code, and usually there is no need to modify the properties manually. To modify the parameters, a better method is to use parameter dialog box so that the properties can be modified manually.

If the user insists on modifying certain properties, function `set_param()` can be called, and details of the function syntaxes will be presented later.

Example 6.21. List all the currently opened models and low-level block names.

Solutions. In the MATLAB command window, the following command is issued:

```
>> h=find_system, h1=h{5}
```

All of the names of the currently opened Simulink models and their low-level blocks are obtained in a cell array:

```
8×1 cell array
  {'newmodel1'                        }
  {'c4mbandc'                         }
  {'c4mbandc/Band-Limited White Noise'}
  {'c4mbandc/Gain1'                   }
  {'c4mbandc/Gain2'                   }
  {'c4mbandc/Sum'                     }
  {'c4mbandc/Transfer Fcn1'           }
  {'c4mbandc/Out1'                    }
```

It can be seen that the block names are absolute names, since apart from the names themselves, the model names are also given. It is known that there are two Simulink models opened: one is `newmodel1` and the other is `c4mbandc`. There is no lower-level block in the former one, while in the latter one, there are six lower-level blocks. With the command h_1=`h{5}`, the handle of the fifth item in the list can be found, that is, the Gain2 block in the `c4mbandc` model, and its handle is passed to variable h_1.

(5) Close the model. Function `close_system()` can be used to close an opened Simulink model. The model name is used in the function call. If the model is modified but unsaved, a warning message is displayed, informing the user to save the model. If commands `close_system('MyModel',1)` or `close_system(`h`,1)` are used, the model is closed, and the modification is discarded.

6.4.2 Copying blocks

To copy a block from one model window into another, the command `add_block()` can be used, with the syntax,

h_1=`add_block(s,t,`name1,property1,name2,property2, ⋯)

where `s` and `t` are strings, representing respectively the source block name and target block name; h_1 is the handle of the block. The names `s` and `t` should be full names, and will be demonstrated later through examples. The full name here means to include model name and block name, joined up with / sign in a string. In the Simulink model library, the original model name can be expressed as `built-in`. The user needs to find out the name of the block himself.

A practical way to find the name is to open a blank window, copy the block to the window and save it into a readable `mdl` file. Then the block name and properties can be found by opening the `mdl` file with an editor. The commonly used property names and values can be found in Table 6.1.

Table 6.1: Commonly used properties in the blocks.

Property names	Description and explanation
`Name`	Model name is given as a string
`BlockType`	Block type is described in a string. It is different with `Name`, `BlockType` is the internal unique name in the block, while `Name` can be assigned arbitrarily by the user
`Position`	Block position represented by a row vector $[x_1,y_1,x_2,y_2]$, where (x_1,y_1) and (x_2,y_2) are the coordinates of the lower-left and upper-right corner of the block. The units are in relative pixels (related to 1280×720 resolution in the screen). Note that make sure $x_2 > x_1$, $y_2 > y_1$
`ShowName`	Determines whether the block name is displayed, with options `'on'`, `'off'` and `'auto'`
`Ports`	Numbers of ports in the block, expressed as $[n,m]$, where n and m are respectively the number of input and output ports
`Orientation`	Block orientation, with default `'right'`, meaning input on the left, and output on the right
`DialogParameters`	All the variable names in the parameter dialog box
block parameters	Different blocks may have different parameters, for instance, Gain block may have `Gain` parameter, whose value of gain is in string

To delete a block h_1, command `delete_block(`h_1`)` can be used directly.

Examples are given next to show how to modify model names and parameters, how to find parameters and how to create blocks.

Example 6.22. To draw a Demux block in a model, find its absolute path of its source block, and create a new model window. At the upper-left corner, draw the Demux block, with width 5, length 50, and with 5 output ports. The new block name is chosen as MyDemux.

Solutions. Open a blank window, and copy a Demux block to the window. Save the model into a mdl file. Open the model file with an editor, and it can be found at the bottom of the file, there is a paragraph below.

```
Block {
    BlockType       Demux
    Name            "Demux"
    SID             "1"
    Ports           [1, 2]
    Position        [35, 16, 40, 54]
    ZOrder          1
    ShowName        off
    Outputs         "2"
    DisplayOption   "bar"
}
```

It can be seen that the block name is Demux, with the original block name being built-in/Demux. Apart from block name, other information is also available; for instance, Position and Ports. Therefore, the following commands can be used to set up a new block in the new window:

```
>> h=new_system('c6mnm1'); open_system(h);
   h1=add_block('built-in/Demux','c6mnm1/MyDemux',...
      'ShowName','on','Position',[10,10,15,60],...
      'Outputs','5')
```

where Ports is a read-only property, and cannot be modified. To change the number of output ports, the Outputs property can be set. Note that the values in strings can be used, not digits.

6.4.3 Linking blocks

Automatic routing function add_line() is provided in MATLAB, with the syntaxes

h_1=add_line(h,s,t)
h_1=add_line(h, [x_1,y_1; x_2,y_2; $\cdots$; x_n,y_n])

where h is the handle of the model, h_1 is the handle of the signal line. As in the syntaxes in other functions, s and t represent the start and end ports in the blocks. Specific format of these ports will be presented later through examples. Under the default setting, the layout of the signal lines are arranged automatically. Sometimes the automatic routing layout may not be satisfactory. The latter syntax can be used by setting directly the matrix composed of turning points (x_i, y_i).

To delete a certain signal line, function `delete_line(`h_1`)` can be used.

Example 6.23. Assume that to build a blank model, three blocks are needed: with a sinusoidal input block, a saturation block and a scope block. The sine block is linked to the saturation block, and the saturation is connected to the scope block. Use MATLAB commands to draw the Simulink model, and save it to a file.

Solutions. Of course, the simple model can easily be established with the modeling technique discussed earlier. If MATLAB command methods are used, the following commands can be used to draw the model:

```
>> h=new_system('c6mnm2'); open_system(h);            % a new model
   h1=add_block('simulink/Sources/Sine Wave','c6mnm2/Sine',...
      'Position',[40, 80, 80, 120]);                   % add sine block
   h2=add_block('simulink/Discontinuities/Saturation',...
      'c6mnm2/NL','Position',[140, 70, 230, 130]); % add saturation block
   h3=add_block('built-in/Scope','c6mnm2/My Scope',...
      'Position',[290, 80, 310, 120]);                 % add scope
   h4=add_line(h,'Sine/1','NL/1');       % connect sine and saturation block
   h5=add_line(h,'NL/1','My Scope/1'); % connect saturation to scope
```

The model drawn with the commands is shown in Figure 6.43. The model drawing commands are easy to understand. A new model is created first, and its position property is set. From Simulink built-in block library, the three blocks are selected and copied into the model window, with assigned new names. Then the three blocks are connected accordingly.

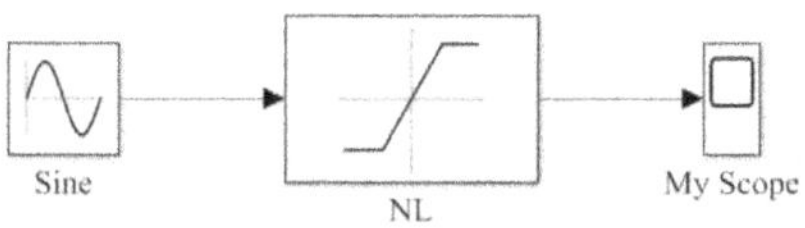

Figure 6.43: The Simulink model drawn with MATLAB commands (file: `c6mnm2.slx`).

Example 6.24. Consider the model established in Example 6.23. If a Mux block is added in between the saturation block and the scope block, let the scope block display simultaneously the input and output signals. Draw the Simulink model again.

Solutions. If automatic routing layout method is used, the following command can be used, where a Mux block is inserted. Connect output ports of the sinusoidal block and saturation block into the two input ports of Mux block. The output port of the Mux block is connected to the scope block, as shown in Figure 6.44(a):

```
>> h=new_system('c6mnm3'); open_system(h);
   h1=add_block('simulink/Sources/Sine Wave','c6mnm3/Sine',...
      'Position',[40, 80, 80, 120]);
   h2=add_block('simulink/Discontinuities/Saturation',...
      'c6mnm3/NL','Position',[140, 70, 230, 130]);
   h2a=add_block('built-in/Mux','c6mnm3/Mux',...
      'Position',[290,80,295,120],'Inputs','2');
   h3=add_block('built-in/Scope','c6mnm3/My Scope',...
      'Position',[320,80,340,120]); % move right the block
   h4=add_line(h,'Sine/1','NL/1');
   h4a=add_line(h,'Sine/1','Mux/2');
   h5=add_line(h,'NL/1','Mux/1');
   h6=add_line(h,'Mux/1','My Scope/1');
```

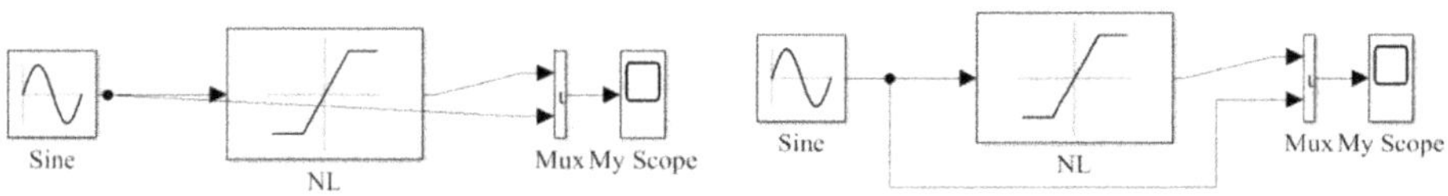

(a) automatic routed model (file: c6mnm3.slx) (b) modified model (file: c6mnm3a.mdl)

Figure 6.44: Simulink models drawn with commands.

It can be seen that the automatic routed signal lines is not ideal in the system model in Figure 6.44(a). Under this situation, the turning points can be assigned, with the following statements:

```
>> delete_line(h4a);    % delete the line
   add_line(h,[100,100; 100,150; 250,150; 250,110; 290,110]);
```

The model shown in Figure 6.44(b) can be constructed. It can be seen that any two blocks can be linked together, but the turning points need to be found. This method is sometimes rather complicated.

6.4.4 Filling block parameters

To set up the Simulink model, the function `set_param()` can be used to set the related parameters. The syntax of the function is

`set_param(`h`,property1,value1,property2,value2,` $\cdots$`)`

where h is the handle or the model name of the Simulink model, and it can also be a block handle or block name. "Property" is the property name supported in the model or block, and "value" is the property value of the object. For the model, the commonly used properties are shown in Table 6.2. For blocks, the property names and values are given in Table 6.1.

Table 6.2: Commonly used properties in Simulink model window.

Property name	Property value and explanation
`AbsTol`	Absolute error tolerance, with default `'auto'`, and the value is 10^{-6}
`RelTol`	Relative error tolerance, with default value of 10^{-3}. Generally, this value is too large, and may lead to large computing errors. It is recommended to set it to smallest possible value of 3×10^{-14}.
`FixedStep`	Fixed-step size, with default of `'auto'`
`InitialState`	Initial states
`StopTime`	Stop time
`SaveFormat`	Returned variable format, with default option `'Dataset'`. It is recommended to modify it to `'Array'`
`OutputVariables`	Returned variable name, default is `'ty'`, indicating the time and output are expected. If states are also expected, it can be set to `'txy'`
`Location`	Window location, as a 1×4 row vector. The first two components are the same as the ones in `Position` properties in the block. In model window, only `Location` can be used, while `Position` cannot be used.
`UserData`	Used to pass parameters
`Solver`	The default algorithm is `'VariableStepAuto'`, and if needed, it can be assigned as `ode45`, `ode15s` and so on.

To extract or modify a certain property in the Simulink model, the simplest way is to open the model and chose the block, then with h=`gcb` command to extract its handle. Or extract the block from `gcb`. To acquire the handle of the model, the command h=`gcs` can be used.

The property values of a model can be extracted with `f1=simget`(file name) command. The model contains various properties, and there is no need to modify them one by one. The modification of the commonly used ones is sufficient.

Example 6.25. A Simulink model was created in the previous example. The property name of the relative error tolerance is RelTol, with default value of 10^{-3}. The value is too large, which should be set to smaller values such as 10^{-8}. Besides, the properties such as MinStep and MaxStep can also be assigned with MATLAB commands.

Solutions. To modify the parameters in a model, the first thing to do is to get its handle of the model. With the handle, the properties can be modified directly, and the model can be saved into a model file:

```
>> f1=gcs;              % get the model handle
   set_param(f1,'MinStep','1e-5','MaxStep','1e-3','RelTol','1e-8');
   save_system(f1) % save the modified model
```

Example 6.26. Continue considering the model in Example 6.23. Set the upper and lower bounds of the saturation element to 3 and −2, respectively.

Solutions. Generally, for an unfamiliar block, if one wants to modify a certain parameter, the crucial step is to acquire the variable name of the parameter. For this particular example, the `c6mnm3.slx` model can be opened first, and select the saturation block. Then issue the following command:

```
>> get_param(gcb,'DialogParameters')
```

The variables names in the parameter dialog box are displayed:

```
      UpperLimit: [1×1 struct]
      LowerLimit: [1×1 struct]
LinearizeAsGain: [1×1 struct]
       ZeroCross: [1×1 struct]
      SampleTime: [1×1 struct]
          OutMin: [1×1 struct]
          OutMax: [1×1 struct]
  OutDataTypeStr: [1×1 struct]
       LockScale: [1×1 struct]
         RndMeth: [1×1 struct]
```

It can be seen that the property names of the upper and lower bounds are respectively UpperLimit and LowerLimit. Note that there is no space in the current version. The original model can be saved into the file `c6mnm3a.slx`. The following commands can be used to set the upper and lower bounds of the NL block:

```
>> set_param('c6mnm3a/NL','LowerLimit','-2','UpperLimit','3')
```

After setting, double click the block, and the parameter dialog box is opened. It can be seen that the two parameters are assigned to the expected values.

6.4.5 Modeling examples

Similar to the direct drawing methods, MATLAB commands can be used to draw Simulink models. The procedures of model drawing are:

(1) **Set up a new model.** A blank model can be created with `new_system()` and `open_system()` functions.

(2) **Add a block.** Function `add_block()` can be used to add necessary blocks to the model window.

(3) **Link blocks.** Function `add_line()` can be called to link blocks.

(4) **Model parameter modification.** The parameters in the model and blocks can be modified directly with function `set_param()`.

(5) **Model save.** Function `save_system()` can be used to save model into a model file.

In this section, examples are given to demonstrate the modeling methods with MATLAB commands.

Example 6.27. Statement modeling method is suitable to draw complicated by regular models. Set up the simulation model in Figure 6.45, and find the equivalent model from port A to port B, with $L = 7$.

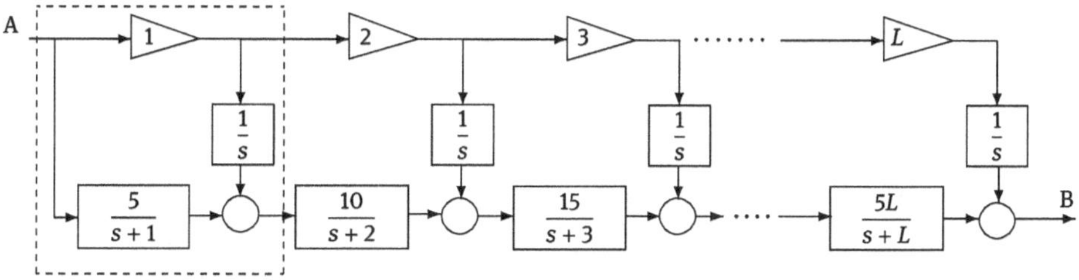

Figure 6.45: Complicated and regular circuit.

Solutions. If L is large enough, it will be very complicated to draw the model manually. Therefore, the command-line driven method should be used to draw the simulation model.

It can be seen from the diagram that the model has composed of L similar substructures. The substructure is shown in the dash box shown in Figure 6.45. The substructure includes a gain block, a transfer function block, an integrator block and a sum block. Therefore, a loop structure should be used to draw the entire simulation model.

If the first gain block is located at $[100, 50, 125, 70]$, and the width of the substructure is 95, the ith gain block is located at $[100 + 95(i-1), 50, 125 + 95(i-1), 70]$. The positions of other blocks in the substructure can be assigned accordingly. The blocks can be drawn and then connected. After the loop execution, the input and output ports can be drawn, and saved into a file. Based on the above ideas, the following MATLAB function can be written:

```
function h=c6drawsys(M,L)
h=new_system(M); open_system(h);
I='built-in/Integrator'; Sm='built-in/Sum'; %simple notation
R='built-in/Gain'; TF='built-in/TransferFcn';
for i=1:L, i1=int2str(i); i0=int2str(i-1);  %positions
   pos1=[90+95*(i-1) 50 115+95*(i-1) 70];
   pos2=pos1+[15 40 15 50];
   pos3=pos1+[-43 95 -20 110];
   pos4=pos1+[20 105 15 100];;
   add_block(R,[M '/gain' i1],'Position',pos1,'Gain',i1)
   add_block(I,[M '/int' i1],...
       'Position',pos2,'Orientation','down')
   add_block(TF,[M '/tf' i1],'Position',pos3,...
       'Numerator',int2str(5*i),'Denominator',['[1 ' i1 ']'])
   add_block(Sm,[M '/sum' i1],...
       'Position',pos4,'IconShape','round','Inputs','++|')
   add_line(h,['gain' i1 '/1'],['int' i1 '/1']) %connect blocks
   add_line(h,['int' i1 '/1'],['sum' i1 '/1'])
   add_line(h,['tf' i1 '/1'],['sum' i1 '/2'])
   if i>1   %if not the first group
      add_line(h,['gain' i0 '/1'],['gain' i1 '/1'])
      add_line(h,['sum' i0 '/1'],['tf' i1 '/1'])
end, end
add_block('built-in/Inport',[M '/in'],'Position',[7 50 27 70])
pos5=pos4+[35 0 35 0];
add_block('built-in/Outport',[M '/out'],'Position',pos5)
add_line(h,'in/1','gain1/1');                   %input port
add_line(h,'in/1','tf1/1')
add_line(h,['sum' int2str(L) '/1'],'out/1'); %output port
save_system(h)  %save the model to a file
```

With the above general function, the following commands can be used to draw the simulation model, as shown in Figure 6.46. When the model drawing process is completed, it can be saved into the `c6mnm5.slx` file.

```
>> L=7; M='c6mnm5'; c6drawsys(M,L)
```

Note that if the above commands are executed twice, the warning message "The model name `'c6mnm5'` is shadowing another name in the MATLAB workspace or path" is displayed. This is a normal phenomenon, since when the first time the code is executed, a model file `c6mnm5.slx` is created. If it is executed again, MATLAB may find the existence of such a model file. Therefore, the warning message is displayed. To solve

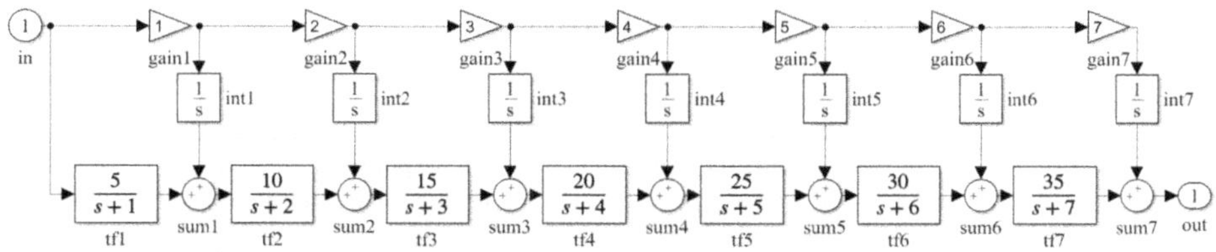

Figure 6.46: The automatically generated Simulink model (file: `c6mnm5.slx`).

the problem, the easiest way is to modify the variable `M`, and select a different model name.

With the simulation model, the linearization method in Chapter 5 can be adopted, to extract the equivalent zero-pole model of the system. Function `minreal()` can be used to find the minimum realized model, that is, to eliminate the pole and zero pairs at the same locations.

```
>> G=linearize(M); zpk(minreal(G))
```

and the zero-pole model obtained is

$$G(s)=\frac{5040(s+0.2987)(s^2+22.66s+138.5)(s^2-0.8458s+13.56)(s^2+10.89s+71.76)}{s^7(s+7)(s+6)(s+5)(s+4)(s+3)(s+2)(s+1)}.$$

6.4.6 Masking blocks with MATLAB commands

Apart from the above presented parameter dialog box design methods, MATLAB commands can be used to mask Simulink blocks. The two types of subsystem masking commands and methods are illustrated.

(1) **Classical method.** The so-called classical method is to use low-level functions such as `set_param()` and `get_param()` to mask the blocks. The commonly used parameter names and values are presented in Table 6.3.

Example 6.28. A subsystem model `c6mpid1.slx` of a PID controller is designed in Example 6.1. Use MATLAB commands to mask the block into a reusable one.

Solutions. Open the existing `c6mpid1a.slx` model. Find the Subsystem block to be masked, and set it to variable `blk`. The variable is the string of the model name. With variable `blk`, function `set_param()` can be used to implement the block masking process. In the commands below, four controls are defined, with the first three as edit boxes, and the final one the listbox. Note that `'popup'` control can be specified as `'popup'`, followed by selectable list and no space is allowed in the string.

```
>> open_system('c6mpid1a'); blk='c6mpid1a/Subsystem';
   p='popup(10|100|1000|10000)'; %set popup items
```

Table 6.3: The commonly used properties in model masking.

Property name	Property values and explanation
MaskTypes	Dialog box control types, such as 'edit', 'popup', 'checkbox', 'radio' and so on. If many parameters are contained in a dialog box, cell arrays are used to represent the controls. For instance, if a parameter dialog box contains two edit boxes and a checkbox, the property value is set to {'edit','edit','check'}.
MaskValues	Default values of the parameters, cell array can be used to specify the values as strings. For instance, if the default values of the above three controls are 1.5, 1.3 and 0, it can be set to '1.5','1.3','0'.
MaskPrompts	Prompt for parameters, specified in a cell array of strings
MaskVariables	Assign the variable names and values, where, 'Kp=@1' command assign the value in the first control to variable Kp, where Kp is not a MATLAB workspace variable, but an internal variable in the masked block.

```
set_param(blk,...                    % set masked parameters
   'MaskStyles',{'edit','edit','edit',p},...
   'MaskVariables','Kp=@1; Ki=@2; Kd=@3; N=@4;',...
   'MaskPrompts',{'Proportional Kp','Integral Ki',...
                  'Derivative Kd','Filter N'},...
   'MaskValues',{'1','1','1','10'});
```

The above block masking commands can be used, where, @i represents the ith parameter in the parameter dialog box.

(2) **Object oriented method.** A class Simulink.Mask is defined in Simulink for block masking. Various methods are supported in this class, and the commonly used methods are shown in Table 6.4.

Table 6.4: Commonly used methods in the class.

Method name	Method value and explanation
create	Create a mask
delete	Relieve a mask
addParameter	Add a parameter item in the dialog box. Commonly used properties and their values are, for Type property, the options 'edit', 'popup', 'checkbox', 'radiobutton' and so on; The Name property has the variable name as its value; 'Prompt' property with prompts in a string; Value property with the default string as the value; Visible property with options 'on' and 'off'. Besides, properties Enable and Callback are allowed. To set 'popup' type, the option list can be provided in the TypeOptions property.
Initialization	Initialization setting
addDialogControl	Add a pane or other control

If the handle of a subsystem or a block is h_1, the following command can be used to create a mask object h:

h=`Simulink.Mask.create(`h_1`)`

If h_1 is unknown, the subsystem block should be selected first. Then the following command can be used to create a block masking handle h:

h=`Simulink.Mask.create(gcb)` or h=`Simulink.Mask.create(gcbh)`

where the `gcb` command is used to find the name string of the selected block, while `gcbh` is used to get the handle of the block. The mask handle can be created, and function `addParameter()` can be used to add dialog box parameter items one-by-one. Parameter dialog box design method is demonstrated next through the following example.

Example 6.29. Use MATLAB command to mask the subsystem in model `c6mpid1a.slx`, with object oriented method.

Solutions. With the object oriented methods, the `Simulink.Mask.create()` command should be used to create a blank model framework. Then function `addParameter()` is called to add parameter items in the parameter dialog box. The following MATLAB commands can be used to mask a subsystem block:

```
>> h0=open_system('c6mpid1a'); mod='c6mpid1a/Subsystem';
   h=Simulink.Mask.create(mod); %create a mask object
   h.addParameter('Type','edit','Name','Kp',...
      'Prompt','Proportional Kp','Value','1.0')
   h.addParameter('Type','edit','Name','Ki',...
      'Prompt','Integral Ki','Value','1.0')
   h.addParameter('Type','edit','Name','Kd',...
      'Prompt','Derivative Kd','Value','1.0')
   h.addParameter('Type','popup','Name','N',...
      'Prompt','Filter N','TypeOptions',{'10','100','1000'})
```

In the object oriented masking method, an individual parameter item can be modified. For instance, an extra option 10000 is to be added to the Filter N listbox, and the following command can be issued:

```
>> h.Parameters(4).TypeOptions={'10','100','1000','10000'};
```

6.5 Control simulation processes with MATLAB commands

The method to start a simulation process presented earlier is to press the Run button in the Simulink interface. In fact, in many applications, MATLAB commands are needed to control the simulation process. For instance, to test the time elapse of a simulation process, or embed the Simulink model in an optimization process, the button clicking method cannot be used. Function `sim()` can be used to execute simulation process once. In fact, this method has been used earlier. In this section, MATLAB commands can be used to control simulation process. The Simulink model control parameter setting method is presented.

6.5.1 Invoke simulation process with MATLAB commands

Function `sim()` has been used earlier. The simple syntax of the function is

[***t***,***x***,***y***]=sim(model name)

where ***t*** returns the time vector, ***x*** is the matrix composed of internal states and ***y*** is the matrix of output signals. The output signals are identified with output ports. In real applications, the statement may lead to error messages “Too many return arguments are specified. Specify only one.” This is because in the new versions of MATLAB single simulation output data format is allowed. Besides, the returned ***y*** is not a matrix, but a structured data variable. The recommended setting is shown in Figure 4.5. For convenience of presentation, the Single simulation output checkbox should be unchecked, and in the Format listbox, the default item should be changed from Dataset to Array. The use of the function under default setting will be presented later.

A more complicated syntax of `sim()` function is

[***t***,***x***,***y***]=sim(model name, [t_0,t_n],name1,property1, name2, property2,⋯)

where, property names and values can be found in Table 6.2. Of course, the name and values in the parameters can be set with `set_param()` function, and then the above mentioned simple syntax of `sim()` function is used. The latter method is recommended in this book.

Example 6.30. Simulate the model in Example 4.1 with commands.

Solutions. Since Single simulation output item is checked in the `c4mmove1.slx` model, function `sim()` may yield error information. Open the model control dialog box, and in the MODELING pane, click the icon in the toolbar, the simulation control dialog box can be opened. Select the Data Import/Export item on the left, the dialog pane is changed to the format shown in Figure 4.5. According to the above mentioned setting, the new model is saved to `c6mmove1.slx`. The following commands can be used to carry out simulation, and the results are exactly the same.

```
>> [t x y]=sim('c6mmove1'); plot(t,y)
```

Of course, the control parameters can also be set. For instance, to set stop time to 20, and set the two error tolerances to 3×10^{-14}, such that the most accurate results can be obtained under double precision data type. Note that when using parameter values, strings should be used, rather than digits:

```
>> set_param('c6mmove1','StopTime','20',...
             'RelTol','3e-14','AbsTol','3e-14');
   [t x y]=sim('c6mmove1'); plot(t,y)
```

Compared with direct simulation method from the toolbar button, the command-line driven method has a potential advantage. That is, the simulation efficiency is higher, and in some examples, the difference is rather significant. An example is given next to compare the two methods. The time elapse of the two methods are compared. Later parallel simulation are also compared.

Example 6.31. Compare the time elapse of the model in Example 4.20 in the two simulation methods.

Solutions. It is rather difficult to measure the time elapse in the direct simulation method. An alternative way is taken here. By using direct method, it implies that the simulation model must be opened in simulation process. Therefore, command-line driven method are used in both cases, one is to open the model, the other is to close the model. When the model is opened, the time elapse is 28.58 seconds, while if the model is closed, the time elapsed is 22.34 seconds. It can be seen that the command-line driven method is more efficient. Besides, in the simulation process, the prompt "Compiling" appears in the status bar of the model. To avoid the phenomenon, the dedicated method will be presented later.

6.5.2 Setting control parameters

With the `set_param()` function discussed earlier, other information can be assigned, to control the simulation process. Examples are given next to show how to get simulation results under different relative error tolerances.

Example 6.32. In Example 4.6, various relative error tolerances are tried, and maximum errors are recorded. Use loop structure to complete the same comparison tasks.

Solutions. To use `set_param()` function to modify the model parameters, the model must be opened; otherwise, the model parameters cannot be found or modified. When the model is opened, loop structure can be used to set different relative error tolerances, and measure the maximum error such that the data can be recorded in a table variable T. Note that to set the parameter, it must be converted first into string, with function num2str(); otherwise, it cannot be assigned:

```
>> open_system('c4mkey2')  % open the model
   hh=[1e-3, 1e-5, 1e-8, 1e-10, 1e-12, 3e-14]; T=[];
   for h=hh                    % test the error with loop structure
      set_param('c4mkey2','RelTol',num2str(h));
      tic, [t0,~,y0]=sim('c4mkey2'); toc
      y1=exp(cos(t0.^2)); y2=exp(sin(t0.^2));
      e1=max(abs(y0(:,1)-y1)), e2=max(abs(y0(:,2)-y2))
      T=[T, [e1; e2]];         % record the result in a table
   end
```

Example 6.33. In Example 3.10, the PreLoadFcn callback function was assigned using the Simulink interface. Set the same action using MATLAB commands.

Solutions. The model c3mpid1.slx can be opened first, then with set_param() function, the PreLoadFcn callback function can be set. The action is the same as the one in Example 3.10. After modification, the model can be saved into the same file:

```
>> open_system('c3mpid1'); % open the Simulink model
   set_param('c3mpid1',... % set the PreLoadFcn callback
         'PreLoadFcn','Kp=0.998; Ki=0.333; Kd=0.0809;')
   save_system('c3mpid1')  % save the model
```

6.5.3 Input and output data structures in Simulink models

The **Single simulation output** option was discussed earlier. If this option is checked, sim() function returns only one argument, with data structure of Simulink.SimulationOutput. Under certain cases, for instance, in parallel simulation to be presented later, this data structure must be used. Similarly, the input information can be described by the Simulink.SimulationInput data type. These data types were introduced in MATLAB R2017a, and the default format in new versions are the direct descriptions of the objects. In this section, the data types and applications are presented.

In Simulink.SimulationInput data structure, one variable is used to described all the information in the entire model. An input object is created with the commands. Examples are given next to demonstrate the creation of the object.

Example 6.34. Describe the Simulink model in Example 4.1 in a simulation input object.

Solutions. Open the `c4mmove1.slx` model, the call `Simulink.SimulationInput()` function directly to create a simulation input object `in`.

```
>> open_system('c4mmove1');                   % open model
   in=Simulink.SimulationInput('c4mmove1')  % create an input object
```

When the model object is created, the `in` variable is displayed as

```
SimulationInput - properties:
        ModelName: 'c4mmove1'
     InitialState: [0×0 Simulink.op.ModelOperatingPoint]
    ExternalInput: []
  ModelParameters: [0×0 Simulink.Simulation.ModelParameter]
  BlockParameters: [0×0 Simulink.Simulation.BlockParameter]
        Variables: [0×0 Simulink.Simulation.Variable]
        PreSimFcn: []
       PostSimFcn: []
       UserString: ''
```

It is easily seen that in the simulation input object, the property `ModelName` is modified, while other properties are unchanged. In the input object, the properties such as `BlockParameters`, `ModelParameters` and `Variables` are provided. They can be reassigned with commands. Also, the properties `InitialStates` and `ExternalInput` can also be modified. Besides, the properties `PreSimFcn` and `PostSimFcn` can be associated to certain callback functions. These properties can be modified or assigned with relevant commands.

If there are variable parameters in the model, function `setVariable()` can be used to modified the parameter values, with the syntax

```
in=in.setVariable(variable name,variable value)
in=in.setVariable(variable name,variable value,'Workspace',model name)
```

In the former command, the variable name in the model can be modified, and the variable value applies to all the models opened, if the same name is used. If the user wants to modify the parameter inside a particular model, the `'Workspace'` option should be assigned, such that the action takes effect within a certain model. Note that function `setVariable()` can only change one parameter at a time.

Similarly, functions `setModelParameter()` and `setBlockParameter()` can be used to modify directly parameters in models and blocks. For instance, `StopTime`. Note that the values can only be specified in strings, rather than specific values; otherwise, error messages are given. On the contrary, `setVariable()` function only accepts specific values, not strings.

Compared with `set_param()` function discussed earlier, these functions do not modify the model, they only tackle the values of the parameters. After simulation, the Simulink model itself are not affected. To modify the model, function `in.applyTo-Model` can be use.

Examples are given next to demonstrate parameter modification methods.

Example 6.35. Consider the simulation problem in Example 4.1. To drop an object on the moon surface, the model must be modified. Set the stop time to 20 seconds, and simulation the dropping problem on the moon.

Solutions. Compared with the dropping problem on the earth, the difference is the gravity accelerator on the moon is $1.63\,\mathrm{m/s^2}$, not $9.81\,\mathrm{m/s^2}$ in the file `c4mmove1.slx`. Of course, one may change the value directly in the model to $1.63\,\mathrm{m/s^2}$. This is not the intension of this example, since the modification only applies to the moon, not suitable for the earth or other planets. Therefore, the specific value in the model is modified as variable K, and save it into a new model `c6mmove1a.slx`. Therefore, the following commands can be used to modify the value of K in the model:

```
>> open_system('c6mmove1a');                        % open the model
   in=Simulink.SimulationInput('c6mmove1a');        % create an input object
   in=in.setVariable('K',1.63);                     % modify variable value
   in=in.setModelParameter('StopTime','20');        % set stop time
```

If the input data structure is modified, command `out=sim(in)` can be used to perform simulation, and the returned result `out` is a `Simulink.SimulationOutput` object. An example is given next to demonstrate the data structure of the simulation results.

Example 6.36. Solve the simulation problem in Example 6.35, and observe the data structure of the simulation results.

Solutions. An input object should be established first, then the following commands can be issued to carry out simulation:

```
>> open_system('c6mmove1a');                        % open model
   in=Simulink.SimulationInput('c6mmove1a');        % create a input object
   in=in.setModelParameter('StopTime','20');        % set stop time
   in=in.setVariable('K',1.63); out=sim(in)         % modify the parameter
```

the result obtained is

```
Simulink.SimulationOutput:
                    tout: [56×1 double]
                    yout: [1×1 Simulink.SimulationData.Dataset]
```

```
SimulationMetadata: [1×1 Simulink.SimulationMetadata]
      ErrorMessage: [0×0 char]
```

It can be seen that `out.tout` is a 56×1 column vector, while `out.yout` is a structure variable. It is rather difficult to visit its particular values. The commands in Example 4.1 can be used:

```
>> plot(out.tout, out.yout{1}.Values.Data)
```

Function `setExternalInput()` can be used to set external input signals. The signal applies automatically to the input port of the model. Some specific problems can be solved in this way, and the readers may explore them.

6.5.4 Parallel simulation

Function `parsim()` is provided in MATLAB to implement parallel simulation of several models. In vector `in` is composed of the input objects of several models, command `out=parsim(in)` can be used to carry out parallel simulation directly, and simulation results are returned in the vector `out`. Examples are given in this section to demonstrate the parallel simulation methods, and execution efficiencies are compared in parallel and traditional series simulation methods.

Example 6.37. Consider the simulation problem in Example 6.35. The gravity accelerators on different planets are provided in Table 6.5. Find the displacement of the object in 20 seconds, when dropped from different planets. Assess also the simulation efficiency.

Table 6.5: The gravity accelerators on different planets.

planets	Earth	Moon	Mercury	Venus	Mars	Jupiter	Saturn	Uranus	Neptune
Gravity accelerator (m/s^2)	9.81	1.63	3.7	8.87	3.69	20.87	7.21	8.43	10.71

Solutions. The entries in the above table can be assigned into a vector $\boldsymbol{K}_0$, then loop structure can be used to set up the nine models and compose the `in` vector. Start the parallel simulation process. It is found that the total time elapse is 5.18 seconds.

```
>> K0=[9.81,1.63,3.7,8.87,3.69,20.87,7.21,8.43,10.71]; tic
   for i=1:length(K0)
      in1=Simulink.SimulationInput('c6mmove1a'); % create object
      in1=in1.setVariable('K',K0(i));              % modify variables
```

```
    in1=in1.setModelParameter('StopTime','20'); in(i)=in1;
  end
  out=parsim(in); toc
```

For this specific example, if the model is opened, then carry out simulation process; the time elapse is as high as 44.38 seconds, much higher than the time lapse demonstrated above. Therefore, it is advised to close the model before simulation process is initiated.

Series simulation can be carried out with the following commands. The total time elapsed is only 1.82 seconds, far much lower than the parallel method. For small-scale problems, there is no need to carry out parallel simulation methods. If the model is opened, the time lapse with series simulation method is 3.60 seconds, almost doubled.

```
>> K0=[9.81,1.63,3.7,8.87,3.69,20.87,7.21,8.43,10.71]; tic
  for i=1:length(K0)
    in=Simulink.SimulationInput('c6mmove1a'); %create object
    in=in.setVariable('K',K0(i));             %modify parameters
    in=in.setModelParameter('StopTime','20'); out=sim(in);
  end, toc
```

Example 6.38. The computational load can be increased, such as using fixed-step algorithm, and set the step-size to $h = 0.00001$, where 2000001 points are computed in each simulation execution. Compare the efficiencies in the series and parallel simulation.

Solutions. The gravity accelerator vector $\boldsymbol{K}_0$ can be established first, then nine models are created with loop structure, and compose the `in` model vector. Start the parallel simulation mechanism, and it is found that the total time elapsed is 56.98 seconds. It should be noted that the simulation task was automatically assigned to four workers, and nine simulation processes are carried out. If eight simulation processes are carried, the efficiency may further increased.

```
>> K0=[9.81,1.63,3.7,8.87,3.69,20.87,7.21,8.43,10.71]; tic
  for i=1:length(K0)
    in1=Simulink.SimulationInput('c6mmove1b'); %create input object
    in1=in1.setVariable('K',K0(i));            %modify variable
    in1=in1.setModelParameter('StopTime','20'); in(i)=in1;
  end
  out=parsim(in); toc
```

Execute the above simulation statements, and the intermediate information in the simulation process is displayed, from which the time distribution can be observed:

```
[14-Jul-2020 16:00:44] Checking for availability of parallel pool
[14-Jul-2020 16:00:44] Starting Simulink on parallel workers...
[14-Jul-2020 16:00:45] Configuring simulation cache folder
[14-Jul-2020 16:00:45] Loading model on parallel workers...
[14-Jul-2020 16:00:48] Running simulations...
[14-Jul-2020 16:01:08] Completed 1 of 9 simulation runs
[14-Jul-2020 16:01:08] Completed 2 of 9 simulation runs
[14-Jul-2020 16:01:09] Completed 3 of 9 simulation runs
[14-Jul-2020 16:01:09] Completed 4 of 9 simulation runs
[14-Jul-2020 16:01:26] Completed 5 of 9 simulation runs
[14-Jul-2020 16:01:27] Completed 6 of 9 simulation runs
[14-Jul-2020 16:01:27] Completed 7 of 9 simulation runs
[14-Jul-2020 16:01:27] Completed 8 of 9 simulation runs
[14-Jul-2020 16:01:36] Completed 9 of 9 simulation runs
[14-Jul-2020 16:01:36] Cleaning up parallel workers...
```

Series simulation can be carried out with the following statements, and the total time elapsed is 94.36 seconds. For this particular example, the time elapse is increased a lot.

```
>> K0=[9.81,1.63,3.7,8.87,3.69,20.87,7.21,8.43,10.71]; tic
   for i=1:length(K0)
      in=Simulink.SimulationInput('c6mmove1b'); % create input object
      in=in.setVariable('K',K0(i));              % modify parameters
      in=in.setModelParameter('StopTime','20'); out=sim(in);
   end, toc
```

6.6 Exercises

6.1 Refer to the enabled subsystem in Example 6.7. If the enabled subsystem is replaced by an enabled and triggered subsystem, observe the behavior of the subsystem.

6.2 In Example 6.31, the time elapsed measurement of the `c4disc.slx` model is carried out with indirect method. In fact, when callback functions are called automatically before and after the simulation process, more accurate time elapse information can be found. Find the accurate time elapsed in the simulation process.

6.3 The Oustaloup filter block used in Section 4.5 is a masked block designed by the author. The block can be used directly in the simulation processes of various fractional-order differential equations. Open the mask editor interface and observe the internal structure and mask parameter design.

6.4 Celsius and Fahrenheit temperature systems are the most widely used systems. The relationships between Celsius temperature c and Fahrenheit temperature f is given by

$$c = \frac{5}{9}(f - 32), \quad f = \frac{9}{5}c + 32.$$

Set up a temperature conversion masked block, where a listbox is used to identify the system, and the temperatures to be converted and given in edit boxes.

6.5 A model reference adaptive control system model is given in Exercise 5.16 (see Figure 5.50), where the sum block, the transfer function $1/(a_2 s + a_3)$ and the integrator follows composed the plant model. Apart from these blocks, the remaining parts are the model reference adaptive controller. Make a subsystem from the blocks, and then mask the subsystem, so that a reusable model reference adaptive controller block can be constructed.

7 Programming implementations of Simulink blocks

In the previous chapters, various low-level modeling methods were presented to create Simulink models. In real applications, if a certain part in the model whose mathematical expression is too complicated, low-level modeling methods may not be suitable. Programming based modeling methods should be introduced. Two kinds of programming based modeling methods are supported in Simulink: M-functions and S-functions. Their uses are different, where the former one is suitable to describe the algebraic relationship between the input and output signals, while the latter is suitable for describing dynamic relationships. The so-called dynamic relationships are those described by state-space expressions. S in S-function stands for "system." In control systems theory, controllers designed with complicated algorithms are always needed, and the complexities in the modeling these algorithms are usually difficult with low-level block diagram construction. Programming based simulation analysis is always needed by the use of S-functions. If the utilization of M- and S-functions are mastered, theoretically speaking, the modeling and simulation of systems of any complexity can be handled successfully in Simulink.

In Section 7.1, the use of M-functions are demonstrated with examples, and the limitations of M-functions are pointed out. The concept of system functions are then introduced. From Section 7.2 onward, the fundamental structure and programming in S-function blocks are discussed. S-function programming and skills are demonstrated with examples. The masking methods of S-function blocks are also presented such that the construction of reusable block are illustrated. In Section 7.3, S-function programming in C language is briefly introduced, with C language code and C compiler setting methods. Compared with MATLAB versions of S-functions, the S-functions in C are more suitable for real-time simulation and control. In Section 7.4, modeling and simulation of active disturbance rejection control based on S-functions are illustrated.

7.1 M-function blocks and implementation

If the static function $f(u)$ is too complicated to be described with low-level block building methods, and programming based methods can be considered. In the User-Defined Functions group in Simulink, many blocks for static function description are provided, as shown in Figure 7.1. In fact, some examples are given to demonstrate the use of M-function blocks in the earlier presentation, including the use of Interpreted MATLAB Fcn and MATLAB Function blocks.

Apart from these blocks, the Fcn block can also be used to process static functions with scalar outputs. Block C Caller allows to describe the function in C language. Besides, there are blocks to control simulation processes, such as Initialize Function, Reset Function and Terminate Function, allowing responses to certain simulation processes. In this section, various static M-function blocks are demonstrated.

https://doi.org/10.1515/9783110734959-007

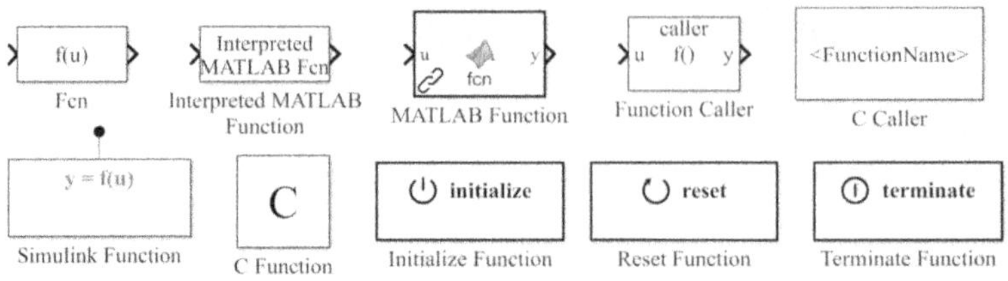

Figure 7.1: M-function related blocks.

7.1.1 Static nonlinearities

Static function are those whose output signal $\boldsymbol{y}(t)$ can be determined uniquely by the current values of the input signal $\boldsymbol{u}(t)$, with the mathematical expression of $\boldsymbol{y}(t) = \boldsymbol{f}(\boldsymbol{u}(t))$, where $\boldsymbol{f}(\cdot)$ can be of any nonlinear functions.

Several simple static function blocks are provided in MATLAB, where the simplest one is the Fcn block. This block was used extensively in the earlier examples. The advantages of using such a block is that there is no need to write a genuine MATLAB function file for it. All the static function expressions can be specified directly in the edit box of its parameter dialog boxes. There are limitations in this block as well. One of them is that the output can be evaluated with one statement, and no intermediate computations are allowed; the other one is that vector input signal $\boldsymbol{u}(t)$ is acceptable, but the output $y(t)$ must be a scalar signal, with no vector output signals allowed in the block.

On the contrary, the blocks Interpreted MATLAB Function and MATLAB Function both support vector output signals, and intermediate statements are allowed. The difference between the two blocks are that, in the former one, a genuine M-function file must be provided, while in the later, the function is embedded in the block, with no actual file required. A simple example is given next to demonstrate the processing of static vector function.

Example 7.1. Generate a cluster of sinusoidal signals $u_i(t) = \sin(\omega_i t + \phi_i)$, $i = 1, 2, \ldots, 5$, with frequencies $\omega_i = i$, and initial phases $\phi_i = i\pi/2$.

Solutions. Of course, several sinusoidal signal blocks can be used to generate the expected signals, but the modeling may be complicated. Now consider using M-function to generate the expected signals. For the expected signal generator, the input signal is t, which can be generated with Clock block. Signal t can be used to generate the M-function block, the outputs of the block are the expected signal generator. Meanwhile, the M-function block accepts scalar input signal, and output one vector signal, with five scalar ones:

```
function y=c7msins(u) %direct relationship with no other parameters
i=[1:5]'; w=i; phi=i*pi/2; y=sin(w*u+phi);
```

With the above M-function, the Simulink model in Figure 7.2(a) can be established directly. Double click the **Interpreted MATLAB Fcn** block, the parameter dialog box is shown in Figure 7.2(b). In the opened dialog box, the name of the M-function file, `c7msins`, the Simulink model is associated with the file `c7msins.m`, so that the file can be called directly. Note that the default value in the **Output dimension** edit box is −1, indicating that the sizes of input and output signals are the same. In this example, the input signal t is scalar, and output is vector signal with 5 scalar ones. Therefore, 5 should be specified in the edit box.

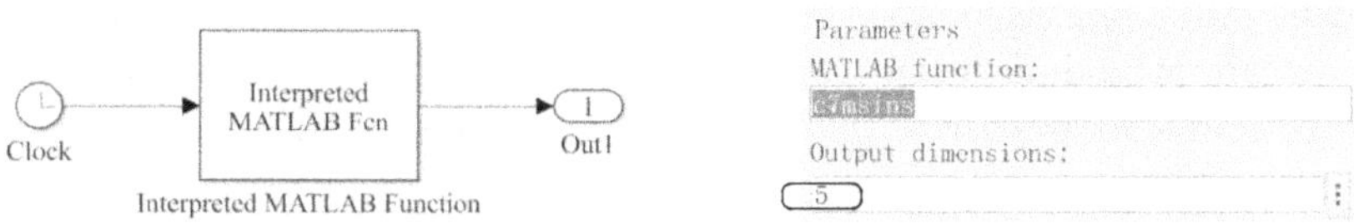

(a) Simulink model (file: `c7mmfun1.slx`) (b) parameter dialog box

Figure 7.2: The use of the M-function block.

Start the simulation process, the output signals are obtained, as shown in Figure 7.3. It can be seen that there are five output signals generated, with different frequencies and initial phases, as expected. Note that since there is no dynamic block in the model, a fixed-step algorithm must be selected. A small step-size such as 0.0001 can be set; otherwise, erroneous results may be obtained.

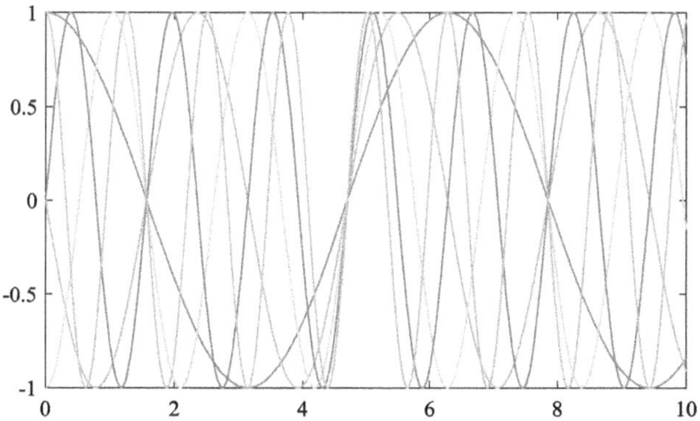

Figure 7.3: Simulation result.

If the **Interpreted MATLAB Fcn** block is replaced by **MATLAB Function** block, the above code can be copied into the internal editor, such that the same results can be obtained. The difference between this block and the **Interpreted MATLAB Function** block is that no genuine M-function file is provided. Besides, there is no need to specify the dimension of the output signal. The dimension of the output signal can be extracted automatically from the internal editor.

7.1.2 Limitations in M-function blocks

It should be noted that the model created by M-function block has a fatal problem, since the block can only be used to process fixed problems. No additional parameters are allowed. For instance, in the sinusoidal signal generator in Example 7.1, the frequencies and initial phases must be given first, and represented in the M-function file. The values cannot be changed and cannot be modified with parameter dialog box. This may bring trouble for the users. The S-function to be presented later can be used to solve the problems.

7.2 Programming of S-functions

In the low-level Simulink modeling method discussed earlier, M-function based methods are also presented. The major restriction of M-function method is that it can only be used to handle static models, while in the modeling of dynamical systems, the description of dynamic models is needed. Dynamic here means that the output signals not only depend upon the input signals, but also upon the states of the system. S-function description is a vigorous means in describing dynamical systems. In the User-Defined Functions group in Simulink, the blocks related to S-functions are shown in Figure 7.4, with the ordinary S-Function block, Level 2 MATLAB S-Function block. Also, a S-Function Builder block provided in the group.

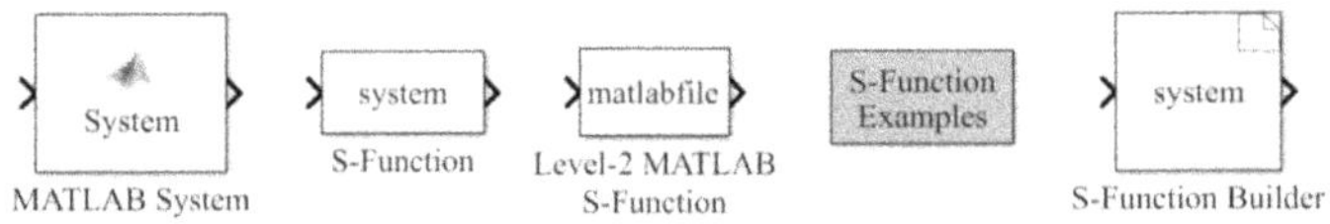

Figure 7.4: S-function related blocks.

In this section, the concept of hybrid system is proposed, and the mathematical models of the hybrid systems are formulated. The fundamental structures and execution mechanisms of S-functions are illustrated. Examples are used to demonstrate programming and masking methods for S-function blocks, such that reusable blocks can be constructed, so as to better complete modeling and simulation problems.

7.2.1 State space models

In the systems discussed earlier, the system contains continuous states only, or discrete states only. In real applications, they may contain continuous and discrete states simultaneously. For instance, in computer control systems, the plant model is usually described by a continuous state space equations, while the controller is usually described by a discrete state space model. Therefore, continuous and discrete states are both contained in the system. The models containing both continuous and discrete states are referred to as hybrid systems.

In fact, hybrid models may refer to a more generalized model format. The discrete states aforementioned are more accurately referred to as a discrete-time system or sample data system. The states of the system are driven by the unified time frame. Apart from discrete-time systems, there is yet another class of systems, referred to as discrete event systems, where the entire system is driven by events, for instance, in the Windows system. If a button is pressed, or a menu item is selected, it is an event from the system viewpoint. When an event happens, it is inserted into the event list, waiting for processing. Discrete event systems are not considered here, and the concept of the hybrid system is restricted to continuous and discrete-time systems.

Definition 7.1. Assume that the state of the system can be composed of $\boldsymbol{x}(t) = [\boldsymbol{x}_\mathrm{c}(t), \boldsymbol{x}_\mathrm{d}(t)]$, where $\boldsymbol{x}_\mathrm{c}(t)$ is the continuous state vector, while $\boldsymbol{x}_\mathrm{d}(t)$ is the discrete state vector. The continuous state space model in the hybrid system is given by

$$\boldsymbol{x}_\mathrm{c}'(t) = \boldsymbol{f}(t, \boldsymbol{x}(t), \boldsymbol{u}(t)), \tag{7.2.1}$$

while the discrete state space model given by

$$\boldsymbol{x}_\mathrm{d}(t+1) = \boldsymbol{g}(t, \boldsymbol{x}(t), \boldsymbol{u}(t)), \tag{7.2.2}$$

with the output equation of the system

$$\boldsymbol{y}(t) = \boldsymbol{h}(t, \boldsymbol{x}(t), \boldsymbol{u}(t)). \tag{7.2.3}$$

S-function block modeling methods are suitable for the Simulink description of these types of hybrid systems.

7.2.2 Fundamental structures of S-functions

Some blocks with complicated algorithms are suitable to be described in the fixed format of S-function format, using MATLAB language. If the original model can be mathematically described as hybrid systems, it can be described directly with S-function blocks. It should be noted that the S-function blocks thus programmed can only be used in Simulink based simulation, and cannot be converted to standalone programs, independent of MATLAB. The S-function in C language can be converted into standalone programs, which will be discussed later.

The leading S-function statement is

`function [sys,`$\boldsymbol{x}_0$`,str,ts]=`function name`(`t`,`$\boldsymbol{x}$`,`$\boldsymbol{u}$`,flag,`p_1`,`p_2`,...)`

where, "function name" is also the file name of an S-function source code; the input arguments t, $\boldsymbol{x}$ and $\boldsymbol{u}$ are respectively the time and the current values of states and input signals; `flag` is an identifier, as shown in Table 7.1, where the `flag` values of 4 and 9 are seldom used. The returned arguments under different `flag` values are also illustrated in the table. Arbitrary number of additional parameters $p_1, p_2, \ldots$ are allowed in the S-function. These parameters can be specified in the S-function parameter dialog box, as demonstrated later in examples.

Table 7.1: The `flag` values and response functions.

Flag	Physical meaning	Response function	The returned arguments
0	initialization	`mdlInitializeSizes`	`sys` is the initialization object, $\boldsymbol{x}_0$, `str` and `ts` are as in definitions
1	continuous states	`mdlDerivatives`	`sys` returns the derivative of the continuous state variables
2	discrete states	`mdlUpdate`	`sys` returns the values of the discrete state at next sample
3	output	`mdlOutputs`	`sys` returns the output
4	next time	`mdlGetTimeOfNextVarHit`	`sys` returns next time instance
9	terminate setting	`mdlTerminate`	`sys` returns an empty matrix

It can be seen from Table 7.1 that the tasks of S-function is to call response functions according to the values of `flag` variable. The main framework is suitable to be implemented with switch structures, and the following main framework is a universal one:

```
function [sys,x0,str,ts]=function name(t,x,u,flag,p1,p2,...)
switch flag
   case 0  % call initialization response function
      [sys,x0,str,ts]=mdlInitializeSizes(input arguments)
   case 1  % compute the derivatives of the continuous states
      sys=mdlDerivatives(t,x,u,input arguments)
   case 2  % compute the next values of discrete states
      sys=mdlUpdate(t,x,u,input arguments)
   case 3  % compute the output signal
      sys=mdlOutputs(t,x,u,input arguments)
```

```
    case 4   % set next simulation instance
       sys=GetTimeOfNextVarHit(t,x,u,input arguments)
    case 9   % complete simulation processing
       sys=mdlTerminate(t,x,u,input arguments)
end
```

Note that the user has to write these response functions according to the practical needs, and the input arguments in the function can be chosen by the users. Besides, the response function names are suggested ones, it is not necessary to select the exact same names, as long as the names the user selected match the ones in the caller. For simple problems, there is no need to write response functions, the simple response commands can be embedded in the main framework of the S-function. This will be demonstrated later in examples.

7.2.3 Response functions in S-functions

In the S-function programming, the response functions to `flag` input arguments are needed. Therefore, the main framework using switch structure is usually used, for the reactions to different values of `flag`. With the program framework, the tasks next are to write programs for response functions in S-functions. Here, a brief explanation is given, and they will be demonstrated later with examples.

(1) **Block initialization.** Command `sizes=simsizes` can be used to retrieve the system variable `sizes`. The variable `sizes` is in fact a structured variable, whose members are given in Table 7.2.

Table 7.2: Commonly used initialization parameters.

Member name	Member variable explanation
`NumContStates`	the number of continuous states in S-function block
`NumDiscStates`	the number of discrete states
`NumInputs`	the number of inputs
`NumOutputs`	the number of outputs
`DirFeedthrough`	indicate whether the input signal appears explicitly in the output, and it is the direct feedthrough discussed earlier, with 1 for direct feedthrough. This parameter is important in automatic sorting. The blocks with no direct feedthrough are ranked at the top, while the one with direct feedthrough can only be computed, if all its input signals are known
`NumSampleTimes`	the number of sample times, and more than one sample times are supported in S-function blocks

According to the request in the `sizes` structured data, `sys=simsizes(sizes)` command is used to load the default parameters in the variable `sys`, and other members can be set directly. Apart from `sys`, the initial state vector $\boldsymbol{x}_0$ should be assigned. Also, the description variable `str` and sample time variable `ts` should also be returned, where, `ts` should be a double-column matrix, with each row corresponds to a sample time. For continuous block or discrete block with one sample time, it is represented as $[t_1, t_2]$, where t_1 is the sample time, and t_2 is the offset, with default value of 0. If $t_1 = -1$, it means that the sample time of the input signal is inherited.

(2) **Update the states.** The derivatives of the continuous states in the block can be evaluated with the `mdlDerivatives` function, while the discrete states at the next sample instance can be computed with `mdlUpdate` function. The results of the states can be returned in the variable `sys`. If hybrid systems are tackled, the two functions should be written to describe respectively the continuous and discrete state variables.

(3) **Output signal evaluation.** Call the `mdlOutputs` function, the output value of the block can be evaluated, and still returned in variable `sys`.

A `sfuntmpl.m` template file is provided in Simulink, and it can be used as the starting point in S-function design. If needed, this file can be copied into the working folder, and from it the S-function can be created. In fact, the structure of S-function is rather simple. The users' own S-functions can be programmed, by closely observing an example.

7.2.4 Execution mechanism of S-functions

The execution of an S-function block in Simulink is as follows: At the start of the simulation process, the `flag` value is set to 0, automatically by the Simulink engine. The initialization process is executed. Then the value of `flag` is set to 3, such that the output signal of the block can be computed. The value of `flag` is then set to 2 and 1, respectively, such that the discrete and continuous states can be computed, and one step in simulation is completed. In the next simulation step, the value of `flag` is set in turn to 3 then to 2 and 1. In each simulation step, the execution like this is executed once, until the end of the simulation.

7.2.5 Writing S-functions with examples

In the previous presentation on S-functions, the structure of S-function framework and programming are introduced. The examples chosen here are representative ones. A continuous block modeling and S-function programming are demonstrated first, then the S-function programming of two discrete blocks are discussed, and finally S-function implementation of a static signal generator is addressed.

Example 7.2. Consider the continuous transfer function model with nonzero initial conditions. Create a universal S-function block for the model.

Solutions. If the transfer function model is given, that is the coefficient vectors num and den are known, and the initial value vectors of output and input signals are given in $\boldsymbol{y}_0$ and $\boldsymbol{u}_0$. These variables can be used as the additional parameters of the S-function block. The main framework of the S-function can be created. In normal cases, the command global is not necessary in S-function blocks, that is, no global variables are necessary. For this particular example, since other types of intermediate variables cannot be efficiently transmitted, global variable is a good choice. In other examples, global variables are not used.

```
function [sys,x0,str,ts]=tf_nzic(t,x,u,flag,num,den,y0,u0)
global G %global variable responsible for transfer function transmission
switch flag
   case 0       %call initialization function
      [sys,x0,str,ts] = mdlInitializeSizes(num,den,y0,u0);
   case 1, sys = mdlDerivative(x,u); %derivatives of continuous states
   case 3, sys = mdlOutput(x,u);       %compute the output signal
   case {2,4,9}, sys = []; %unused flag, with empty matrix returned
   otherwise                  %flag values not supported
      error(['Unhandled flag = ',num2str(flag)]);
end
```

It can be seen from the framework that three response functions should be written by the user:
- initialization function mdlInitializeSizes()
- function mdlDerivative() to update continuous state variables
- function mdlOutput() to compute the output signals

Since there is no discrete state variable in this example, it is not necessary to write mdlUpdate() function.

Look at the initialization function. The tasks in the initialization are to declare some information about the block. The simsizes() function can be called first to load the S-function template, and then some members of the template can be modified accordingly, and finally call simsizes() function again to confirm the settings. For this particular example, the algorithm presented in Section 5.1.4 can be used to find the equivalent state space model and the equivalent initial state vector. Assume that the length of the den vector is m, the number of continuous state is $n = m - 1$ and number of discrete state is 0. The block has one input port, and $n + 1$ output signals, with the first one being the output signal, and the remaining n the state variables.

Since the system is assumed to be strictly proper, there is no direct feedthrough. The following initialization function can be written:

```
function [sys,x0,str,ts] = mdlInitializeSizes(num,den,y0,u0)
global G; sizes = simsizes; %load the template
[x0,G]=find_x0(tf(num,den),y0,u0); n=length(x0);
sizes.NumContStates = n;      %number of continuous states
sizes.NumDiscStates = 0;      %no discrete states
sizes.NumOutputs = n+1;       %number of outputs
sizes.NumInputs = 1;          %one channel of input
sizes.DirFeedthrough = 0;     %no direct feedthrough
sizes.NumSampleTimes = 1;     %a single sample time
sys = simsizes(sizes);        %confirm the initialization parameters
str = [];                     %set str to empty
ts = [0 0];                   %sample time of continuous block are all 0's
```

In contrast, the computation of the derivatives of the state variables and the output signals are much easier. The derivative of the states can be evaluated with $\boldsymbol{Ax}+\boldsymbol{Bu}$, and the output signal can be found from $\boldsymbol{Cx}$. There is even no need to write response functions for them. The computation can be embedded in the switch framework. For this particular problem, the output returns not only the output signal, but also the state variables:

```
case 1, sys = G.a*x+G.b*u; %find derivatives of the continuous states
case 3, sys = [G.c*x; x];  %compute the output
```

With the S-function file, the problem in Example 5.6 can be solved directly. The following commands can be used to input the additional parameters into MATLAB workspace:

```
>> num=[2,1]; den=[1,5,9,7,2]; G=tf(num,den);
   u0=[0;1;0]; y0=[1;0;0;0];
```

Then the simulation model in Figure 7.5(a) can be established, and the commands above can be written into the PreLoadFcn property of the model. Open the S-function block, the parameter dialog box is opened as shown in Figure 7.5(b). In the S-function name edit box, the file name of S-function file can be specified. In the S-function parameters edit box, the names of the additional variables can be specified. The S-function block can the be used directly in the simulation model. Click the Edit button, and the S-function editor interface can be opened. If the model is executed, the simulation results obtained are exactly the same as the one in Example 5.6.

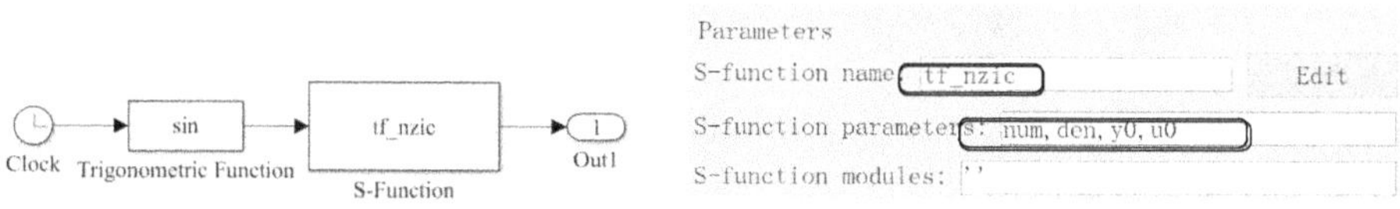

(a) simulation model (file: `c7mtf.slx`) (b) S-function parameter dialog box

Figure 7.5: Setting and simulation of S-function models.

Example 7.3. Consider the fractal tree model in Example 6.9, whose mathematical model is given again as

$$\begin{cases} x_1 = 0,\ y_1 = y_0/2, & \text{if } y_i < 0.05, \\ x_1 = 0.42(x_0 - y_0),\ y_1 = 0.2 + 0.42(x_0 + y_0), & \text{if } 0.05 \leqslant y_i < 0.45, \\ x_1 = 0.42(x_0 + y_0), y_1 = 0.2 - 0.42(x_0 - y_0), & \text{if } 0.45 \leqslant y_i < 0.85, \\ x_1 = 0.1x_0,\ y_1 = 0.2 + 0.1y_0, & \text{otherwise.} \end{cases}$$

The conditional subsystem was used to describe the model and the modeling process is complicated, and unfortunately the model was not successfully built. Use S-function block to build a Simulink model for the fractal tree, and draw the simulation results.

Solutions. Select the state variables $z_1(k) = x$ and $z_2(k) = y$, and the above iterative relationship can be expressed by the discrete state space model:

$$\begin{bmatrix} z_1(k+1) \\ z_2(k+1) \end{bmatrix} = \boldsymbol{A} \begin{bmatrix} z_1(k) \\ z_2(k) \end{bmatrix} + \boldsymbol{B}.$$

Note that the discrete state space model is different from the conventional discrete state space models. The $\boldsymbol{A}$ and $\boldsymbol{B}$ matrices in the four formulas can be written

$$\boldsymbol{A}_1 = \begin{bmatrix} 0 & 0 \\ 0 & 1/2 \end{bmatrix},\quad \boldsymbol{B}_1 = \begin{bmatrix} 0 \\ 0 \end{bmatrix},\quad \boldsymbol{A}_2 = 0.42 \begin{bmatrix} 1 & -1 \\ 1 & 1 \end{bmatrix},\quad \boldsymbol{B}_2 = \begin{bmatrix} 0 \\ 0.2 \end{bmatrix},$$

$$\boldsymbol{A}_3 = 0.42 \begin{bmatrix} 1 & 1 \\ -1 & 1 \end{bmatrix},\quad \boldsymbol{B}_3 = \begin{bmatrix} 0 \\ 0.2 \end{bmatrix},\quad \boldsymbol{A}_4 = 0.1 \begin{bmatrix} 1 & 0 \\ 0 & 1 \end{bmatrix},\quad \boldsymbol{B}_4 = \begin{bmatrix} 0 \\ 0.2 \end{bmatrix}.$$

Random input signal y_i can be used as the input signal of the S-function block, and the output signals are the two state variables. It can be seen from definition that there is one input signal, two output signals. Also, there is no continuous state, and with two discrete states, since the outputs are the states, there is no direct feedthrough. These information can be specified in the initialization process. The state update and output computation can also be programmed for the block. The S-function for this problem can be written as

```
function [sys,x0,str,ts]=c7stree(t,x,u,flag)
switch flag
   case 0                          % initialization function
      [sys,x0,str,ts] = mdlInitializeSizes;
   case 2                          % Update the discrete states
      sys = mdlUpdates(x,u);
   case 3, sys = x;                % compute the output directly
   case {1,4,9}, sys = []; % unused flag values
   otherwise                       % handle processing
      error(['Unhandled flag = ',num2str(flag)]);
end
% when flag is 0, perform initialization for the system
function [sys,x0,str,ts] = mdlInitializeSizes
sizes = simsizes;               % read in initialization template
sizes.NumContStates = 0;        % no continuous states
sizes.NumDiscStates = 2;        % two discrete states
sizes.NumOutputs = 2;           % two output signals
sizes.NumInputs = 1;            % one input signal
sizes.DirFeedthrough = 0;       % no direct feedthrough
sizes.NumSampleTimes = 1;       % single sample time
sys = simsizes(sizes);          % confirm the initial parameter setting
x0 = [0; 0];                    % zero initial conditions
str = [];                       % set str to an empty string
ts = [-1 0];                    % sample time and offset
% when flag=2, update the discrete state variables
function sys = mdlUpdates(x,u)
if u<0.05,      A=[0 0; 0 1/2]; B=[0; 0];
elseif u<0.45, A=0.42*[1 -1; 1 1]; B=[0; 0.2];
elseif u<0.85, A=0.42*[1 1; -1 1]; B=[0; 0.2];
else,          A=0.1*eye(2); B=[0; 0.2]; end
sys = A*x+B;
```

Since the output equation is simple, the states can be passed directly to the output, there is no need to write a `mdlOutputs()` response function, and it can be embedded directly in the switch structure.

With the S-function, the simulation model can be established, as shown in Figure 7.6, where the input signal is generated with the **Uniform Random Number** block, with value range of $[0, 1]$, rather than the default $[-1, 1]$. The sample time can be specified as any value, for instance, 0.01. In fact, the execution results in this example is independent of the sample time.

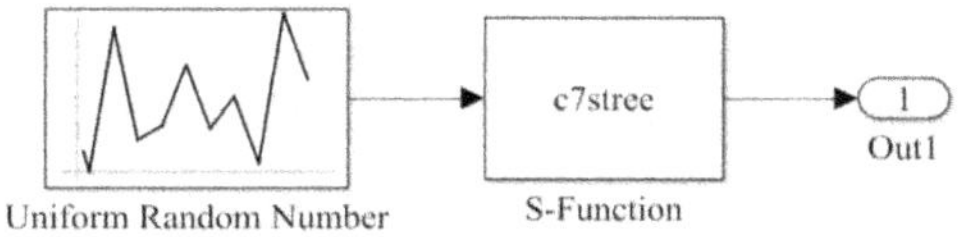

Figure 7.6: The Simulink model for the fractal tree (file: `c7mtree.slx`).

Start the simulation process, and the scattered plot of the fractal tree can be obtained as shown in Figure 7.7. Note that the option `'.'` must be specified to draw the scattered plot; otherwise, the plot obtained is not of any use.

```
>> plot(yout(:,1),yout(:,2),'.')
```

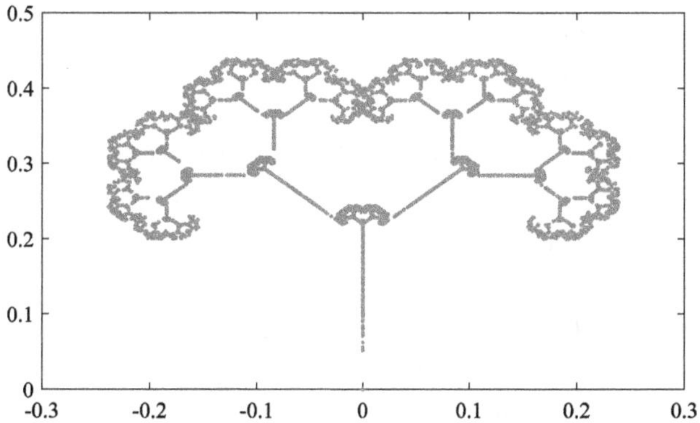

Figure 7.7: Scatter points of the fractal tree.

Of course, if the users do not want to summarize the unified linear state space description, the following commands can be used directly to implement the iterative part, and define the new subfunction `mdlUpdates()`:

```
function sys = mdlUpdates(x,u)
x0=x(1); y0=x(2);   % save temporarily the initial states
if u<0.05,       sys=[0; 0.5*y0];
elseif u<0.45, sys=[0.42*(x0-y0); 0.2+0.42*(x0+y0)];
elseif u<0.85, sys=[0.42*(x0+y0); 0.2-0.42*(x0-y0)];
else,            sys=[0.1*x0; 0.1*y0+0.2]; end
```

Example 7.4. Now consider the mathematical description of the tracking and differentiator[32] studied in Example 4.31. For convenience, it is given again here:

$$\begin{cases} x_1(k+1) = x_1(k) + Tx_2(k), \\ x_2(k+1) = x_2(k) + T\mathrm{fst}(x_1(k), x_2(k), u(k), r, h), \end{cases} \tag{7.2.4}$$

where T is the sample time, $u(k)$ is the input signal at the kth instance, r determines the tracking speed and h is the filter constant, when the input signal is corrupted by noise. Function fst can be computed from the following expressions:

$$\delta = rh, \quad \delta_0 = \delta h, \quad y_0 = x_1(k) - u + hx_2(k), \quad a_0 = \sqrt{\delta^2 + 8r|y_0|}, \tag{7.2.5}$$

$$a = \begin{cases} x_2(k) + y_0/h, & |y_0| \leqslant \delta_0, \\ x_2(k) + 0.5(a_0 - \delta)\,\mathrm{sign}(y_0), & |y_0| > \delta_0, \end{cases} \tag{7.2.6}$$

$$\mathrm{fst} = \begin{cases} -ra/\delta, & |a| \leqslant \delta, \\ -r\,\mathrm{sign}(a), & |a| > \delta. \end{cases} \tag{7.2.7}$$

Two low-level methods were tried previously. The models are rather complicated, but the expected results were not obtained. Use the S-function to build up a simulation model.

Solutions. It was stated in Example 4.31 that, in order to update the states, the correct computational order is to compute in turn δ, δ_0, y_0 and a_0. Then find a, and finally find fst. Since piecewise functions are involved here, the low-level modeling process is rather complicated and error-prone. The manipulation of piecewise functions can be implemented easily in MATLAB programming. Therefore, the model in S-function can easily be established.

It can be seen from the original problem that the model has two discrete states and no continuous states. There is one input signal $u(k)$ and two outputs. The model is with one sample time T, and no direct feedthrough. Three additional parameters, r, h and T, are used in the model. Based on the given mathematical model, it is not hard to write the following S-function program:

```
function [sys,x0,str,ts]=han_td(t,x,u,flag,r,h,T)
switch flag
  case 0                % call initialization function
    [sys,x0,str,ts] = mdlInitializeSizes(T);
  case 2                % update the discrete state variables
    sys = mdlUpdates(x,u,r,h,T);
  case 3, sys = x; % compute the output directly
  case {1, 4, 9}, sys = []; % unused flag values
  otherwise             % handling errors
    error(['Unhandled flag = ',num2str(flag)]);
```

```
end
% when flag is 0, process initialization information
function [sys,x0,str,ts] = mdlInitializeSizes(T)
sizes = simsizes;             % read in template
sizes.NumContStates = 0;      % no continuous states
sizes.NumDiscStates = 2;      % with two discrete states
sizes.NumOutputs = 2;         % two outputs
sizes.NumInputs = 1;          % one input signal
sizes.DirFeedthrough = 0;     % no direct feedthrough
sizes.NumSampleTimes = 1;     % single sample time
sys = simsizes(sizes);        % confirm the initialization parameters
x0 = [0; 0];                  % set up zero initial conditions
str = [];                     % set str to empty string
ts = [T 0];                   % sample time and offset
% in the function when flag=2, update the state variables
function sys = mdlUpdates(x,u,r,h,T)
sys = [x(1)+T*x(2); x(2)+T*fst(x,u,r,h)];
function f=fst(x,u,r,h)
delta=r*h; delta0=delta*h; b=x(1)-u+h*x(2);
a0=sqrt(delta*delta+8*r*abs(b));
a=x(2)+b/h*(abs(b)<=delta0)+0.5*(a0-delta)*sign(b)*(abs(b)>delta0);
f=-r*a/delta*(abs(a)<=delta)-r*sign(a)*(abs(a)>delta);
```

Since the output signal is rather easy to compute, it can be embedded directly in the main framework. No extra response function is needed. Since the piecewise function is implemented with the `if` structure, it is easy to debug.

The test signal is defined in two parts: one is in the $(0, 2\pi)$ interval, and the sinusoidal signal is used, while in the $(2\pi, 4\pi)$ interval, the triangular signal is used. The following M-function can be used to generate the test signal:

```
function y=c7han_fun(x)
if x<=2*pi, y=sin(x);                      % sinusoidal signal in the first part
elseif x<=2.5*pi, y=2*(x-2*pi)/pi;  % triangular signal in three parts
elseif x<=3.5*pi, y=1-2*(x-2.5*pi)/pi;
elseif x<=4*pi, y=-2+2*(x-3*pi)/pi; end
```

Therefore, the simulation model for the tracking and differentiator can be constructed as shown in Figure 7.8. The simulation results are given in Figure 7.9. It can be seen that the tracking signal and its first-order derivative can be obtained. The tracking signal is the same as the input signal, and the tracking of the first-order derivative is rather rapid. The results are ideal.

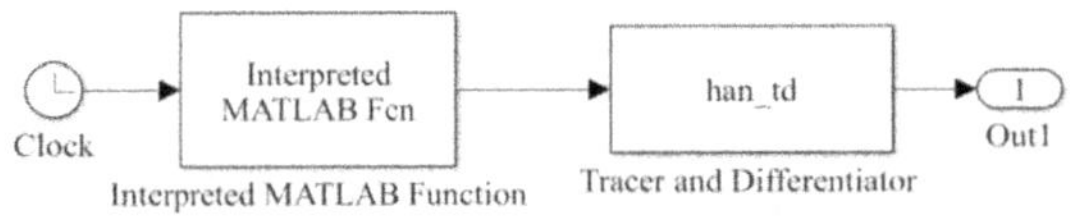

Figure 7.8: Tracking and differentiator model (file: `c7mtd.slx`).

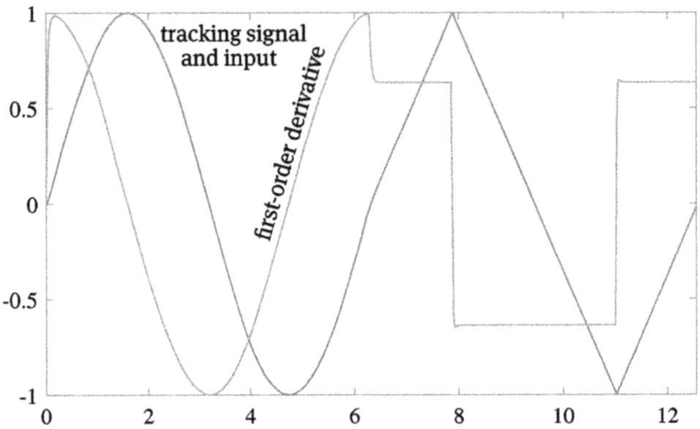

Figure 7.9: Tracking and derivative signals.

Compared with the modeling methods tried earlier, the S-function modeling and simulation are quite simple and straightforward. The ease of modeling and program reliability are much better than the other low-level modeling methods. Therefore, in the modeling of complicated state space models, the S-function modeling method is recommended.

Example 7.5. A periodic staircase signal generator studied in Example 2.2 can be implemented with the **Repeating Sequence** block. Use the S-function to construct a reusable periodic staircase signal generator.

Solutions. No input signal is needed in such a signal generator. It is no continuous and discrete states. The block has one output signal. There is no direct feedthrough in the block. Therefore, in the initialization process, this information can be specified. Two additional parameters, `tTime` and `yStep` vectors, can be used to describe the turning points in the signal. The following S-function can be written. Since there is no continuous nor discrete states, the `case 1` and `case 2` are not needed, and they can be arranged in `case 4,9`, to set `sys` to an empty matrix:

```
function [sys,x0,str,ts]=multi_step(t,x,u,flag,tTime,yStep)
switch flag
   case 0       % call the initialization process
      [sys,x0,str,ts] = mdlInitializeSizes;
   case 3, sys=mdlOutputs(t,tTime,yStep); % compute the outputs
```

```
    case {1,2,4,9}, sys = []; % unused flag values
    otherwise % handling errors
        error(['Unhandled flag = ',num2str(flag)]);
end
% when flag is 0, carry out initialization
function [sys,x0,str,ts] = mdlInitializeSizes
sizes = simsizes; % call the initialization template
sizes.NumContStates=0; sizes.NumDiscStates=0;
% no continuous/discrete states
sizes.NumOutputs = 1; sizes.NumInputs = 0;      % input and output signals
sizes.DirFeedthrough = 0; sizes.NumSampleTimes = 1;
sys = simsizes(sizes);              % initialization
x0 = []; str = []; ts = [0 0]; % block is continuous
% when flag is 3, compute the output
function sys = mdlOutputs(t,tTime,yStep)
i=find(tTime<=t); sys=yStep(i(end));
```

7.2.6 Masking S-function blocks

The S-function block can be masked directly to build up reusable blocks. Generally speaking, if the S-function block is clicked, the shortcut menu can be obtained by right click. Select the Mask item, and the method in Chapter 6 can be used to start masking process for S-function blocks. Since usually the S-function has its own additional parameters, these parameters can be designed in the parameter dialog box. The S-function block masking methods will be demonstrated through examples.

Example 7.6. Convert the S-function block in Example 7.2 into a reusable block, such that the transfer function models with nonzero initial conditions can be constructed.

Solutions. It can be seen from the S-function that, the model has four additional parameters, as `num`, `den`, `y0` and `u0`, where the first two are the coefficient vectors of the numerator and denominator, while the latter ones are the initial values of the input and output. Therefore, four edit boxes can be designed for the parameters, as shown in Figure 7.10.

Command `dpoly()` provided in MATLAB can be used to draw transfer function block icons. The commands can be provided in the Icon & Ports pane, in the masking editor interface. The masked `num` and `den` parameters cannot be passed to the Icon & Ports pane, by default, since the item Run Initialization is by default set to Off. It can be modified to On. Under the Icon & Ports pane, fill in `dpoly(num,den)` command, the transfer function icon can be displayed directly with the masked S-function model set up as shown in Figure 7.11. It can be seen that the transfer function icon can be displayed normally.

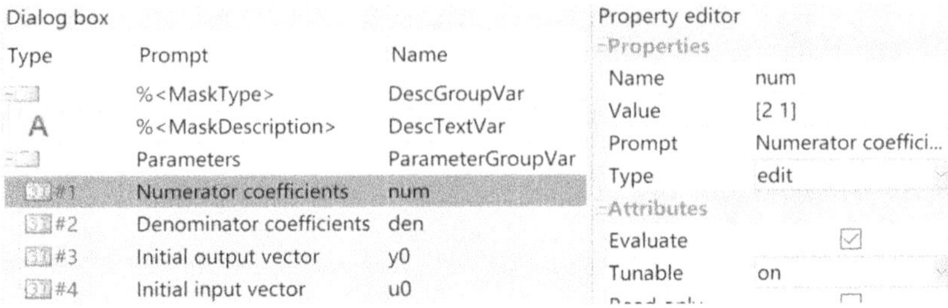

Figure 7.10: S-function block masking.

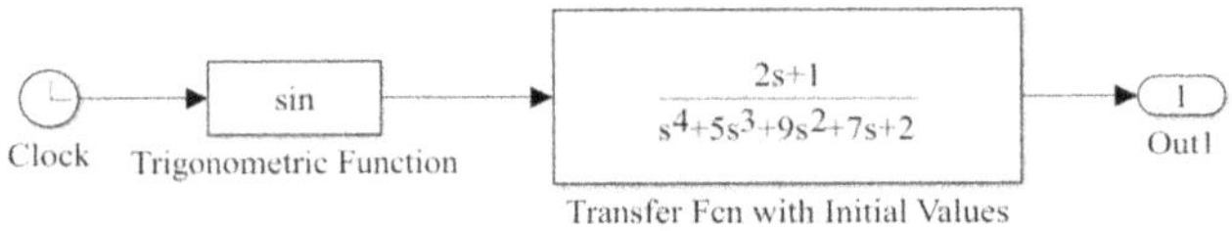

Figure 7.11: Masked S-function (file: `c7mtf1.slx`).

Example 7.7. Consider the masked block in Example 7.6, where there is one output port. The actual output of the block is composed of system output and n state variables. Modify the block such that the block has two output ports: one for the system output and the other for the states. Make a new masked block.

Solutions. The new Simulink model can be constructed, as shown in Figure 7.12 (a), where a Demux block is used, with parameters set to `[1, n]`. Then the model has two output ports, which can be made into a subsystem. Then mask the block. When using the demux block, the number n of states should be measured first. Therefore, in the Initialization pane, a command n=`length(den)` can be specified, while other masked parameters are the same as the one in Figure 7.10. With the new masked block, the model in Figure 7.12(b) can be executed, and the system has only one output signal, the states are transmitted directly to the Terminator block.

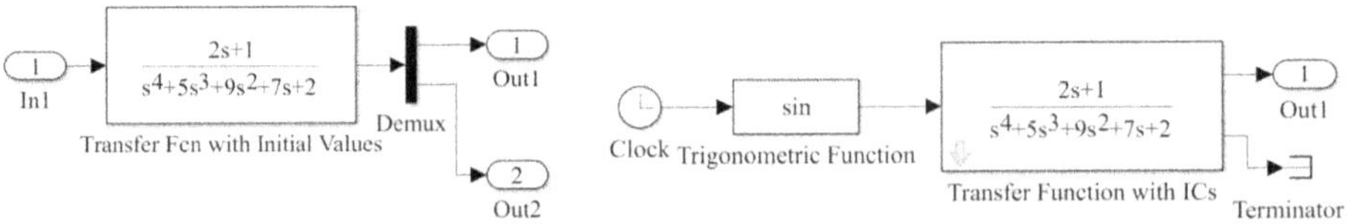

(a) with two output ports (b) the new model (file: `c7mtf2.slx`)

Figure 7.12: New masked block.

Example 7.8. Mask the staircase signal generator block in Example 7.5, and design an icon for the block.

Solutions. Two additional parameters, `tTime` and `yStep`, are used in the block. The parameter dialog box design can be made, as shown in Figure 7.13 for the masked block.

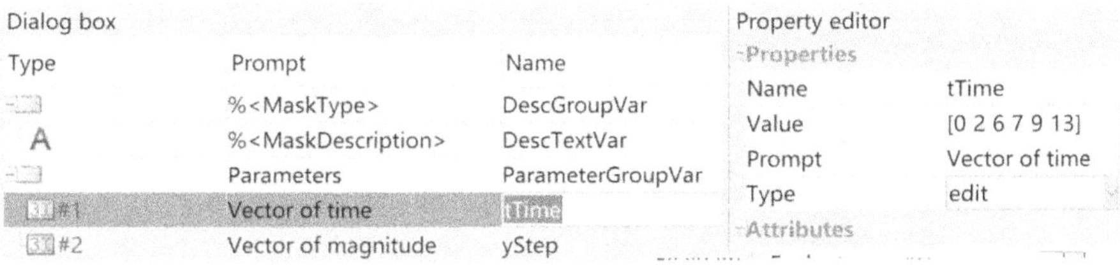

Figure 7.13: Masking editor interface.

In the Initialization pane, the following data can be prepared for latter use in icon plotting:

```
ee=1e-5; n=length(tTime); tt=tTime(2:n)+ee;
y0=[yStep(1) yStep(1:n-1) yStep(2:n)];
[t0,ii]=sort([tTime,tt]); y0=y0(ii);
T=diff(t0); if isempty(T), T=10; else, T=T(end-1); end
t0=[t0 t0(end)+T]; y0=[y0 y0(end)];
```

The vectors $\boldsymbol{t}_0$ and $\boldsymbol{y}_0$ can be used to draw directly the staircase plot. In the Icon & Ports pane, set Run Initialization to On, and in the Icon drawing commands edit box, the following code can be filled in, such that the icon of the staircase signal can be displayed:

```
plot(t0,y0) % the commands below are to add margins in the icon
dX=max(t0)-min(t0); dY=max(y0)-min(y0);
plot(min(t0)-0.08*dX,min(y0)-0.08*dY)
plot(max(t0)+0.08*dX,max(y0)+0.08*dY)
```

The Simulink model shown in Figure 7.14(a) can be created. Double click the Multi-Staircase Signal Generator block, and the staircase data can be specified in the parameter dialog box, as shown in Figure 7.14(b). The icon of the staircase signal can be automatically displayed.

7.2.7 Level-2 S-functions

The S-function style studied so far is referred to as Level-1 S-functions. Apart from this type of ordinary S-functions, Level-2 S-function can also be used to describe the

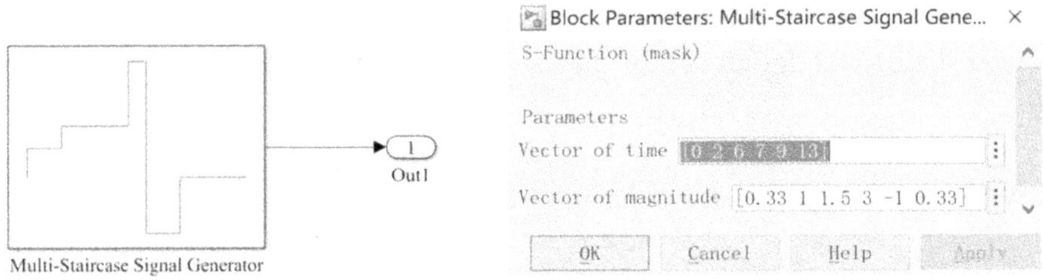

(a) model (file: `c7mmstep.slx`) (b) parameter dialog box

Figure 7.14: The dialog box design for the staircase signal generator block.

system blocks. Object oriented programming ideas are used in Level-2 S-functions. The `switch-case` structure based on the values of `flag` is not used. Multiple input port and multiple output ports are directly supported in Level-2 S functions. Complex matrix output signals are also supported in the programming style, such that the application field is much wider than the ordinary S-functions. The reference template file `msfuntmpl.m` can be used for Level-2 S-function programming. An example for Level-2 S-function programming is demonstrated next.

Example 7.9. Consider the tracking and differentiator block studied in Example 7.4. Implement the model with the Level-2 S-function program.

Solutions. If Level-2 S-function is used, the following S-function can be written. Details of the Level-2 S-function implementation is listed below, with no further explanations:

```
function han_td_L2(block)
setup(block); % the main program
%---- initialization function
function setup(block)
block.NumDialogPrms=3; block.NumInputPorts=1;
block.NumOutputPorts=1; block.NumContStates=0;
block.SetPreCompInpPortInfoToDynamic;
block.SetPreCompOutPortInfoToDynamic;
block.InputPort(1).Dimensions=1;
block.InputPort(1).DirectFeedthrough=false;
block.OutputPort(1).Dimensions=2;
T=block.DialogPrm(3).Data; block.SampleTimes=[T 0]; % sample time
% define the response functions
block.RegBlockMethod('PostPropagationSetup', @DoPostPropSetup);
block.RegBlockMethod('InitializeConditions', @InitConditions);
block.RegBlockMethod('Outputs',              @Output);
block.RegBlockMethod('Update',               @Update);
```

```
%---define discrete state variables
function DoPostPropSetup(block)
block.NumDworks=1; block.Dwork(1).Name='x0';
block.Dwork(1).Dimensions=2; block.Dwork(1).DatatypeID=0;
block.Dwork(1).Complexity='Real';
block.Dwork(1).UsedAsDiscState=true;

%--- initial values of the states
function InitConditions(block)
block.Dwork(1).Data=[0; 0];

%--- compute the output
function Output(block)
block.OutputPort(1).Data = block.Dwork(1).Data;

%--- discrete state update
function Update(block)
r=block.DialogPrm(1).Data; h=block.DialogPrm(2).Data;
T=block.DialogPrm(3).Data; u=block.InputPort(1).Data;
x=block.Dwork(1).Data;
block.Dwork(1).Data=[x(1)+T*x(2); x(2)+T*fst2(x,u,r,h)];

%--- subfunction
function f=fst2(x,u,r,h)
delta=r*h; delta0=delta*h; y=x(1)-u+h*x(2);
a0=sqrt(delta*delta+8*r*abs(y));
if abs(y)<=delta0, a=x(2)+y/h;
else, a=x(2)+0.5*(a0-delta)*sign(y); end
if abs(a)<=delta, f=-r*a/delta; else, f=-r*sign(a); end
```

Note that the discrete state variables are stored in the `block.Dwork` property, which should be defined with the method `DoPostPropSetup()`. In this example, for simple problems, unless absolutely necessary, this programming style is not recommended.

7.3 Programming S-functions with C

Apart from using MATLAB language to write S-functions, computer languages such as C, C++, Fortran and Ada can also be used to write S-functions. In this section, the installation and settings of C language compiler are demonstrated. Then an S-function programming in C language is demonstrated.

7.3.1 C compiler setting

In recent versions of MATLAB, the TDM-GCC MinGW compiler is supported, which can be downloaded from

https://sourceforge.net/projects/tdm-gcc/?source=typ_redirect

The installation can be made in default paths, and the following commands must be issued to set up the C compiler. Note that each time when MATLAB is invoked, the commands should be executed. Alternatively, the command can be written into the `startup.m` file:

```
>> setenv('MW_MINGW64_LOC','C:\TDM-GCC-64')
```

The following command can be used to compile the C source code:

```
>> mex filename.c
```

The file can be compiled directly. If the process is successfully completed, the executable file "filename.mexw64" can be generated. The file can be called directly in the same way as other MATLAB functions.

7.3.2 Examples in S-functions

A great amount of S-function examples are provided in Simulink. Open the **S-function demos** group, and various examples can be referenced. The relevant examples can be studied to learn more in S-function programming in other languages. Here, only one example is given to illustrate S-function programming in C.

Example 7.10. Write S-function for the tracking and differentiator in C.

Solutions. The C version of the S-function is given below, for reference only. The lengthy comments and blank lines in the source code are not listed here:

```
#define S_FUNCTION_NAME  sfun_han /* the actual S-function name*/
#define S_FUNCTION_LEVEL 2        /* Level-2 S-function*/
#include "simstruc.h"
#include "math.h"      /* mathematical computing library*/
double sign(double x) /* defined sign function*/
{
   double f; f=1; if (x<=0) f=-1; return(f);
}

double fst(real_T *x, const real_T *u,
           const real_T *p1, const real_T *p2)
```

```
{  /* fst function*/
   double delta, delta0, a0, a, y, r, h;
   r=p1[0]; h=p2[0]; delta=r*h; delta0=delta*h; y=x[0]-u[0]+h*x[1];
   a0=sqrt(delta*delta+8.0*r*fabs(y)); /* absolute value with fabs*/
   if (fabs(y) <= delta0) a=x[1]+y/h;
   else a=x[1]+0.5*(a0-delta)*sign(y);
   if (fabs(a) <= delta) return (-r*a/delta);
   else return(-r*sign(a));
}

static void mdlInitializeSizes(SimStruct *S) /* initialization*/
{
    ssSetNumSFcnParams(S, 3);   /* number of additional parameters*/
    ssSetNumContStates(S, 0);   /* number of continuous states*/
    ssSetNumDiscStates(S, 2);   /* number of discrete states*/
    if (!ssSetNumInputPorts(S, 1)) return;
    ssSetInputPortWidth(S, 0, 1); /* set input signals*/
    ssSetInputPortDirectFeedThrough(S, 0, 0); /* direct feedthrough*/
    if (!ssSetNumOutputPorts(S, 1)) return;
    ssSetOutputPortWidth(S, 0, 2);     /* output signal numbers*/
    ssSetNumSampleTimes(S, 1);         /* sample time*/
    ssSetNumRWork(S, 3);               /* 3 additional parameters*/
    ssSetNumIWork(S, 0); ssSetNumPWork(S, 0); ssSetNumModes(S, 0);
    ssSetNumNonsampledZCs(S, 0); ssSetOptions(S, 0);
}

static void mdlInitializeSampleTimes(SimStruct *S)
{   /* sample time etting*/
    ssSetSampleTime(S,0,*mxGetPr(ssGetSFcnParam(S,2)));
    ssSetOffsetTime(S, 0, 0.0);     /* define 3 additional parameters*/
}

static void mdlOutputs(SimStruct *S, int_T tid)
{   /* system o*/
    const real_T *x = ssGetRealDiscStates(S); /* get x and y pointers*/
    real_T       *y = ssGetOutputPortSignal(S,0);
    y[0] = x[0]; y[1] = x[1];     /* write output function*/
}

static void mdlUpdate(SimStruct *S, int_T tid)
{  /* discrete state update*/
    real_T       *x = ssGetRealDiscStates(S);
```

```
    const real_T *u = (const real_T*) ssGetInputPortSignal(S,0);
    const real_T *r = mxGetPr(ssGetSFcnParam(S,0));
    /* additional parameters*/
    const real_T *h = mxGetPr(ssGetSFcnParam(S,1));
    const real_T *T = mxGetPr(ssGetSFcnParam(S,2));
    real_T tempX[2] = {0.0, 0.0}; /* save to temporary data*/
    tempX[0] = x[0] + T[0]*x[1];  /* command x[0]=... cannot be used*/
    tempX[1] = x[1] + T[0]*fst(x,u,r,h);
    x[0] = tempX[0]; x[1] = tempX[1]; /* save the data to x*/
}
```

7.4 Programming practice of S-functions – active disturbance rejection controllers

Active disturbance rejection control (ADRC) is a very effective industrial control strategy. ADRC was pioneered by Chinese scholar, Jingqing Han (1937–2008) and it is an active control strategy in the control community.[31] The typical ADRC controller has three components, including the tracking and differentiator discussed earlier, an extended state observer (ESO) and ADRC controller itself. The structures with the three components of ADRC controller are illustrated in Figure 7.15, where the one in the dash box is the ADRC controller.

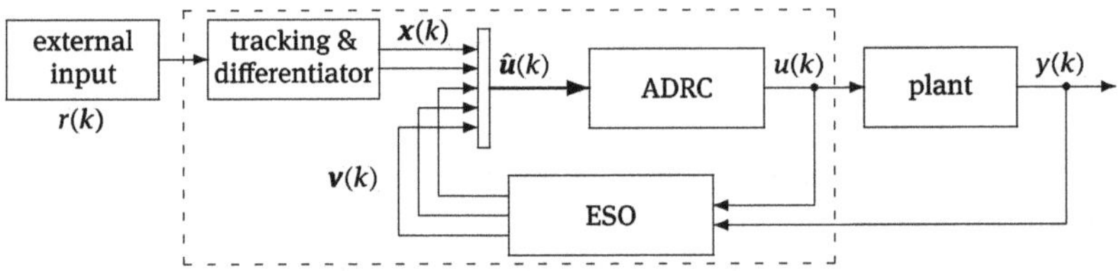

Figure 7.15: Block diagram of ADRC.

The three components are all suitable for S-function implementation. Therefore, the control strategy is used as the background to demonstrate modeling and simulation of ADRC systems with S-functions. The mathematical forms and S-function implementation are presented for the extended state observer and ADRC controller. The ADRC model masking is also presented. Examples are provided to demonstrate control results for different plants.

7.4.1 Modeling of extended state observers

The concept of the extended state observer presented here is different from the traditional state observer in state space control theory. Concentrations are made on the extraction of some features in the plant model, not for genuine states. In extended state observers, the name of the state observer was borrowed, and the disturbances affecting the output of the plant are defined as the extended state variables, and special feedback mechanism is established to observe the extended state variables. In certain cases, ESO is in fact the generalized and practical disturbance observer. Second-, third- or even fourth-order extended state observers are studied in[31]. In this section, the third-order extended state observer is used as an example. The mathematical form and the S-function modeling of ESO are covered. The names of state variables are systematically handled with other parts of the controller.

Definition 7.2. The mathematical model of third-order ESO is[31]

$$\begin{cases} v_1(k+1) = v_1(k) + T[v_2(k) - \beta_{01}e(k)], \\ v_2(k+1) = v_2(k) + T[v_3(k) - \beta_{02}\,\mathrm{fal}(e(k), 1/2, \delta) + bu(k)], \\ v_3(k+1) = v_3(k) - T\beta_{03}\,\mathrm{fal}(e(k), 1/4, \delta), \end{cases} \tag{7.4.1}$$

where $e(k) = v_1(k) - y(k)$, and

$$\mathrm{fal}(e(k), a, \delta) = \begin{cases} e(k)\delta^{a-1}, & |e(k)| \leqslant \delta, \\ |e(k)|^a\,\mathrm{sign}(e(k)), & |e(k)| > \delta. \end{cases} \tag{7.4.2}$$

Example 7.11. Design an S-function for the third-order extended state observer.

Solutions. It can be seen from the mathematical model that this is a discrete block, with three discrete states and no continuous states. It can be seen from Figure 7.15 that the block has two input signal and three output signals, which are the three states $v_i(k)$. Therefore, the block has no direct feedthrough. The sample time can be set to −1. The additional parameters are δ, $\boldsymbol{\beta}$, b and T, where the vector is $\boldsymbol{\beta} = [\beta_{01}, \beta_{02}, \beta_{03}]$. The input signal to the block is $\tilde{\boldsymbol{u}}(k) = [u(k), y(k)]$.

With this in mind, the following S-function can be written for the extended state observer:

```
function [sys,v0,str,ts]=han_eso(t,v,u,flag,d,bet,b,T)
switch flag
   case 0, [sys,v0,str,ts] = mdlInitializeSizes;
   case 2, sys = mdlUpdates(v,u,d,bet,b,T);
   case 3, sys = v;
   case {1, 4, 9}, sys = [];
   otherwise, error(['Unhandled flag = ',num2str(flag)]);
end
```

```
%--- when flag is 0 carry out initialization process
function [sys,v0,str,ts] = mdlInitializeSizes
sizes = simsizes;
sizes.NumContStates=0; sizes.NumDiscStates=3;
sizes.NumOutputs=3; sizes.NumInputs=2;
sizes.DirFeedthrough=0; sizes.NumSampleTimes=1;
sys = simsizes(sizes);
v0 = [0; 0; 0];   % set zero initial state
str = [];         % set str to an empty string
ts = [-1 0];      % sample time to inherits its input signal
%--- whenflag is 2, the discrete states are updated
function sys = mdlUpdates(v,u,d,bet,b,T)
e=v(1)-u(2);
sys=[v(1)+T*(v(2)-bet(1)*e);
     v(2)+T*(v(3)-bet(2)*fal(e,0.5,d)+b*u(1));
     v(3)-T*bet(3)*fal(e,0.25,d)];
%--- subfunction fal
function f=fal(e,a,d)
if abs(e)<d, f=e*d^(a-1); else, f=(abs(e))^a*sign(e); end
```

7.4.2 Modeling of ADRC controllers

With the tracking and derivative signal, and when the extended states signals are measured, they can be used to compute the actual input signal, so as to control the plant directly. In this section, the mathematical form and its S-function implementation of the ADRC controller are given.

Definition 7.3. The mathematical form of the ADRC controller is

$$\begin{cases} e_1(k) = v_1(k) - x_1(k), \\ e_2(k) = v_2(k) - x_2(k), \\ u_0(k) = \beta_1 \operatorname{fal}(e_1(k), a_1, \delta_1) + \beta_2 \operatorname{fal}(e_2(k), a_2, \delta_1), \\ u(k) = u_0(k) - v_3(k)/b, \end{cases} \tag{7.4.3}$$

where from external input and its derivative, also with the extended observer states, the two error signals $e_1(k)$ and $e_2(k)$ can be computed, from which the controller signal $u(k)$ can be obtained through nonlinear operation.

Example 7.12. Write an S-function program for the ADRC controller.

Solutions. It can be seen from Definition 7.3 that the ADRC controller has five input signals $\hat{\boldsymbol{u}}(k) = [\boldsymbol{x}(k), \boldsymbol{v}(k)]$, with the former two, $\boldsymbol{x}(k)$, the output of the tracking and differentiator block, and the latter three, $\boldsymbol{v}(k)$, the output of the ESO block. There is one output signal, which is the actual control signal $u(k)$. There is no continuous states nor discrete states. The additional parameters are $\boldsymbol{a}$, $\boldsymbol{\beta}$, b and δ_1, where $\boldsymbol{a} = [a_1, a_2]$ and $\boldsymbol{\beta} = [\beta_1, \beta_2]$. Since the input $\hat{\boldsymbol{u}}(k)$ is directly reflected in output signal $u(k)$, there is direct feedthrough in this block.

Based on the above information, the S-function can be written:

```
function [sys,x0,str,ts]=han_ctrl(t,xn,u,flag,aa,bet1,b,d1)
switch flag
   case 0, [sys,x0,str,ts] = mdlInitializeSizes(t,u,xn);
   case 3, sys = mdlOutputs(t,xn,u,aa,bet1,b,d1);
   case {1,2,4,9}, sys = [];
   otherwise, error(['Unhandled flag = ',num2str(flag)]);
end
% --- when flag is 0, carry out initialization
function [sys,x0,str,ts] = mdlInitializeSizes(t,u,xn)
sizes = simsizes;
sizes.NumContStates=0; sizes.NumDiscStates=0;
sizes.NumOutputs=1; sizes.NumInputs=5;
sizes.DirFeedthrough=1; sizes.NumSampleTimes=1;
sys = simsizes(sizes);
x0=[]; str=[]; ts=[-1 0]; % other parameters
% --- when flag is 3 compute the output signal
function sys = mdlOutputs(t,xn,u,aa,bet,b,d1)
x=u(1:2); v=u(3:5); e1=x(1)-v(1); e2=x(2)-v(2);
u0=bet(1)*fal(e1,aa(1),d1)+bet(2)*fal(e2,aa(2),d1);
sys=u0-v(3)/b;
% --- user defined function fal, the same as the earlier ones
function f=fal(e,a,d)
if abs(e)<d, f=e*d^(a-1); else, f=(abs(e))^a*sign(e); end
```

7.4.3 Simulation of ADRC systems

Simulation study on the ADRC controller is given in this section. Refer to the block diagram representation in Figure 7.15, and use the S-functions, — `han_td()`, `han_eso()` and `han_ctrl()`, for the three blocks given earlier, the Simulink model of the ADRC controller can be constructed, as shown in Figure 7.16.

The ADRC controller can be made into a subsystem first, then through masking process, a reusable ADRC controller block can be created. In the masking editor in-

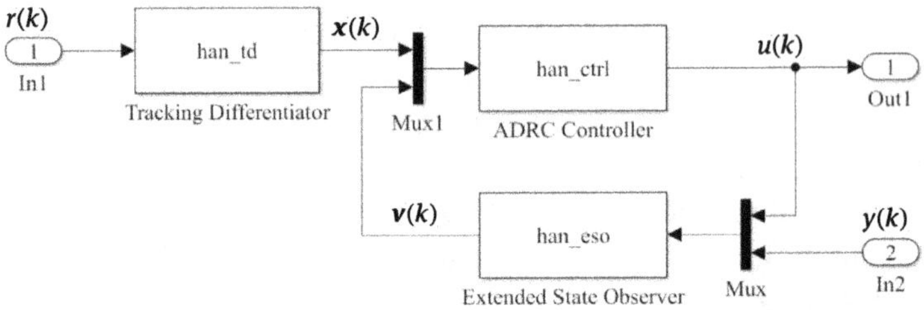

Figure 7.16: ADRC controller model (file: `c7madrc.slx`)stem.

terface, the edit box in the Icon & Ports pane can be expressed as `disp('ADRC\nController')`, such that a simple icon is designed. The key issues to be presented here is how to design the parameter dialog box. Since the ADRC controller is composed of three components, three panes can be designed in the dialog box, and each correspond to a component, as shown in Figure 7.17(a). After masking process, a reusable ADRC controller is constructed, as shown in Figure 7.17(b).

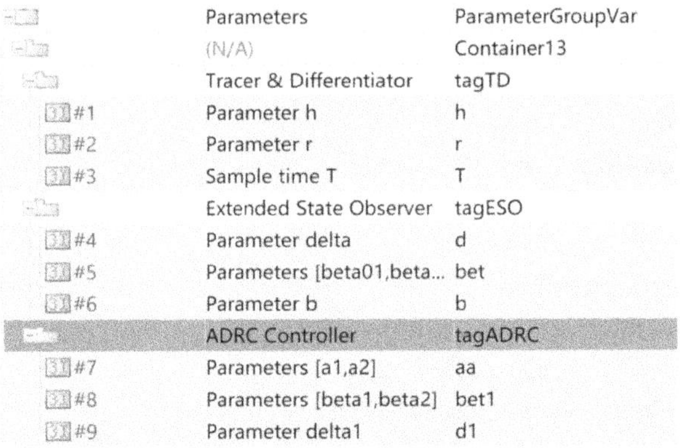

(a) parameter dialog box design

(b) file: `c7madrc1.slx`

Figure 7.17: Design of ADRC controller block.

Double click the controller block, and the parameter dialog box is opened, as shown in Figure 7.18. It can be seen that there are three panes in the dialog box, the same as expected. The controller parameters in different components can be specified in different panes of the dialog box. The control behavior of the ADRC controller will be demonstrated through an example.

Example 7.13. For the nonlinear continuous plant given by

$$\begin{cases} x_1'(t) = x_2(t), \\ x_2'(t) = \operatorname{sign}(\sin t) + u(t), \end{cases}$$

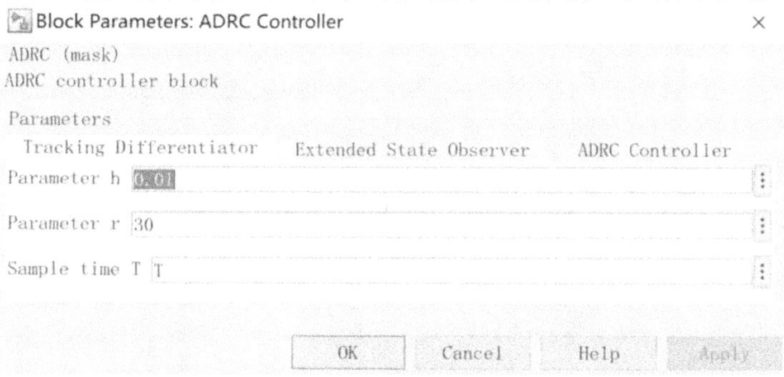

Figure 7.18: Parameter dialog box of ADRC controller.

selecting ADRC controller parameters as $r = 30$, $h = 0.01$, $T = 0.01$, $\boldsymbol{\beta} = [100, 65, 80]$, $\boldsymbol{\beta}_1 = [100, 10]$, $\boldsymbol{a} = [0.75, 1.25]$, $\delta = \delta_1 = 0$ and $b = 1$, find the output signal of the closed-loop system.

Solutions. It can be seen from the plant model that it is essentially a double-integrator system, subject to the periodic square wave disturbance, of amplitude 1. The plant model can be established, and ADRC controller can be connected to the plant. The input signal to follow is a square wave generated by the **Signal Generator** block, The simulation model thus established is as shown in Figure 7.19. Start the simulation process, the input and output signals can be obtained, as shown in Figure 7.20. It can be seen that the output follows well the input signal.

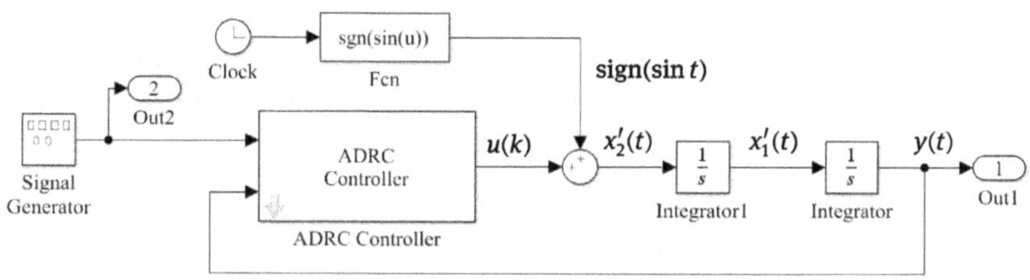

Figure 7.19: ADRC system model (file: `c7madrc3.slx`).

If there are variations in the plant, for instance, the disturbance on $x_2'(t)$ is changed from $\operatorname{sign}(\sin t)$ to $-\operatorname{sign}(\sin t)$, and the simulation model can be executed again. Almost no change can be witnessed from the output signal. If the disturbance is changed to $-\operatorname{sign}(\sin 8t)$, that is, the frequency of the disturbance is increased, the system output can be obtained, as shown in Figure 7.21. It can be seen that there are variations with small magnitude of the system output, and the control effect is satisfactory.

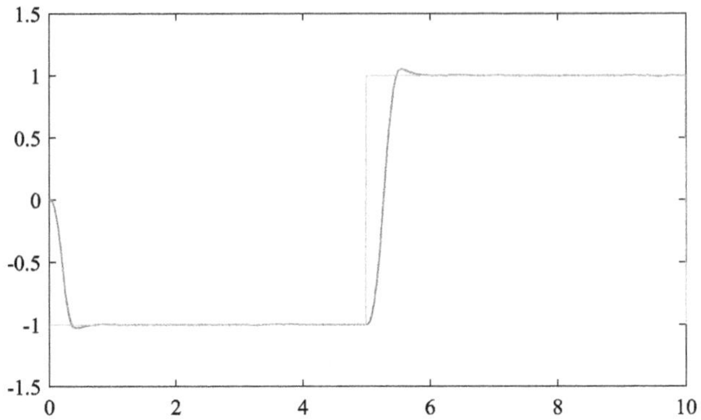

Figure 7.20: Control signal and output of the system.

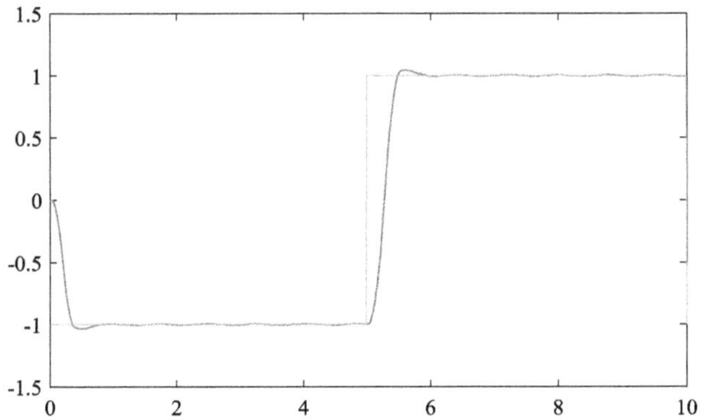

Figure 7.21: System output when disturbance changes.

If the parameter r of the controller is set to $r = 100$, the system response obtained is shown in Figure 7.22. It can be seen that when r increases, the speed of the response also increases.

7.5 Exercises

7.1 In Example 7.1, an M-function block is designed, where a cluster of sine waves are generated. The frequencies and initial phases must be specified in advance in the M-function file. Design an equivalent S-function block, such that three vectors of frequency, initial phase and magnitude can be accepted as additional parameters. If the lengths of the three vectors are different, adjust them according to the shortest one.

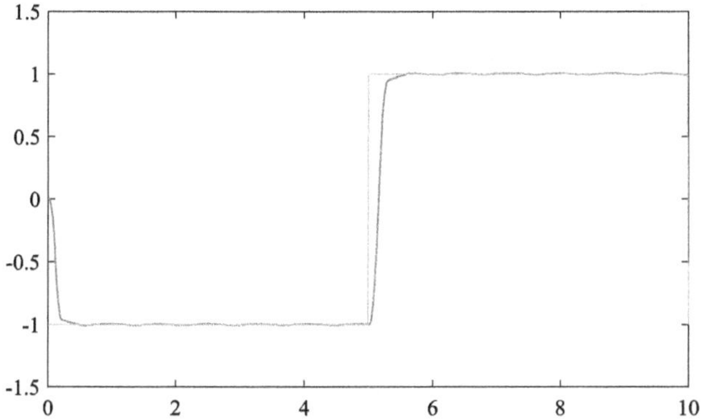

Figure 7.22: System output at $r = 100$.

7.2 For the van der Pol differential equation, $y''(t)+\mu(y^2(t)-1)y'(t)+y(t)=0$, if μ and $\boldsymbol{x}(0)$ are selected as additional parameters, design an S-function, and mask it into a reusable Simulink block.

7.3 For the mathematical model of Lorenz equation,

$$\begin{cases} x_1'(t) = -\beta x_1(t) + x_2(t)x_3(t), \\ x_2'(t) = -\rho x_2(t) + \rho x_3(t), \\ x_3'(t) = -x_1(t)x_2(t) + \sigma x_2(t) - x_3(t), \end{cases}$$

if β, ρ, σ and $\boldsymbol{x}(0)$ are the additional parameters, design an S-function, and mask it into a reusable Simulink block.

7.4 For the piecewise linear nonlinearities, if in the i piece, that is, $e_i \leqslant x < e_{i+1}$, the output signal is modeled as $y(x) = k_i x + b_i$. If the break points $e_1, e_2, \dots, e_{N+1}$ are given, and the parameters $k_1, b_1, k_2, b_2, \dots, k_N, b_N$ are also known, design an S-function, and mask it into a reusable Simulink block.

7.5 For a programmable logic device (PLD), there are six input signals, A, B, W_1, W_2, W_3 and W_4, and W_i are encoding signals, whose values depends the logic relationship of Y. The logic relationship is described in Table 7.3. It may be rather complicated to design the PLD block, using a low-level modeling method. Write an S-function to implement the block.

7.6 For a given linear discrete state space model,

$$\begin{cases} \boldsymbol{x}(k+1) = \boldsymbol{F}\boldsymbol{x}(k) + \boldsymbol{G}u(k), \\ y(k) = \boldsymbol{C}\boldsymbol{x}(k) + \boldsymbol{D}u(k), \end{cases}$$

Table 7.3: Logic relationship in Exercise 7.5.

W_1	W_2	W_3	W_4	Y	W_1	W_2	W_3	W_4	Y
0	0	0	0	0	1	0	0	0	$A\overline{B}$
0	0	0	1	AB	1	0	0	1	A
0	0	1	0	$\overline{A+B}$	1	0	1	0	$\overline{B}$
0	0	1	1	$AB+\overline{AB}=A\odot B$	1	0	1	1	$A+\overline{B}$
0	1	0	0	$\overline{A}B$	1	1	0	0	$\overline{A}B+A\overline{B}=A\oplus B$
0	1	0	1	B	1	1	0	1	$A+B$
0	1	1	0	$\overline{A}$	1	1	1	0	$\overline{A}+\overline{B}=\overline{AB}$
0	1	1	1	$\overline{A}+B$	1	1	1	1	1

if the sample time is T, write a general purpose S-function block. If the matrices are given

$$F=\begin{bmatrix}0.2769 & 0.8235 & 0.9502\\ 0.0462 & 0.6948 & 0.0345\\ 0.0971 & 0.3171 & 0.4387\end{bmatrix},\quad G=\begin{bmatrix}0.3816\\ 0.7655\\ 0.7952\end{bmatrix},\quad C=[1,0,0],\quad D=0.3,$$

find the step response of the system. If in the design of S-function, the value of `DirectFeedthrough` is set to 0, and observe whether normal simulation can be carried out, why? Use the discrete state space block provided in Simulink to validate the results.

7.7 Rewrite the S-function for the extended state observer in C.

7.8 If the plant model is an unstable one $1/s(s-1)$, see whether ADRC controller can still yield good control results. If the plant model is changed into a third-order model, $1/(s^3+3s^2+2s+4)$, see whether the ADRC controller is still effective.

7.9 Consider the plant model with large time delay, $G(s)=10\mathrm{e}^{-20s}/(2s+1)$. See whether the ADRC controller can be used directly the plant model. If it cannot be used, adjust the controller parameters and observe the control results, such that a suitable controller can be found.

8 Promising combinations of Simulink and numerical optimization techniques

In the early chapters, the powerful modeling and simulation facilities of Simulink were illustrated. If the system models to be studied are complicated, the simulation block diagram can be established by Simulink, and the behaviors of the complicated systems can be acquired. With the powerful tools provided in Simulink, system behaviors of models of any complexity can be handled, theoretically.

Optimization ideas and solution methods are very important in tackling scientific and engineering research. It is not exaggerating to say that when equipped with ideas and methodologies of optimization, the practical research level of the researcher may be boosted into a higher level. If the users are not using optimization concepts, they will be satisfied if a solution in science and engineering can be acquired. If optimization ideas are introduced, targets of the users may be upgraded, and they may not be happy to find merely a solution. They may begin to pursuit for the best solution. Therefore, the objective functions and constraints can be deliberately chosen, so as to find the optimum solutions of the problems, by using numerical optimization techniques. In Volume IV of the series, the MATLAB based optimization problem solutions are fully addressed.[83]

If the two powerful tools can be cleverly used together, some unsolved problems can be explored under the new framework, and new methods or even new research fields can be developed, such that the practical research capabilities of the researchers can be boosted to an upper level.

Talking about introducing optimization techniques into scientific research, an inevitable question must be answered: "What is the most important factor in the application of optimization techniques?" The answer to this question may be differ in many ways, depending on the viewpoints and research background of the researcher. According to the author, the selection of the objective function is the most crucial in optimization. Only the objective function is selected reasonably, the solution is meaningful. With the powerful software and hardware support of modern computers, naturally no matter how complicated the objective is selected, the results needed can be obtained by the contemporary computer resources.

In the research in the past, since there were no powerful tools in carrying out optimization research, indirect methods were chosen in objective functions. The target of selecting objective functions is not because they are really useful, but because they can be computed. From contemporary viewpoints, the way to select a computable objective function, regardless of its physical meaning, is questionable. Therefore, to get really useful results, the truly meaningful objective functions should be chosen, and there is no need to worry about whether the objective function is too hard to compute.

https://doi.org/10.1515/9783110734959-008

In this chapter, explorations are made on how to combine together the powerful simulation facilities in Simulink and the numerical optimization technique provided by MATLAB to tackle application problems. Some of the problems to be explored do not have solution algorithms in the traditional aspect. With the tools provided in numerical optimization techniques, the problems may be solved easily. Two types of problems are presented in this chapter: one is to solve boundary value problems in ODEs, and the other, optimal controller design. In Section 8.1, boundary value problems to differential equations are studied, with concentration on how to directly convert the problems into numerical optimization problems. A simple example is used to illustrate the idea and implementation method, and also fast restart methods of the Simulink model are introduced. Then the solutions of other problems are considered, such as boundary value problems of fractional-order ODEs and delay differential equations. These equations are usually not solvable by traditional methods. In Section 8.2, the APP design method using recent versions of MATLAB are introduced, which can be regarded as the foundation for the interface design methods in the subsequent sections. In Section 8.3, the conventional optimal control methods are objectively commented, and the hidden problems are pointed out. The Servo control system is used as the background, to explore the application of optimization techniques to the controller design problem. Finally, a universal optimal controller design interface is developed. It is shown by optimization that for certain plants that PID controllers are the best second-order controllers. In Section 8.4, another APP designed by the author, OptimPID, for the design of optimum PID controllers is presented. The low-level Simulink model and APP design are illustrated. Examples are used to demonstrate the optimum PID controller methods with the APP.

8.1 Boundary value problems in ODEs

It has been shown in the previous chapters that Simulink is useful in finding numerical solutions of differential equations. The methods discussed earlier are the solutions on differential equations with initial values, therefore, the problems are referred to as initial value problems. With tools such as Simulink, the numerical solutions of differential equations can be derived step by step, from the given initial values. In contrast, the solutions of boundary value problems are much more complicated. In this section, the boundary value problems of various differential equations are explored. The mathematical form of boundary value problems are presented first, and then converted into optimization problems. With the numerical optimization solution facilities provided in MATLAB, the boundary value problem solutions can be found. The methods are tried to find boundary value problem solutions of fractional-order differential equations and delay differential equations, where no other solvers are known.

8.1.1 Shooting algorithms in boundary value problems

The mathematical form of boundary value problems of differential equations is presented first in this section, then the basic ideas of a commonly used method, the shooting method, are introduced.

Definition 8.1. The general form of the first-order explicit differential equations containing parameters is given below:

$$\boldsymbol{y}'(t) = \boldsymbol{f}(t, \boldsymbol{y}(t), \boldsymbol{\theta}), \tag{8.1.1}$$

where $\boldsymbol{y}(t)$ is the state vector, $\boldsymbol{\theta}$ is the vector containing undetermined parameters. The boundary values of the equations are known as

$$\boldsymbol{\phi}(\boldsymbol{y}(a), \boldsymbol{y}(b), \boldsymbol{\theta}) = \boldsymbol{0}. \tag{8.1.2}$$

As a special case for Definition 8.1, consider first a special second-order differential equation, whose explicit form is

$$y''(t) = f(t, y(t), y'(t)). \tag{8.1.3}$$

The values $y(a)$ and $y(b)$ are known. Find the numerical solution $y(t)$ of the equation over the interval (a, b).

The ordinary idea for solving a boundary value problem is to assume that the initial value $\hat{y}'(a)$ is known, and the solver `ode45()` or Simulink environment can be used to solve the initial value problem. The value of $\hat{y}(b)$ in the solution must exists errors with the given boundary value $y(b)$. The error information can be used to modify the initial value $\hat{y}'(a)$ and solve the problem again. An iterative process can be used until a consistent solution is found, such that $|\hat{y}(b) - y(b)| < \epsilon$. With the consistent initial values obtained, the initial value problem can be solved again, which can be regarded as the numerical solution of the boundary value problem. This type of algorithm is known as the shooting method.

In fact, the above shooting method can easily be extended to handle the boundary value problems in Definition 8.1. Select randomly a set of initial values, and the differential equation solutions can be found. The value of the algebraic equation function in (8.1.2) can be found. Of course, the value of the function has differences with zero value, known as errors. The error can be used to adjust the initial values, and an iterative method can be used to reduce the error. When the error reaches zero, the initial values obtained can be used to find the solutions of the boundary value problems.

8.1.2 Converting shooting methods into optimization problems

It is obvious that the above presented idea can be used to solve the algebraic equations in (8.1.2). In fact, the equation solution methods are too complicated to implement in real applications. An alternative method is to convert the original problem into an optimization problem.

Definition 8.2. The state variables $y_i(a)$ in Definition 8.1 can be selected as the decision variables, such that the original boundary value problem can be converted into the following optimization problems:

$$\min_{\boldsymbol{x}} \|\boldsymbol{\phi}(\boldsymbol{x}, \hat{\boldsymbol{y}}(b), \boldsymbol{\theta})\|, \tag{8.1.4}$$

where $\boldsymbol{x} = [y_1(a), y_2(a), \dots, y_n(a)]^{\mathrm{T}}$, and $\hat{\boldsymbol{y}}(b)$ is the initial value problem solution at time b.

Since theoretically speaking, MATLAB and Simulink can be used to solving initial value problems of any complexity, if it is combined with numerical optimization techniques, and the boundary value problem of any complexity can be obtained.

As a special case, in the boundary value problem in (8.1.3), some of the initial values are known, while others are not. Therefore, in optimization problems, it is not necessary to select all the initial states as decision variables. The unknown initial values must be chosen as decision variables. The total number of the decision variables can be reduced, so as to solve the original problem effectively.

8.1.3 Examples of simple boundary value problems

In ordinary cases, traditionally, Simulink cannot be used to solve boundary value problems. If powerful simulation facilities in Simulink are used to solve initial value problems, together with the optimization problem solution capabilities, the solutions of various complicated boundary value problems can be found. Some of the problems may not be solved with any other known algorithms. In this section, examples are used to demonstrate the ideas and solution methods for boundary value problems.

Example 8.1. For the differential equation $y''(t) = 2y(t)y'(t)$, if the boundary values are $y(0) = -1$ and $y(\pi/2) = 1$, and the exact solution is $y = \tan(t - \pi/4)$, use the Simulink to solve the boundary value problem.

Solutions. An integrator chain with two integrators can be introduced, so that the Simulink model in Figure 8.1 can be established. Since the initial value of $y(0) = -1$ is known, while $y'(0)$ is unknown, it can be set to variable a. Let the decision variable $x = a$, and the simulation process can be invoked for the given value of x, such that the numerical solution of the initial value problem can be found. The final value of $y(t)$ can be compared with the given $y(\pi/2) = 1$, and an error can be found. The absolute value of the error can be selected as the objective function, such that the boundary value problem can be converted into an optimization problem. The following M-function can be written to describe the objective function. The value x of the decision variable can be written to MATLAB workspace with function `assignin()` to assign to variable a. Therefore, the simulation process can be invoked, and the absolute value of the error is assigned to y. If simulation process is not successful, `try-catch` structure can be used, such that the value of y can be set to a large value of 10.

```
function y=c8mbvp1(x)
assignin('base','a',x(1))
try, [t,~,y0]=sim('c8mbvp1a',[0,pi/2]);    y=abs(y0(end,1)-1);
catch, y=10; end
```

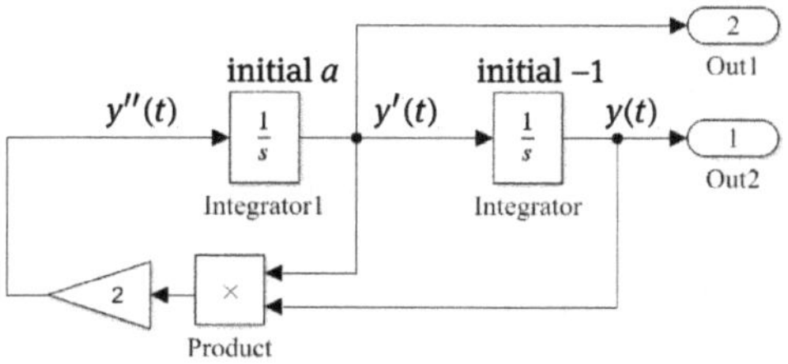

Figure 8.1: Simulink model (file: `c8mbvp1a.slx`).

With the objective function, the following commands can be used to solve the numerical optimization problem, and the solutions found are shown in Figure 8.2. The circled point in the plot is the given boundary value. It is found that $a = 1.999999999998268$, whose theoretical value is 2. The total time elapse is 7.51 seconds. It can be seen that the maximum error in the solution is 8.4745×10^{-12}. The error is too small to be witnessed from the curves.

```
>> ff=optimset; ff.TolX=eps; ff.TolFun=eps;
   tic, [x0 f0]=fminsearch(@c8mbvp1,0,ff), toc %solve optimization
   [t,~,y]=sim('c8mbvp1a',[0,pi/2]);
   plot(t,y,t,tan(t-pi/4),[0 pi/2],[-1 1],'o')
   max(abs(y(:,1)-tan(t-pi/4)))
```

If the above commands are used, but the `ff` option is not used, the maximum error obtained is 8.0335×10^{-5}. For this specific example, if the `ff` control is used, the time elapsed is 9.26 seconds, while if it is not used, the time elapsed is 4.09 seconds. It can be seen that the time elapsed is reduced by half, but the maximum error is increased

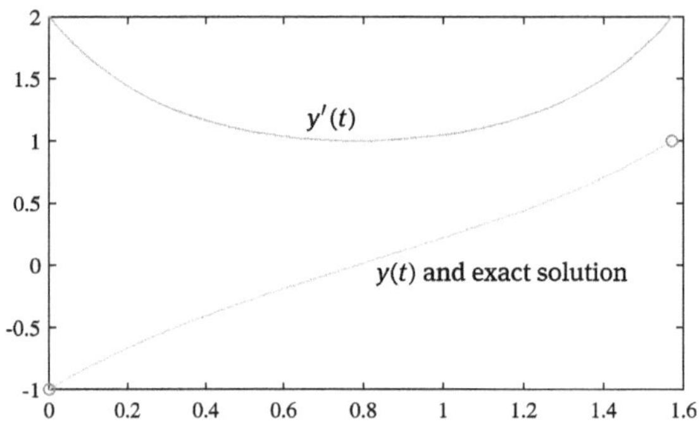

Figure 8.2: Numerical and exact solutions.

by 10^7 times. It can be seen that in simulation and optimization models, the tough error tolerances should be used, to reduce the total error.

Example 8.2. Solve the boundary value problem of the ODE:[2]

$$(t^3u''(t))'' = 1, \quad 1 \leqslant t \leqslant 2,$$

where $u(1) = u''(1) = u(2) = u''(2) = 0$, and the analytical solution is known

$$u(t) = \frac{1}{4}(10\ln 2 - 3)(1 - t) + \frac{1}{2}\left[\frac{1}{t} + (3 + t)\ln t - t\right].$$

Solutions. The original form of the differential equation cannot be solved directly in Simulink. Although the second-order derivative of the $t^3u''(t)$ can be found, it is not suitable to represent in integrators, since the initial values of the integrator are hard to decide. Therefore, Symbolic Math Toolbox can be used to further derive the formula, such that the explicit form can be obtained.

```
>> syms c u(t)
   diff(t^3*diff(u,2),2), latex(ans) % derive the left side expression
```

The following form can be found:

$$t^3u^{(4)}(t) + 6t^2u'''(t) + 6tu''(t) = 1.$$

Since the interval is $t \in (1,2)$, then $t \neq 0$, and the explicit form can be immediately found:

$$u^{(4)}(t) = -\frac{6}{t}u'''(t) - \frac{6}{t^2}u''(t) + \frac{1}{t^3}.$$

An integrator chain is used to define key signals $u(t)$, $u'(t)$, $u''(t)$, $u'''(t)$ and $u^{(4)}(t)$. It is not hard to construct the Simulink model shown in Figure 8.3. With the Mux block, the key signals and t can be grouped into a vector signal. The block Fcn can be used to express the right side of the equation, and the output can be fed into the $u^{(4)}(t)$ signal, to close the simulation loop.

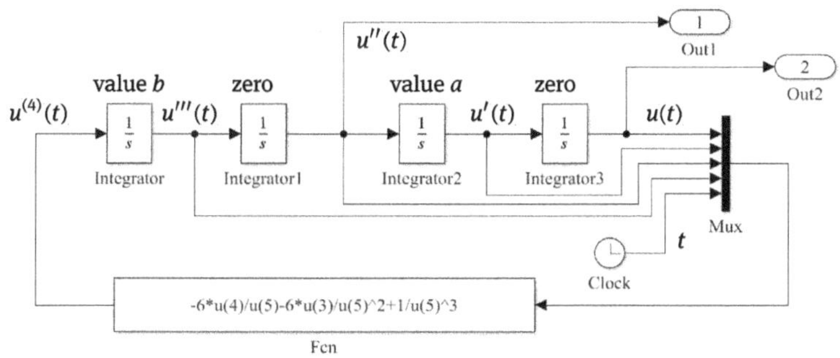

Figure 8.3: Simulink model (file: c8mbvp2a.slx).

It can be seen from boundary conditions that $u(1) = u''(1) = 0$ are initial values, indicating that the initial values of $u(t)$ and $u''(t)$ integrators are zeros, while the initial values of the other two integrators are unknown. They must be determined with the optimization process. The initial values of the two integrators can be set to a and b, as shown in Figure 8.3. Select the decision variable as $\boldsymbol{x} = [a, b]$, Simulink process can be invoked, such that the final values of $u(t)$ and $u''(t)$ can be found. With the Simulink model, the following objective function can be written, for the condition $|\hat{u}_1(2)|+|\hat{u}_3(2)|$. If there are errors in the simulation process, the value of the objective function is set to 10.

```
function y=c8mbvp2(x)
assignin('base','a',x(1)), assignin('base','b',x(2))
try
   [t,~,y0]=sim('c8mbvp2a',[1,2]); y=abs(y0(end,1))+abs(y0(end,2));
catch, y=10; end
```

With the objective function, the optimization process is activated. The consistent initial values are $a = 0.01713$ and $b = -0.5$, with time elapsed of 58.04 seconds. Substituting the initial values back to the model, the simulation results $u(t)$ and $u''(t)$ curves are shown in Figure 8.4. The maximum error between the numerical and analytical solution is 8.7931×10^{-12}. The two curves are almost identical, the errors cannot be identified from the curves.

```
>> ff=optimset; ff.TolX=eps; ff.TolFun=eps;
   tic, [x0 f0]=fminsearch(@c8mbvp2,[1,1],ff), toc %optimization
   [t,~,y]=sim('c8mbvp2a',[1,2]);
```

```
u=1/4*(10*log(2)-3)*(1-t)+1/2*(1./t+(3+t).*log(t)-t);
plot(t,y,t,u,[1,2],[0,0],'o'), max(abs(y(:,2)-u))
```

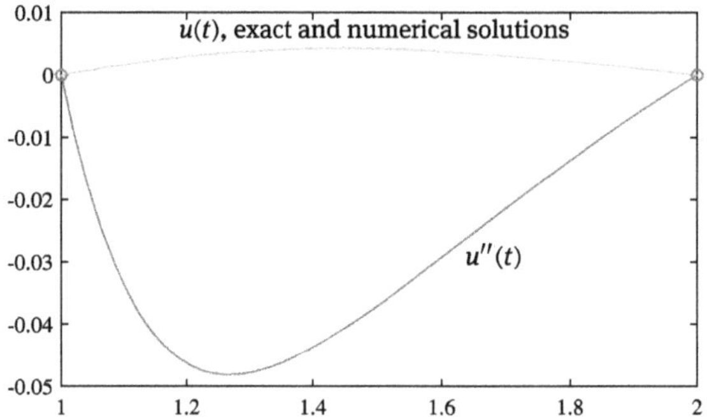

Figure 8.4: Simulation results.

8.1.4 Fast restart mode of Simulink models

In the optimization process, when the Simulink model is opened, it can be found that in each simulation step, in the Simulink model status bar, the "Compiling" message is displayed. This is normal, since under the default setting, the model is compiled, each time the model is executed. The compilation process is rather time consuming. How can we avoid the recompiling process? "Fast restart" simulation mode is supported in Simulink. When the mode is activated, if the structure and the unadjustable parameters are not changed, there is no need to recompile it. Direct simulation can be carried out. To activate fast restart mode, the following changes should be made in the optimization process:

(1) The variables in model workspace rather than the MATLAB workspace should be set with the function `assignin()`; otherwise, the variable change is not accepted.

(2) The `sim()` function call is only allowed to return one argument, even if the **Single simulation output** item is unchecked. Otherwise, error messages are displayed and the members `tout` and `yout` in the returned argument are time vector and output matrix.

(3) The parameters such as **Start Time** and **Stop Time** are not allowed to be set in the `sim()` function call. These parameters must be set within the model. Even though different intervals are specified in the `sim()` function, they do not take any effect.

The fast restart mode is demonstrated next through an example. In the remaining part of the chapter, the optimization problem is involved. The fast restart model is adopted, so as to increase the solution efficiency.

Example 8.3. Solve the problem again in Example 8.2 with fast restart mode, and assess the time consumption.

Solutions. To use the fast restart mode, the following modifications must be made to the objective functions:

```
function y=c8mbvp2b(x)
W=get_param(gcs,'ModelWorkspace'); %model workspace
assignin(W,'a',x(1)), assignin(W,'b',x(2))
try, txy=sim('c8mbvp2a'); %only one returned argument allowed
   y=txy.yout; y=abs(y0(end,1))+abs(y0(end,2));
catch, y=10; end
```

Therefore, the fast restart mode can be set with the following statements, and the optimization process can be carried out. Under the new simulation mode, the time elapsed is 13.80 seconds, which is significantly less the 58.04 seconds time in normal simulation mode. The result obtained is exactly the same, with the maximum error of 2.8677×10^{-17}. It can be seen from the example that when using the fast restart mode, the simulation results obtained are exactly the same, and the solution efficiency is significantly increased:

```
>> c8mbvp2a; set_param(gcs,'FastRestart','on');        %fast restart
   ff=optimset; ff.TolX=eps; ff.TolFun=eps;
   tic, [x0 f0]=fminsearch(@c8mbvp2b,[1,1],ff), toc %optimization
   txy=sim('c8mbvp2a',[1,2]); t=txy.tout; y=txy.yout;
   u=1/4*(10*log(2)-3)*(1-t)+1/2*(1./t+(3+t).*log(t)-t);
   plot(t,y,t,u,[1,2],[0,0],'o'), max(abs(y(:,2)-u))
```

In the MODELING pane in the model toolbar, a button Fast Restart is provided. If it is pressed down, many of the buttons in the toolbar turn gray, meaning that they cannot be changed. If the button is unpressed, the parameters can be modified. When the optimization process is used, make sure that this button is pressed; otherwise, error messages may appear in computing the objective function.

8.1.5 Boundary value problems with parameters

In some differential equations, the mathematical forms are given, and they may contain undetermined parameters. If there are a sufficient number of boundary conditions, such that the number of algebraic equations equal the number of unknowns, the optimization problem can be tried to solve the problem directly. An example is given here to demonstrate the conversion and the solution process.

Example 8.4. Solve the following differential equation:

$$\begin{cases} x'(t) = 4x(t) - \alpha x(t)y(t), \\ y'(t) = -2y(t) + \beta x(t)y(t), \end{cases}$$

with the conditions $x(0) = 2$, $y(0) = 1$, $x(3) = 4$ and $y(3) = 2$. Find the solutions of the differential equation, and find the consistent parameters α and β.

Solutions. From the given differential equation, the Simulink model shown in Figure 8.5 can be established. Since $x(0) = 2$ and $y(0) = 1$ are given, they can be filled accordingly into the two integrators.

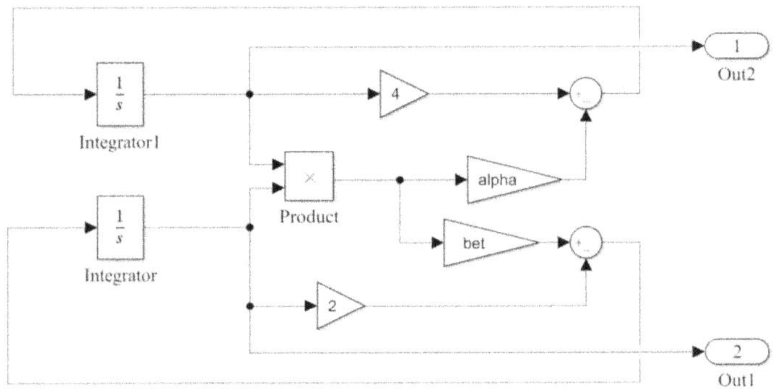

Figure 8.5: Simulink model with parameters (file: `c8mab1a.slx`).

With the given boundary conditions $x(3) = 4$, $y(3) = 2$, the following objective function can be written. The target is to find the undetermined parameters α and β, such that the final value conditions are satisfied:

```
function y=c8mab1(x)
W=get_param(gcs,'ModelWorkspace'); %model workspace
assignin(W,'alpha',x(1)), assignin(W,'bet',x(2))
try
   [t,~,y0]=sim('c8mab1a');
   y=abs(y0(end,1)-4)+abs(y0(end,2)-2);
catch, y=10; end
```

Solving the optimization problem, the two undetermined parameters are found, as $\alpha = 0.7557$, $\beta = 1.7016$, and time elapsed is 8.76 seconds. The value of the objective function is 1.3323×10^{-15}. The solution of the differential equation is shown in Figure 8.6.

```
>> c8mab1a; set_param(gcs,'FastRestart','on');
   ff=optimset; ff.TolX=eps; ff.TolFun=eps;
   tic, [x0 f0]=fminsearch(@c8mab1,rand(2,1),ff), toc %optimization
   txy=sim('c8mab1a');
   plot(txy.tout,txy.yout,[0 0 3 3],[2 1 4 2],'o')
```

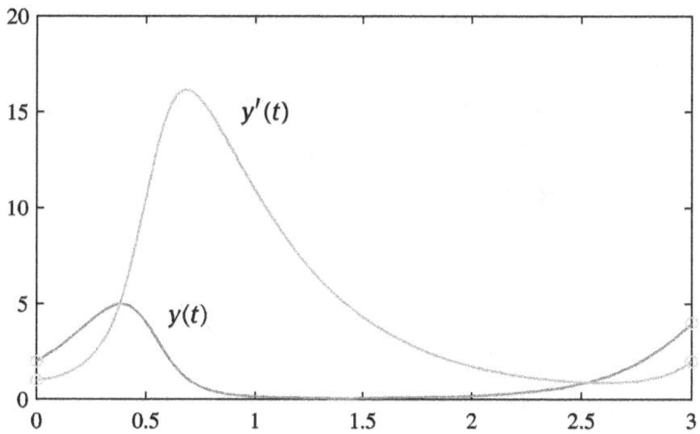

Figure 8.6: Solutions of the differential equations.

8.1.6 Solutions of multiple-point problems

In the examples studied earlier, the boundary value problems are all two-point problems. In real applications, problems with multiple given points are encountered. Therefore, if several function values at point $t_1, t_2, \dots, t_m$ are known, and the number of conditions equals to the number of unknowns, the given points can be sorted. It is safe to assume that $t_1 < t_2 < \cdots < t_m$, the simulation interval can be divided into subintervals $(t_0, t_1), (t_1, t_2), \dots, (t_{m-1}, t_m)$. The initial values of each subinterval are the final values of the previous one. Therefore, the problem can also be converted into numerical optimization problems, such that the final solutions can be found.

In fact, when calling function `sim()`, the following syntax can be used:

`[`$\boldsymbol{t}$`,`$\boldsymbol{x}$`,`$\boldsymbol{y}$`]=sim(`model name`, [`t_0`,`t_1`,`t_2`,…,`t_m`])`

Then the solution results on these points are returned only. There is no need to divide the interval into several subintervals. Note that the first point in the time vector should be t_0, not t_1; otherwise, the function may misunderstand t_1 as the initial time, for the given initial values, such that wrong results are obtained.

The following example is given to demonstrate the idea, and the solution to the problem can be found.

Example 8.5. Consider the following linear differential equation:

$$y^{(4)} + 10y'''(t) + 35y''(t) + 50y'(t) + 24y(t) = e^{-5t}\cos(2t+1) + 5.$$

If the given values are $y(0) = 0.5$, $y'(\pi) = 0$, $y'(4) = 0$ and $y''(2\pi) = 0$, find the numerical solutions with optimization method, and draw the solution curves.

Solutions. Since the values at the four points $t = 0, \pi, 4$ and 2π are known, where only $y(0)$ is the known initial value, and the other initial values are unknown. Let $y'(0) = a$, $y''(0) = b$ and $y'''(0) = c$, the Simulink model for the differential equation can be established, as shown in Figure 8.7. Select the decision variable vector $\boldsymbol{x} = [a, b, c]$, and the original problem can be converted into a numerical optimization one. The objective function is given below. To avoid the unnecessary troubles, the fast restart mode is not used in this example. The conventional mode is used instead.

```
function y=c8mmult1(x)
assignin('base','a',x(1))
assignin('base','b',x(2)), assignin('base','c',x(3))
[t1 x1 y1]=sim('c8mmult1b',[0,pi 2*pi]);
y=abs(y1(2,2))+abs(y1(3,2))+abs(y1(3,3));
```

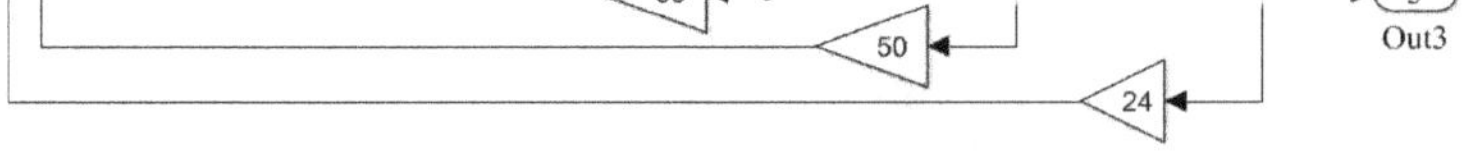

Figure 8.7: Simulink model (file: c8mmult1b.slx).

Through the following statements, the optimization problem can be found, and the simulation results are shown in Figure 8.8. It can be seen that the given points are all on the curves:

```
>> ff=optimset; ff.TolX=eps; ff.TolFun=eps;
   tic, [x0 f0]=fminunc(@c8mmult1,rand(3,1),ff), toc %optimization
   [t,x,y]=sim('c8mmult1b',[0,2*pi]);
   plot(t,y,[0 pi 4 2*pi],[0.5 0 0 0],'o')
```

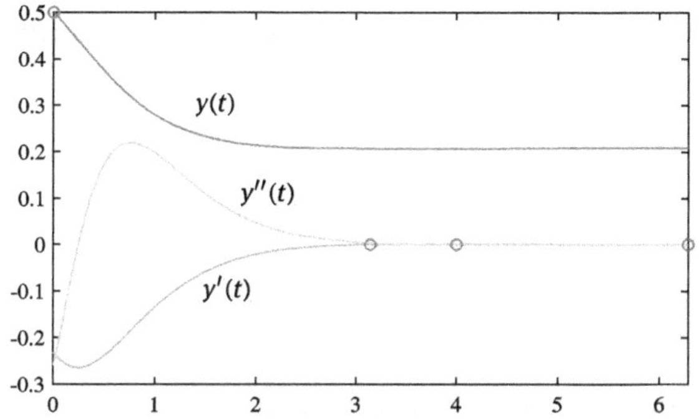

Figure 8.8: Solutions of the differential equation.

8.1.7 Boundary value problems of fractional-order ODEs

To the author's knowledge, there is no available methods suitable of solving boundary value problems of fractional differential equations. The initial value problem solvers for nonlinear fractional-order differential equations are systematically studied in [79], where the block diagram based method can be used to simulate fractional-order differential equations of any complexity, with high precision. Therefore, the method is combined with optimization methods, such that the boundary value problems of nonlinear fractional-order differential equations can be found.

Example 8.6. Consider again the nonlinear Caputo fractional-order differential equation studied in Example 4.28:

$$ {}_0^{\rm C}\mathscr{D}_t^{1.455}y(t) = -t^{0.1}\frac{\mathrm{E}_{1,1.545}(-t)}{\mathrm{E}_{1,1.445}(-t)}\mathrm{e}^t y(t){}_0^{\rm C}\mathscr{D}_t^{0.555}y(t) + \mathrm{e}^{-2t} - \left[y'(t)\right]^2. $$

If the boundary values are $y(0) = 1$ and $y(1) = \mathrm{e}^{-1}$, solve the boundary value problem. It is known that the analytical solution is $y(t) = \mathrm{e}^{-t}$, and assess the accuracy of the solution.

Solutions. From the given problem, since $y(0)$ is known, the decision variable should be selected as $x = y'(0)$. Therefore, the numerical solution for the initial value problem can be represented as $\hat{y}(1)$, where the absolute value of the difference between $\hat{y}(1)$ and the given $y(1)$ can be used as the objective function. The Simulink model `c4mfode.slx` in Figure 4.28 can be used. If the initial value of $y'(t)$ integrator, the left one in Figure 4.57, is changed from −1 into a, and saved into model `c8mfode.slx`, the model can be used to solve initial value problem for this example.

For certain initial values of a, the problem may become singular, such that an abnormal interrupt may occur; the trial structure can be used here to bypass the singularities, and set the object function value to large ones, and keep the optimization process going on. Based on the considerations, the following objective function can be written:

```
function y=c8mfode1(x)
W=get_param(gcs,'ModelWorkspace'); assignin(W,'a',x(1))
try
   txy=sim('c8mfode',1); y0=txy.yout;  y=abs(y0(end)-exp(-1));
catch, y=10; end
```

With the objective function, with suitable Oustaloup filter parameters, the following commands can be used to solve the boundary value problem. The time elapsed is 22.51 seconds, and the consistent initial value is $x_2(0)$, $a = -0.99997037$, very close to the exact one of −1. The maximum error of the solution is 1.6593×10^{-7}. It can be seen that the solution process is successful for this problem. Note that the filter parameters should be set first, before setting the fast restart mode; otherwise, the parameters cannot be used.

```
>> ww=[1e-5,1e5]; N=30; c8mfode, format long %filter parameters
   set_param(gcs,'FastRestart','on');
   opt=optimset; opt.TolX=1e-13;
   tic, [x0,f0]=fminsearch(@c8mfode1,rand(1),opt), toc  %optimization
   txy=sim('c8mfode'); max(abs(txy.yout-exp(-txy.tout)))
```

8.1.8 Boundary value problems of delay differential equations

The initial value problems of delay differential equations can normally be solved with the solvers in MATLAB, and also with Simulink. However, the solutions of boundary value problems are not thus simple, since there is no existing solvers provided in MATLAB, nor in Simulink. Similar to the above problems, optimization methods can be combined with the initial value facilities provided in Simulink, and the boundary value problem of delay differential equations can be solved directly. In this section, the delay differential equation with nonzero history functions is used as an example to demonstrate the combined Simulink and optimization solution method.

Example 8.7. Consider the following delay differential equations with constant delay times:

$$\begin{cases} x'(t) = 1 - 3x(t) - y(t-1) - 0.2x^3(t-0.5) - x(t-0.5), \\ y''(t) + 3y'(t) + 2y(t) = 4x(t), \end{cases}$$

where when $t \leqslant 0$, the history functions are constants. It is known that $x(0) = 0$, $y(10) = 0.3289$ and $x(10) = y'(10) + 0.163$. Find the numerical solution of the equation.

Solutions. Since $x(0) = 0$, for $t \leqslant 0$, $x(t) = 0$, that is, the history value of the state is zero. The initial values of the other two key signals are not known. It is safe to assume that the initial value of $y(t)$ is a, and the initial value of $y'(t)$ is b. Since the delay signal of $y(t)$ is needed, the method in Example 4.25 can be followed, and set the history value $y(t)$ to a with a switch block. Therefore, the simulation model is established as shown in Figure 8.9, where the Mux block is composed of five scalar signals. The former three are the key signals, and the remaining two are respectively the delay signals $x(t-0.5)$ and $y(t-1)$. The two differential equations can be expressed by the Fcn and Fcn1 blocks.

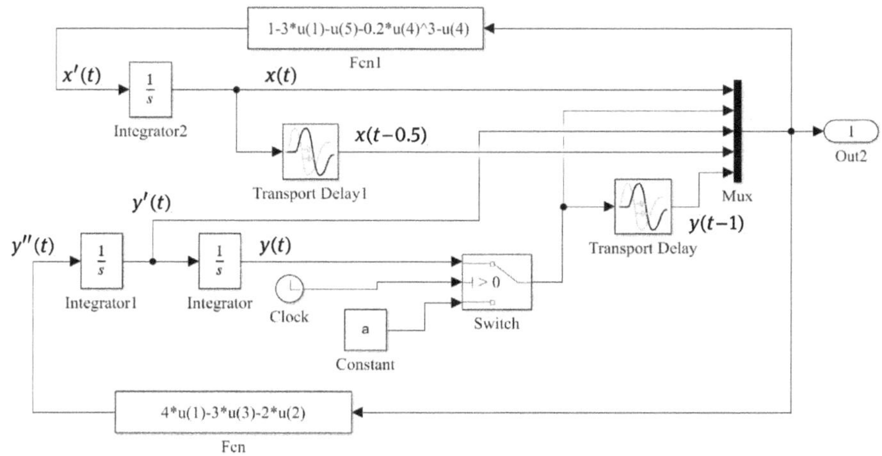

Figure 8.9: Delay model (file: `c8mdde1a.slx`).

Select the decision variables $\boldsymbol{x} = [a, b]$, from the boundary condition $y(10) = 0.3289$, $x(10) = y'(10) + 0.163$, the following MATLAB function can be written for the objective function:

```
function y=c8mdde1(x)
W=get_param(gcs,'ModelWorkspace'); %model workspace
assignin(W,'a',x(1)), assignin(W,'b',x(2))
try
   txy=sim('c8mdde1a',[0,10]); y0=txy.yout;
   y=abs(y0(end,2)-0.3289)+abs(y0(end,1)-y0(end,3)-0.163);
catch, y=10; end
```

With the following commands, the optimization problem can be solved, and it is found that $a = 1.9682$ and $b = 1.0849$. The time elapsed is 8.25 seconds. The solution of the differential equation obtained is shown in Figure 8.10. The error in the boundary value is $f_0 = 4.8209 \times 10^{-8}$. Since for delay systems, the error tolerance cannot be assigned to too small values; the one set in this example is 10^{-8}. It can be seen that the solutions are rather accurate.

```
>> c8mdde1a, set_param(gcs,'FastRestart','on');
   opt=optimset; opt.TolX=1e-13; % optimization control parameters
   tic, [x0 f0]=fminsearch(@c8mdde1,[2 2],opt), toc
   txy=sim('c8mdde1a'); t=txy.tout; y0=txy.yout;
   plot(t,y0(:,[1 2 3]),[0,10],[0 0.3289],'o')
```

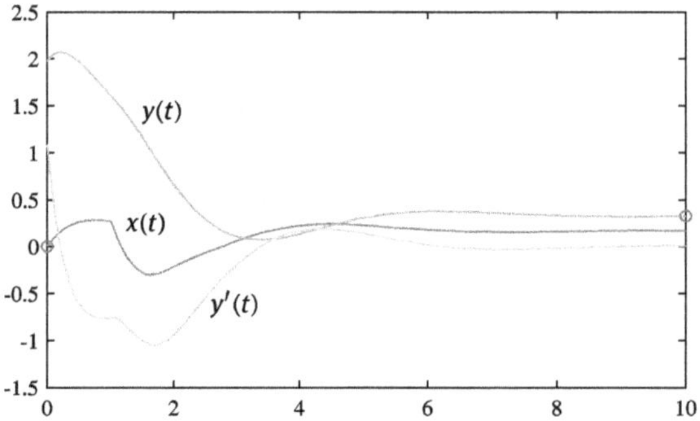

Figure 8.10: Simulation results of differential equations.

8.1.9 Global optimum solutions

In optimization theory, the global optimum solutions are always welcome, and the local optimum solutions can be neglected. In the solutions of boundary value problems, why are the global optimum solutions so meaningful? A counterexample is given in this section to demonstrate the solutions of boundary value problems with multiple solutions, and the global optimum solutions are explored in this example.

Example 8.8. Solve the differential equation given below. Also, since the analytical solution can be found, compare the accuracies of the numerical solutions:

$$(y''(t))^2 = 4(ty'(t) - y(t)) + 2y'(t) + 1, \quad y(0) = 0, \quad y(1) = 1.$$

Solutions. Since the differential equation given here is not in explicit form, it cannot be handled directly with MATLAB and Simulink. It must be converted first to explicit form. The simplest way is to take square roots to the expression on the right side. It can be seen that

$$\begin{cases} y''(t) = \sqrt{4(ty'(t) - y(t)) + 2y'(t) + 1}, \\ y''(t) = -\sqrt{4(ty'(t) - y(t)) + 2y'(t) + 1}. \end{cases}$$

A unified Simulink model can be established as shown in Figure 8.11, where a variable key can be set to -1 or 1, indicating that there are two branches in the solutions. With the model, the following MATLAB function can be written, to set up the objective function of the boundary value problem:

```
function y=c8msqrt1(x)
W=get_param(gcs,'ModelWorkspace'); assignin(W,'a',x)
try
   txy=sim('c8msqrt1a'); y=abs(txy.yout(end,1)-1);
catch, y=10; end
```

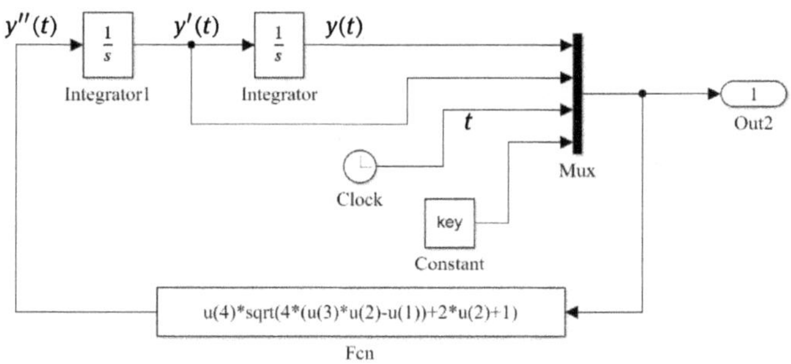

Figure 8.11: Differential equation model (file: c8msqrt1a.slx).

Now consider the first branch in the differential equation. Set key to 1, and the following commands can be used to solve the optimization problem. The consistent value obtained is $a = 0.1708$, and the time elapsed is 2.27 seconds. The value of the objective function is 3.5083×10^{-14}, and the solution is shown in Figure 8.12. It can be seen that the boundary values are satisfied, thus that it is the solution of the original problem.

```
>> key=1; c8msqrt1a, set_param(gcs,'FastRestart','on');
   opt=optimset; opt.TolX=1e-13;
   tic, [x0 f0]=fminsearch(@c8msqrt1,1,opt), toc
   txy=sim('c8msqrt1a'); t=txy.tout; y0=txy.yout;
   plot(t,y0(:,[1 2]),[0 1],[0 1],'o')
```

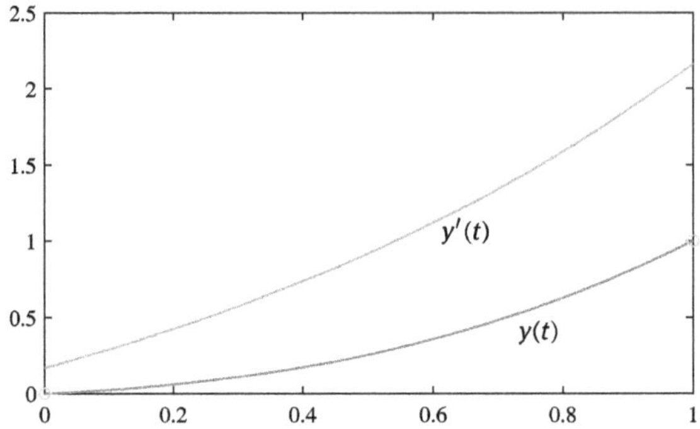

Figure 8.12: Solution of the first branch.

Now consider the second branch. Set key to −1. The following commands can be used to solve optimization problem. The consistent initial value is $a = 1.8292$, and the time elapse is 2.56 seconds. The value of the objective function is 2.6423×10^{-14}. The solution of the differential equation is shown in Figure 8.13. It can be seen that the solution also satisfies the original differential equation:

```
>> key=-1; opt=optimset; opt.TolX=1e-13;
   tic, [x0 f0]=fminsearch(@c8msqrt1,1,opt), toc
   txy=sim('c8msqrt1a'); t=txy.tout; y0=txy.yout;
   plot(t,y0(:,[1 2]),[0 1],[0 1],'o')
```

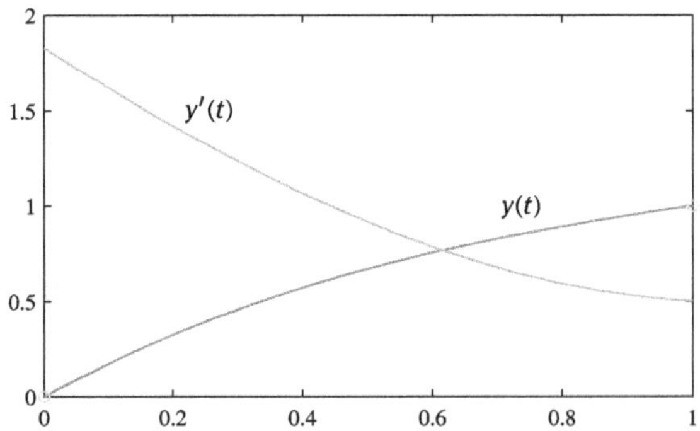

Figure 8.13: The numerical solution of the second branch.

It can be seen that the value of objective function in the second branch is slightly smaller than the first one. From the conventional viewpoints, this solution can be regarded as the global optimum value, and this solution is better than the first one. In fact, it is not. The two solutions are all the solutions of the original problem, and there is no need to say which is better. In this particular application, the conventional idea on global minimum solution should not be used; otherwise, misleading interpretations are obtained.

It can be seen from the example that sometimes many solutions can be found with numerical optimization methods. All of the solutions such that the value of the objective function close to zero can be regarded as the solutions of the original problem. A solution cannot be regarded as a better one, by merely know the value of the objective function is smaller than the other one.

The optimization problems studied here are essentially the algebraic equation solution problems. The targets are to find the algebraic equations solutions of the boundary value conditions. If a solution is found, it is the solution to the original value problem. Compared with equation solution methods, finding the numerical optimization solutions have many advantages. For instance, it can be used to handle the case, where the number of unknowns are different from the number of equations. Therefore, the conversion of the boundary value problem into numerical optimization problems is meaningful. It is a promising attempt, since many of the problems cannot be solved by any other methods.

8.2 APP interface design

In the earlier versions of MATLAB, the `guide` tool was recommended in the design of graphical interfaces.[84] From R2016a on, a new APP designer is recommended in user interface design, or design of application programs (APPs). With command `appdesigner()`, the design process can be launched. The main tasks of APP designer is to provide the controls in graphical user interface design, and generate the program framework. The response functions to the controls can also be written in the framework.

The command `appdesigner` can be used, and an initial design interface is shown in Figure 8.14. If the user has already a graphical user interface, it can be converted into the new framework and can be maintained in the APP designer. A new APP can also be created. Help information for quick start is provided in the interface.

Click Blank App icon, a design interface is opened as shown in Figure 8.15. Three portions in the interface are shown. The left one is the COMPONENT LIBRARY, the middle one is the prototype design for the APP and the right one is the COMPONENT BROWSER. In the toolbar, an App Details button is given, which allows the display of basic information about the APP.

Expand the component library, and all the controls provided are listed. The controls are shown in many parts, such as the commonly used ones in Figure 8.16, and the container and figure tools in Figure 8.17. The commonly used controls such as edit boxes and buttons are given. Also, the containers and various dashboard controls

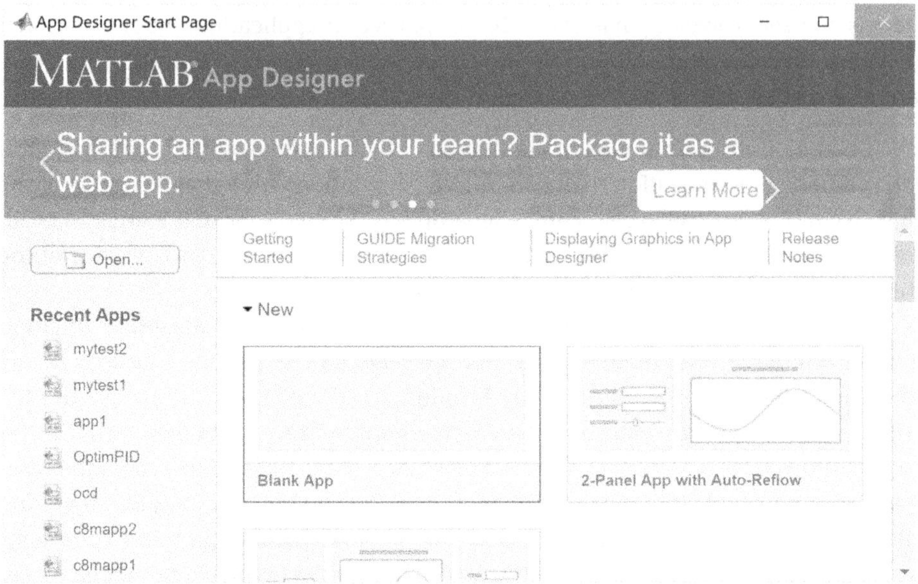

Figure 8.14: APP designer main interface.

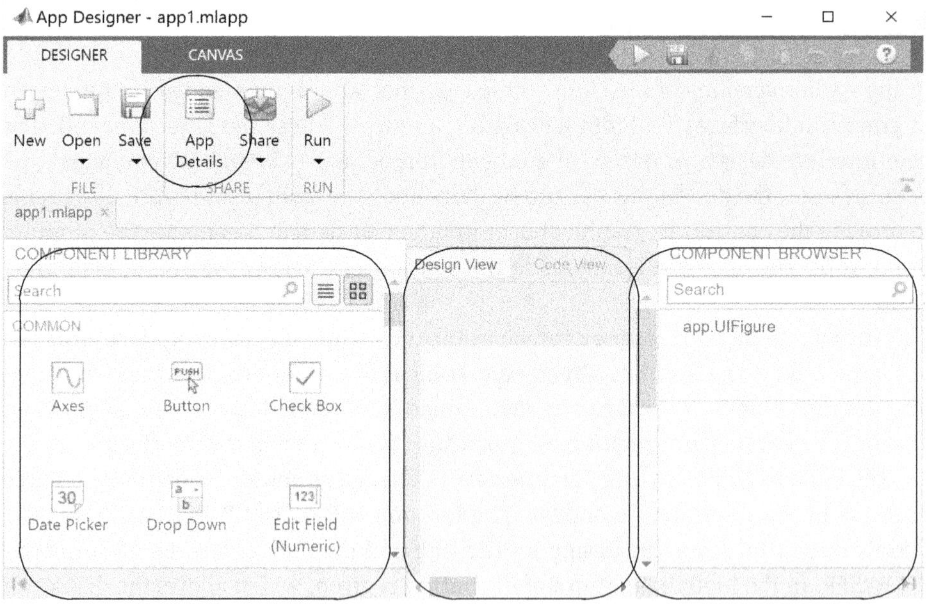

Figure 8.15: Blank APP design interface.

are provided. These controls can be drawn in the prototype window, so that practical interface can be designed. The properties of the controls can be assigned in the right

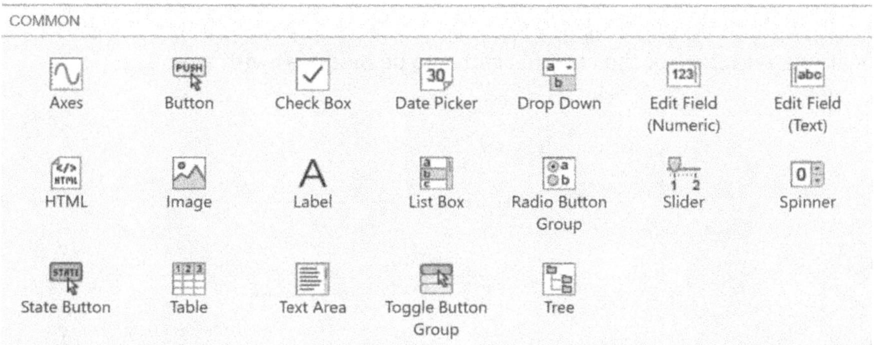

Figure 8.16: Commonly used control library.

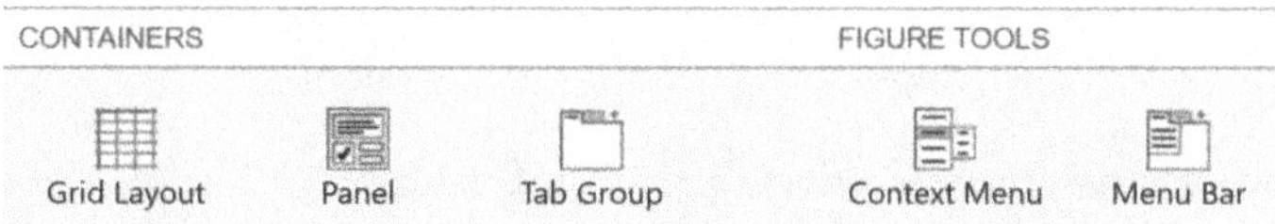

Figure 8.17: Containers and figure tools.

portion of the interface, and finally the program framework can be created automatically.

Two simple examples are given next in the section to demonstrate the design process of APPs. The interface drawing, interface design considerations and program implementation are illustrated.

Example 8.9. Design a simple program. Open a graphics window, with two controls. One is a button, the other is a label. If the button is pressed, the text "Hello World!" is displayed on the label.

Solutions. The object oriented programming method is different from classical programming methods. The essential task in program design is to write response functions to certain events. In this interface, there are two controls. The button is an active object. When it is pressed, it informs the interface to activate a callback function. The task of the callback function is to modify the properties of the label object. The label object is a passive one, and accepts actions only from other controls. It usually has no callback functions. Therefore, the key points in program design are to assign tasks and write callback functions. In this particular example, the tasks for the button are to find the label object, and modify the value of its Text property to "Hello World!".

With the `appdesigner` command, the APP designer interface is launched. The user may drag the icon on the lower-right corner of the prototype window, so as to adjust the size of the prototype window. Then the button and label controls can be dragged from the component library to the prototype window, as shown in Figure 8.18. The locations and sizes of the controls can also be modified with a mouse.

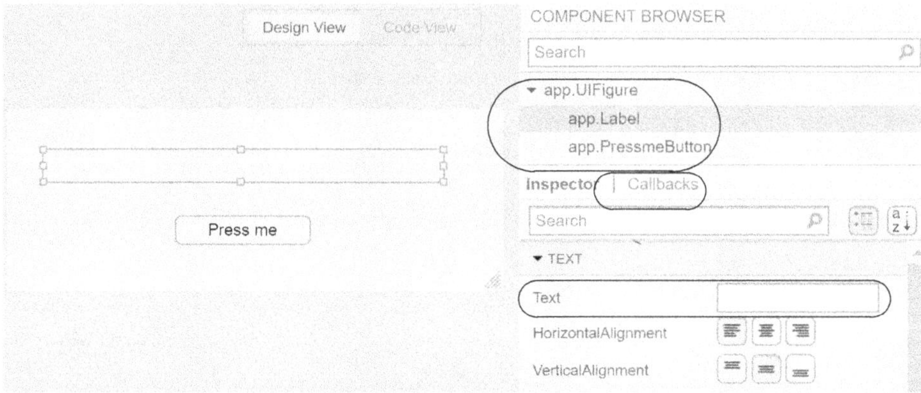

Figure 8.18: Prototype windows and component browser (file: `c8mapp1.mlapp`).

It can be seen that two controls are drawn on the prototype window. One is a label and the other is a button. For the label object, the component browser can be used to assign the initial value of its Text property to empty. The name of the label object is automatically assigned to `app.Label` in MATLAB, as listed in the component browser on the right. Other properties such as font and size of the controls can also be assigned.

Another control is the button, whose Text property can be set to Press me. A control name `app.PressmeButton` is then assigned in MATLAB. The properties of the two controls are displayed in the component browser. An upper-level object, app.UIFigure, is also displayed, which is the window of the program.

When completing drawing the prototype window, it can be saved, for instance, to `c8mapp1.mlapp` file. The suffix `mlapp` is a new one for MATLAB APP designer. The file can be executed directly in MATLAB command window, in the same way as an ordinary `.m` file.

Click the `app.PressmeButton` list item in the component browser, and click Callbacks button, a callback function item can be created. Let us recall now the task assigned to the button. That is, to find the Label control, and change the Text property of the control to “Hello World!”. How to find the control? The control name is `app.Label`. How do we modify the properties? Modify properties are simple, and it is not necessary to use `set()` function in the old versions. In the framework of the automatically opened editor, the following command can be written, and the program design process is completed:

```
function PressmeButtonPushed(app, event)
   app.Label.Text='Hello World!';
end
```

When the program is saved, command c8mapp1 can be issued in MATLAB command window, such that the expected interface appears. If the **Press me** button is clicked, the message "Hello World!" is displayed in the interface.

It can be seen that the APP interface design is rather simple. Three tasks —draw interface, task assign and callback function programming—should be completed, so that a program can be designed.

Example 8.10. Design a control system analysis program. The sketch for the interface is given in Figure 8.19. In the two edit boxes labeled **Numerator** and **Denominator**, the numerator and denominator coefficient vectors of the transfer function can be specified. Click **Model** button, and the transfer function object can be composed. The items in the listbox labeled **list** can be selected, such that the corresponding analysis plot of the system can be drawn. Write a MATLAB program to implement all the functions.

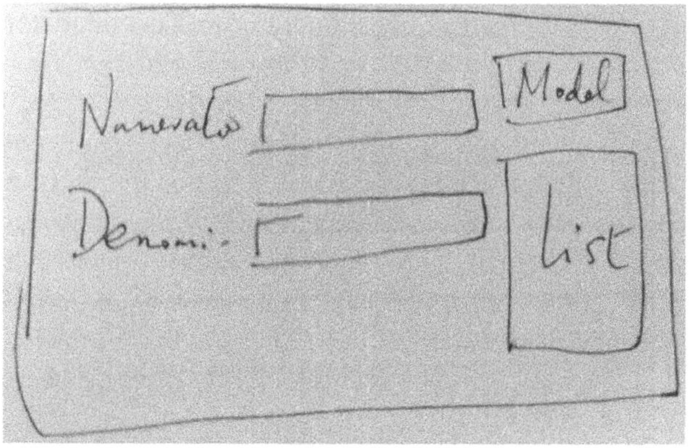

Figure 8.19: Sketch of the expected user interface.

Solutions. The command appdesigner can be used to invoke APP designer interface. A blank prototype window can be designed. Based on the requests in Figure 8.19, two edit boxes, a button and a list box can be drawn on the prototype window, as shown in Figure 8.20. Also, the component browser is also given in the figure. The names for each controls can be assigned automatically in MATLAB. For instance, the name of the numerator edit box is app.NumeratorEditField, as seen in the component browser. The default values of the numerator and denominator polynomials can also be specified, in their Value properties.

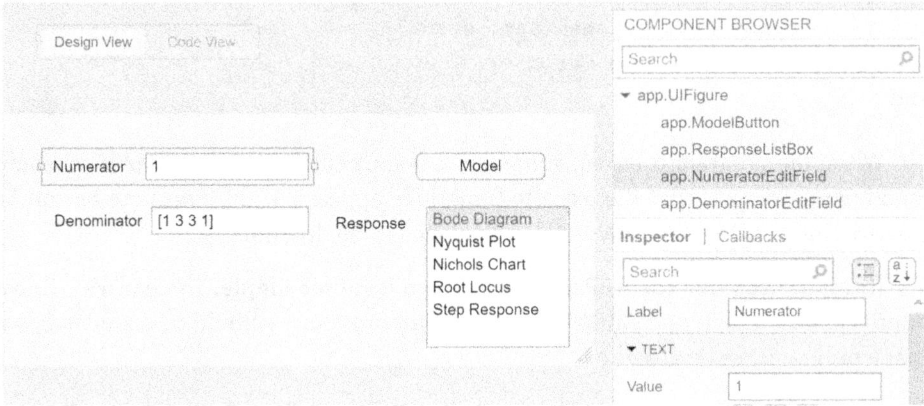

Figure 8.20: The designed interface and component browser (file: `c8mapp2.mlapp`).

It can be seen from the problem that the two edit boxes are passive controls, since they only accept actions from others, and not manipulating others. The button and listbox controls are active objects, and tasks should be assigned to them individually. The specific tasks are:

(1) **Button control**. The tasks are ① to read in the numerator and denominator parameter vectors from the corresponding edit boxes, in strings `num` and `den`. They should be converted into numeric vectors, in the same names; ② Use G=`tf(num,den)` command to compose the transfer function object; ③ The variable G is normally not visible from other callback functions. To make it visible, it can be stored in the **UserData** property of the window. Besides, the conversion from string to numeric vector may lead to errors, if unrecognizable characters are typed into the edit box. Fault tolerant processing should be made, for instance, by set an error trap with `try-catch` structure. The callback function below can be written:

```
function ModelButtonPushed(app, event)
   try
      num=str2num(app.NumeratorEditField.Value);
      den=str2num(app.DenominatorEditField.Value);
      G=tf(num,den); app.UIFigure.UserData=G;
   catch
      errordlg('Data type fault in the edit box','ERROR');
end, end, end
```

(2) **Listbox control**. The tasks assigned to the listbox object are: ① retrieve the temporarily saved model G from **UserData** property of the window; ② draw the analysis curve from the selected item in the listbox. The callback of this object is suitable to be implemented with switch structure.

```
function ResponseListBoxValueChanged(app, event)
   G=app.UIFigure.UserData;
   switch app.ResponseListBox.Value
      case 'Bode Diagram', bode(G)
      case 'Nyquist Plot', nyquist(G)
      case 'Nichols Chart', nichols(G)
      case 'Root Locus', rlocus(G)
      case 'Step Response', step(G)
end, end
```

8.3 Design of optimum controllers

The so-called "optimal control" means to complete a certain task, the control such that the preselected performance index is maximized or minimized. The performance index is in fact the objective function in optimization. The commonly used objective functions are the integral type of performance index

$$J_{\mathrm{ISE}} = \int_0^\infty e^2(t)\mathrm{d}t, \quad J_{\mathrm{ITAE}} = \int_0^\infty t|e(t)|\mathrm{d}t, \tag{8.3.1}$$

and the shortest time, the least energy consumption and so on. Similar to optimization problems, optimal control problems are also classified as constrained and unconstrained optimal control problems. Unconstrained optimal control problems can be solved with calculus of variations,[13, 74] where analytical solutions for very small-scale problems can be found. For instance, optimal linear quadratic controller design has closed-form formulas for direct solutions. The constrained optimal control problems are difficult to deal with, sometimes maximum principles of Pontryagin should be applied.

In this section, the hidden problems in conventional optimal control are pointed out. Then, for ordinary servo control problems, the reasonable objective functions are explored. The Simulink facilities and optimization technique are combined to design optimal controllers. A general-purpose APP for design optimal controllers is given. Examples are given to show that for some kinds of plants models, optimal PID controllers are the best second-order controllers.

8.3.1 Hidden problems in classical optimal control methods

In conventional optimal control problems, in order to let the problem solvable analytically, various unnecessary constraints and conditions are introduced, and many unexplainable indirect factors are introduced, such that the performance indexes are subjective and artificial ones. Linear quadratic optimal control and $\mathcal{H}_2$ norm index are used as examples, to illustrate the possible mistaken ideas in conventional optimal control theory and research.

Example 8.11. For a time invariant state space model,

$$\begin{cases} \boldsymbol{x}'(t) = \boldsymbol{A}\boldsymbol{x}(t) + \boldsymbol{B}\boldsymbol{u}(t), \\ \boldsymbol{y}(t) = \boldsymbol{C}\boldsymbol{x}(t) + \boldsymbol{D}\boldsymbol{u}(t), \end{cases} \tag{8.3.2}$$

an performance index is introduced, to design an input $\boldsymbol{u}(t)$, such that

$$J = \frac{1}{2}\boldsymbol{x}^{\mathrm{T}}(t_{\mathrm{n}})\boldsymbol{S}\boldsymbol{x}(t_{\mathrm{n}}) + \frac{1}{2}\int_{t_0}^{t_{\mathrm{n}}}\left[\boldsymbol{x}^{\mathrm{T}}(t)\boldsymbol{Q}(t)\boldsymbol{x}(t) + \boldsymbol{u}^{\mathrm{T}}(t)\boldsymbol{R}(t)\boldsymbol{u}(t)\right]\mathrm{d}t \tag{8.3.3}$$

is minimized, where $\boldsymbol{Q}$ and $\boldsymbol{R}$ are weighting matrices for the state variables and input signals. Time t_{n} is the stop time in control. Matrix $\boldsymbol{S}$ is a weight matrix for the final value of control. The control problem is referred to as optimal linear quadratic control problem.

It is known from linear quadratic control theory[1] that, in order to minimize J, the control signal should be

$$\boldsymbol{u}^*(t) = -\boldsymbol{R}^{-1}\boldsymbol{B}^{\mathrm{T}}\boldsymbol{P}(t)\boldsymbol{x}(t), \tag{8.3.4}$$

where $\boldsymbol{P}(t)$ is a symmetrical matrix, satisfying the famous Riccati differential equation

$$\boldsymbol{P}'(t) = -\boldsymbol{P}(t)\boldsymbol{A} - \boldsymbol{A}^{\mathrm{T}}\boldsymbol{P}(t) + \boldsymbol{P}(t)\boldsymbol{B}\boldsymbol{R}^{-1}\boldsymbol{B}^{\mathrm{T}}\boldsymbol{P}(t) - \boldsymbol{Q}, \tag{8.3.5}$$

where the final value of $\boldsymbol{P}(t)$ matrix is $\boldsymbol{P}(t_{\mathrm{n}}) = \boldsymbol{S}$. Even from contemporary computers and tools, the solutions of Riccati differential equations are hard to solve, therefore, the controller design based on such a solution is even more complicated. Thus only the static problem is considered. Assume that $\boldsymbol{P}(t)$ is a constant matrix $\boldsymbol{P}$, the Riccati differential equation can be simplified as

$$\boldsymbol{P}\boldsymbol{A} + \boldsymbol{A}^{\mathrm{T}}\boldsymbol{P} - \boldsymbol{P}\boldsymbol{B}\boldsymbol{R}^{-1}\boldsymbol{B}^{\mathrm{T}}\boldsymbol{P} + \boldsymbol{Q} = \boldsymbol{0}. \tag{8.3.6}$$

The equation is referred to as Riccati algebraic equation. The control problem is changed to a linear quadratic regulator problem. Assume that $\boldsymbol{u}^*(t) = -\boldsymbol{K}\boldsymbol{x}(t)$, where $\boldsymbol{K} = \boldsymbol{R}^{-1}\boldsymbol{B}^{\mathrm{T}}\boldsymbol{P}$, the closed-loop state space model under state feedback control can be written as $(\boldsymbol{A} - \boldsymbol{B}\boldsymbol{K}, \boldsymbol{B}, \boldsymbol{C} - \boldsymbol{D}\boldsymbol{K}, \boldsymbol{D})$.

It seems that the mathematical presentation above is rigorous, and beautiful mathematical formulas can also be found. Is there any hidden problems in linear quadratic optimal control?

Solutions. Linear quadratic control is an important milestone in the development of control theory. However, from the critical viewpoint of today, there are many hidden problems. First, in industrial servo control, it is expected that the output signal follow very well with the input signal; there is no overshoot or very small overshoot; the response speed is fast. These expectations have nothing to do with the quadratic performance index. The only reason why quadratic performance index was chosen is that the derivative can be found, such that the closed-form formula of optimal control can be derived.

Besides, two weighting matrices $\boldsymbol{Q}$ and $\boldsymbol{R}$ should be introduced, if final value is considered, $\boldsymbol{S}$ matrix is also needed. Although very beautiful mathematical form under state feedback can be derived, there is to date no widely accepted methods in choosing the matrices. There are too much artificial factors in the whole performance indexes, and it is too subjective. If the matrices can be reasonably chosen, the optimal control is meaningful; otherwise, it is meaningless, or even harmful, if not chosen reasonably.

With the objective function whose physical interpretation is questionable, when the Riccati differential equation is obtained, it is found that the differential equation cannot be solved, the solution matrix is artificially assigned to a constant matrix $\boldsymbol{P}$, and the corresponding equation is simplified as Riccati algebraic equation. Although after the simplification process, a good mathematical formula is obtained, the solutions is biased away from the original problem. The optimal controller obtained is not the controller of the original problem, but a controller of another artificial problem. Therefore, the controller obtained is questionable.

Example 8.12. An example in real life is used here for contrast analysis, trying to explain why there are possible mistaken ideas in linear quadratic optimal control. In Figure 8.21, the map of Liaoning Province, China, is given. If you want to travel from Shenyang to Dalian, the description similar to optimization language is to travel to southwest to the furthest place, the optimum point is Dalian. This is an objective description. If you find there is no road to the southwest, and there is one to the south (to chose a quadratic criterion), so you travel to the south. At some point, you find that there is no road traveling on (unable to solve Riccati differential equation), there is one slightly biased to the east (assume it a constant matrix and change it to Riccati algebraic equation). You can then follow this road to the furthest point (find the optimum solution). You may find you are in Dandong, not Dalian. If a biased direction is chosen (unsuitable objective function is selected), the final solution may become meaningless. Travel southeast may not be the worst thing in real life. If you find that the there is no road to the southwest, there is an immediate road heading north (wrong weighting matrices), you choose to follow that road, you may still get the optimum solution, but the solution is harmful.

For optimal control problem, you find no road, because you are unable to set up an differentiable objective function such that no beautiful mathematical formula

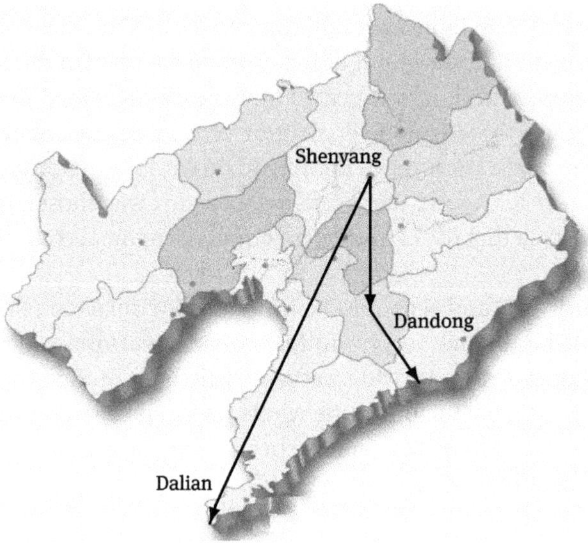

Figure 8.21: Map of Liaoning Province.

can be derived. There is not really no road. The correct road to follow is numerical optimization. It may create you a road to the genuine destination.

Example 8.13. Assume that the plant model is given by $G(s) = 1/(s+1)^5$, design an optimal PID controller, such that the ISE criterion is minimized. Draw the step response of the closed-loop system.

Solutions. Consider the PID control structure, shown in Figure 8.22. The saturation nonlinear element is temporarily neglected here. For the tracking error signal $e(t)$, the ISE criterion in (8.3.1) can be introduced, such that the area enclosed by the squared error is minimized. It seems to be a reasonable performance index. An alternative way to evaluate the criterion is to find the Laplace transform of the error signal $e(t)$, and when $E(s)$ is stable, the ISE criterion can be computed directly by evaluating the $\mathcal{H}_2$ norm of $E(s)$, with function `norm()`. The following MATLAB function is written to compute the objective function:

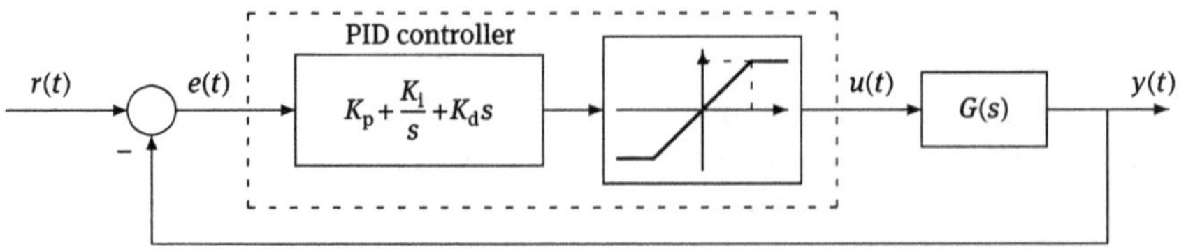

Figure 8.22: Block diagram of PID control system.

```
function y=c8mise(x,G,s)
R=1/s; Gc=x(1)+x(2)/s+x(3)*s;
E=minreal(feedback(1,G*Gc)*R); y=norm(E);
```

With the objective function designed, the following statements can be used to search directly the parameters of the optimal PID controller. After 3.32 seconds of waiting, the optimal parameters can be found as $\boldsymbol{x} = [1.5375, 0.6908, 4.6574]$, with the objective function 1.3329. Under such a controller, the step response of the closed-loop system can be found, as shown in Figure 8.23.

```
>> s=tf('s'); G=1/(s+1)^5;
   f=@(x)c8mise(x,G,s); % objection function without additional parameters
   opt=optimset; opt.TolX=1e-13;
   tic, [x,f0]=fminsearch(f,rand(3,1),opt), toc
   Gc=x(1)+x(2)/s+x(3)*s; step(feedback(G*Gc,1))
```

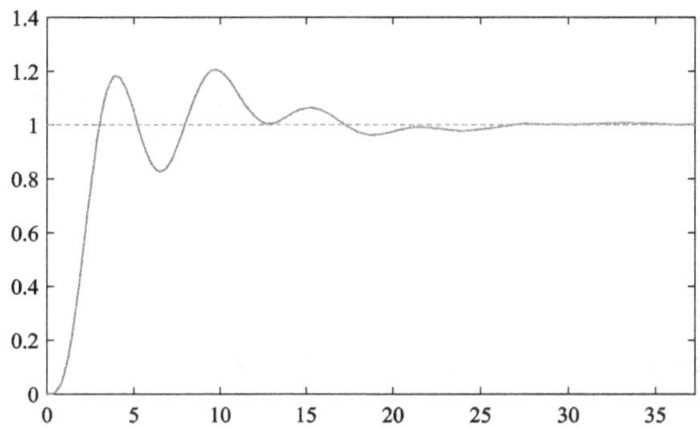

Figure 8.23: Step response of the closed-loop system.

It can be seen that the quality of the PID controller thus designed is not satisfactory. This indicates that the selection of ISE criterion is questionable. Besides, if the control signal is expected, it is not possible to draw it directly, since with pure derivative term, the controller is not physically realizable. An approximate derivative term can be used instead. The control signal obtain is as shown in Figure 8.24. For this particular example, the instant value of the control signal is too large, and it may damage the hardware system. Therefore, in real control systems, the saturation nonlinearity cannot neglected in reality. Unfortunately, if the saturation nonlinearity is considered in the design process, the system is a nonlinear one, and the above mentioned method cannot be used, since it can only be used in linear systems.

```
>> Gc=x(1)+x(2)/s+x(3)*s/(0.0001*s+1);
   [y t]=step(feedback(Gc,G),0.01); plot(t,y)
```

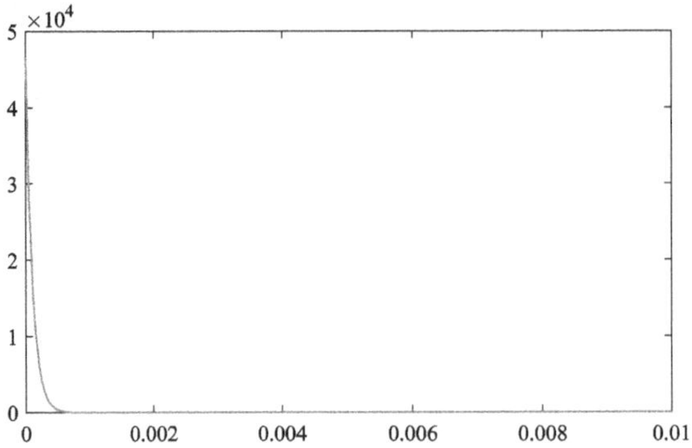

Figure 8.24: The curve of control signal $u(t)$.

The final target in optimal control in engineering is not how beautiful the mathematical formulas are, but the selection of the truly meaningful objective function. If a nonsense objective function is chosen, the beautiful mathematical formulas obtained may become meaningless. In real applications, the objective functions must not be chosen because although it is easy in computation, and neglects its reasonableness. A truly meaningful objective function must be chosen.

8.3.2 Performance indices and optimum design in servo control systems

With popularity of the powerful computer tools such as MATLAB, the solutions of optimization problems are made much easier. Besides, with the powerful simulation facilities in Simulink, the computation of really meaningful objective functions are much easier. Therefore, many optimal control problems can be converted to ordinary optimization problems, and numerical optimization techniques can be used to solve easily optimal control problems, although no beautiful mathematical formulas are found, and truly meaningful results can be obtained. Examples are given later on the design of optimal servo control systems with numerical optimization methods.

In this section, a discussion is made on how to select a reasonable objective function in servo control systems. In the ISE criterion, the error signal at any time is treated equally, which leads to strong oscillations when t is large. This is the phenomena to be avoided in servo control design. The weights can be imposed on the absolute values of the error signal, then carried out with integral action, such that the ITAE criterion in (8.3.1) can be introduced. It can be seen from the mathematical formula that when t increases, the weight also becomes large, the objective function may force the error signal to reduce to zero as soon as possible. At least, when t is rather large, the oscillation phenomenon like those in ISE criterion is avoided.

How to evaluate the ITAE criterion? Since the upper bound of the integral is infinity, it is not possible to carry out the evaluation by simulation method. In other words, it cannot be evaluated. Assume that from a certain time t_n on, the error signal approaches to zero, and the ITAE criterion can be replaced by a finite-time ITAE criterion:

$$I \approx \int_0^{t_n} t|e(t)|dt. \tag{8.3.7}$$

If t_n is large enough, the approximation is reasonable. The finite-time ITAE criterion can be evaluated with the Simulation method. In fact, in Example 2.10, a Simulink model is designed for the ITAE integral. Examples are given next to demonstrate the ITAE criterion based controller design method.

Example 8.14. Consider the plant model in Example 8.13. Design an optimal PID controller under the ITAE criterion.

Solutions. The PID Controller block in Simulink can be used directly, with three parameters assigned to variables `Kp`, `Ki` and `Kd`. The simulation is established, as shown in Figure 8.25. The ITAE integral is returned in the **Out1** port. Select a stop time $t_n = 20$, and the final value is an approximate value of the ITAE criterion.

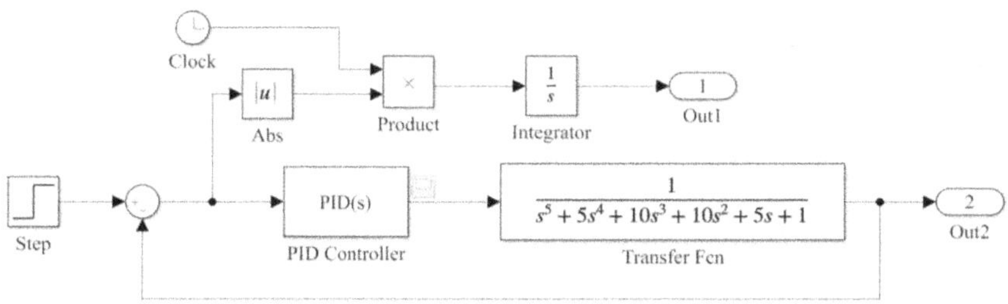

Figure 8.25: Simulink model of PID control system (file: `c8mpid1a.slx`).

Based on the model, the following objective function is written:

```
function y=c8mpid1(x)
W=get_param(gcs,'ModelWorkspace'); assignin(W,'Kp',x(1))
assignin(W,'Ki',x(2)); assignin(W,'Kd',x(3));
txy=sim('c8mpid1a'); y=txy.yout(end,1); % the final value
```

With the objective function, the following commands can be used to start the numerical optimization process. Since the robustness of PID controller is rather strong, there is no need to specify very tough error tolerance in optimization. The time elapse is 16.46 seconds, with the optimal parameter of the controller $\boldsymbol{x} = [1.2469, 0.3569, 1.6889]$, and the value of the objective function 6.9254. Since the objective function is different from the one in Example 8.13, it is meaningless to compare the two values. If the controller in Example 8.13 is converted into ITAE criterion, the value if 17.80, much larger:

```
>> c8mpid1a, set_param(gcs,'FastRestart','on');
   tic, [x,f0]=fminsearch(@c8mpid1,rand(3,1)), toc
   txy=sim('c8mpid1a'); plot(txy.tout,txy.yout(:,2))
```

The step response of the closed-loop system obtained is shown in Figure 8.26, superimposed by the control behavior in Figure 8.13. It can be seen that the response is slightly slower than the one in Example 8.13, but the overshoot is significantly reduced, and the settling time is significantly shortened. It is seen that in engineering viewpoint, the PID controller designed under ITAE criterion is much superior than the one with ISE criterion. Double click the virtual scope block, and it can be seen that the instant control signal is reduced to a value less than 80, also much more reduced than the other controller.

Note that for this particular example, if fast restart mode is not used, the time elapse is as high as 169.95 seconds, almost 10 times more than the fast restart mode. Therefore, when optimization is involved, since a significant runs in Simulink are executed, it is recommended to use fast restart mode whenever possible.

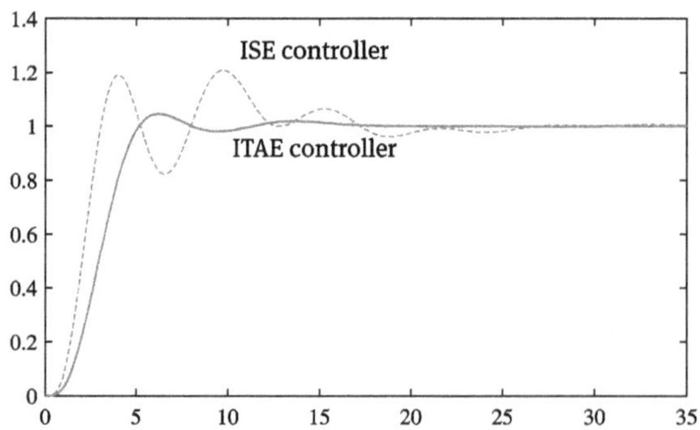

Figure 8.26: Step responses of the closed-loop systems.

Example 8.15. In the design process in Example 8.14, there are still many questions yet to be answered: Why the stop time is set to 20, not some other value? If different values are selected, what is the impact to the control behavior? Try different values of t_n's, and draw possible conclusions.

Solutions. The stop time can be set in the model to variable tn, and set different values to tn in a loop structure. The controller parameters and time elapses are measured and given in Table 8.1. The ITAE integral curve is also drawn, as shown in Figure 8.27:

```
>> c8mpid1a, set_param(gcs,'FastRestart','on');
   for tn=[15 20 25 30 50 80 120 500]
      tic, [x,f0]=fminsearch(@c8mpid1,[1.2 0.4 1.7]), toc
   end
```

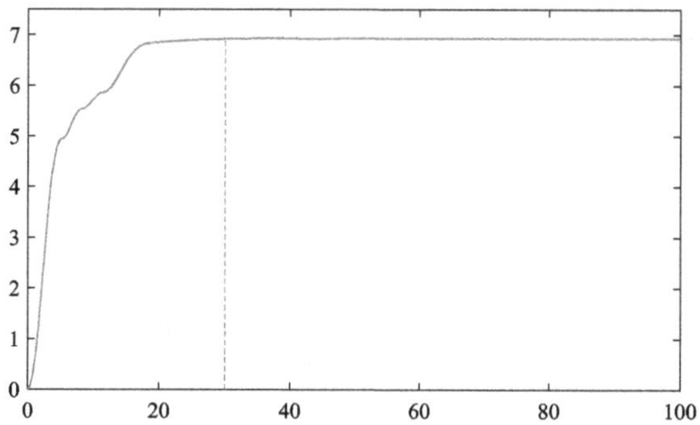

Figure 8.27: ITAE integral curve.

Table 8.1: Comparisons under different stop time.

Stop time t_n	15	20	25	30	50	80	120	500
K_p	1.2126	1.2469	1.2421	1.2335	1.2338	1.2338	1.2338	1.2338
K_i	0.3496	0.3569	0.3584	0.3541	0.3542	0.3542	0.3542	0.3542
K_d	1.6576	1.6890	1.6697	1.6397	1.6376	1.6376	1.6376	1.6376
ITAE value	6.4329	6.8435	6.8844	6.9188	6.9273	6.9273	6.9273	6.9273
Time elapse	11.51	11.01	11.05	12.19	14.31	16.26	19.95	78.53

Since the integrand in ITAE integral is $t|e(t)| \geqslant 0$, therefore, the integral signal is a monotone increasing one. It can be seen that from $t_n = 30$ on, the ITAE integral is getting flat. This is because that when t is large enough, the value of the error signal $e(t)$ tends to zero, such that the integral value is unchanged. It can be seen that there is no change witnessed in the subsequent controller parameters. The time elapsed is increasing, since the simulation interval is enlarged.

It can be seen from the example that, if t_n is large enough, the controller designed is reasonable. How to check whether t_n is selected large enough? In fact, the validation is rather simple. After the controller is designed, the ITAE integral curve can be drawn. If the curve is already flat, the controller can be accepted. If it is still increasing, the value of t_n should be increased, and design and validation processes can be carried again. The value $t_n = 20$ selected in Example 8.14 is slightly smaller, since the ITAE integral curve is still slightly increasing. It can be seen from the control results that the closed-loop step response has no significant differences with the ones designed for $t_n = 30$ or even larger values. Therefore, the requirement for the selection of t_n is not quite rigorous. If the above design and validation guidelines are followed, the designed controllers are acceptable.

Although the magnitude of the control signal at the initial time instance is reduced to about 80 under ITAE optimal PID controller, it is still too large in certain applications. Therefore, a restriction to the control signal $u(t)$ is usually needed. For instance, an actuator saturation can be introduced, such that $|u(t)| \leqslant 15$. Due to the introduced actuator saturation, the entire system is a nonlinear one, and most of the design methods may fail. The above optimization based controller design method can still be used, and there is almost no impact on the solution speed in the design process.

Example 8.16. If the controller in Example 8.14 contains actuator saturation, such that the control signal satisfies $|u(t)| \leqslant 15$, design the controller again, and observe the control behaviors.

Solutions. The PID Controller block provided in Simulink allows the use of saturation nonlinearities in the model. Open the parameter dialog box, and select the **Output saturation** pane, as shown in Figure 8.28. If the checkbox **Output limit** is checked, the values 15 and −15 can be filled respectably to **Upper limit** and **Lower limit** edit boxes, a ⎽∫ is marked on the controller icon, indicating the presence of saturation in the block.

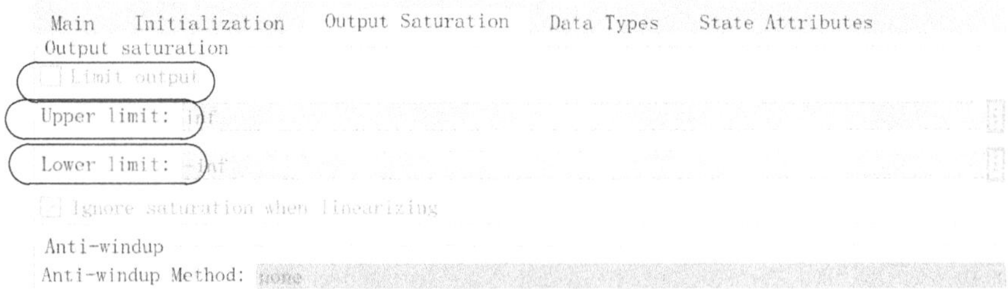

Figure 8.28: PID controller parameter dialog box.

Set the stop time to $t_n = 50$, and turn on the fast restart mode, then begin the controller design process, the result obtained is $\boldsymbol{x} = [1.3203, 0.2807, 1.7651]$, with objective function value of 9.0551, and time elapsed is 16.21 seconds. The step response under new controller is shown in Figure 8.29, with the controller designed earlier, with no actuator saturation superimposed. It can be seen that there is only a very small sacrifice in the control behavior, and the benefit is that the control signal satisfies $|u(t)| \leqslant 15$. Compared to the time elapsed, the inclusion of nonlinear element adds almost no extra computational load.

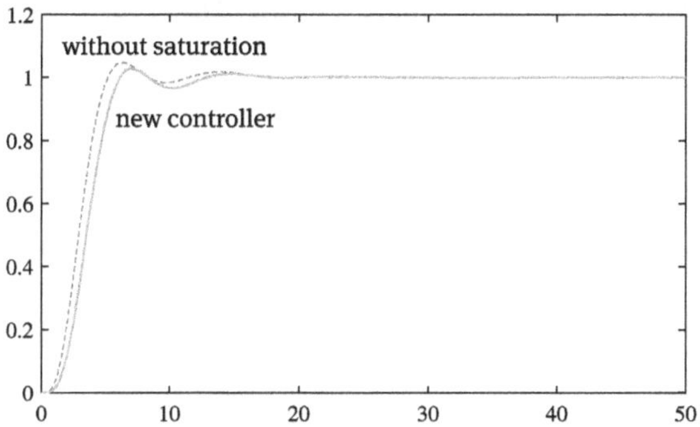

Figure 8.29: The behavior under the new controller.

```
>> tn=50; c8mpid1a, set_param(gcs,'FastRestart','on');
   tic, [x,f0]=fminsearch(@c8mpid1,rand(3,1)), toc
   txy=sim('c8mpid1a'); plot(txy.tout,txy.yout(:,2))
```

8.3.3 Optimal control design interface

It can be seen from the above demonstration that the optimization based optimal controller design methods are not restricted to the conventional optimal control patterns. The objective functions can be chosen arbitrarily. It may have better application expectations than the conventional ones.

The general optimal servo control patterns are summarized by the author, and an tracking error based optimal controller design APP, named OCD, for optimal controller designer.[85]

With the combinations of MATLAB and Simulink, the meaningful optimal controller parameters can be found. The APP allows the user to describe the whole control system model with Simulink, where the controller can be of any form, with tunable parameters. An objective function can be generated automatically in MATLAB files, from which the optimal control problem can be solved, and the optimal controller parameters can be found.

Due to the update in Simulink versions, the models should be compiled first each time it is executed in normal cases, therefore, with the fast restart facilities, the design process is significantly improved. The author rewrote the whole OCD program, with the APP designer, such that the APP can be executed in fast restart mode, so as to increase the efficiency of the program.

The calling procedures of the OCD program is as follows:

(1) Under the MATLAB prompt, issue the command `ocd`, and the interface shown in Figure 8.30 can be opened. The facilities provided in MATLAB and Simulink can be used to design optimal controllers.

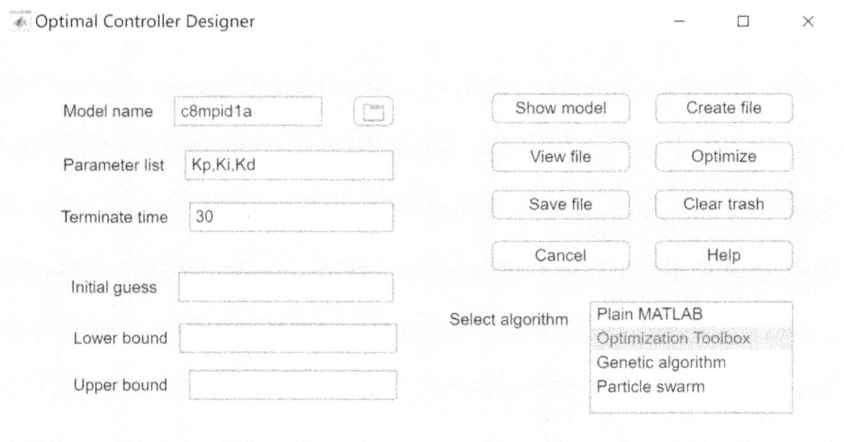

Figure 8.30: OCD interface.

(2) Set up a Simulink model, which contains at least two factors. One is the controller must have undetermined parameters, such as variables `Kp` in `Ki` in the PI controller block. The stop time of the model must be set to the variable `tn`. The second one is an output port to describe the error-based criterion signal, for instance, the ITAE integral can be expressed and connected to an output port **Out1**, the first channel of output signal.

(3) The corresponding Simulink model name can be specified in **Model name** edit box.

(4) The decision variable names can be specified in the **Parameter list** edit box, separated by commas.

(5) The stop time should be specified by the user, in the **Terminate time** edit box. The designed optimal controller must be validated to make sure the error tends to zero by the end of the time interval.

(6) Click the **Create file** button, a MATLAB file describing the objective function can be automatically generated, with file name `opt_fun*.m` automatically assigned

by OCD program. Click the **Clear trash** button, and all of these objective function files are deleted.

(7) Click the **Optimize** button, the optimization process is invoked, to search for the specified parameters. The parameters of the variables are returned to MATLAB workspace, with the same name as specified in the above edit box. In the actual design process, to ensure satisfactory controllers, the button may be clicked again, to get a better controller. In the program, the solvers such as `fminsearch()`, `fminunc()` or `ga()` can be called according to the selection, to find optimum parameters. If constraints are given, the `fmincon()` solver can be called.

(8) The upper and lower bounds of the decision variables can be assigned, and initial search points can also be specified in the corresponding edit boxes. Different simulation algorithms can be selected in the Simulink model.

Example 8.17. Solve the optimal PID controller design problem in Example 8.14 with the OCD program.

Solutions. Launch the OCD program. Set the parameters according to the ones in the interface in Figure 8.30. Click the **Create file** button to automatically generate an MATLAB file for the objective function. Press the **Optimize** button to initiate the design process. The same result in Example 8.14 can be obtained directly with the interface. For this particular example, if the **Optimization Toolbox** item is selected from the listbox, the low-level function `fminunc()` can be called by the program, and the time elapse is significantly reduced than the use of the `fminsearch()` function in Example 8.14.

Example 8.18. Consider a plant model $G(s) = 1/[s(s+1)^4]$. Use OCD program to design an optimal PD controller.

Solutions. The model can be modified, and the new model obtained is as shown in Figure 8.31.

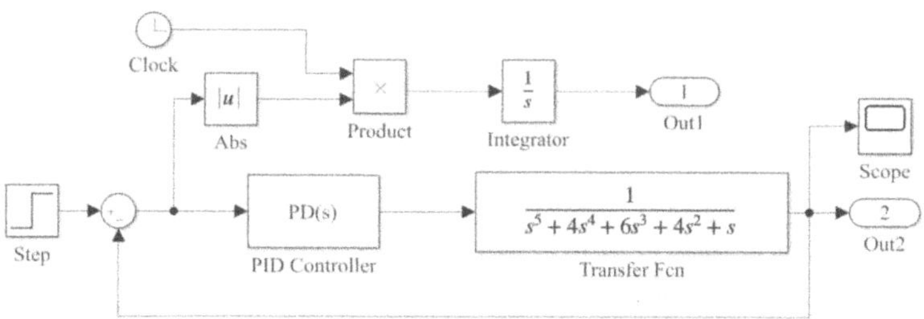

Figure 8.31: Simulink model of PD control system (file: `c8mpid2a.slx`).

Under a MATLAB prompt, input the command `ocd`. In the interface displayed, file name `c8mpid2a` is specified in the **Model name** edit box. `Kp,Kd` is specified in the

Parameter list edit boxes, while in Terminate time edit box, fill in the number 40. Click the Create file button, and the automatically generated MATLAB file for the objective function can be written:

```
function y=optfun_2(x)
W=get_param(gcs,'ModelWorkspace');
assignin(W,'Kp',x(1));
assignin(W,'Kd',x(2));
try, txy=sim('c8mpid2a'); y=txy.yout(end,1); pause(1e-5);
catch, y=1e10; end
```

where, in the first statement, the interface of the objective function is specified. In the second statement, the model workspace W is acquired, and in the third and fourth statements, the decision variables are assigned to the K_p and K_d in Simulink model workspace W, followed by a `try-catch` structure. The ITAE value is assigned to output argument y, to complete the objective function evaluation. If the system is unstable, a large value is returned, such that the search in the unstable system direction can be detoured. A `pause()` statement is executed for 10 μs of pause, such that the scope block displays in real-time the simulation results. To better observe the design process, the menu item File →Open at Start of Simulation in the scope block is chosen, such that it automatically opens the scope.

Click the Optimize button, and the optimization process is invoked. After the searching process, the PD controller parameters $G_c(s) = 0.2592+0.6982s/(0.006982s+1)$ can be found, such that the ITAE criterion is minimized. The output signal of the system is shown in Figure 8.32. It can be seen that the control behavior is satisfactory.

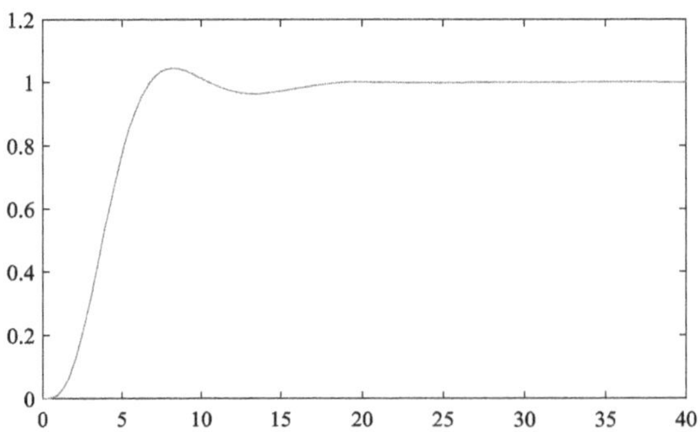

Figure 8.32: The output signal in PD control system.

Example 8.19. The application of OCD program is not restricted to the design of simple PID controllers. If there are more complicated structures, such as the cascade PI controllers design in Figure 8.33, with the conventional method, the inner loop is designed first, then the outer one. Use OCD to design the two controllers simultaneously.

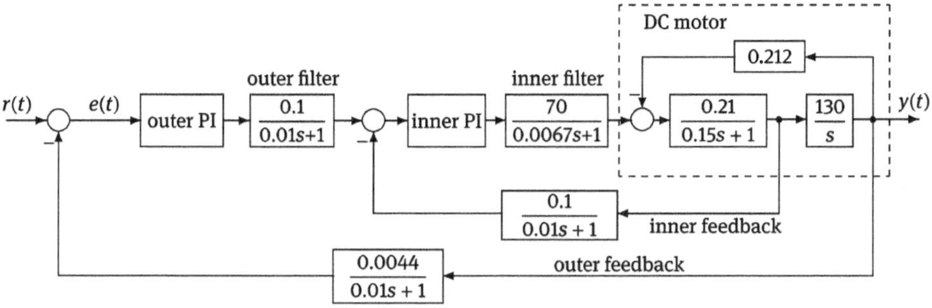

Figure 8.33: DC motor drive system with two loops.

Solutions. To solve the problem, the Simulink model in Figure 8.34 should be established first. Note that four undetermined variables, `Kp1`, `Ki1`, `Kp2` and `Ki2` are included in the model. The ITAE integral is designed and connected to **Out1** block.

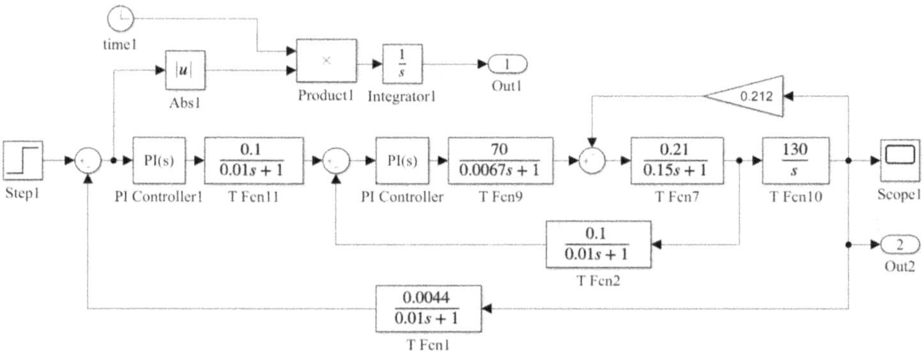

Figure 8.34: Cascade control model (file: `c8model2.slx`).

To launch the OCD interface, fill in name `c8model2` in the **Model name** edit box; and in the **Parameter list** edit box, fill in the undetermined variable names `Kp1,Ki1,Kp2,Ki2`; fill in 0.6 at the **Terminate time** edit box. Now click the **Create file** button, the objective function file in MATLAB can be generated. Click the **Optimize** button, and the optimal controller parameters $K_{p_1} = 11.3195$, $K_{i_1} = 9.8829$, $K_{p_2} = 39.2670$ and $K_{i_2} = 4.8072$ can be found. That is, the outer loop controller is $G_{c_1}(s) = 39.2670 + 9.8829/s$, while the inner one is $G_{c_2}(s) = 11.3195 + 4.8072/s$. Under these controllers, the step response of the system is shown in Figure 8.35, and it is a satisfactory one.

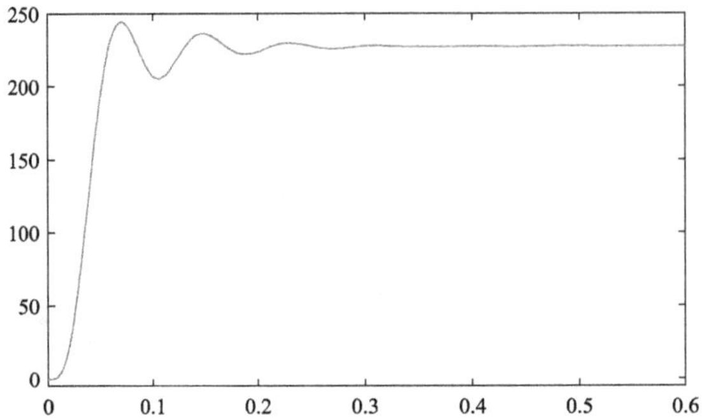

Figure 8.35: Step response of the drive system.

8.3.4 Other applications of OCD interface

OCD program cannot only be used in optimal control design, it can be used in solving other problems, for instance, in the optimal model reduction. If the error signal or error criterion can be described in Simulink, the OCD interface can be used to find the optimum parameters.

In this section, the application of OCD interface in model reduction problem is demonstrated.

Example 8.20. Now consider using OCD interface to study optimal model reduction problems. Before using the OCD interface, an error-based criterion signal can be designed, based on it optimization process can be carried out. OCD interface can be used to find the optimal reduced model. Assume that the original model is $G(s) = 1/(s+1)^6$, and find an approximate first-order model with delay $G_r(s) = Ke^{-Ls}/(Ts + 1)$.

Solutions. Establish a Simulink model `c8model3.slx`, as in Figure 8.36. Here, the ITAE criterion is used to define the error signal, to carry out model reduction tasks.

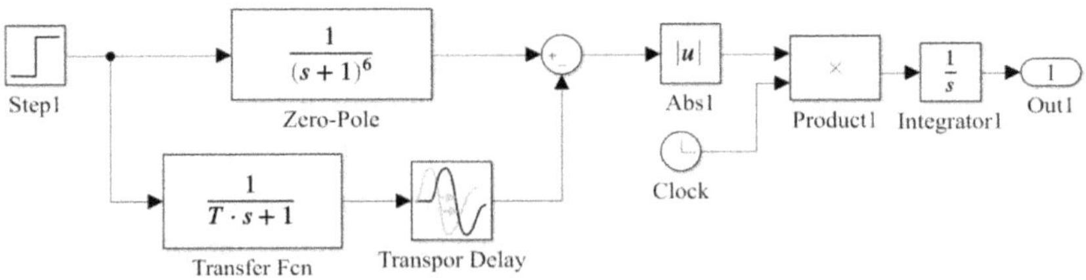

Figure 8.36: Simulink model for model reduction (file: `c8model3.slx`).

For convenience, there is no need to find K, since the steady-state value can be used to find it, with $K = 1$. Therefore, only two parameters, T and L, are needed.

Invoke OCD interface, fill `c8model3` in the **Model name** edit box; in **Parameter list** edit box, fill in L, T; and in **Terminate time** edit box, fill in 10. Click **Create file** button to generate the objective function, then press the **Optimize** button and the optimal parameters under ITAE criterion can be found as $L = 3.6683$ and $T = 2.6605$. That is the optimal reduced model is $G^*(s) = \mathrm{e}^{-3.6683s}/(2.6605s + 1)$.

The OCD interface can be used directly in many other field. Nonlinear factors in the controller design can be considered easily, which are difficult to handle in other controller design methods.

8.3.5 Open framework of the program

Open framework are adopted in the design of OCD program. The users are allowed to modify or extended the existing program. For instance, in the optimization algorithm selection part, switch structure is adopted in calling the corresponding solvers. The user may insert his own solver in the framework, to compare optimization algorithms.

```
switch app.SelectalgorithmListBox.Value
   case 'Plain MATLAB'
      Prob.solver='fminsearch'; ctrl_pars=fminsearch(Prob)
   case 'Optimization Toolbox'
      Prob.solver='fminunc'; ctrl_pars=fminunc(Prob)
   case 'Genetic algorithm'
      Prob.solver='ga'; ctrl_pars=ga(Prob)
   case 'Particle swarm'
      Prob.solver='particleswarm'; ctrl_pars=particleswarm(Prob)
end
```

8.3.6 PID-type controllers – the best second-order controller?

PID controllers are the most widely used controllers in industrial control, because it is flexible and the parameters in the controllers are easy to tune. In fact, with simulation methods, two other second-order controllers can be explored:

$$G_1(s) = \frac{a_1 s^2 + a_2 s + a_3}{s(a_4 s + 1)}, \quad G_1(s) = \frac{a_1 s^2 + a_2 s + a_3}{s^2 + a_4 s + a_5}, \tag{8.3.8}$$

and see whether better performance can be achieved. The two controllers have their own characteristics. In the former model, an integrator is contained to eliminate steady-state errors. The latter one is suitable for the plant models containing their own integrators. It can be seen that the second controller is a generalized lead–lag controller, since complex zeros and poles are allowed in the controller. An example is used next to demonstrate that PID controller is the best second-order controller.

Example 8.21. Consider the controller design problem in Example 8.14, where the plant model is $G(s) = 1/(s+1)^5$. Since the integrator is not provided in the plant, the first type of controller should be tried. Design an optimal controller and compare its behaviors with optimal PID controller designed in Example 8.14.

Solutions. Similar to model `c8mpid1a.slx`, if the new controller is used to replace the original PID controller, the simulation model in Figure 8.37 can be established. With the help of the OCD program, it is found that the optimal controller parameters are $\boldsymbol{a} = [1.6551, 1.2425, 0.3558, 0.0000]$. It can be seen that since $a_4 = 0$, the controller thus designed is the standard PID controller. The parameters are very close to the ones in Example 8.14, indicating that the PID controller is the best second-order controllers.

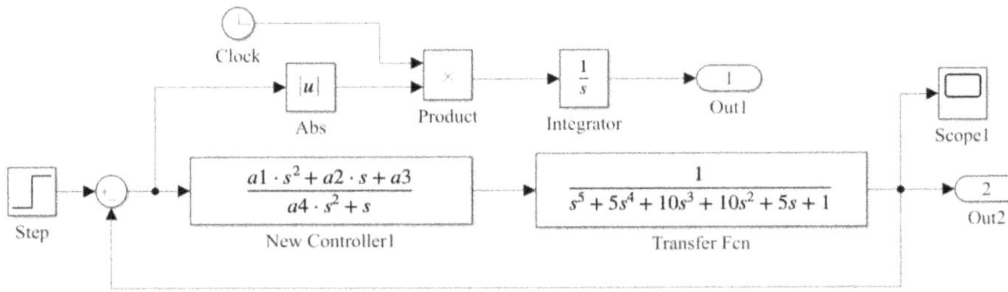

Figure 8.37: Simulink model with the new controller (file: `c8mpid1b.slx`).

Many examples are tried using this method, and the conclusions obtained are consistent. That is, the parameter $a_4 = 0$ is found by using the searching method. It can be seen that PID controller is the best second-order controller in terms of performance.

Example 8.22. Consider the plant model $G(s) = 1/[s(s+1)^4]$ in Example 8.18. Since an integrator is already contained in the plat model, the second structure can be tried. Find the best controller parameters and compare the control quality with the optimal PD controller in Example 8.18.

Solutions. If the Simulink model `c8mpid2a.slx` is modified, by substituting the PD controller by the new controller, the Simulink model in Figure 8.38 can be created.

With the OCD program, with stop time set as $t_n = 40$, the optimal controller obtained is

$$G_1(s) = \frac{78.632(s^2 + 0.8974s + 0.3)}{(s^2 + 0.2491s + 41.66)}.$$

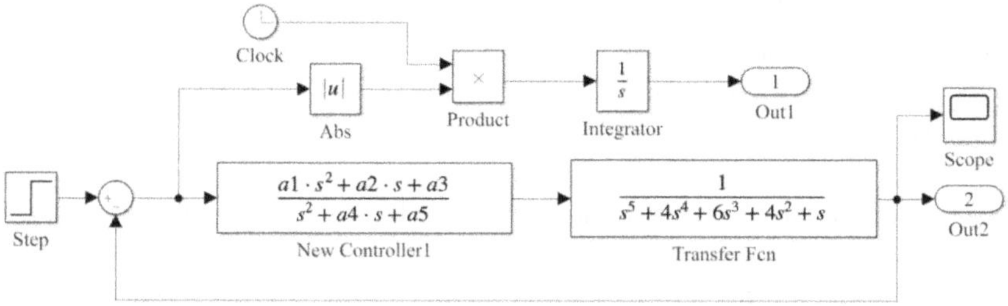

Figure 8.38: Simulink model with the new controller (file: `c8mpid2b.slx`).

The step response under the new controller is shown in Figure 8.39, superimposed by the PD controller response in Example 8.18. It can be seen that the controller is better than the PD controller. Of course, it is not fair for such a comparison, since the new controller has two more tunable parameters. Besides, the poles and zeros in the new controller are complex ones, without clear physical interpretations. Theoretically, the PD controller is just a special case of such a controller, therefore, it is true that the controller found is not worse than the PD controller. In normal cases, the better controller can be designed in the new structure.

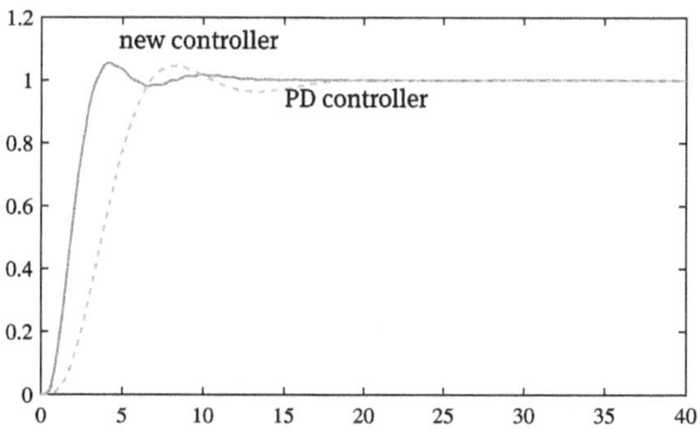

Figure 8.39: Comparisons in step responses.

8.4 OptimPID – optimal PID controller design APP

PID controller is one of the earliest proposed control strategies.[10] The structures and design methods are simple. Besides, the exact plant model is not expected in PID control strategy, and the PID control behaviors are satisfactory. Therefore, PID controllers are widely used in industrial control. Over the recent 20 to 30 years, PID controllers receive again the attention from the control researcher and engineers, since the intelligent PID with automatic tuning strategies introduced by Swedish scholar, Karl Åström, which have better performance than classical PID controllers.[6] Automatic tuning PID controllers have hardware products, such that they are widely used in industrial control.[5, 63, 38]

The OCD program developed in the previous section combined successfully the numerical optimization techniques and powerful simulation facilities of Simulink together, and an effective solution method is provided to design optimal controllers. It was also shown with examples that the PID controller is the best second-order controller. Therefore, another APP dedicated to design optimal PID controllers, named OptimPID,[85] is written. The user needs only to express the plant model in Simulink. With such a tool, optimal PID controllers can be designed directly. Compared with the existing parameter setting interface, OptimPID is easier to use, and the control performance is more reasonable and objective. For unstable and nonlinear plants, the OptimPID program can be used directly, and the control performance is better than other PID controller design methods.

Since the old version of OptimPID is not quite compatible in the new versions of MATLAB, the fast restart mode in Simulink is not supported, and a new OptimPID program is designed with APP designer, the readability and efficiency of the program are enhanced. Demonstrations on OptimPID are given, and low-level simulation model for PID control systems is briefly discussed. Programming details are not given in the book. Interested readers are advised to open `OptimPID.mlapp` file in APP designer, to appreciate the low-level programming and implementation.

8.4.1 Low-level simulation model of the control systems

Low-level Simulink model in OptimPID is shown in Figure 8.40. If the plant model is specified, the model will be embedded automatically to the Model block. The low-level model is set to invisible mode by default, and it is not suggested to modify the model. Exiting normally from OptimPID, all of the modifications are abandoned.

Double click the Subsystem block icon, the internal structure of the subsystem block is displayed, as shown in Figure 8.41. Six commonly used integrands in the objective function are designed, with the Multiport Switch block structure. An external constant `keyCriterion` is used to control the switch to select suitable objective function.

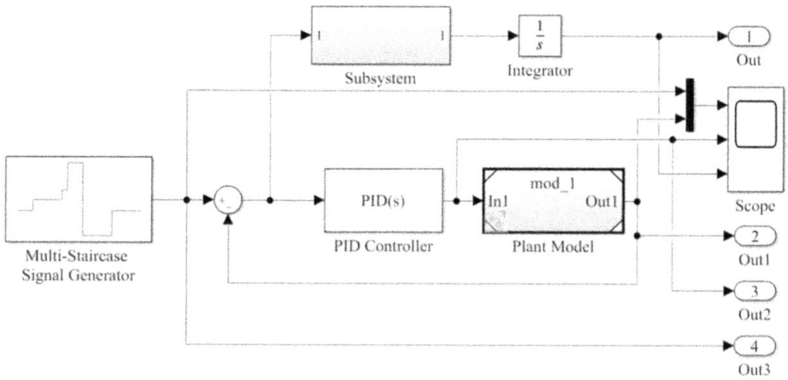

Figure 8.40: Low-level Simulink model (file: `pidctrl_model.slx`).

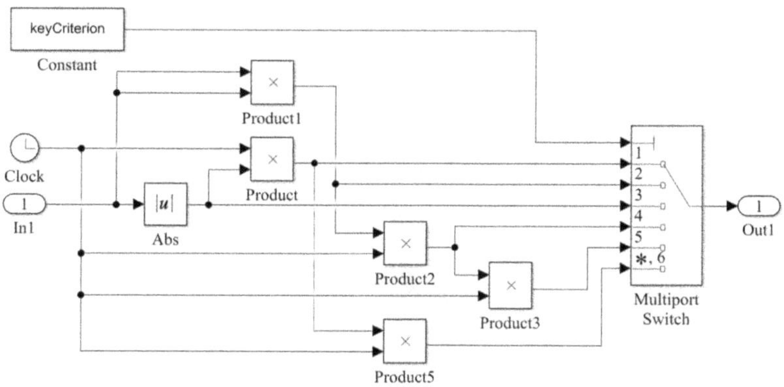

Figure 8.41: Simulink subsystem.

8.4.2 OptimPID APP examples

The use of the program OptimPID is demonstrated in this section. The PI controller design method for a composite transfer function model is demonstrated first, then a PD controller for a time-varying plant is illustrated. The controllers for the two plants are difficult to be designed with any other methods, but can be implemented with OptimPID easily, with satisfactory control performance.

Example 8.23. Consider the composite transfer function model in Example 2.7. A Simulink model `c2mtf2.slx` was already created. If the model is used as the plant model, and saturation actuator with $|u(t)| \leqslant 15$ is expected, design an optimal PI controller.

Solutions. At the MATLAB prompt, type command `OptimPID`, and the program interface in Figure 8.42 is opened. Fill the name `c2mtf2` to the **Plant model name** edit box.

Figure 8.42: Main OptimPID interface.

Try a stop time of 20 in the **Terminate time** edit box. Select **PI controller** item from the listbox. Now click **Create file** button to create an objective function file. Then press **Optimize** button to launch the optimization process. In the design process, a scope is opened automatically, display "real-time" the three signals, that is, the input/output signal, the control signal and the ITAE integral signal. If the design is completed, the ITAE integral is a horizontal one, then the design is successful; otherwise, increase the stop time, and design and validate again.

After design, a PI controller $G_c(s) = 0.8995 + 0.3642/s$ is obtained. The scope display is shown in Figure 8.43(a). It can be seen that in about 15 seconds, the ITAE integral in the bottom plot is already in horizontal. Therefore, the PI controller design is successful. The upper waveforms are the input and output signals, and the control behavior is satisfactory. The signal in the middle is the control signal, and it can be seen that the signal is kept within the range of ± 15.

If the staircase signal is specified by the two vectors at the bottom-right corner in the interface, then click **Simulation** button, and the staircase response can be ob-

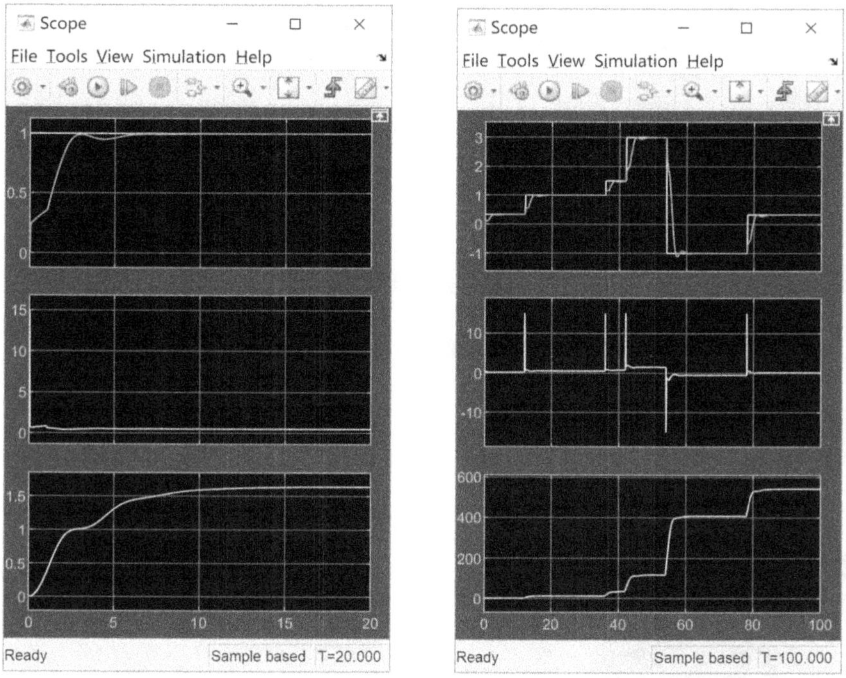

(a) designed responses

(b) response to a staircase input

Figure 8.43: Scope display of the control behaviors.

tained, as shown in Figure 8.43(b). It can be seen that the tracking of the output to the jumping signal is also satisfactory, and it ensures that $|u(t)| \leqslant 15$.

Example 8.24. Consider the following time-varying plant model. Design an optimal PD controller:

$$y''(t) + \mathrm{e}^{-0.2t}y'(t) + \mathrm{e}^{-5t}\sin(2t+6)y(t) = u(t).$$

Solutions. The interactive design interface provided in MATLAB was used to design directly the PD controller for this plant in [85], but the behavior of the controller is rather poor. Now consider using OptimPID interface to design the optimal PD controller. The relationship between the control signal $u(t)$ and the output signal $y(t)$ can be expressed with a Simulink model shown in Figure 8.44. Assume that the saturation actuator is $|u(t)| \leqslant 4$, and the stop time 10. An optimal PD controller can be designed under ITAE criterion that can be found, and the output signal is shown in Figure 8.45. The controller parameters found are $K_\mathrm{p} = 484.0281$, $K_\mathrm{d} = 91.3725$. The control behavior is much better than the one in [85], and is ensured that the control signal satisfies $|u(t)| \leqslant 4$.

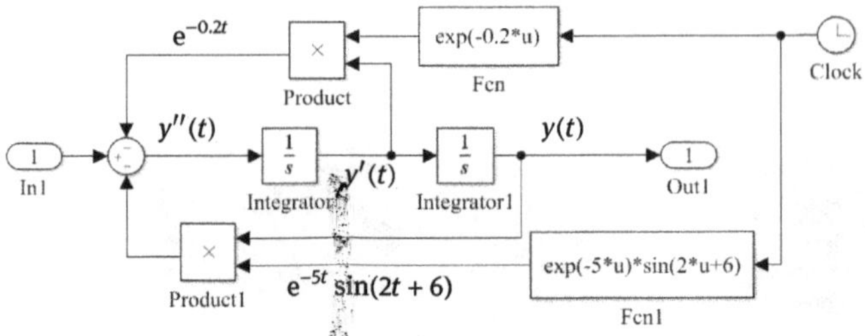

Figure 8.44: A time-varying plant model (file: `c8mod_3.slx`).

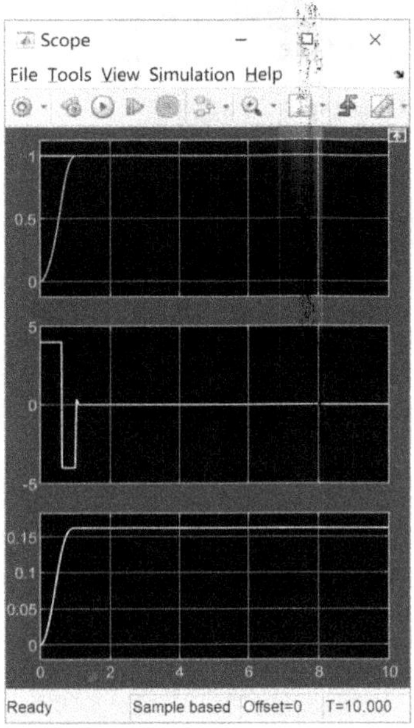

Figure 8.45: Scope display of the control results.

The intelligent optimization algorithms such as **Optimization algorithms** and **Particle swarm optimization** can be selected from the listbox, such that the intelligent solvers can be adopted in the design process. The same results can be obtained. It can be seen that these algorithms are very time consuming, and usually tens of time are needed. If not absolutely necessary, these methods are not recommended.

It can be seen that there is no programming needed in the optimal controller design. The Simulink description to the plant model is needed, then the interface manipulations can be carried out to design directly the optimal controllers. The two APPs are provided in the package for the book. The interested readers can reference the source code in the APPs to better learn MATLAB and Simulink simulation methods.

8.4.3 Open frameworks and extensions

As in the cases in OCD program, open framework is also adopted in the design of OptimPID program. The programs can be easily extended to solve more problems. For instance, if a meaningful integral-type criterion is to be introduced, an extra switch input port can be added in the subsystem block in Figure 8.41 to define the objective function. Then in the interface design, the corresponding item can be added to the Optimization criterion listbox, and the callback function in `switch-case` structure can be modified, if necessary.

Apart from objective functions, optimization algorithms and PID controller types can be added.

In fact, the original OptimPID interface was extended to design optimal fractional-order PID controllers. More about this interface can be found in [79, 80], and they are not discussed in this book.

8.5 Exercises

8.1 Understand why the `try-catch` structures are used in computing the objective functions. If the `try-catch` structure is deleted from the objective function in Example 8.1, what kind of phenomenon may happen? Execute the commands several times to observe the results.

8.2 Solve the following boundary value problem:[56]

$$\begin{aligned} y''(t) &= -0.05y'(t) - 0.02y^2(t)\sin t + 0.00005\sin t\cos^2 t \\ &\quad - 0.05\cos t - 0.0025\sin t, \end{aligned}$$

where $y(0) = y(2\pi)$ and $y'(0) = y'(2\pi)$. The exact solution is $y(t) = 0.05\cos t$.

8.3 Solve the following boundary value problem:

$$y^{(4)}(t) = 6\mathrm{e}^{-4y(t)} - \frac{12}{(1+t)^4},$$

where $y(0) = 0$, $y(1) = \ln 2$, $y''(0) = -1$, $y''(1) = -0.25$, and the analytical solution is $y(t) = \ln(1+t)$.

8.4 Solve the following boundary value problem:[56]

$$y''(x) - 400y(x) = 400\cos^2\pi x + 2\pi^2\cos 2\pi x,$$

where $y(0) = y(1) = 0$. The analytical solution is

$$y(x) = \frac{\mathrm{e}^{-20}}{1+\mathrm{e}^{-20}}\mathrm{e}^{20x} + \frac{1}{1+\mathrm{e}^{-20}}\mathrm{e}^{-20x} - \cos^2\pi x.$$

8.5 The boundary value problem of magnetic monopoles mapped into the interval $(0,1)$ is given below[2]

$$\begin{cases} y_1''(x) = \dfrac{2y_1'(x)}{1-x} + \dfrac{y_1(x)}{x^2(1-x)^2}\left[y_1^2(x) - 1 + \dfrac{y_3^2(x) - y_2^2(x)}{(1-x)^2}\right], \\ y_2''(x) = \dfrac{2y_2(x)y_1^2(x)}{x^2(1-x)^2}, \\ y_3''(x) = \dfrac{y_3(x)}{x^2(1-x)^2}\left[y_1^2(x) + \dfrac{\beta(y_3^2(x) - x^2)}{(1-x)^2}\right], \end{cases}$$

where $y_1(0) = 1$, $y_2(0) = y_3(0) = 0$, $y_1(1) = 0$, $y_2(1) = \eta$, $y_3(1) = 1$, $0 \leqslant \beta \leqslant 20$, $0 < \eta < 1$ are all constants. Let $\eta = 0.5$, and solve the differential equation for $x \in (0,1)$.

8.6 Solve the following boundary value problem:[56]

$$\begin{cases} y_1'(t) = ay_1(t)(y_3(t) - y_1(t))/y_2(t), \\ y_2'(t) = -a(y_3(t) - y_1(t)), \\ y_3'(t) = [b - c(y_3(t) - y_5(t)) - ay_3(t)(y_3(t) - y_1(t))]/y_4(t), \\ y_4'(t) = a(y_3(t) - y_1(t)), \\ y_5'(t) = -c(y_5(t) - y_3(t))/d, \end{cases}$$

where $a = 100$, $b = 0.9$, $c = 1000$, $d = 10$, and the given boundary values are $y_1(0) = y_2(0) = y_3(0) = 1$, $y_4(0) = -10$, $y_3(1) = y_5(1)$.

8.7 Solve the following boundary value problem:[57]

$$\begin{cases} y_1'(t) = y_2(t), \\ y_2'(t) = y_3(t), \\ y_3'(t) = -(3-n)y_1(t)y_3(t)/2 - ny_2^2(t) + 1 - y_4^2(t) + sy_2(t), \\ y_4'(t) = y_5(t), \\ y_5'(t) = -(3-n)y_1(t)y_3(t)/2 - (n-1)y_2(t)y_4(t) + s(y_4(t) - 1), \end{cases}$$

where $n = -0.1$, $s = 0.2$, and the boundary conditions are $y_1(0) = y_2(0) = y_4(0) = y_2(b) = 0$, $y_4(b) = 1$, $b = 11.3$.

8.8 Solve the boundary value problem of semiinfinite interval:[25]

$$\begin{cases} 3y(t)y''(t) = 2(y'(t) - z(t)), \\ z'''(t) = -y(t)z'(t). \end{cases}$$

If the given boundaries are $y(0) = z(0) = 1$, $y'(\infty) = z'(\infty) = 0$. Note that infinite stop time cannot of course be used in simulation. A relatively large stop time can be assigned. Also, a larger one can be tried. If the two results are not consistent, a larger stop time can be tried.

8.9 Consider the following boundary value problem:[2]

$$\begin{cases} f'''(t) - R[(f'(t))^2 - f(t)f''(t)] + AR = 0, \\ h''(t) + Rf(t)h'(t) + 1 = 0, \\ \theta''(t) + Pf(t)\theta'(t) = 0, \end{cases}$$

where $R = 10$, $P = 0.7R$. A is an undetermined constant. If the boundaries are given as $f(0) = f'(0) = 0, f(1) = 0, f'(1) = 0, h(0) = h(1) = 0, \theta(0) = 0$ and $\theta(1) = 1$, solve the differential equation. If $R = 10000$, solve the differential equation again.

8.10 Observe the subsystem in Figure 8.41. If the value of `keyCriterion` is 4, what is the mathematical form of the output signal?

8.11 Approximate the following models with a first-order model with time delay, and compare the time domain and frequency domain responses with the original model:

① $G(s) = \dfrac{12(s^2 - 3s + 6)}{(s+1)(s+5)(s^2+3s+6)(s^2+s+2)}$,

② $G(s) = \dfrac{-5s + 2}{(s+1)^2(s+3)^3}\mathrm{e}^{-0.5s}$.

8.12 If pure delay terms are included in the plant model, design optimal PID controllers under ITAE, IAE and ISE criteria and compare the results:

① $G_\mathrm{a}(s) = \dfrac{1}{(s+1)(2s+1)}\mathrm{e}^{-s}$, ② $G_\mathrm{b}(s) = \dfrac{1}{(17s+1)(6s+1)}\mathrm{e}^{-30s}$.

8.13 If the plant model is given by a delay differential equation,

$$\frac{\mathrm{d}y(t)}{\mathrm{d}t} = \frac{0.2y(t-1)}{1 + y^{10}(t-1)} - 0.1y(t) + u(t),$$

and a PI controller is used. Convert the controller design problem into an optimization problem, and find the optimal PI controller parameters. Draw the step response of the closed-loop system. If the overshoot thus obtained is too large, constraints can be introduced to convert the problem into a constrained optimization problem. Solve the problem with MATLAB.

8.14 For an unstable plant model,

$$G(s) = \frac{s+2}{s^4 + 8s^3 + 4s^2 - s + 0.4},$$

with saturation actuators $|u(t)| \leqslant 5$, select a suitable stop time and design an optimal PID controller. Observe the closed-loop step response. If ordinary optimization methods failed to get convergent controllers, a genetic algorithm or particle swarm optimization solvers can be tried to design controllers.

8.15 Consider the plant with large time delay

$$G(s) = \frac{\mathrm{e}^{-20s}}{(s+1)^3}.$$

Use the `pidtune()` function in MATLAB and OptimPID program to design PID controllers, and compare the control performances.

8.16 Design optimal continuous and discrete PID controllers for the following plant models:[34]

① $H(z) = \dfrac{7}{z^4 - 1.31z^3 + 1.21z^2 - 0.287z - 0.0178}$, $T = 0.01$ seconds.

② $H(z) = \dfrac{3z^2 - 1}{z^5 - 0.6z^4 + 0.13z^3 - 0.364z^2 + 0.1416z - 0.288}$, $T = 0.01$ seconds.

8.17 For the fractional-order plant model $G(s) = 1/(0.8s^{2.2} + 0.5s^{0.9} + 1)$, design an optimal PID controller and observe the results.

9 Processing and simulation of multimedia information

Multimedia is a collection of different medias. The so-called medias here include text, sound, images and videos. Of course, generally speaking, more medias are contained, such as storage media and prorogation media and so on. The multimedia technique uses computers to process and manage text, data, sound and videos, such that the users can have real-time communications with computers through different sensory organs.

The multimedia technique is the product of the information age. They may be seen anywhere in human daily life, scientific research and engineering practices. As the major forms of expression in multimedia processing, the processing of the signals such as sound, images and videos receive significant attention. In this chapter, MATLAB and Simulink based multimedia information processing methods are illustrated.

The input and output methods for audio signals are presented in Section 9.1. Concentrations are made on introducing the low-level tools provided in MATLAB Audio System Toolbox and its blockset to handle audio information. An introduction is given on how to load audio information into computers, and how to send the processed audio information back to the user. In Section 9.2, various audio information processing methods, such as filtering, dynamic range control, reverberation in sound effect processing and so on is provided. Voice activity detection method is also demonstrated. In Section 9.3, a brief introduction to the Computer Vision Toolbox and its applications in image and video processing is given. In Section 9.4, introductions are given on how to load images and videos into MATLAB, and how to show and play the information. In Section 9.5, commonly used color spaces are introduced, and image and video conversion methods are presented, including color space conversion, image scale conversion, Gamma correction and so on. In Section 9.6, examples are given to demonstrate fundamental digital image processing methods, such as the histogram equalization method, edge detection method and image composition and so on. Morphological based image processing methods such as dilation, erosion and skeleton extraction are also demonstrated.

9.1 Input and output of audio information

Audio information usually comes from audio files and audio equipment, such as a microphone. In this section, the methods to load audio information into MATLAB and Simulink environment are illustrated. Besides, the audio information play is illustrated, and the audio information can also be saved into audio files. In this section, the MATLAB and Simulink based audio waveform display and audio file recording methods are presented.

https://doi.org/10.1515/9783110734959-009

9.1.1 Introduction to Audio System Toolbox

The major tool for audio information processing provided in MATLAB is the Audio System Toolbox.[48] In the toolbox, two categories of audio information processing methods are supported. One is the command-line driven methods, and the other, block diagram based methods. Command-line driven means that a significant amount of functions provided in the toolbox, by calling the functions, the audio information processing tasks are implemented. Block diagram format means that with the powerful blockset, the Simulink model can be composed to implement audio signal processing.

In the Simulink model library window, shown in Figure 2.6, double click the Blocksets & Toolboxes icon, and all the toolboxes and blocksets are automatically shown in Figure 2.41. Select from the list on the left, and the Audio Toolbox item, the Audio System Toolbox blockset is opened. An alternative way to open the Audio System Toolbox blockset is by the command `audiolib`. The opened blockset is shown in Figure 9.1. The Sources, Sinks groups are provided in the blockset. Also, the groups such as Dynamic Range Control, Effect, Filters and Measurement are provided. With the blocks provided in the groups, Simulink models can be established to implement audio signal processing tasks.

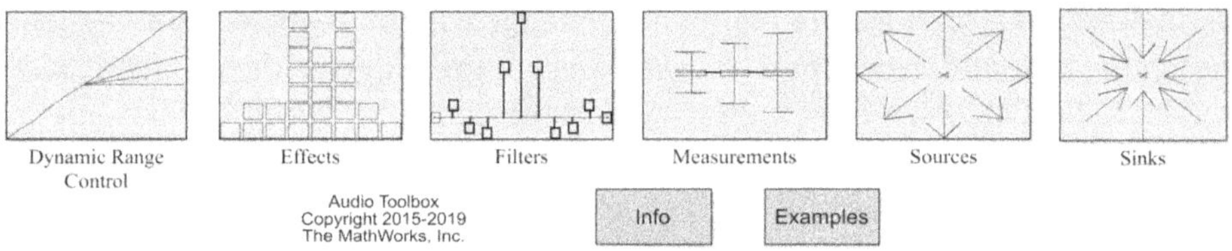

Figure 9.1: Audio System Toolbox blockset.

9.1.2 Input and output of audio information

In the blockset in Figure 9.1, it can be seen their own Sources and Sinks groups are presented, where the input and output audio blocks are provided, respectively.

The blocks of the Sources group is shown in Figure 9.2. The commonly used groups are summarized as follows:

(1) Audio Device Reader block, which can be used to connect the audio device in the computer as the signal source. The audio information can be loaded into the Simulink model. In this way, real-time processing of audio information is possible.

(2) From Multimedia File block can be used to load audio signal in a multimedia file directly into Simulink, as in input signal. In this way, the audio information saved in the multimedia file can be introduced.

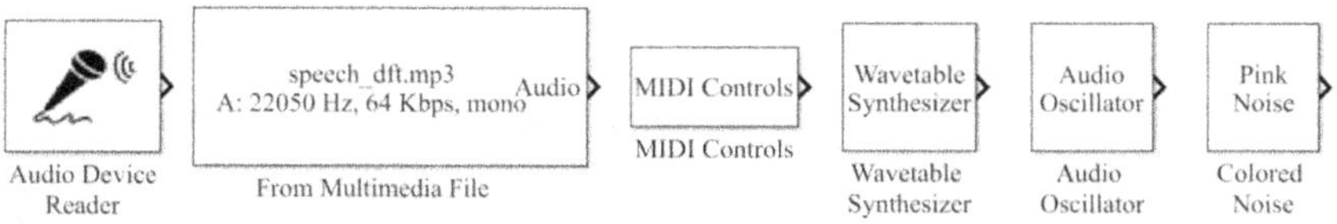

Figure 9.2: Audio source group.

(3) Other signal sources include **Audio Oscillator**, **MIDI Control** and **Wavetable Synthesizer** blocks. The block **Colored Noises** can be used to introduce noise signals into audio signals, such that the denoise algorithms can be tested.

The blocks provided in the **Sinks** group are shown in Figure 9.3, where the **Audio Device Writer** block allows the user to play the audio signal generated in Simulink model on the speakers connected to the computer. The **To Multimedia File** block can be used to write the Simulink generated audio signal into a multimedia file. If the specified file already exists, it will be automatically overridden by the block.

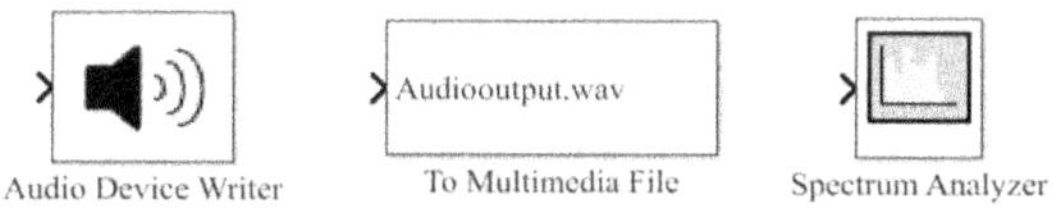

Figure 9.3: Audio sinks group.

The **Spectrum Analyzer** block allows to carry out spectrum analysis to the audio signal. Apart from this block, the signal can also be connected to the blocks provided in the **DSP System Toolbox** blockset, such that professional signal analysis can be implemented. The blockset can be opened by the command `dsplib`, where the blocks in the group are shown in Figure 9.4. Especially, the **Time Scope** block in the **Sinks** group can be used to display the waveform of the audio signal.

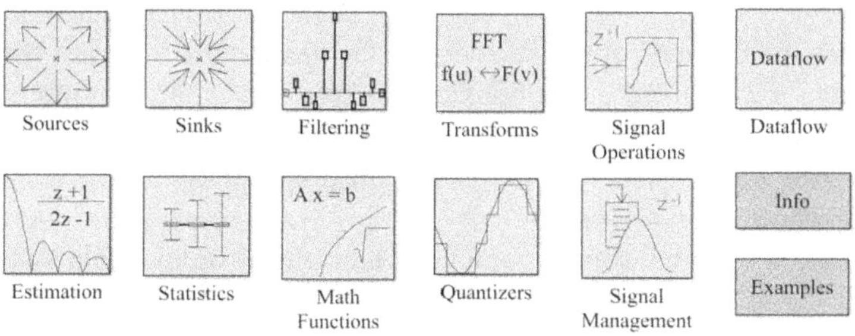

Figure 9.4: Digital signal processing blockset.

Example 9.1. Create an audio file playback system in Simulink.

Solutions. Consider using the From Multimedia File block to load audio file, and connect the block directly to the Audio Device Writer block. The Simulink model in Figure 9.5 can be established, where the parameters of Audio Device Writer are automatically assigned, as shown in Figure 9.6. The signal can be sent directly to the speaker connected to the computer. There is no extra setting needed.

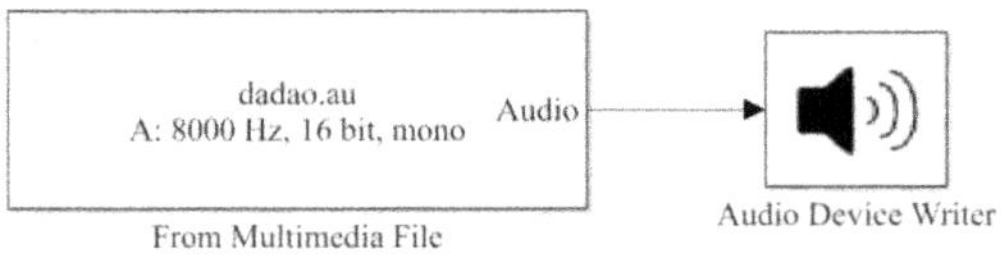

Figure 9.5: A simple audio file playback system (file: `c9maud1.slx`).

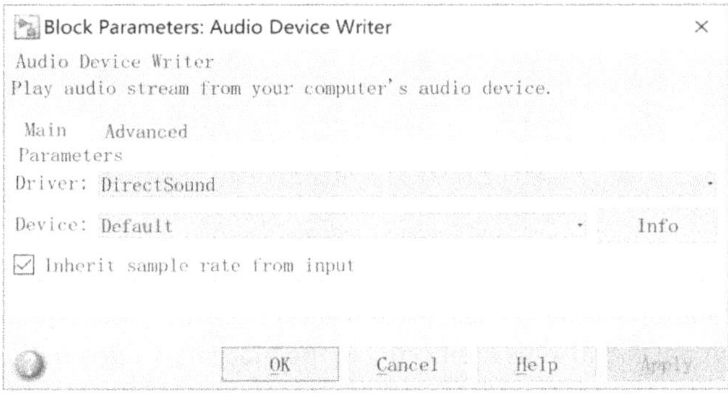

Figure 9.6: The parameter dialog box of the Audio Device Writer block.

Double click the From Multimedia File block, and the multimedia file name selection dialog box can be opened, as shown in Figure 9.7. It is seen that this is a common file name dialog box. The user needs to select an audio file name in the File Name edit box. The user can click the Browse button, and select a file name is a visual way. For instance, the file `dadao.au` provided in the book can be selected. Different types of audio files are supported, such as `wav`, `mp3`, `wma` and so on. After selection, the audio file is associated directly with the block. If simulation process is invoked, the audio file is played on the speaker.

When the audio file is selected, some information is displayed on the From Multimedia File block icon automatically. It can be seen that the `dadao.au` file is of the sound information in 8000 Hz, 16 bit monophonic format.

Click the Browse button in the dialog box shown in Figure 9.7, and a standard file name selection dialog box is opened. An audio file can be selected from the dialog

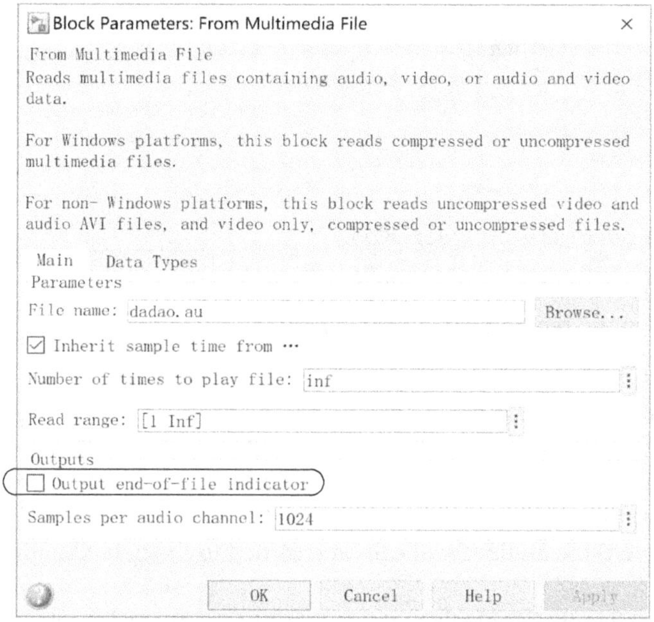

Figure 9.7: Parameter dialog box of From Multimedia File block.

box. Video files can also be chosen from the dialog box. It can be seen that the block is rather powerful, because it not only extracts audio information, but also video information, such that the audio information in the file is played automatically.

It is worth mentioning that the easily designed audio file playback system has certain defects, because the actual length of play cannot be made depends on the real length of the audio file. The Stop Time value is assigned by the user. If the actual length of the audio file is longer than the value in Stop Time, the simulation process is terminated automatically at Stop Time. If the actual length of the audio file is shorter than the value in Stop Time, the audio file is played again and again, until the end of the Stop Time.

Example 9.2. Modify the audio file playback system designed earlier, such that a complete audio file can be played. After play, the simulation process is terminated automatically.

Solutions. Various methods can be used to complete the tasks like this. Of course, double click the From Multimedia File block, at the bottom in the dialog box shown in Figure 9.7, and there is an Output end-of-file indicator checkbox. If it is checked, an extra output port EOF is automatically added to the From Multimedia File block. The signal can be connected to the Stop Simulation block in the Sinks group in Simulink, such that the new block diagram in Figure 9.8 can be created. The Stop Time edit box in Simulink can be set to `inf`.

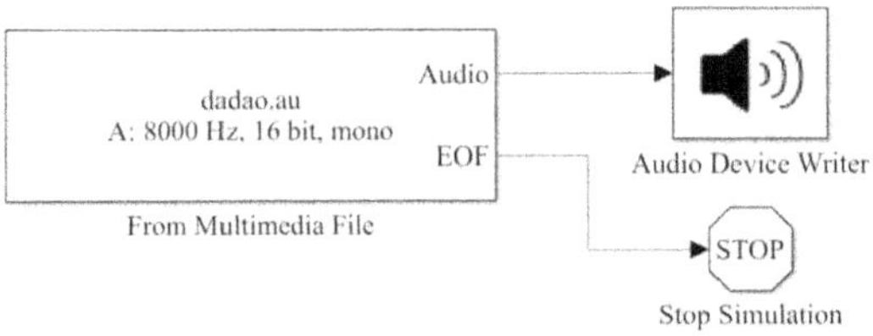

Figure 9.8: A modified playback system (file: `c9maud1a.slx`).

Example 9.3. Construct a sound recording system in Simulink.

Solutions. To the contrary in the previous examples, the microphone connected to the computer can be used as the input block, which can be implemented with the **Audio Device Reader** block. The block can be connected to the **To Multimedia File** block in the **Sinks** group, as shown in Figure 9.9, where the **Audio Device Reader** block extracts audio information from the microphone connected to the computer. If the simulation model is executed, the audio signal can be recorded in the default audio file `output.wav`.

Double click the **To Multimedia File** block, and the dialog box is opened as shown in Figure 9.10, allowing the user to select a file name to save. Uncompressed `wav` format can be selected. Also, in the **File type** and **Audio compressor** listboxes, the file names

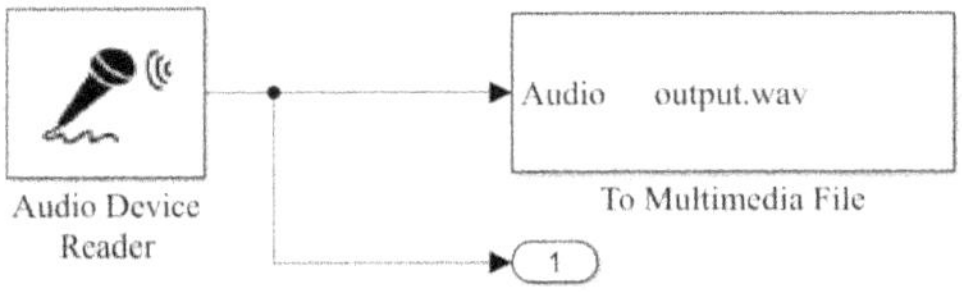

Figure 9.9: A simple sound recording system (file: `c9maud2.slx`).

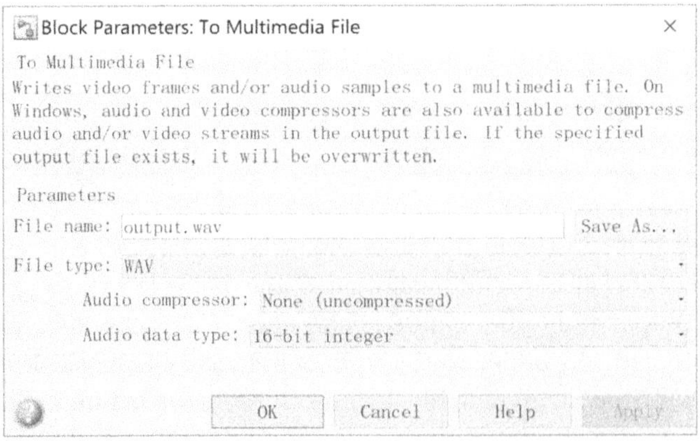

Figure 9.10: Parameter dialog box of To Multimedia File block.

such as wav, flac and wma and so on can be selected, to save the audio signal into audio files. The compression format may not be arbitrarily chosen, for instance, the commonly used mp3 file format cannot be selected.

Note that apart from the file-saving block, an ordinary **Out1** output port block is given. It is not used in this example, and it is prepared for later use.

9.1.3 Data structure of audio signals

The audio signals travel in Simulink models along signal lines. What are the data types of the signal line like this? The data type of the audio signal is measured next through examples, to allow the user to better understand the audio signal processing tasks with Simulink.

Example 9.4. Observe the redundant **Out1** block in Figure 9.9. If the sound recording process is activated, observe the output signal.

Solutions. Start the simulation model, and after a while, the simulation process can be stopped. An output.avi file in MATLAB working path is created. Since the existence of the **Out1** block, the simulation results are returned to MATLAB workspace in the variables tout and yout. The following commands can be used to extract the simulation results:

```
>> size(tout), size(yout), diff(tout)
```

where the returned tout is a 423×1 column vector, and yout is a $1024 \times 1 \times 423$ three-dimensional array, where 1 in the second dimension indicates that it is a monotonic signal. The difference vector is equally spaced in time, and each step-size is 0.0232 seconds.

9.1.4 Audio input and output with MATLAB commands

Functions audioread(), audiowrite() and audioinfo() are provided in MATLAB to read, write and extract the audio information. The sound($\boldsymbol{W}$,f,r) function can be used to playback sound, where $\boldsymbol{W}$ is the extracted audio vector, f is the frequency and r is the bit rate of the audio signal. If default parameters are used to play the sound, the latter two parameters can be omitted.

Example 9.5. Consider the dadao.au file used earlier. Read it into the MATLAB workspace, and draw the waveform of the audio signal.

Solutions. Function audioread() can be used to load audio file into the MATLAB workspace, and the frequency is $F_s = 8000\,\text{Hz}$, the same as explained earlier. The audio signal is returned in a column vector $\boldsymbol{W}$. Function plot() can be used to draw its waveform, as shown in Figure 9.11, the horizontal axis unit is seconds.

```
>> [W,Fs]=audioread('dadao.au'); plot([1:length(W)]/Fs,W)
```

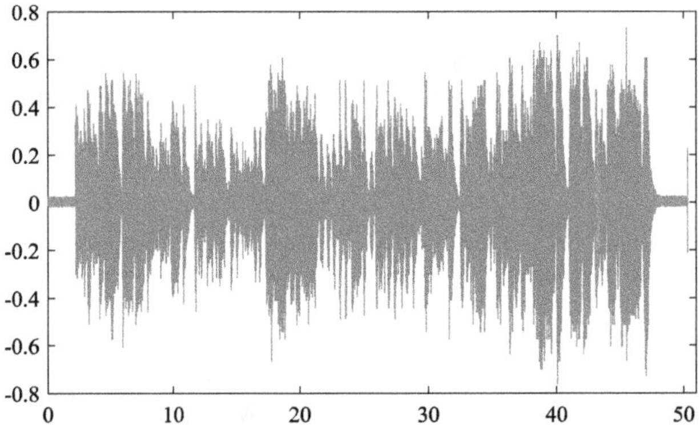

Figure 9.11: Waveform of the audio signal.

If function sound() is called, the audio file can be played, If double speed is expected, the following commands can be used. Besides, the function audioinfo() can be used to display the audio file information.

```
>> sound(W), sound(W,2*Fs), T=audioinfo('dadao.au')
```

where T is a structured variable, containing all the information in the audio file. The members include Filename, SampleRate, Title and so on.

```
         Filename: 'C:\xue.dy\BOOKS\...\dadao.au'
CompressionMethod: 'u-law'
      NumChannels: 1
       SampleRate: 8000
     TotalSamples: 402636
         Duration: 50.3295
            Title: []
          Comment: []
           Artist: []
    BitsPerSample: 8
```

Example 9.6. It can be seen from the waveform in Figure 9.11 that at the beginning and the end of the plot, there is a small period each the sound with very small volume. These periods can be regarded as noise signals. Cut off these parts, and extract the effective audio signal and save it into a new audio file.

Solutions. Zooming the waveform in Figure 9.11, the subscripts for the start and terminate points for the noise signal can be found. The submatrix extraction method can be used to cut off the unnecessary parts. The threshold method can be used to find the points whose volume value is smaller than 0.05. It can be seen from the waveform that there is a sudden change at the very end of the signal, therefore, the final 100 points are not considered. With the following statements, the effective audio signal, in between the points $i_1 = 17622$ to $i_2 = 382498$, can be extracted and saved into the new audio file `dadao1.wav`. The extracted new waveform is shown in Figure 9.12, and it can be seen that the effective waveform is extracted, as expected:

```
>> [W,Fs]=audioread('dadao.au'); W=W(1:end-100);
   ii=find(abs(W)>0.05); i1=ii(1), i2=ii(end)
   W1=W(i1:i2); audiowrite('dadao1.wav',W1,Fs)
   plot([1:length(W1)]/Fs,W1)     % draw the sound waveform
```

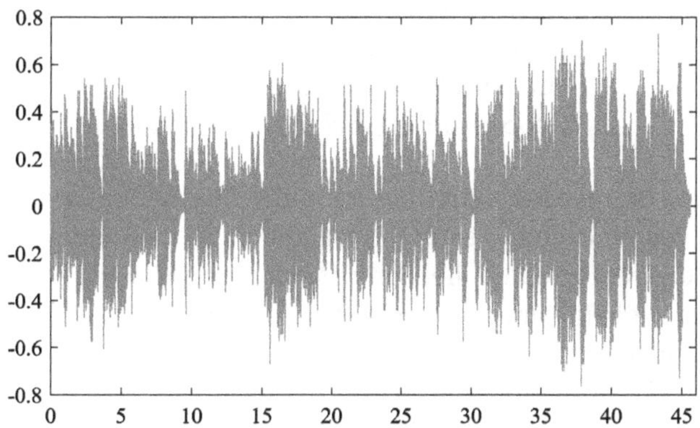

Figure 9.12: The waveform of the newly recorded signal.

Note that since the vector $\boldsymbol{W}_1$ stores the original audio signal, which is uncompressed, it can only be saved into uncompressed audio files, such as `wav`, `flac` and so on. It cannot be saved into compressed files such as `au` or `mp3`. Certain compression should be made first to the vector $\boldsymbol{W}_1$, in order to save to such files.

In the recent versions, object oriented programming technique is fully supported. Functions `audioplayer()` and `audiorecorder()` can be used to read or write audio information. The former can be used to replace the `sound()` function, and in the latter, an audio object *p* is opened. The `record()` function can then be called to record sound, or use `play()` function to play the sound. In the recording and play process, the command `pause()` can be called, or use `stop()` command to stop the play process. Function `resume()` can be used to restore record and play process. All of these commands can be used to manipulate directly the audio object *p*.

Besides, the `recordblocking` method in object *p* can be used to activate recording process, and assigns no response to other actions is allowed, within a certain time. The use of the audio object and recording methods are demonstrated next with an example.

Example 9.7. Record a sound for 10 seconds using command-line driven methods, and play the recording.

Solutions. Function `audiorecorder()` can be called first to create a recording object *p*, with sample rate of 44100 Hz, and assign that it is a 16 bit, stereo sound. The `record()` function can be used directly to record the sound, or the following format can be adopted to record sound, that is to call the `recordblocking` method in *p* object for recording, and ensure that other actions are suspended for 10 seconds. After 10 seconds, the command `stop()` can be used to terminate the recording process. When the recording process is completed, function `play()` can be called to playback the recorded object.

```
>> p=audiorecorder(44100,16,2); % create a recording object with parameters
   recordblocking(p,10)         % in 10 seconds, recording is made
   stop(p); play(p)             % stop recording and playback sound
```

The audio object *p* cannot be saved directly into an audio file. Before saving it, function `getaudiodata()` can be called to extract the audio data, then function `audiowrite()` can be called to save the audio file. For instance, the following commands can be used to generate audio file directly, where in the design of the *p* object, it is assumed that the sample rate is 44100 Hz:

```
>> W=getaudiodata(p,'int16');  % convert to audio data
   audiowrite('ttt.wav',W,p.SampleRate)
```

If the waveform of the audio information is to be displayed in the Simulink model, the existing Audio System Toolbox blockset cannot be used. The Time Scope block in the Sinks group of the DSP blockset can be used instead. This will be demonstrated later in examples.

9.2 Audio signals processing

The input and output blocks are discussed earlier in the previous section. With the audio signal input, the blocks in the Audio System Toolbox blockset can be used to further process the audio signals, such as filtering, dynamic range control and sound effect handling. In this section, the demonstrations of audio signal processing with MATLAB and Simulink are presented.

9.2.1 Audio signal filtering

Double click the Filters group icon in the blockset shown in Figure 9.1, the contents of the group is as shown in Figure 9.13, with different filter blocks provided. The blocks in the group are summarized below:

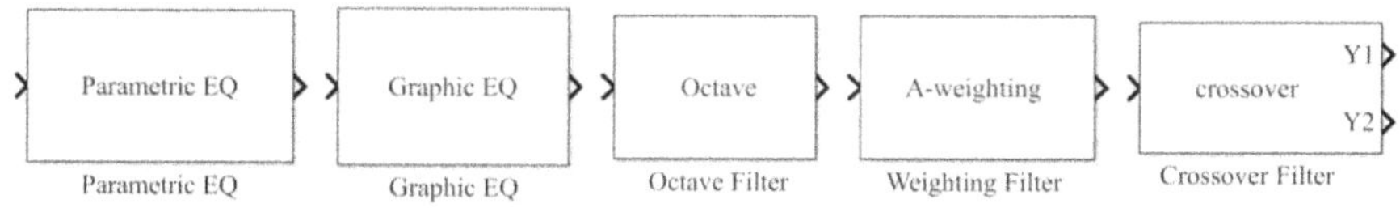

Figure 9.13: Audio filters group.

(1) Crossover Filter block can be used to divide the audio signal into two or more bands. Each band has its own output port.

(2) Graphic EQ block provides an interface, such the gains at the key frequencies in the low- mid- and high-frequency ranges can be adjusted graphically.

(3) Besides, other blocks such as Octave Filter, Parametric EQ and Weighting Filter are provided in the group. Examples are used to demonstrate the former two blocks. The use of these blocks are similar, and the users may try the other blocks and observe the processed signals.

Example 9.8. Divide audio file `dadao.au` into two bands with crossover frequency 2000 Hz.

Solutions. The Crossover Filter block can be used to divide an audio signal into two or more bands. The divided signals can be connected to a Manual Switch block, whose output is connected to the speaker block. The Simulink model thus created is as shown in Figure 9.14.

Double click the Crossover Filter block, and its parameter dialog box is opened, as shown in Figure 9.15, where the default value of Number of crossovers is 1, which divide the audio signal into two bands, such that two output ports are added to the block. Other values can be assigned in the edit box, such that the signal is divided

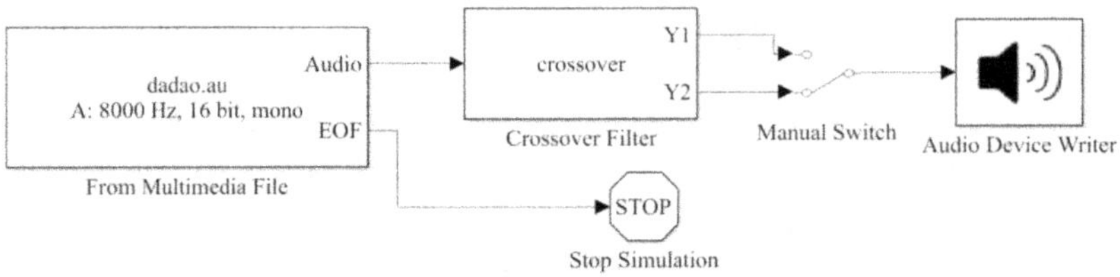

Figure 9.14: Signal dividing system (file: `c9maud3a.slx`).

into more bands. The edit box or tuning knob can be used to adjust the value of the Crossover frequency frequency, for instance, set it to 2000 Hz.

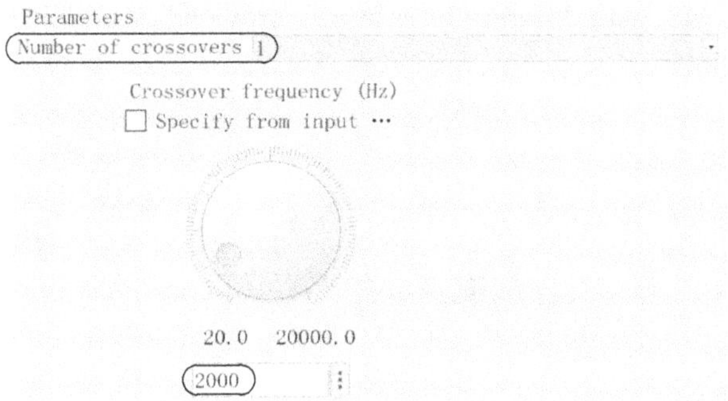

Figure 9.15: Parameter dialog box of the Crossover Filter block.

Example 9.9. Carry out necessary processing to the `dadao.au` audio file with graphical equalizer.

Solutions. Refer the Simulink model in Figure 9.8. A **Graphic EQ** blocked is inserted in between the source and speaker blocks, the simulation model can be created, as shown in Figure 9.16.

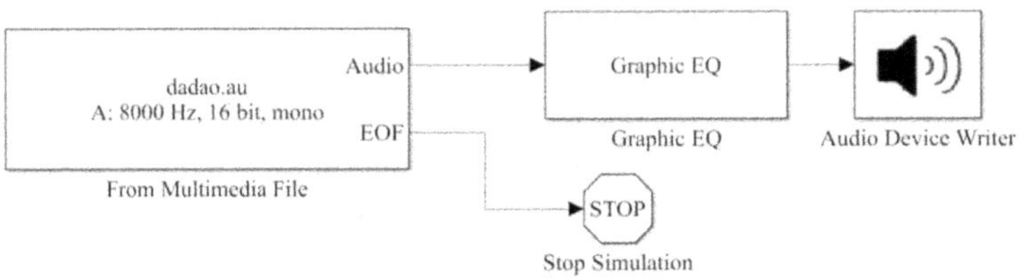

Figure 9.16: Sound processing with graphical equalizer (file: `c9maud3.slx`).

Double click the Graphic EQ icon, and a graphical equalizer interface is opened, with the controls on its Low Frequency parts is shown in Figure 9.17. The gains at three specific frequencies can be adjusted with the sliders. Visual methods are allowed to change the sound effect. Of course, the gains in the middle and high frequency ranges can be adjusted in the same way.

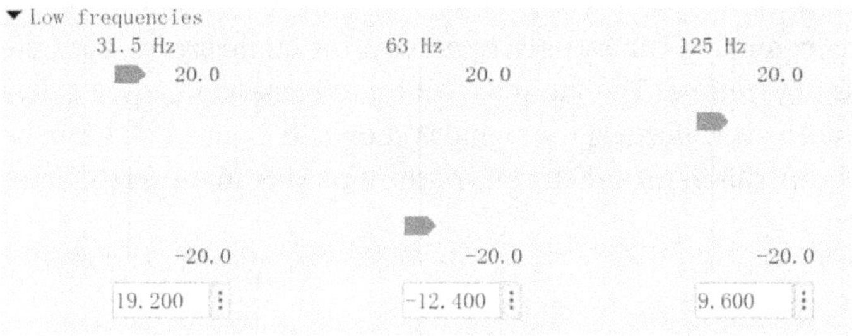

Figure 9.17: Graphical equalizer interface – low-frequency part.

9.2.2 Dynamic range control

Dynamic Range Control group is provided in the Audio System Toolbox, as shown in Figure 9.18. The dynamic range of the audio signal refers to the difference between the maximum and minimum volumes in the waveform. Generally speaking, the larger the range, the higher the sound resolution. If in a certain audio signal, if part of it is with higher volume, while the other is lower volume, it is necessary to adjust automatically the dynamic ranges so that the volumes in the two parts are equalized. The audio signal adjustment method is referred to as dynamic range control. Simply speaking, the different parts are multiplied by different values of the gains, to equalize the different parts. Naturally, the blocks like Compressor and Expander can be used. Besides, the Limiter and Noise Gate blocks can also be used to process dynamic ranges.

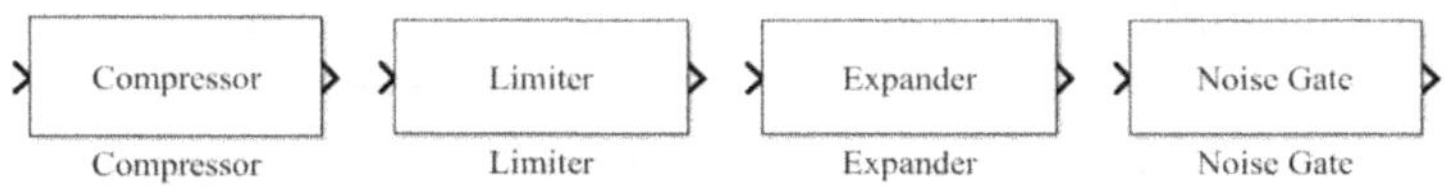

Figure 9.18: Dynamic range control group.

In simple words, the functions of the compressor is to compress the volumes over the prespecified threshold such that the volumes are reduced. Of course, other factors can be adjusted in the compressor block. Here, only the simplest threshold adjustment methods are demonstrated. To learn other adjust methods in dynamic range control, the readers are advised to refer to [64].

Example 9.10. Consider again the dadao.au file demonstrated earlier. If the volumes in the latter part of the audio signal is reduced deliberately, correct the audio signal via compression method such that the entire audio file can be played in an equalized volume.

Solutions. The following command can be used to process the audio signal, with the latter half reduced to its 1/3 volume. The modified audio information can be saved into the file dadao1a.wav. The waveform of the signal is shown in Figure 9.19. It can be seen that there are significant differences in the two parts. If played, there are different audio effects:

```
>> [W,Fs]=audioread('dadao.au'); N=length(W);
   W(round(N/2):end)=0.333*W(round(N/2):end);
   audiowrite('dadao1a.wav',W,Fs), plot([1:N]/Fs,W) % new waveform
```

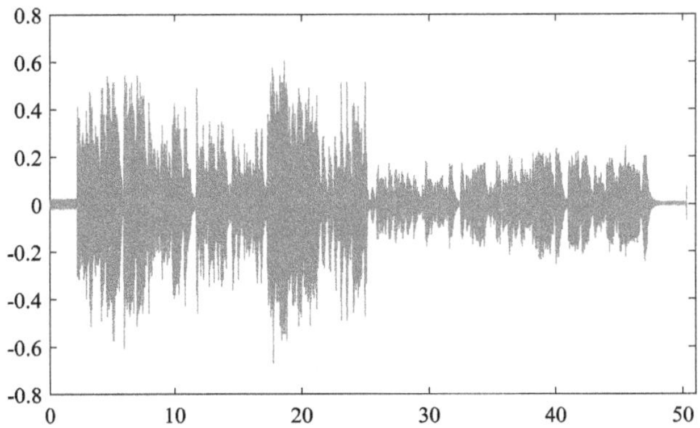

Figure 9.19: The waveform of the newly edited audio signal.

Now, in the c9maud3a.slx model, if the Graphic EQ block is replaced by the Compressor block, the audio signal can be compressed manually. While if it is directly played, the total volume is rather low. Therefore, a three time gain can be multiplied to the signal, to restore to the expected volume. The Simulink model thus created is as shown in Figure 9.20. In order to display the waveform, the Time Scope block in the DSP blockset can be used. The time range in the block can be selected, such that the waveform in the 0~52 seconds interval can be displayed.

Double click the Compressor block, and the parameter dialog box shown in Figure 9.21 is opened. Different compression methods can be selected. After some trials, the value of −37 dB can be assigned in the Threshold edit box, such that the volumes

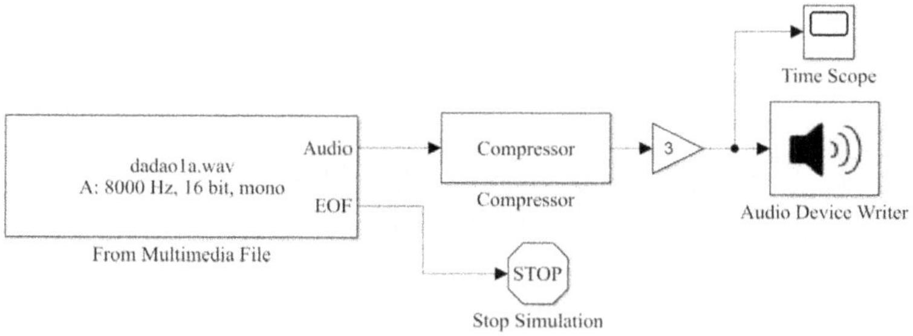

Figure 9.20: Audio compression model (file: `c9maud4.slx`).

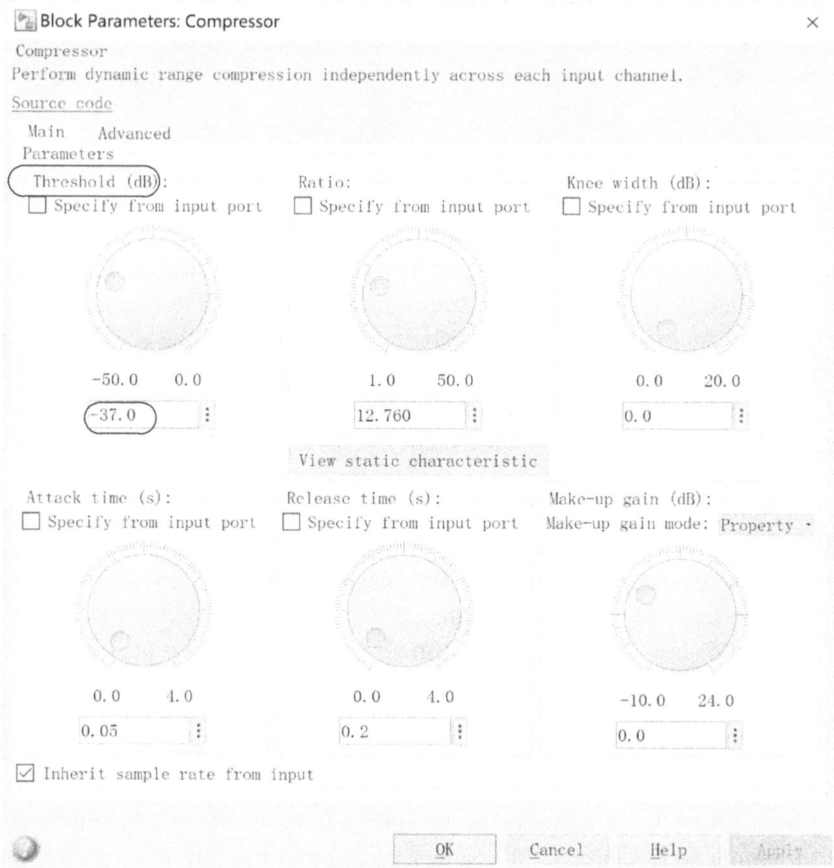

Figure 9.21: Parameter dialog box of the Compressor block.

in the two parts are equalized. Starting the simulation process, the processed signal audio signal is obtained, and the waveform shown in Figure 9.22 is obtained. Since the gain block is used in the model, the noise signal is also amplified.

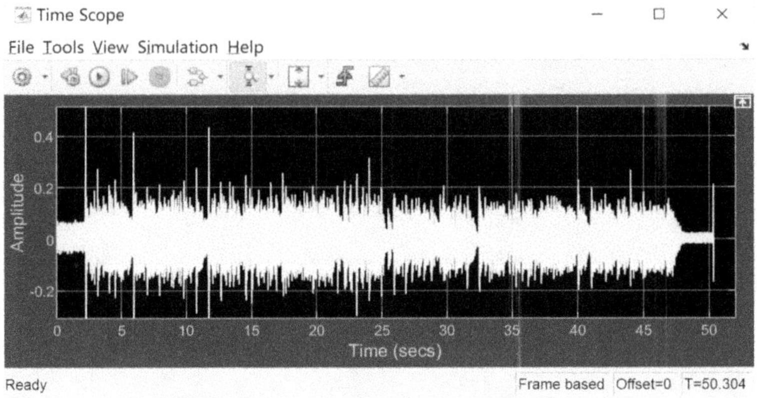

Figure 9.22: The waveform of the processed audio signal.

9.2.3 Audio effect processing

In some specific fields in audio signal processing, such as music processing, the sound effect of audio signals can also be processed. The basic concepts in sound traveling and sound effect are illustrated first. The sound travels from the voice source to the receiver is referred to as the direct sound. The music received by the audience in a certain environment include the direct sound, and also the sounds generated by the physical reflection of the sound in the environment. In the reflected sound, the sound effect such as decay and delayed echo are included. These sounds are referred to as reverberation.

The sound effect processing can be implemented by the blocks provided in the Audio System Toolbox. Double click the **Effect** group icon, and the group can be opened. In fact, there is only one block, the **Reverberator** block, in the group. The block can be used to simulate a different reverberation effect. Double click the reverberation block, and the parameter dialog box in Figure 9.23 can be obtained. The users are allowed to adjust different reverberation modes and parameters.

Example 9.11. Set up a sound reverberation system in Simulink.

Solutions. The sound source in Example 9.24 is adopted, and the **Graphic EQ** block is replaced by the **Reverberator** block; the Simulink model in Figure 9.24 can be established to handle sound effect. The readers may adjust the tuning knobs in the interface, and assign different parameters, to experience and appreciate the changes in the sound effect.

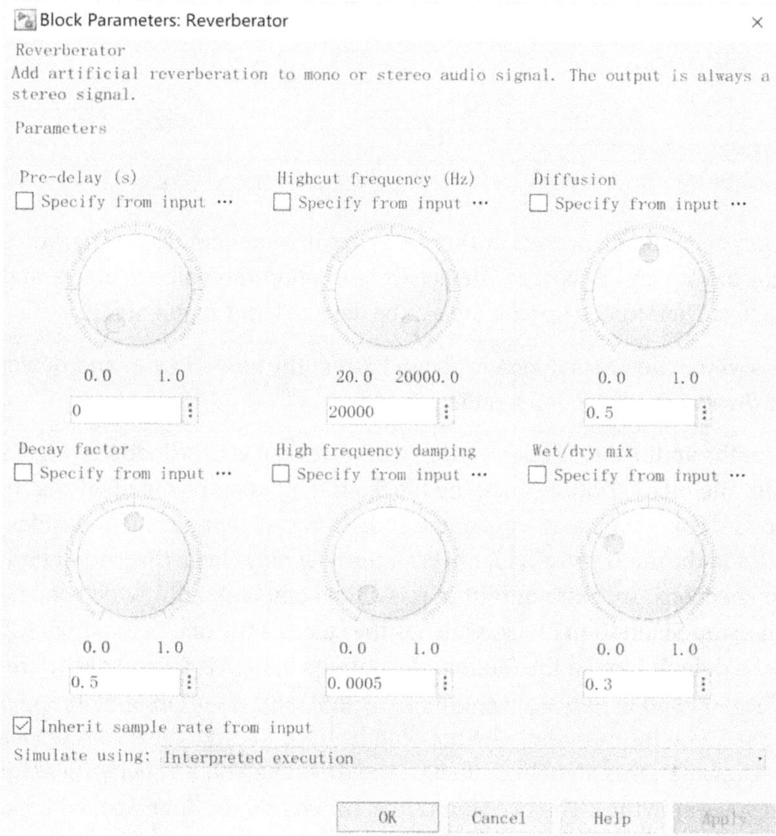

Figure 9.23: Parameter dialog box of a reverberation block.

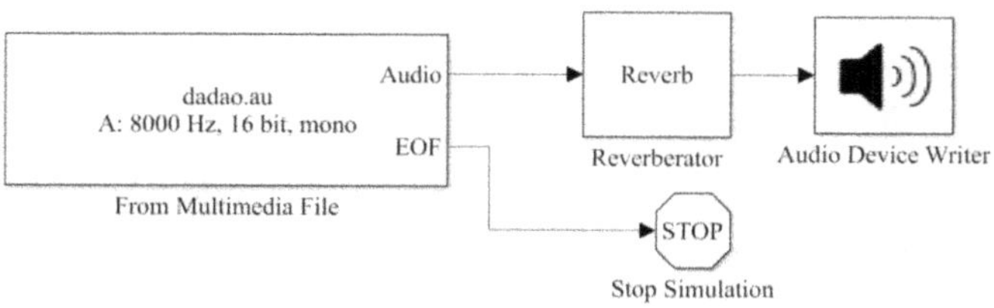

Figure 9.24: Sound effect processing model (file: `c9maud5.slx`).

9.2.4 Detection of audio information

Double click the Measurement group icon, and the blocks in Figure 9.25 are displayed. The commonly used blocks are Voice Activity Detector, Loudness Meter and Cepstral Feature Extractor.

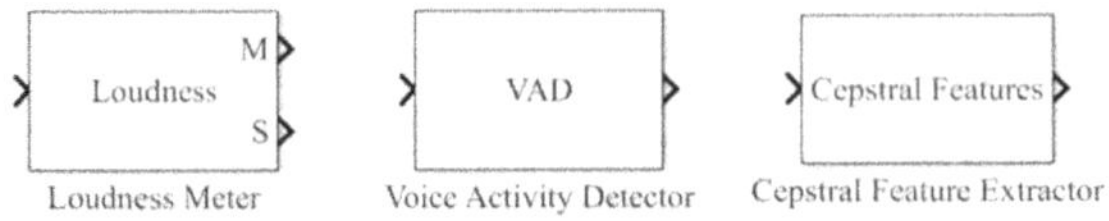

Figure 9.25: The measurement group.

For instance, for speech signals, normally there are gaps in sentences. Therefore, Voice Activity Detector block can be used, to detect which period the voice is active and which period is not. Therefore, a speech file can be detected and segmented.

Example 9.12. A video file `speech.mp4` is given. Extract the audio signal and detect its activity, and display the detection results.

Solutions. As for the audio files studied earlier, video files can also be loaded with the From Multimedia File block. Double click the block, and a dialog box is displayed to select the video file. For instance, the file name `speech.mp4` can be selected. Besides, at the bottom part in the dialog box in Figure 9.26, the user may check the Output end-of-file indicator checkbox. An extra output port is added automatically, which can be connected to the Stop Simulation block, which is the same as the one in Example 9.2. Besides, since the default item in the Multimedia outputs listbox is video only, therefore, it should be modified to Audio only option. The first output port in the block can be used to output audio information. The port can be used to drive the Audio Device Writer block. Meanwhile, this signal can be used to drive the Voice Activity Detector block, to detect voice activities. The two signals can be fed into the Time Scope block, and set up the Simulink model in Figure 9.27.

Figure 9.26: File name selection dialog box (bottom part).

Executing the simulation model, the waveform of the sound and activity detection methods can be obtained, as shown in Figure 9.28. It can be seen from the results, when the magnitude of the sound signal is large, the activity detection result is close to 1; if the magnitude is smaller, the detection value is also smaller. In this way, the gaps in the voice signal can be detected, and the method can be used to segment effectively voice signal effectively. This block can also be used in monitor systems, such that the activity detection facilities can be used to playback multimedia files, when abnormal phenomenon is detected.

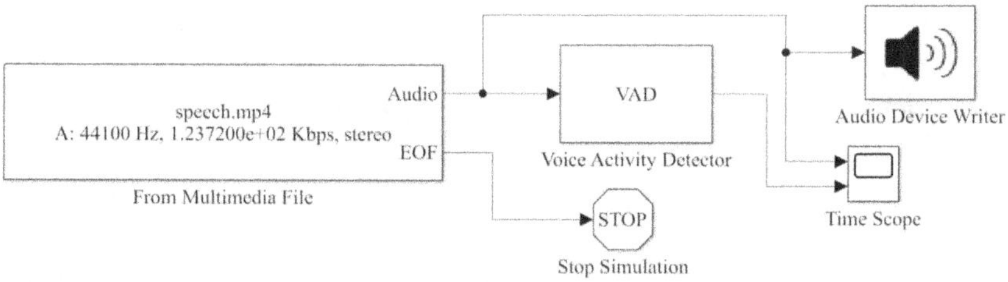

Figure 9.27: Voice activity detection model (file: `c9maud5.slx`).

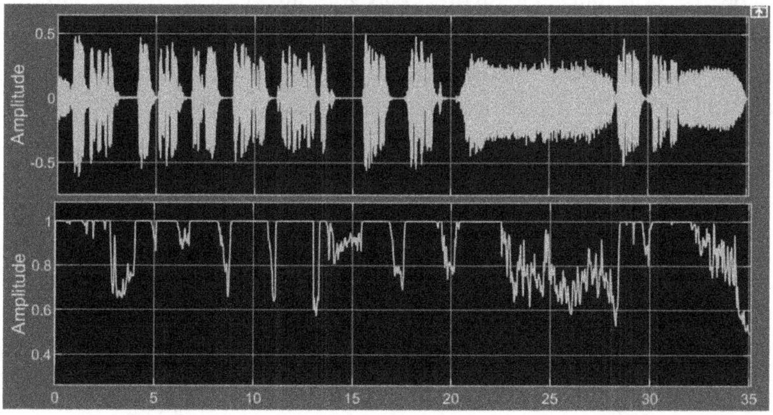

Figure 9.28: Sound waveform and activity detection results.

It should be pointed out that since there is the existence of the Audio Device Writer block, the model can be executed in real time. If the block is removed, the whole Simulink model can be executed faster.

9.3 Video information processing and Computer Vision Toolbox

In signal processing, the audio signal can be regarded as a one-dimensional signal processing, while image signals can be regarded as two-dimensional signals. An extra time axis is introduced in video signals, which can be regarded as three-dimensional signal processing. Two- and three-dimensional signal processing can be manipulated directly with the tools provided in MATLAB and Simulink.

A great amount of image and video processing blocks are provided in the Computer Vision Toolbox. Two categories of image processing methods are supported in MATLAB: one is command-line driven methods, and the other is the block diagram based methods. These two categories of methods have their own advantages.

Command-line driven methods are usually carried out with MATLAB functions provided in the Image Processing Toolbox, or by other third-party toolboxes. An image variable is processed by a MATLAB function, and the result is returned to the caller. These types of methods are easy to implement. It is also easy for the users to write certain MATLAB functions for image processing, and compare their own results with the existing methods. These methods are suitable to process a single image file, and not easy to process video files. To process a video, the whole image file must be loaded first into MATLAB as a variable. Loop structure can then be used to process the video files, frame by frame. The processing methods like this are rather demanding on hardware resources. For large-scale video files, these methods are difficult to implement.

If block diagram based methods are used, the blocks in the Computer Vision Toolbox blockset can be used. Double click the **Blocksets & Toolboxes** icon in the Simulink block library, then in the list provided, double click the **Computer Vision Toolbox** icon, or type in MATLAB command window the command `visionlib`, and the blockset is opened, as shown in Figure 9.29. There are **Sources, Analysis & Enhancement, Sinks, Text & Graphics, Conversions, Transforms, Morphological Operations** and **Filtering** groups in the blockset. A **Demos** group with many practical demonstration models is also provided.

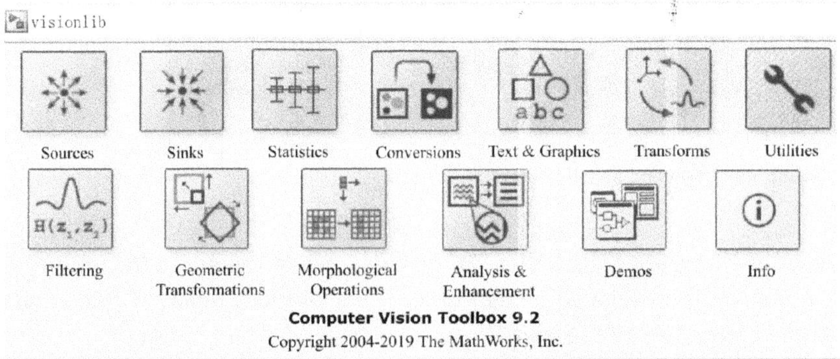

Figure 9.29: Computer Vision Toolbox blockset.

Similar to the Audio System Toolbox, the images or videos can be fed into Simulink as a stream media. The processed results can be returned with output blocks. The image and video processing framework is rather clear. It can be used to process the whole video file, but it is not necessary to load everything into MATLAB workspace first. The processing of stream media looks like real-time video processing, and the processing results can be obtained directly.

In the subsequent presentations in this chapter, computer implementation of image and video processing are introduced, with concentrations on using Computer Vision Toolbox blockset. With the correct tools, the images and videos can be processed easily.

9.4 Input and output of images and videos

It can be seen from the groups provided in the Computer Vision Toolbox that the **Sources** and **Sinks** groups are provided, and the input and output of images and videos can be handled with the blocks. Besides, other toolboxes can be used to output images or videos. In this section, examples are given to demonstrate the input and output of images and videos.

9.4.1 Reading of images and videos

Various image and video input functions and blocks are provided in MATLAB. The input blocks in the Sources group are shown in Figure 9.30. The input methods are summarized below:

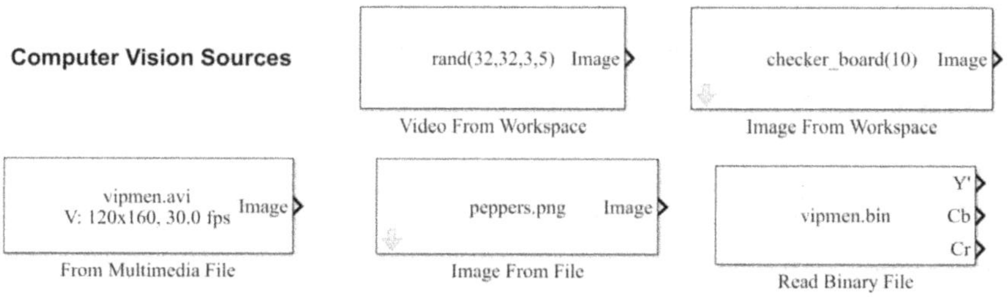

Figure 9.30: Image and video input blocks.

(1) **Image file input.** Image from file block can be used to feed image files into the Simulink model. The currently supported image formats are `bmp`, `jpg`, `jpeg`, `png`, `tif`, `tiff` and so on. Also, the functions in the Image Processing Toolbox can be used to load images into MATLAB workspace. The input of image file can be implemented with the `imread()` function.

(2) **Multimedia file input.** Audio and video files are all supported by the From Multimedia File block. The currently supported file types include `avi`, `mp4`, `wmv` and so on. Audio file formats are `wav`, `wma`, `mp3` and so on. The video and audio files can be loaded to Simulink model for further processing. With MATLAB function `mmreader()`, the multimedia file can be loaded into the MATLAB workspace. Note that the entire multimedia file can be loaded with a command. For large-scale multimedia file, a significant amount of memory is needed, while in the block diagram format, the video is loaded frame by frame.

(3) **Real-time input from camera.** Apart from the blocks in the group, the Image Acquisition Toolbox is also provided in MATLAB, where there is one Simulink block in the toolbox, used to acquire real-time videos from the camera. With such a block, the

real-time videos can be handled in the same way as in reading video files. Therefore, real-time video processing system can be constructed.

9.4.2 Display and output of images and videos

Double click the Sinks group icon, and the blocks in the groups are displayed as shown in Figure 9.31. Various output blocks are provided. For instance, To Multimedia File block can be used to save video signal into a multimedia file. Block Video To Workspace can be used to write video information to MATLAB workspace. Video Viewer or To Video Display blocks can be used to playback video signals. Besides, To Binary File block can be used to save videos in a binary file.

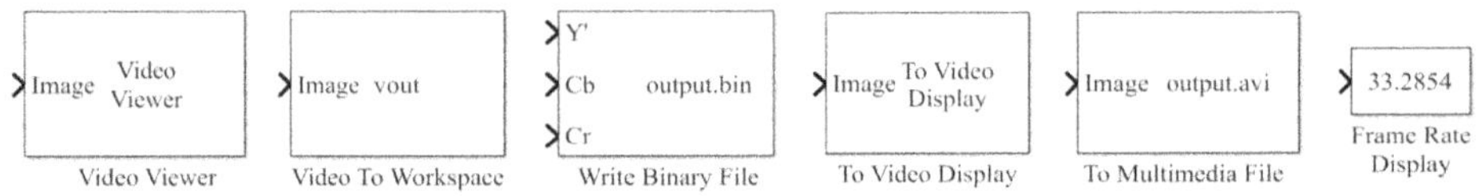

Figure 9.31: Sinks group.

Example 9.13. Establish an image file display system with Simulink.

Solutions. The Image From File block in the Source group in Computer Vision Toolbox can be used to extract image file, and connect it to the Video Viewer block in Sinks group directly, as shown in Figure 9.32. It can be used to display an image in an image file.

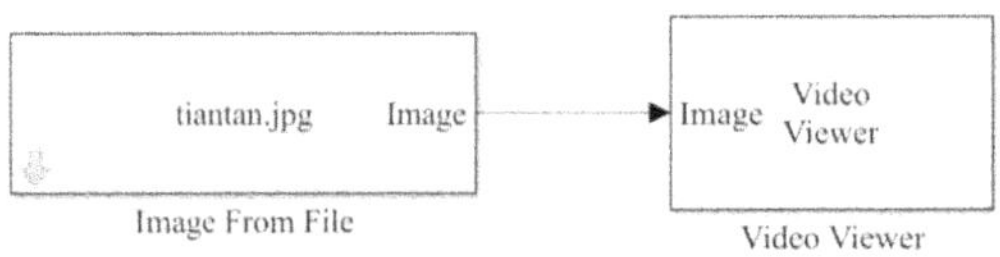

Figure 9.32: Simple image display system (file: `c9mvis0.slx`).

Example 9.14. Establish a simple video file player with Simulink.

Solutions. Refer to the previous example, the From Multimedia File block in the Sources group can be used to read in the multimedia signal. The signal can be connected directly to the Video Viewer block in the Sinks group. The Simulink model in Figure 9.33 can be established. Double click the From Multimedia File block icon, and the multimedia file name can be selected from the parameter dialog box, such that the user can select any multimedia file from the dialog box. For instance, the `Wildlife.wmv` file provided in Microsoft Windows can be selected.

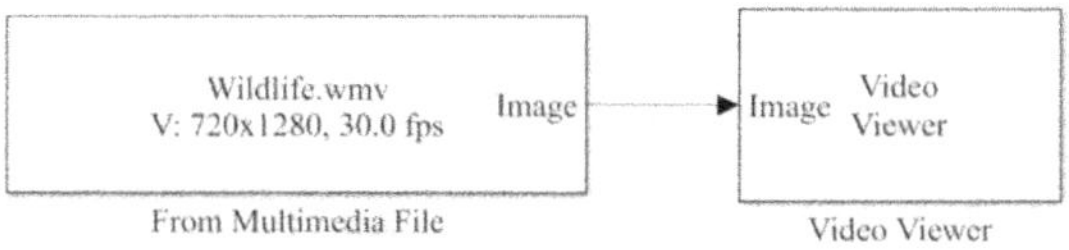

Figure 9.33: A simple video playback system (file: `c9mvis1.slx`).

In fact, in the image display system in Figure 9.32, the From Multimedia File block can also be used to provide the image file. With the Computer Vision Toolbox, the multimedia file play is as simple as the display of an image file.

Example 9.15. The Simulink model constructed in the previous example contains at least two questions. The first one is that the stop time should be defined manually rather than determined by the actual length of the video. The second one is that there is no sound played. Improve the multimedia player.

Solutions. Refer to the model in Example 9.12, and it can be seen that the video is provided with the From Multimedia File block. Double the block, at the bottom of the parameter dialog box shown in Figure 9.26, and the option Video and audio can be selected. An extra Audio output port can be added to the block automatically, which can be used to drive the Audio Device Writer block, to play the sound. This is the solution of the second question. The first question is easy to solve. Check the Output end-of-file indicator checkbox, and use the new output port to derive the Stop Simulation block, and the Simulink model can be established, as shown in Figure 9.34.

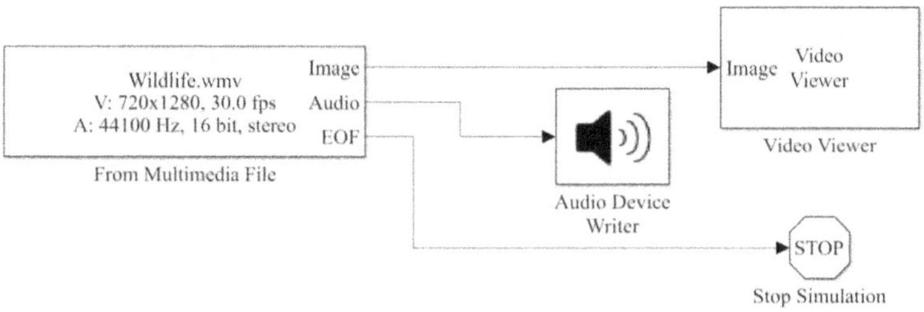

Figure 9.34: Modified movie playback system (file: `c9mvis2.slx`).

9.4.3 Video play with commands

Function `imtool()` provided in MATLAB Image Processing Toolbox can be used to display image files. It can also be used to display other information, and zooming the processing of the images.

Example 9.16. The `tiantan.jpg` file or any other file in MATLAB search path can be loaded into MATLAB workspace with `imread()` function. Display the image.

Solutions. With the following commands, the image file can be loaded into MATLAB workspace. The image can be displayed with the function `imtool()`, as shown in Figure 9.35(a).

```
>> W=imread('tiantan.jpg'); imtool(W)
```

It can be seen that the display window has its own toolbar, and can be used to simply process the image. For instance, click the button, the tool bar is displayed in Figure 9.35(b). In the currently selected zone, the color values of the selected pixels can be directly displayed. The user may move the display box to update the color values of the pixels. With such a tool, some information can be found. For instance, based on the color value changes, such as in the edges of the image, useful information can be provided.

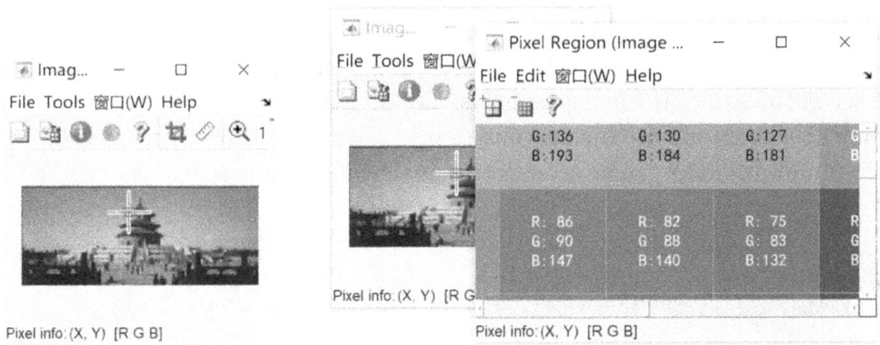

(a) image display (b) pixel display tool

Figure 9.35: Image display window.

It is seen from the above example that the command-line driven method can be used to display an image easily. It may be difficult to display video files with command-line driven methods. The MATLAB function `VideoReader()`, or `mmreader()` in the earlier versions, can be used to load the entire movie file into MATLAB workspace. For large-scale video files, this method requires a significant amount of memory space. After the file is loaded in MATLAB workspace, the structured variable can be created. The members of the structured variable include `Name` (for file name), `Path`, `Duration`, `NumFrames`, `Width`, `Height`, `FrameRate`, `BitsPerPixel` and `VideoFormat`. Loop structure can be used to play the video file frame by frame. The specific play method will be demonstrated with examples.

Example 9.17. Play the `Wildlife.wmv` movie with command-line driven method.

Solutions. The movie playback commands are given next. The movie file is opened first, and assigned to the structured variable p. Select the current axis, and assign the handle to the `currAxes` variable. Some members in p can be displayed such as frame rate 29.97 fps, and number of frames is 901, and duration is 30.093 seconds. The main body of the program is implemented in a `while` loop structure. In the `while` loop, each time `hasFrame()` function is called, and judge whether there are frames remaining. If there are, the `readFrame()` function can be used to extract one frame, then call `image()` function to display it, and set the scales of the axis invisible. When an image is displayed, it is maintained for a period time (the reciprocal of frame rate). In this way, the video can be displayed directly:

```
>> p=VideoReader('Wildlife.wmv');           % read movie file
   currAxes=axes;                           % set current axis
   p.FrameRate, p.NumFrames, p.Duration % display certain information
   while hasFrame(p)   % process each frame with loop
      h=readFrame(p); % extract an image
      image(h,'Parent',currAxes); currAxes.Visible = 'off';
      pause(1/p.FrameRate);   % Each frame is paused for a certain time
   end
```

Compared with the video playback system in Simulink discussed earlier, the use of command-line driven methods are far more complicated, since too many low-level details are to be considered. Besides, the sound in the video file cannot be played in this format. Therefore, it is not recommended to use low-level command-line driven methods to process multimedia files. The block diagram format in Simulink is recommended.

9.5 Simple transformation of images

9.5.1 Color space representation of images

Images are represented in pixels. The image can be divided into mesh grid form in n rows and m columns. Each such grid is referred to as a pixel. Each pixel is assigned to a color value. In this way, the image can be represented directly. The format defining color values are referred to as color space of the image. The density of the mesh grids defines the resolution of the image. In television sets, 2K resolution means that the mesh grid of $1920 \times 1024 = 1966080$ pixels, while 4K resolution is composed of $3840 \times 2048 = 7864320$ pixels. The commonly used color spaces are as follows:

(1) **Binary image.** Binary image are also known as black and white images. A binary image can be expressed by an $n \times m$ matrix, whose elements are 0's and 1's, where 1

for white, and 0 for black. The data type of a binary image is a logic matrix, with each element represented by a binary number. If a binary image is represented by matrix $\boldsymbol{W}$, then its inverse color is obtained with ~$\boldsymbol{W}$ command.

(2) Grayscale image. Grayscale images can be represented by an $n \times m$ matrix, and the data type of the matrix element is `uint8`, that is, 8 bit unsigned integers, ranging 0~255, Therefore, each grayscale value may has 256 values. In very special cases, `uint16` data type can be used for grayscale images.

(3) RGB color image. RGB color space is the most natural and easiest color space for color images. For each pixel, the actual color added up by three primary color, on red, green and blue, to describe the color. In MATLAB, a three-dimensional array can be used to describe an RGB image. A three-dimensional array is in fact piled up by three $n \times m$ matrices, with the first lay for red components, and the second and third layers for green and red components, respectively. The data type of the three-dimensional array is `uint8`, ranging 0×255. Eight commonly used RGB colors are defined in Table 9.1.

Table 9.1: Commonly used RGB components.

Components	black	blue	green	cyan	red	magenta	yellow	white
red	0	0	0	0	255	255	255	255
green	0	0	255	255	0	0	255	255
blue	0	255	0	255	0	255	0	255

(4) CMY color image. CMY are respectively cyan, magenta and yellow as primary colors. It is always used in painting and color printing. Three-dimensional array $n \times m \times 3$ of `uint8` data type can also be used to describe a CMY color image.

(5) HSV color image. HSV, namely hue, saturation and value, is another color space used to describe the direct features of color images. The HSV color space is described by $n \times m \times 3$ array, with data type of `double`. The value of each component is 0~1.

Besides, there are also commonly used color spaces as YUV, USTC, YCbCr, pseudocolor and so on. They are not covered in this book. The interested readers may refer [39] for more information.

9.5.2 Color space conversion

The blocks in the Conversions group is shown in Figure 9.36. In the group, the Color Space Conversion block can be used to convert the image or video from one color space into another one. Apart from color space conversion, other conversions are allowed, such as Chroma Resampling block, Autothreshold block and so on. The Image Complement works for binary or grayscale images. If the original image is a binary one, the action is similar to ~$\boldsymbol{W}$, while if the image is a grayscale one, it is equivalent to 255−$\boldsymbol{W}$. Block Image Data Type Conversion can be used to convert data type of an image signal from one into another. The Gamma Correction block is used to implement Gamma correction, and it will be demonstrated later.

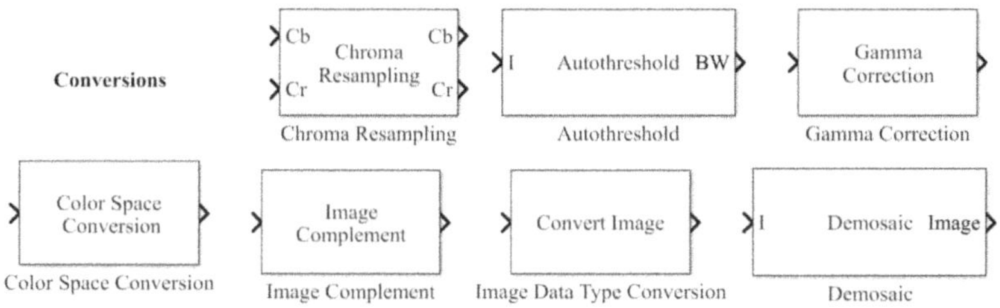

Figure 9.36: Image conversion group.

It can be seen from the blocks that some ports are labeled I, indicating the intensity (or brightness) of grayscale image. If the port is labeled BW, it means that black and white image is connected. Some are labeled Cb, corresponding to a YCbCr color space. In block connection, the input or output signals must be the same as the marked text in Simulink modeling. It can be seen from the given port that the Autothreshold block requires that the input is grayscale signal, while its output is binary signal.

Some conversion functions are available in Image Processing Toolbox. For instance, `rgb2gray()` can be used to change RGB color image into grayscale ones, while `rgb2hsv()` convert RGB into HSV ones. An example is given next to demonstrate color space conversion.

Example 9.18. Convert the video file `Wildlife.wmv` into grayscale and binary ones, and play the converted video.

Solutions. The multimedia playback system in Example 9.15 can be used to study color space. If the expected color video is converted to a grayscale video, in between the From Multimedia File and Video Viewer blocks, a Color Space Conversion block can be inserted, such that the video can be converted into grayscale videos. The Simulink model thus created is shown in Figure 9.37.

Note that the default setting in Color Space Conversion block is R'G'B' to Y'CbCr, which is not expected in this example. Double click the block, and the R'G'B' to intensity item in the listbox in Conversion block can be selected, as shown in Figure 9.38. It can be seen that the conversion system can be used to play grayscale videos.

If the obtained intensity video signal is fed into the Autothreshold block, then the result is connected to the video player block, the video processing system is con-

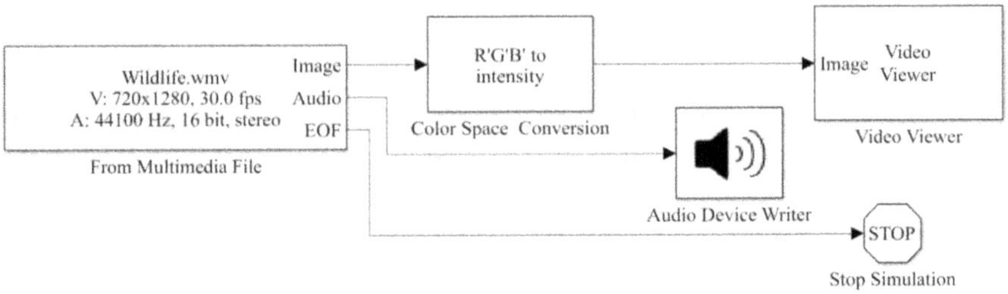

Figure 9.37: Grayscale video play system (file: `c9mvis3.slx`).

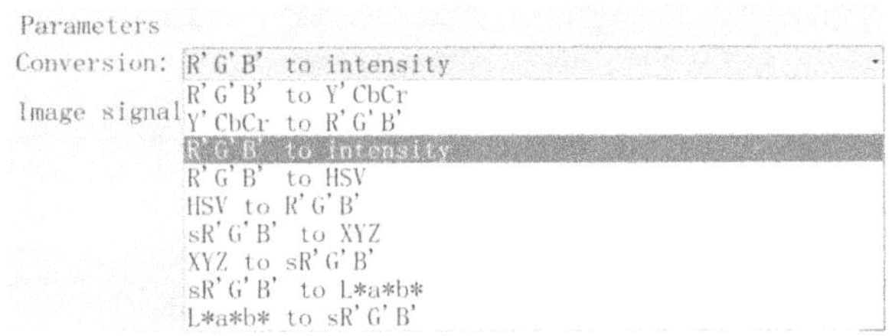

Figure 9.38: Color space dialog box.

structed in Figure 9.39. The new system can play simultaneously the grayscale and black and white videos.

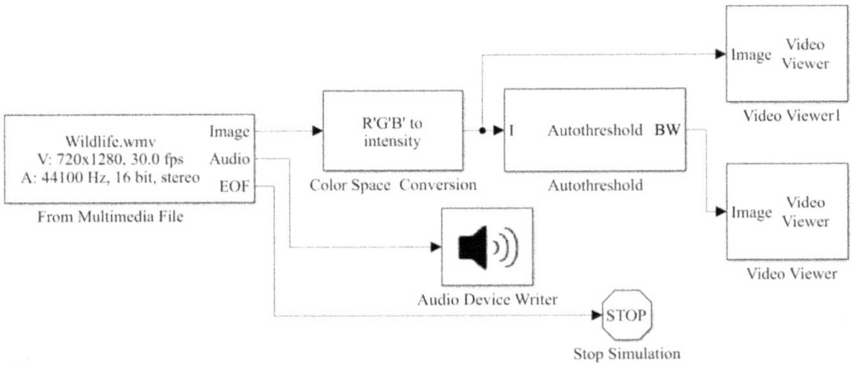

Figure 9.39: The new video players (file: `c9mvis3a.slx`).

For an image file, the functions `rgb2gray()` and `rgb2hsv()` can be used in color space conversion. Function `im2bw()` can be used to convert color images into black and white ones, with fixed threshold. The functions such as `imadjust()` can be used in image correction, including Gamma correction. These functions can only be used to process images, not videos. For video processing, the methods in Example 9.17 can be used, such that loop structures can be used to process the videos frame by frame.

9.5.3 Geometric transformation of images

The Geometric Transformation group is provided in the Computer Vision Toolbox, as shown in Figure 9.40. It can be seen that image rotation, translation and scale transformation can be made with the low-level blocks.

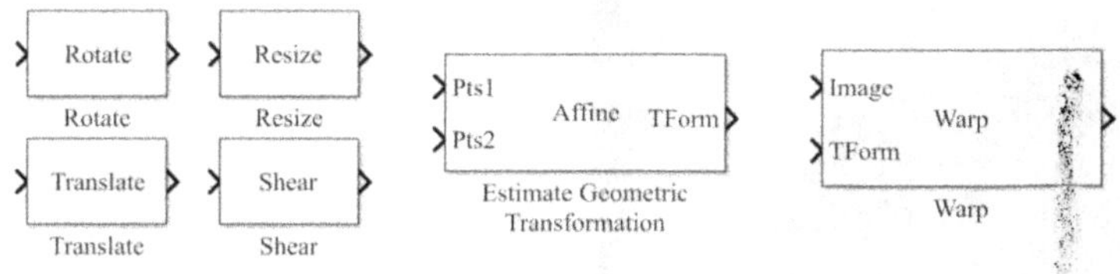

Figure 9.40: Geometric transformation group.

9.5.4 Gamma correction of images

Image contrast is a very important concept in digital image processing. The concept of contrast is introduced based on grayscale images, It concentrates on the description of the difference between the maximum and minimum values in a grayscale image. The larger the difference, the more variations in the grayscale from white to black. Normally, more detailed information can be revealed in the image. Nonlinear transformation can be carried out on the grayscale values in each pixel such that the image contrast can be increased. The commonly used transformations include Gamma correction, logarithmic transformation, exponential transformation and so on. The concept of Gamma correction is introduced first, and an example is given to illustrate Gamma correction results.

Definition 9.1. Power-law operations can be made to the grayscale value of each pixel in Gamma correction.

$$J(i,j) = cI^{\gamma}(i,j), \tag{9.5.1}$$

where $I(i,j)$ is the grayscale value of the (i,j)th pixel before transformation, and $J(i,j)$ is the value after the transformation. The variables c and γ are constants. Normally, when $\gamma > 1$, the contrast in the smaller values in the pixels get contrast increased, and contrast are reduced in the pixels with larger grayscale values. If $\gamma < 1$, the contrary actions are made. Normally, $c = 1$.

Example 9.19. Consider the picture shown in Figure 9.41, from [26]. It is a picture of the Mars surface. For better typesetting, the picture used here is rotated. Since in the lower-left corner is rather dark, details cannot be observed from the picture. Image processing is needed for the image. Use Gamma correction to increase the contrast and see whether more details in the picture can be revealed.

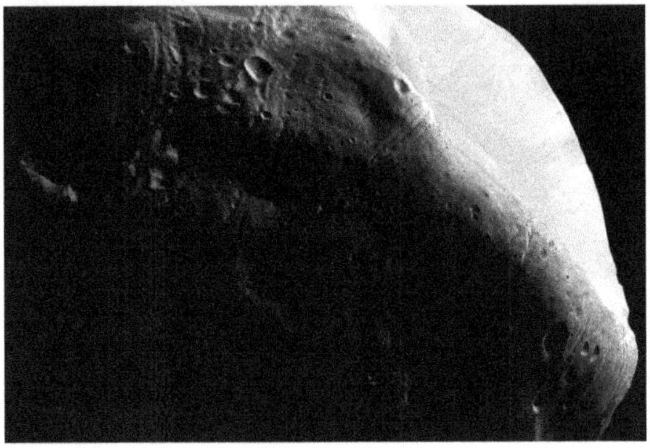

Figure 9.41: Original image.

Solutions. Since in the grayscale images, the grayscale value 0 for black and 255 for white. A relatively larger value of γ can be chosen, such the contrast in the dark regions can be increased. The Simulink model shown in Figure 9.42 can be established, for Gamma correction. The users may try different values of γ. For instance, $\gamma = 5$, such that the corrected picture is shown in Figure 9.43. It can be seen that the lower-left corner is processed very well and some of the detailed information is revealed in the picture. Unfortunately, the details at the upper-right corner are lost.

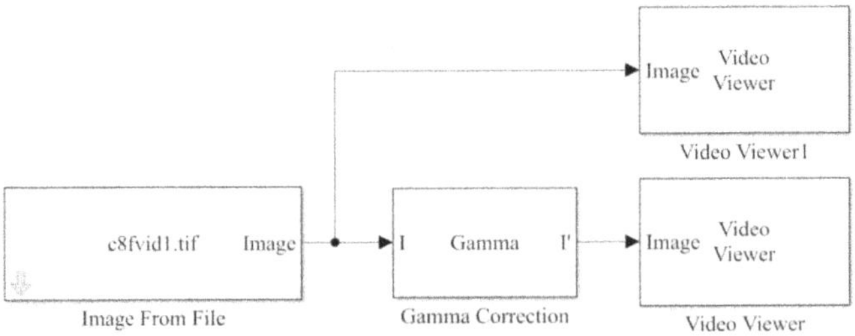

Figure 9.42: Gamma correction model (file: `c9mvis3b.slx`).

It should be noted that the **Gamma Correction** block only supports the correction for $\gamma > 1$.

The correction method used here correct the entire image in a uniform way, such that some parts the quality of the image is improved, but other part may be sacrificed.

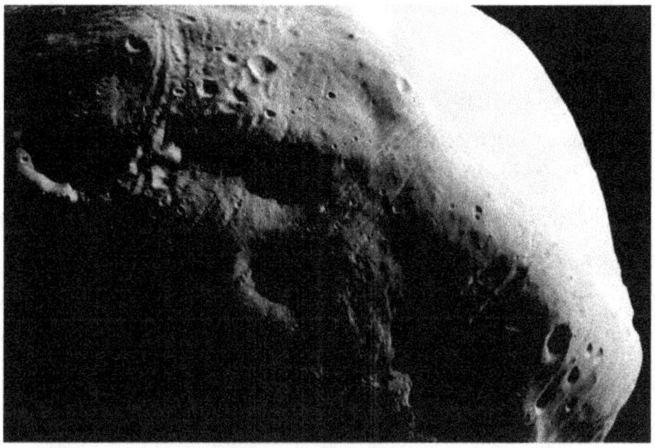

Figure 9.43: Gamma correction result for $\gamma = 5$.

Therefore, this correction method is not recommended. Later more meaningful correction and enhancement methods are discussed.

9.5.5 Frequency domain transform and filters

Many image transformation blocks are provided in the Transform group in the MATLAB Computer Vision Toolbox, as shown in Figure 9.44, including 2D FFT, 2D-IFFT, 2D-DCT, 2D-IDCT and various Hough transform blocks. Filters are used mainly to process images with frequency response methods. Filters can be used to remove noises in the images, while Hough transform is used to identify lines in black and white images.

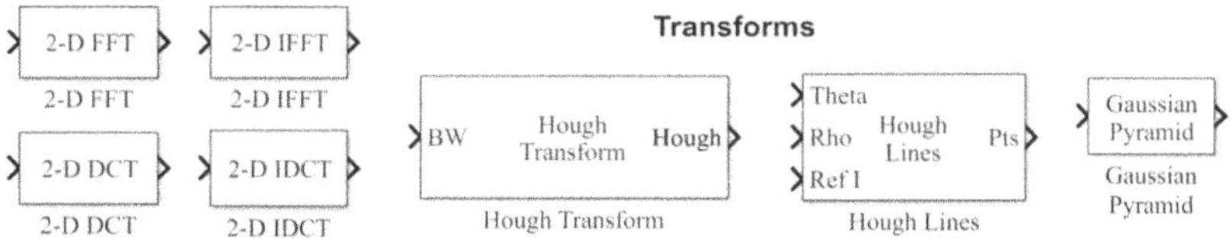

Figure 9.44: Image transform group.

Some functions are provided in the Image Processing Toolbox of MATLAB, such as `fft2()`, `ifft2()`, `dct2()` and `idct2()`, mainly using fast Fourier transforms. Hough transform and line extraction can be implemented with `hough()` or `houghlines()` functions.

9.6 Image Analysis and Enhancement

Many blocks are provided in the Analysis & Enhancement group in Computer Vision Toolbox. Double click the group, the group is opened in Figure 9.45, including commonly used Histogram Equalization, Template Matching, Edge Detection, Median Filter, Corner Detection blocks and so on. These blocks can be used to directly handle the images or video signals. These blocks are essential in the area of digital image processing. Besides, in each block, different parameters can be set, and different algorithms can be chosen. Some of the facilities can be implemented with the functions in the Image Processing Toolbox, such as `edge()`, `histeq()`, `medfilt2()`, `imadjust()` and so on.

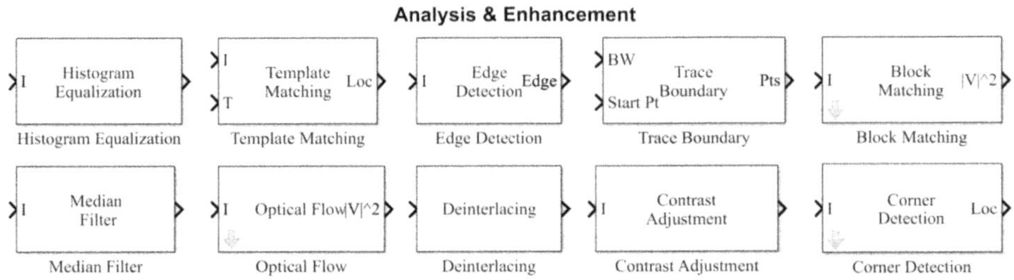

Figure 9.45: Enhancement and analysis blocks.

The blocks in the Text & Graphics group is shown in Figure 9.46, where Draw Shapes, Insert Text, Draw Marker blocks allow the user to superimpose objects on the image, while the Compositing block allows the an image embedded into the other one.

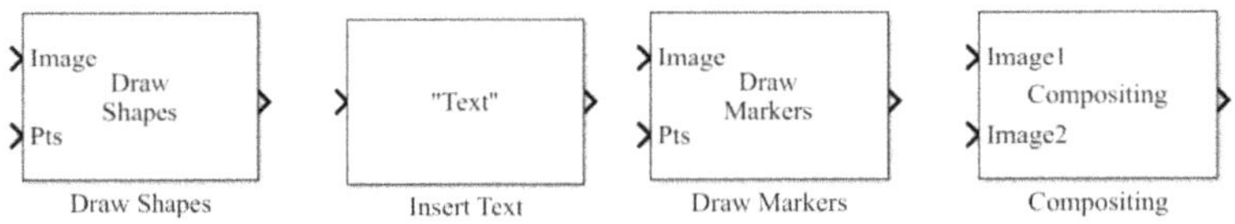

Figure 9.46: Text and graphics group.

9.6.1 Histogram equalization

Image histogram is an important criterion to describe the grayscale value distributions in an image. Since the grayscale values of pixels are in the interval 0~255, all of the pixel values in the image can be extracted, and thrown into 255 bins. For instance, if the value of the (i,j)th pixel is k, then the k counter is added by 1. In this method, the numbers of counts in each bin is found, and represented in histogram form. If the grayscale matrix of the image is $\boldsymbol{W}$, the command `imhist(W)` can be used to draw directly the histogram of the image.

In photography and movie making, the phenomena of overexposure or underexposure are often encountered. The problem is caused and the histograms of the images are not well distributed. The histogram equalization method can be used to correct the images. Even though some of the problems are fairly distributed, the histogram equalization method may still be used to show certain details.

Example 9.20. Consider again the picture in Example 9.19. Use the histogram equalization method to analyze the image histogram, and use the histogram equalization method to process the image, and observe processing results.

Solutions. The image histogram can be drawn with the `imhist()` function in the Image Processing Toolbox, as shown in Figure 9.47. It can be seen that most of the pixel values are clustered in the 0~30 interval, which makes the image very dark, such that it is hard to observe the details in the lower-left corner. Histogram equalization processing is needed for the picture.

```
>> W=imread('c8fvid1.tif'); imhist(W)
```

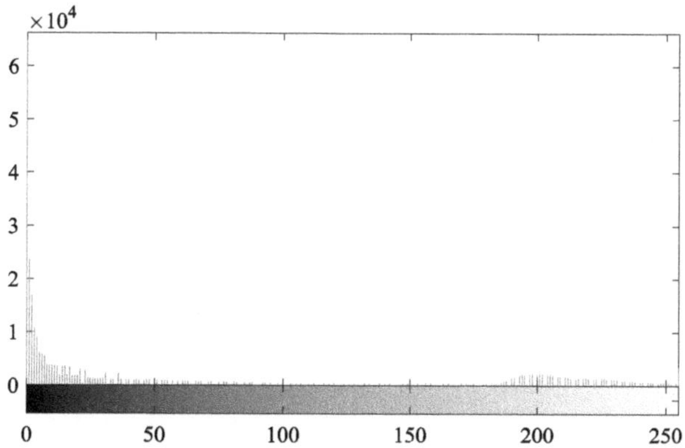

Figure 9.47: Histogram of the original image.

If Computer Vision Toolbox is used, the Simulink model shown in Figure 9.48 can be created. In the model, the image file can be used as an image signal source followed directly by the histogram equalization block, and then connect to an image display block. The image after such processing is obtained as shown in Figure 9.49. It can be seen that after processing, more details can be revealed. The parameters in the Histogram Equalization block can also be adjusted so as to get better results.

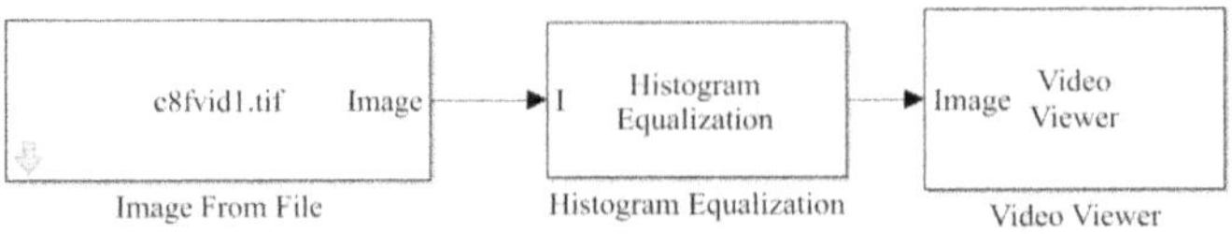

Figure 9.48: Histogram equalization system (file: `c9mvis4.slx`).

Figure 9.49: Image after processing.

The same result can be found with the `histeq()` function in the Image Processing Toolbox. The histogram of the new image is shown in Figure 9.50. It can be seen that the histogram of the new image is extended to a wider interval, while the colors are relatively shallow, and some details cannot be displayed satisfactorily.

```
>> W1=histeq(W); imhist(W1)
```

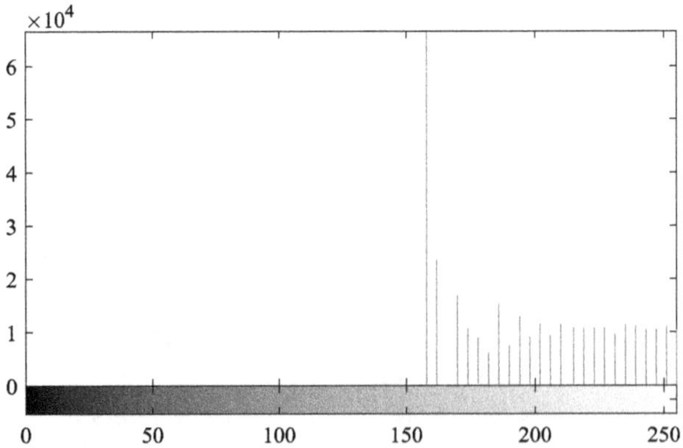

Figure 9.50: Histogram of the new image.

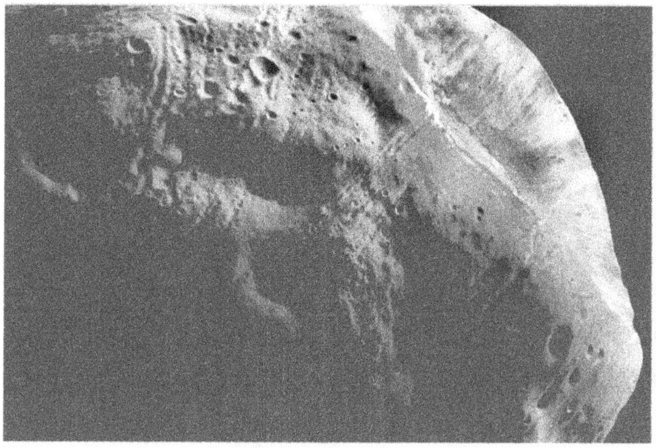

Figure 9.51: Result with adaptive histogram equalization method.

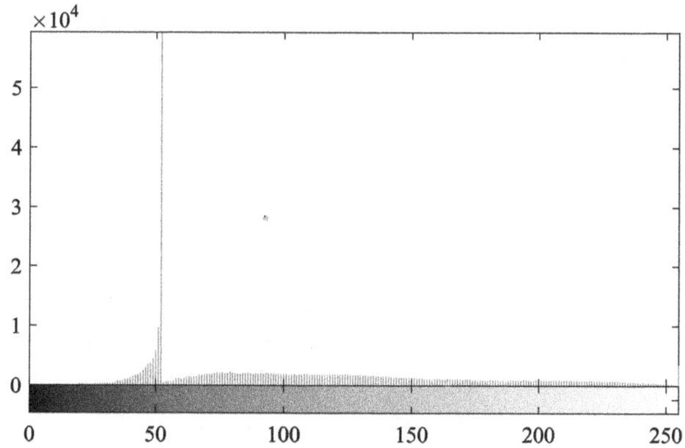

Figure 9.52: Histogram after processing.

An adaptive histogram equalization function `adapthisteq()` can also be used to process the original image, and the equalized result is shown in Figure 9.51. The processing result is much superior to the Gamma correction one in Figure 9.43, and the histogram equalization one in Figure 9.52. It can be seen from the results that the original features of the image are retained, and more details are better expressed. The histogram of the new image is extended to a wider interval, and the quality of the image is significantly improved. The hidden details in the lower-left corner can be displayed. Therefore, the adaptive histogram equalization method can usually exhibit better results:

```
>> W2=adapthisteq(W,'Range','original', ...
                 'Distribution','uniform','clipLimit',0.2);
   imtool(W2); figure; imhist(W2)
```

It can be seen from the result that clipLimit is an important adjustable parameter. Selecting the value 0.2 means that the new histogram is distributed uniformly in the $(0.2, 1)$ interval. Such a parameter can be tuned so as to get good results. It can be seen that the new histogram indeed has a good distribution in the $(0.2, 1)$ interval, that is, the $(51, 255)$ in uint8 data type.

It is a pity that there is no such corresponding Simulink blocks, and it cannot be used to handle videos. The user may consider writing such a Simulink block, or use the function in a loop structure, such that video information can be processed frame by frame. This method is rather complicated.

9.6.2 Edge detection

Edge detection is a very important topic in digital image processing and computer vision. The target of edge detection is to extract all the points such that there are sudden changed on the brightness of the pixels in an image. An edge detection block is provided in the Computer Vision Toolbox in MATLAB to detect edges from video signals. Function edge() is also provided in the Image Processing Toolbox to extract the edge from a grayscale image. The commonly used edge detection algorithms are all for grayscale images. If a color image is to be processed, normally, it should be converted into grayscale images first to extract the brightness information. The information can be regarded as a grayscale image so that edge detection algorithms can be used to extract the edges of an image. In this section, examples are used to detect edges of a video file, and then the composition of videos is introduced.

Example 9.21. Consider the image file tiantan.jpg. Use different operators to detect edges in the image, and compare the effect of the edges.

Solutions. The image file can be loaded into MATLAB workspace first, then the brightness information can be extracted from its HSV color space expression. Take edge detection to the converted image, the edge detection results can be obtained, as shown in Figure 9.53. It can be seen that different operators may yield different results. Suitable operators can be selected for different images, so as to extract useful edge information.

```
>> W=imread('tiantan.jpg');          % real in the image file
   W1=rgb2hsv(W); W2=W1(:,:,3); % extract the intensity components
   W11=edge(W2,'sobel'); W12=edge(W2,'canny');
   W21=edge(W2,'prewitt'); W22=edge(W2,'roberts');
   subplot(221), imshow(~W11), set(gca,'Visible','off');
   subplot(222), imshow(~W12), set(gca,'Visible','off');
   subplot(223), imshow(~W21), set(gca,'Visible','off');
   subplot(224), imshow(~W22), set(gca,'Visible','off');
```

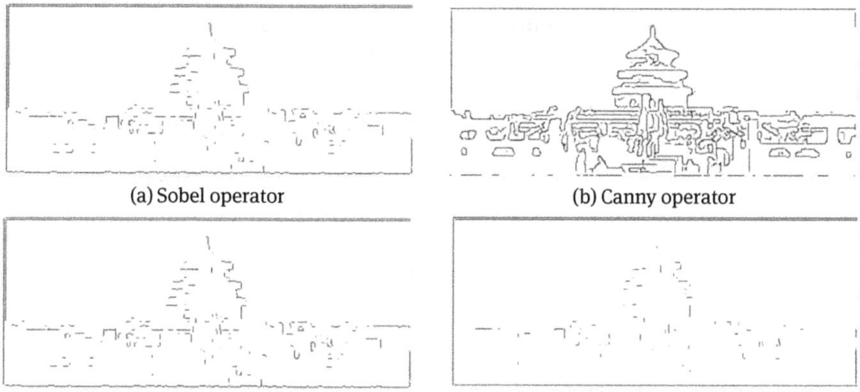

(a) Sobel operator (b) Canny operator

(c) Prewitt operator (d) Roberts operator

Figure 9.53: Edges extracted by different operators.

Example 9.22. Consider the video file `Wildlife.wmv`. Extract the edges for the video file with a Simulink model.

Solutions. The method discussed earlier applies only to individual images, and cannot be used to handle videos directly. If videos are to be processed, a Simulink model is recommended. The video file can be loaded into Simulink first. Since the video is a colored one, color space conversion can be made, and the converted video signal can be fed into the Edge Detection block, such that a Simulink model in Figure 9.54 can be constructed. Executing the model, the result is the edge detection results. Note that the parameters and operators can be assigned in the parameter dialog box of the Edge Detection block.

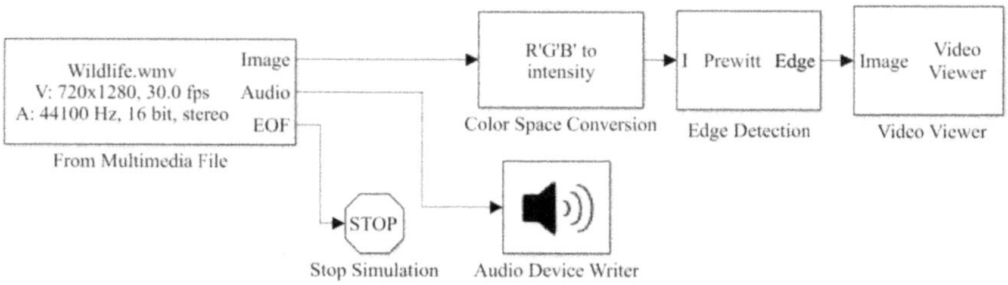

Figure 9.54: Edge detection system (file: `c9mvis6.slx`).

Example 9.23. The edge detected is a binary video, while the original video is in color. Use the two videos to make a composite one, such that the edges detected are highlighted on the color video.

Solutions. The two video signals can be fed into the Compositing block. The Simulink model in Figure 9.55 can be established. Double click the Compositing block, and the parameter dialog box is shown in Figure 9.56. In the Operations listbox, the Highlight selected pixels option can be selected, such that the edges are highlighted, so that a new video signal is constructed.

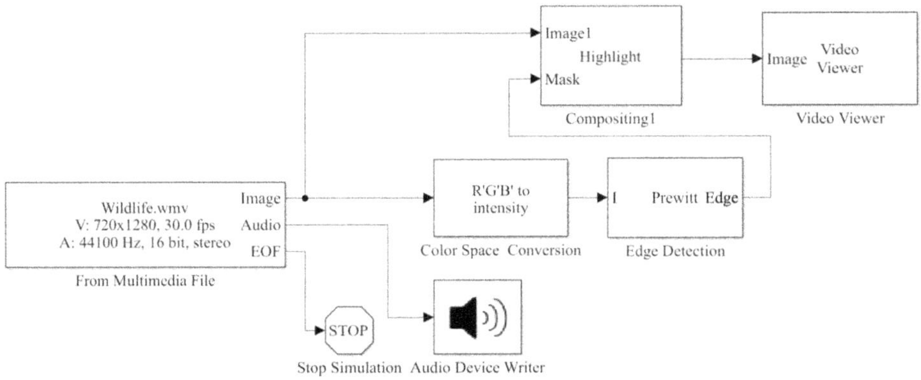

Figure 9.55: Video composite system (file: `c9mvis7.slx`).

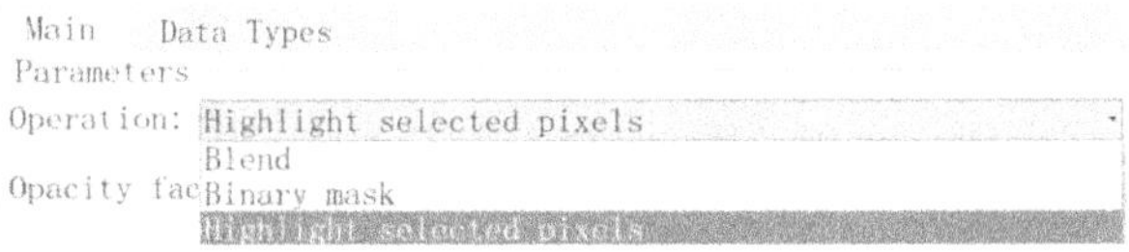

Figure 9.56: Parameter dialog box of the Compositing block.

9.6.3 Morphological processing

Mathematical morphology is widely used in digital image processing. Four operators are defined in morphology: the erosion, dilation, opening and closing operators. A series of practical morphological algorithms for digital signal processing can be derived from the four operators. The application examples are image segmentation, skeleton extraction and image enhancement.

A Morphological Operations group is provided in the Computer Vision Toolbox. Double click the group icon, the group is opened, as shown in Figure 9.57. The four operators in morphology are provided in the group.

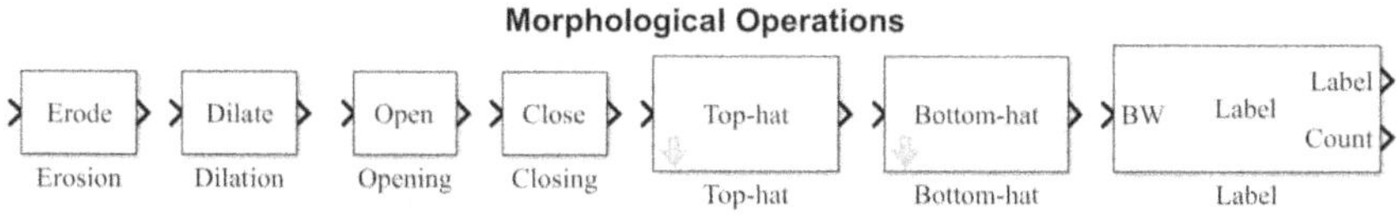

Figure 9.57: Morphological group.

If the image is not quite clear, certain geometric structure can be used to dilate the image, and then the erosion method can be used to extract the skeleton of the image.

For binary images, functions such as `imdilate()` and `imerode()` can be used to carry out morphological operations. Color images may also be processed using corresponding methods. Morphological methods for image processing are demonstrated next through examples.

Example 9.24. Consider the image shown in Figure 9.58(a),[26] which is not very clear. Use the dilation method to process the image and observe the results.

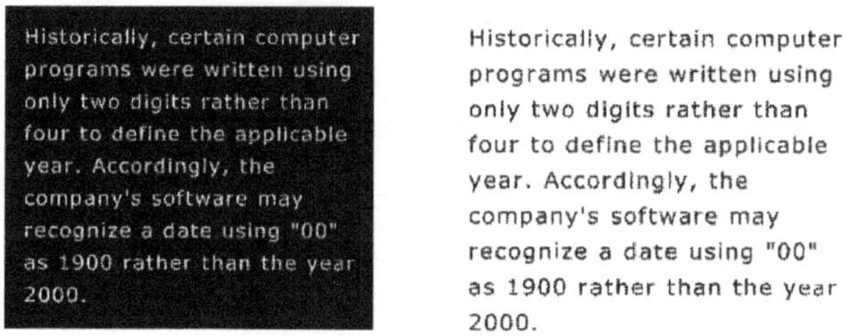

(a) original image (b) color inverted

Figure 9.58: Original image and its inverse one.

Solutions. The original image file is given in `c8fvid2.tif`, which has white characters on the black background. The color can be inverted, such that the image in Figure 9.58(b) can be found. The Simulink model with dilation method is shown in Figure 9.59(a), where the **Neighborhood or structuring elements** matrix is set to `[0,1,0; 1,1,1; 0,1,0]`. The dilation result is shown in Figure 9.59(b). It can be seen that with dilation manipulation, the image gets clearer. Different transformation matrices can be tried, and observe the dilation results.

The image dilation and erosion can also be processed with the `imdilate()` and `imerode()` functions provided in the Image Processing Toolbox. If the dilation matrix is selected as $\boldsymbol{A} = [0,1,0;1,1,1;0,1,0]$, the following commands can be used to get the dilation result, which is the same as the one shown in Figure 9.59(b). It can be seen that the inverted image can be dilated.

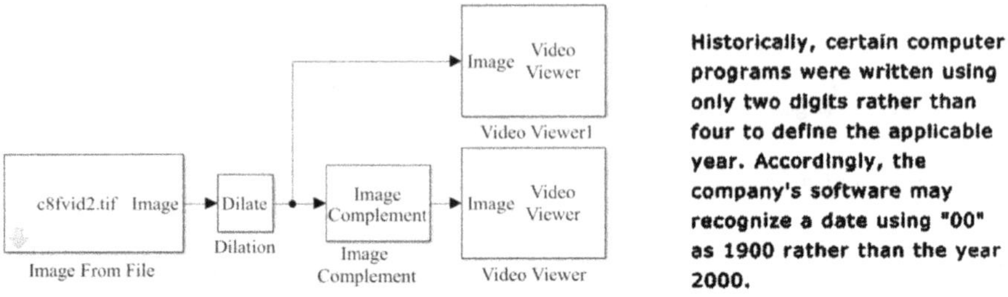

(a) image processing system (file: `c9mvis5.slx`) (b) the dilation results

Figure 9.59: Dilation model and result.

```
>> W=imread('c8fvid2.tif'); imtool(~W); figure;
   A=[0 1 0; 1 1 1; 0 1 0]; W1=imdilate(W,A); imtool(~W1)
```

As in those functions discussed earlier, these functions can only be used to handle image files. Loop structures can be used to process videos in frame by frame format, but it is much more complicated than Simulink models.

The morphological function `bwmorph()` is also provided in the Image Processing Toolbox, which can be used to extract outlines and skeletons in images. An example is given next to demonstrate these tasks and effect.

Example 9.25. Consider the binary image shown in Figure 9.60(a). The image is provided in file `c8fvid3.bmp`. Extract the outline and skeleton of the image.

Solutions. It can be seen from the image file that the file is already a binary image file. There is no need to carry out color space conversion. Function `bwmorph()` can be used to extract the outline and skeleton, as shown in Figures 9.60(b) and 9.60(c). It can be seen that there are some burs in the skeleton extraction results shown in Figure 9.60(c), due to the disturbances on the boundary of the image:

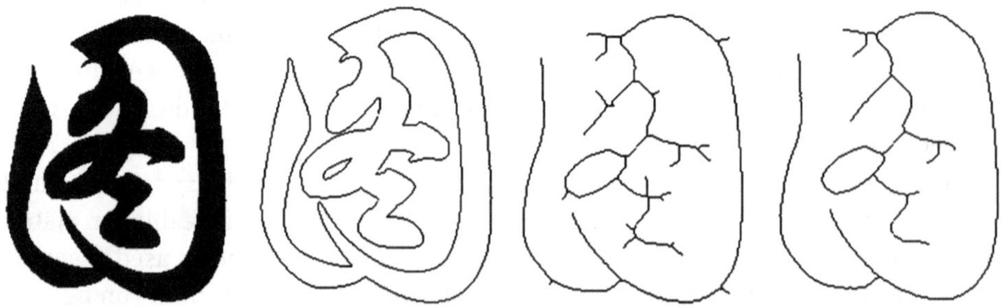

(a) original image (b) outline (c) skeleton extraction (d) refined image

Figure 9.60: Image processing effect based on morphological functions.

```
>> W=imread('c8fvid3.bmp'); imtool(W)          % original image
   W1=bwmorph(W,'remove');  imtool(~W1)        % boundary extraction
   W2=bwmorph(~W,'skel',inf); imtool(~W2)      % skeleton extraction
```

If the 'thin' option in the bwmorph() function is used, a thinner boundary line can be obtained. If this function is repeated 50 times in a loop, a better skeleton can be extracted, as seen in Figure 9.60(d). The result is relatively stable, since there is almost no change if it is iterated more times. Function bwmorph() is many more options, and they can be displayed with the doc bwmorph command:

```
>> W3=~W; % get refined skeleton with a loop
   for i=1:50, W3=bwmorph(W3,'thin'); end, imtool(~W3)
```

9.6.4 Real-time video processing

The Image Acquisition Toolbox is provided in MATLAB, with only one block, the Image Acquisition block. This block can be used to acquire real-time video signals from the camera, connected to the computer. The signal can be used as the input signal, to load into Simulink the videos connected from the camera, as the input signal, and process the signal with Simulink models, so as to implement real-time video processing. An example is used next in this section.

Example 9.26. The video from the camera can be acquired in real-time. Detect the edges in real time with Simulink model.

Solutions. A Simulink model c9mvis7.slx in Figure 9.55 can be used initially, such that block From Multimedia File can be substituted by the Image Acquisition block, with data type set to uint8. See the new Simulink model in Figure 9.61. The videos can be processed in real time with the model.

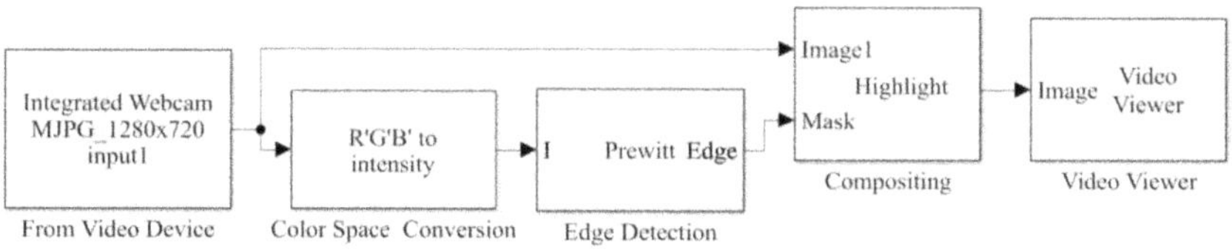

Figure 9.61: Real-time edge detecting system (file: c9mvis8.slx).

9.7 Exercises

9.1 All of the Simulink models constructed in the book are reuseable ones. If the audio and video files are substituted into other files, the same models can be used to carry out the works. Try different audio and video files with the Simulink models, and experience the manipulations of multimedia files.

9.2 Multimedia files are almost everywhere from the internet. Use the methods presented in the book and process the multimedia files with MATLAB and Simulink.

9.3 In Example 9.7, the object `audiorecorder` is demonstrated. Display all of the members of such an object. Display all the usable methods of the object with `help` commands.

9.4 In the earlier versions, many compressed format of audio files, such as `mp3` files, were supported with the `wavwrite()` function. In the new versions, the function is not allowed, and in the new file `audiowrite()`, these formats are not supported. Search from internet and see whether there are files in MATLAB, which can be used in dealing with `mp3` files. The author downloaded and tested several solvers named `mp3write()`, but the low-level support is with the `wavwrite()` function, and they cannot be correctly used.

9.5 Lena image is the most commonly used testing image in digital image processing. The size of the picture is of 512×512, given in the file `lena512color.tif`. Extract the edges of such a picture using different algorithms, and compare the results.

9.6 Consider making the `adapthisteq()` function into a block. The basic idea is load the image as input signals, and process it in S-function. In the S-function, the function `adapthisteq()` can be processed, and feed the result to the output port.

9.7 In the site [7], a great amount of testing images are provided for digital image processing. Select some of the images and carry out analysis tests.

9.8 A significant amount of blocksets and functions are provided in the MATLAB Computer Vision Toolbox. Also, many demonstrative examples are provided, such as resampling, target tracking, line extraction and panoramic image creation. Read the information and execute the models to better understand the applications of digital image processing.

10 Modeling and simulation of engineering systems

Engineering systems can of course be modeled and simulated with the methods discussed in previous chapters. That is for engineering systems, the mathematical models can be constructed first, then low-level modeling methods can be used to express the obtained mathematical models. For complicated engineering systems, the low-level modeling methods may be too complicated to implement. Adequate theoretical knowledge and low-level block diagram modeling methods must be equipped, such that the models can be created in this way. If the modeling of a certain component is not made correctly, imprecise or even misleading simulation results may be obtained. On the other hand, the low-level models constructed in this way are always far too complicated to validate and maintain. Therefore, for many practical engineering systems, the low-level modeling technique is not always the best modeling method.

Multidomain physical modeling ideas are proposed and implemented in Simulink. A new methodology is proposed for the modeling and simulation of engineering and many nonengineering systems. Many reusable professional blocksets are developed by the leading scholars in their own fields, and some of the blocksets can established upon the Simscape framework. Many blocks are designed according to the corresponding hardware components, such that the simulation model can be connected as if they were assembled in hardware systems. The physical models created in this way can be used to automatically generate the mathematical model of the system in Simulink, which can be used directly in numerical simulation. The benefits of the methodology are that building block methods can be used to construct the simulation model, and the model thus created is easy to validate. If there are errors, it can be found easily. Modification of system structures are often made in a simple manner. Besides, no solid background knowledge is required to the users, such that the cross-disciplinary modeling and simulation are made easy, if they know how to use the components. A more important issue is that the whole simulation model can be made in unified Simulink frameworks. Therefore, there are advantages in modeling and simulation.

In Section 10.1, the concept of multidomain physical modeling is presented first, followed by a brief introduction to Simscape. In Section 10.2, the blockset for electrical systems is presented, and simulation model construction upon such a blockset is illustrated. In Sections 10.3 to 10.4, the modeling and simulation of electrical systems and mechanical systems are presented, and illustrated with practical examples.

10.1 Introduction to physical modeling blockset Simscape

The concept and necessity of physical modeling was demonstrated earlier in Chapter 1, with an example. In the section, the requirements of physical modeling are

https://doi.org/10.1515/9783110734959-010

demonstrated, and then the fundamental facilities and components in Simscape are briefly introduced. The physical modeling and simulation of an ordinary network is illustrated with examples.

10.1.1 Limitations in mathematical modeling methods

All the simulation systems are carried out based on the assumption that the system models are already known. These models are established by the conventional methods, step by step. In conventional modeling, physical laws are required to write out the mathematical equations. For instance, in writing equations for the circuits, Kirchhoff's laws are always required, while in mechanical systems, Newton's laws of motion are usually required to write out mathematical models. With these mathematical models, the Simulink model for simulation can be constructed, such that simulation analysis can be carried out.

Therefore, in order to carry out the simulation studied for practical systems, very strong background knowledge of the specific fields are required. For instance, for an electrical engineer, he may need a lot of time to get familiar with mechanical systems and their modeling methods, before he can start effective simulation studies for the systems. This is a time-consuming job to carry out simulation for mechanical systems. On the other hand, since he is not an expert in the related fields, the simulation models thus established may not be very reliable. Besides, in the mathematical modeling process, many of the "unimportant factors" are neglected. If these factors are not negligible, they may induce big modeling errors, and sometimes erroneous results can be obtained.

Example 10.1. Consider the resistor, inductor and capacitor circuit shown in Figure 10.1. Set up its mathematical model.

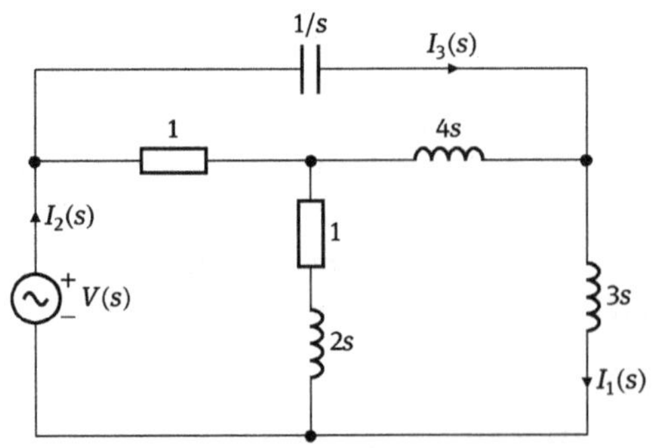

Figure 10.1: Simple circuit composed of resistor, inductor and capacitor.

Solutions. The parameters given here are those used in Laplace transform. With Kirchhoff's laws, three loop current equations can be established, respectively. Since the capacitor and inductor can be expressed as integrator and differentiator, the following equations in Laplace transform can be created:[18]

$$\begin{cases} (2s+2)I_1(s) - (2s+1)I_2(s) - I_3(s) = V(s), \\ -(2s+1)I_1(s) + (9s+1)I_2(s) - 4sI_3(s) = 0, \\ -I_1(s) - 4sI_2(s) + (4s+1+1/s)I_3(s) = 0. \end{cases} \tag{10.1.1}$$

The input source is a voltage signal with voltage is 220 V, and frequency of 1 Hz, such that $V(t) = 220\sin 2\pi t$. Since the original system is linear, the output $v(t)$ is also a sinusoidal signal of different magnitude and initial phase. Due the modeling and solution process, the model and result obtained are to be validated. Of course, the circuit here is rather simple, and the modeling method of complicated circuit may be very complicated. If a circuit is with tens of loops, tens of equations can be established and the modeling and solution process may become extremely complicated. If there are slight errors in the modeling phase, a certain loops can be neglected, such that the simulation result may be erroneous. Therefore, better modeling and simulation methods should be introduced, such as with the physical modeling method demonstrated in Chapter 1.

Example 10.2. Consider the spring–damper system in Figure 10.2, where $x(t)$ is the displacement of the mass, $f(t)$ is the external force. Write out the mathematical model of displacement $x(t)$.

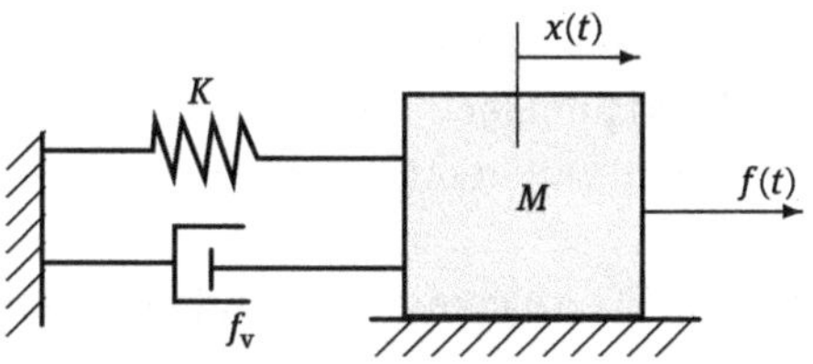

Figure 10.2: Spring–damper system.

Solutions. In the model, the resistant force is proportional to the displacement and its velocity, in the spring and damper. Therefore, based on Newton's second law, the mathematical model below can be immediately written as

$$Mx''(t) + f_v x'(t) + Kx(t) = f(t), \tag{10.1.2}$$

where f_v is the damping ratio of the damper, K is the elastic coefficient of the spring.

The mathematical models for simple systems can of course be created like this. If the systems with many springs and dampers are involved, the modeling of the system is becoming more and more complicated. If an incorrect composite force is used in the model, the simulation of the whole system may fail. Therefore, the modeling and simulation for complicated mechanical problems are also very complicated. Special knowledge are required to build the correct models. Meanwhile, a more powerful modeling and simulation tool is expected in solving modeling and simulation problems for complicated systems. Later, Simulink-based advanced modeling and simulation methods are illustrated.

Due to the above reasons, new ideas should be introduced for modeling complicated engineering systems. For instance, use multidomain physical modeling methods, the creation of simulation models for complicated systems can be completed with components, just as in assembling hardware systems. The Simscape software and related blocksets provided in MathWorks are the ideal tools in multidomain physical modeling processes. With such tools, the simulation models can be established in the unified Simulink framework, such that the entire system can be simulated easily. This method is hard to be implemented in other software platforms.

10.1.2 Simscape introduction

Simscape is a new object oriented multidomain physical modeling tool provided in MathWorks in 2005. After many generations of evolution, the new facilities are more and more powerful, and the coverage fields are becoming wider and wider. In earlier versions of Simscape, many of the tools are self-contained, and somewhat isolated from each other. For instance, the independent blocksets such PowerSystems (earlier versions SimPowerSystems or Power System Toolbox) SimElectronics and so on were provided, while in the new versions, these components are reorganized and unified, such that the new Electrical blockset is formulated.

The command `simscape` can be issued in the MATLAB command window, or select the Simscape item from Simulink model library, the Simscape blockset can be opened directly, as shown in Figure 10.3.

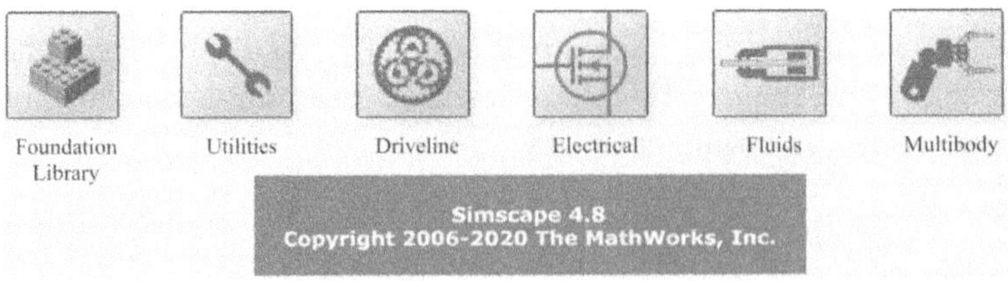

Figure 10.3: Simscape blockset.

In the current versions, the blocksets in Simscape include the Foundation Library, covering the fields such as electrical, magnetic, mechanical, heat, hydraulic and so on. More professional and highly integrated blocksets such as Electrical, Driveline, Multibody, Fluids and tools alike are also provided. The targets of the blocksets are to provide a series component blocks, such that the entire simulation model can be created in the same way as in assembling hardware systems. The mathematical model used in simulation can be generated automatically from the system models constructed. Simscape and related tools in Simulink are meaningful attempts in physical modeling. When building models, it is not quite necessary for the users to have very solid background and experience in mathematical modeling in the field studied. Therefore, direct modeling and simulation can be carried out for the users, even through they are not very familiar in the related fields.

Besides, Simscape language is proposed by MathWorks, allowing the users to define their own reusable blocks in a syntax very similar to MATLAB, in the object oriented form. This may enrich the facilities of multidomain physical modeling facilities in Simulink.

10.1.3 Foundation library in Simscape

Click the Foundation Library blockset icon in Figure 10.3, and the foundation library in Simscape can be opened as shown in Figure 10.4. It can be seen that the blockset includes the subgroups such as Electrical, Mechanical, Hydraulic, Pneumatic, Magnetic, Thermal and so on. A subgroup Physical Signals contains the blocks to make conversions between physical and Simulink signals.

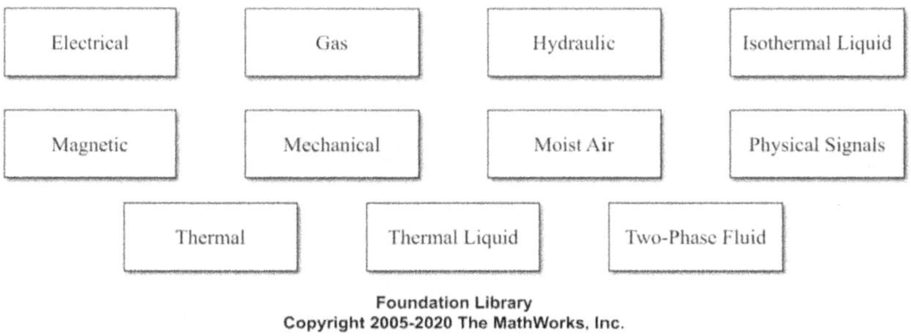

Figure 10.4: Foundation library in Simscape.

Double click the electrical subgroup icon, and the blockset can be opened as shown in Figure 10.5. It can be seen that there are three subgroups: Electrical Elements, Electrical Sources and Electrical Sensors subgroups.

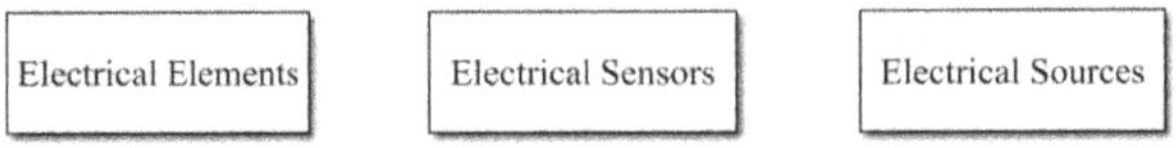

Figure 10.5: Electrical element group in foundation library.

Double click the **Electrical Elements** icon, and the electrical element subgroup shown in Figure 10.6 can be opened. It can be seen that the blockset includes the commonly used electrical components such as **Resistor, Inductor, Capacitor, Mutual Inductor, Ideal Transformer, Variable Resistor** and so on. Also, the electronics components such as **Diode, Op-Amp** (operational amplifier), **Gyrator** and **Memristor** are provided. The other components such as **Rotational Electromechanical Converter** and **Translational Electromechanical Converter** are also provided. Besides, the reference block such as **Electrical Reference** is provided in every blockset.

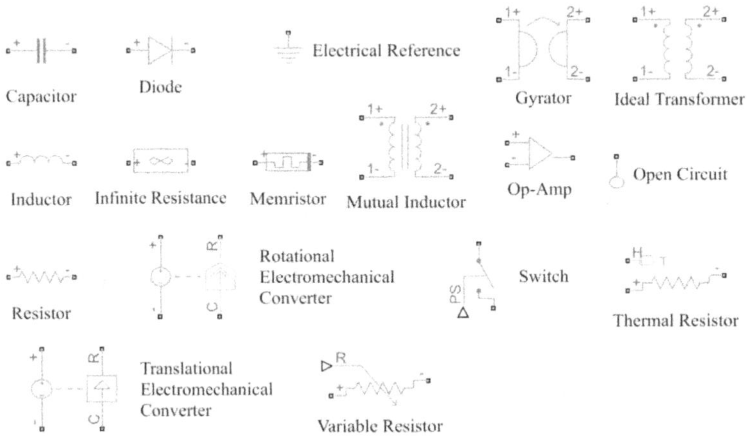

Figure 10.6: Electrical element subgroup.

It can be seen that the display of the port block is different from the ordinary ones in Simulink. In these circled ports, the bidirectional physical signals are accepted, while in the arrow ports in Simulink, unidirectional signals are allowed. The two types of ports cannot be connected directly, without conversions.

The blocks provided in the **Electrical Sources** group are shown in Figure 10.7. The blocks include **AC Voltage Source, AC Current Source, DC Voltage Source, DC Current Source** and so on. Various controlled sources such as **Voltage Controlled Current Source** are also provided.

The blocks provided in the **Electrical Sinks** group are as shown in Figure 10.8, where two blocks are provided: the **Current Sensor** and **Voltage Sensor**, equivalent to the current and voltage meters. The two blocks can be connected in the circuits in

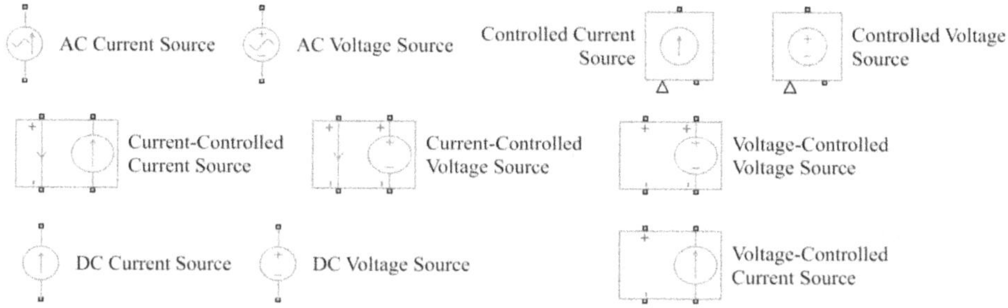

Figure 10.7: Electrical source group.

the way in normal hardware connection. The measured signals are physical ones, and cannot be connected directly to scopes.

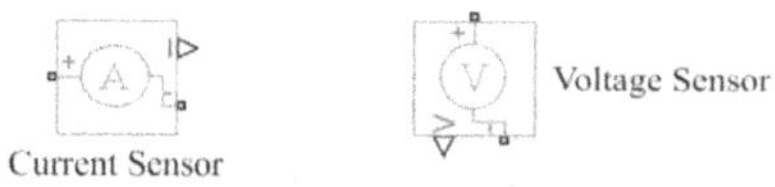

Figure 10.8: Electrical sinks group.

In the foundation library in Simscape, other important subgroups and blocks are provided:

(1) Mechanical. Five subgroups are included in this group, as shown in Figure 10.9, specifically, Mechanisms, Translational Elements, Rotational Elements, Mechanical Sources and Mechanical Sensors.

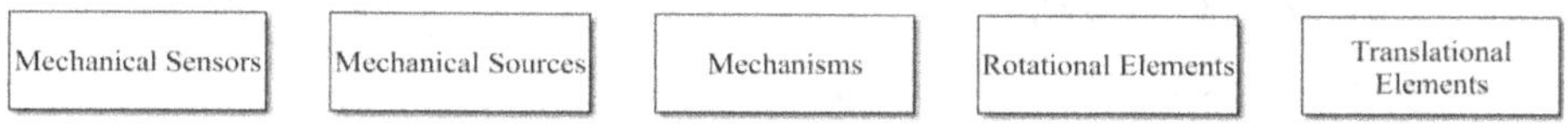

Figure 10.9: Mechanical group.

Translational element subgroup includes the blocks shown in Figure 10.10, including Mass, Translational Friction, Translational Damper, Translational Spring, Translational Hard Stop and so on. As in other component groups, Mechanical Translational Reference block is provided. Therefore, the simulation model in Example 10.2 can be modeled directly with the relevant blocks.

The blocks in the rotational element group are shown in Figure 10.11, including the blocks such as Inertia, Rotational Spring, Rotational Friction, Rotational Damper, Rotational Hard Stop and so on.

The mechanism group includes Gear Box, Lever, Wheel and Axle and so on.

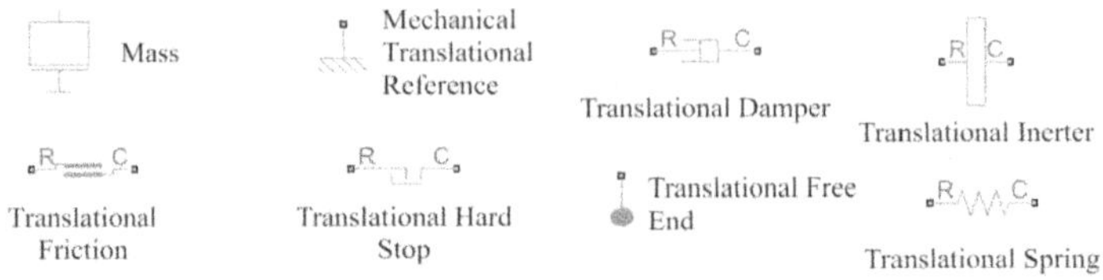

Figure 10.10: Translational elements group.

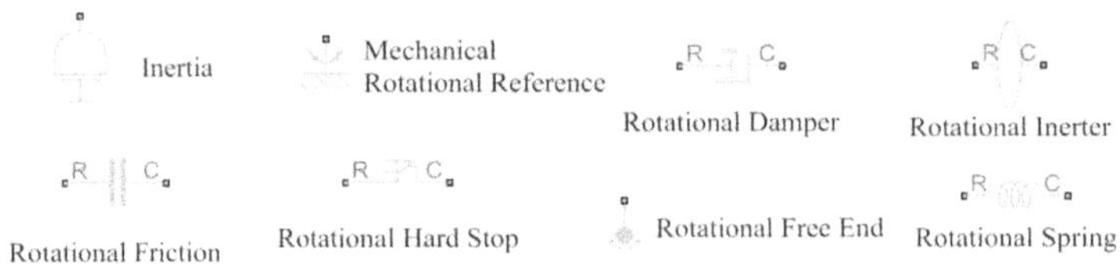

Figure 10.11: Rotational element group.

The blocks in the mechanical source group are shown in Figure 10.12(a), including Ideal Force Source, Ideal Torque Source, Ideal Angular Velocity Source, Ideal Translational Velocity Source. The mechanical sensor group contains the blocks as shown in Figure 10.12(b), including Ideal Force Sensor, Ideal Rotational Motion Sensor, Ideal Torque Sensor, Ideal Translational Motion Sensor.

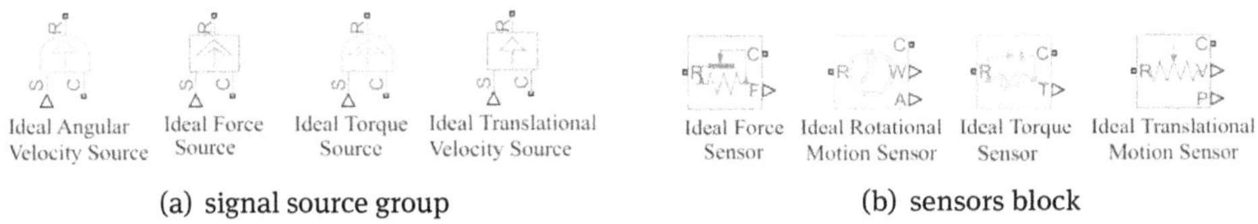

Figure 10.12: Input and sensors groups in mechanical group.

(2) Magnetic. The subgroups are shown in Figure 10.13,

where three subgroups are included: the Magnetic Elements, Magnetic Sources and Magnetic Sensors subgroups. The blocks such as Reluctance, Electromagnetic Converter are provided in the group. In the subgroup in Figure 10.15, the blocks such as Flux Source, MMF Source, Controlled Flux Source, Controlled MMF Source, Flux Sensor, MMF Sensor and so on are also provided in Figure 10.14.

(3) Thermal. The blocks such as Conductive Heat Transfer, Convective Heat Transfer, Radiative Heat Transfer, Thermal Mass and so on are provided in the Heat Element group. The input and output blocks in the thermal group include the Ideal Heat Flow Source, Ideal Temperature Source, Ideal Heat Flow Sensor and Ideal Temperature Sensor. In the remaining parts of this section, the screenshots are not provided.

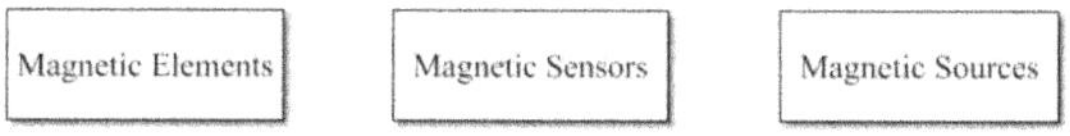

Figure 10.13: Magnetic group.

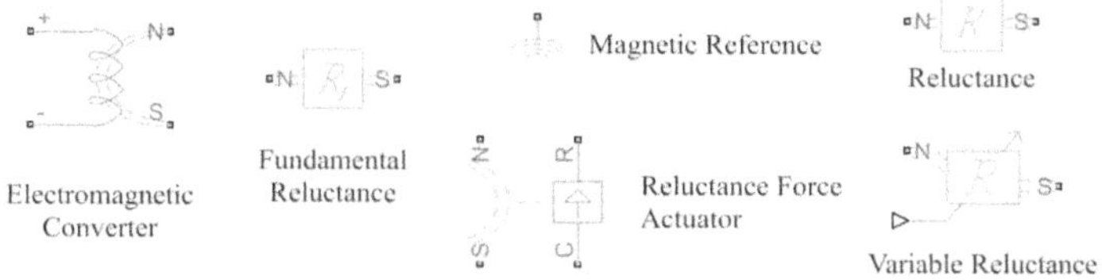

Figure 10.14: Magnetic element blocks.

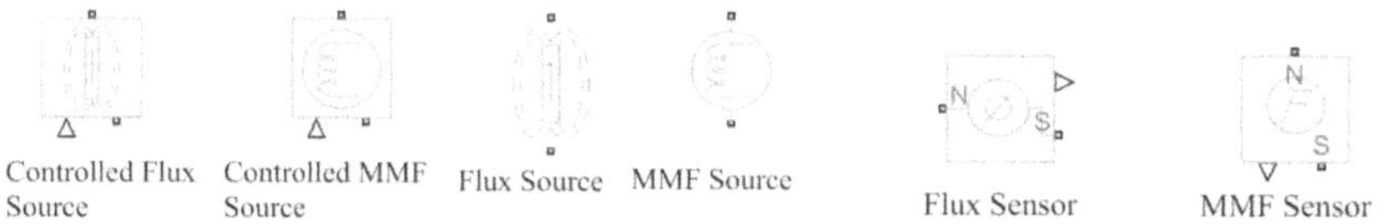

(a) signal source (b) sensors

Figure 10.15: Input and sensors in the magnetic group.

(4) Hydraulic, including Constant Area Hydraulic Orifice, Variable Area Hydraulic Orifice, Hydraulic Resistive Tube, Linear Hydraulic Resistance, Translational Hydro-mechanical Converter, Rotational Hydro-mechanical Converter, Variable Hydraulic Chamber, Constant Volume Hydraulic Chamber, Hydraulic Piston Chamber and so on. Also, the blocks such as Hydraulic Pressure Source, Hydraulic Flow Rate Source, Hydraulic Flow Rate Sensor, Hydraulic Pressure Sensor and so on are provided.

10.1.4 Conversions of two types of signals

If the physical simulation framework is established, two types of signals are involved, with one of the conventional Simulink signals, and the other, the physical signals. These two types of signals always exist in a physical model, and they cannot be connected each other. Conversions must be made to connect them together.

Double click the Utilities group icon in Simscape blockset, the blocks contained in the group can be displayed, as shown in Figure 10.16, where the direct conversion blocks between Simulink and physical signals are provided in PS-Simulink Converter and Simulink-PS Converter. Apart from these two blocks, the Solver Configuration, Two-Way Connection and Connection Port blocks are also provided, where Solver Configuration block is an essential one in the simulation model. If there is no such a block in a Simscape model, simulation processes cannot be executed, otherwise an error message is displayed.

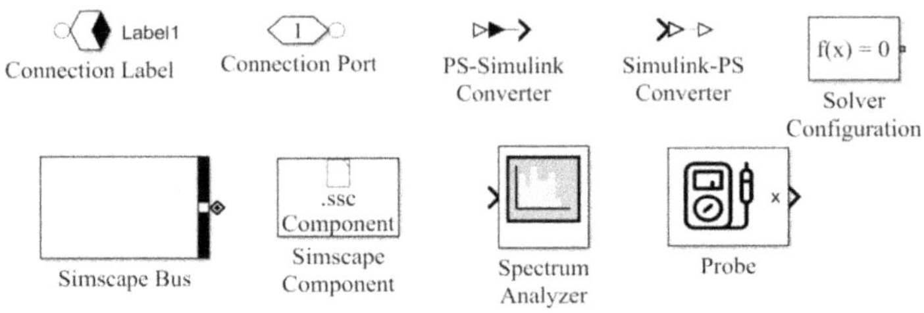

Figure 10.16: Utilities group.

10.1.5 Creating simple Simscape models

Blank Simscape models can of course be created directly with File→New→Model command, however, the essential blocks such as Solver Configuration must be added manually. It is rather complicated to create a model in this way. If the blocks are not added, the simulation process cannot be executed.

It is recommended to use `ssc_new` command to start up modeling process. A blank model framework shown in Figure 10.17 can be created. The blank model contains four essential blocks to start with: Solver Configuration, Simulink-PS Converter, PS-Simulink Converter, and the latter is connected directly to a Scope block. The paragraph displayed in the model framework is about the use of Simscape modeling resources, and it can be removed. In order to connect to Simulink blocks, the signal converters must be used. Examples are given next to demonstrate the modeling process in Simscape.

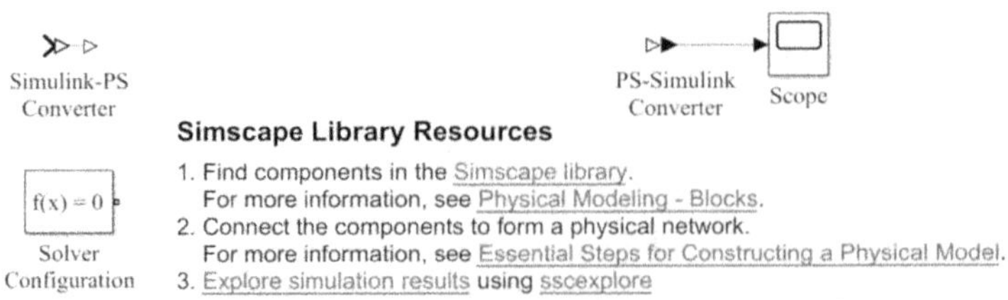

Figure 10.17: A blank Simscape model framework.

Example 10.3. Consider again the circuit in Figure 10.1. Build up a physical model with Simscape, and compute the voltage across the capacitor, and the current signal $i_1(t)$.

Solutions. To measure the voltage and current I_1, the voltage and current meters should be connected in the corresponding places. According to the knowledge in physical experiments, the voltage meter should be connected in parallel to the capacitor, while the current meter should be connected in series in the $I_1(s)$ branch. Therefore, starting from the blank framework in Figure 10.17, the simulation model shown in Figure 10.18 can be established. For the convenient analysis latter, the input and output ports are reserved in the model. It can be seen that the kernel part of the model is exactly the same as the one in the circuit.

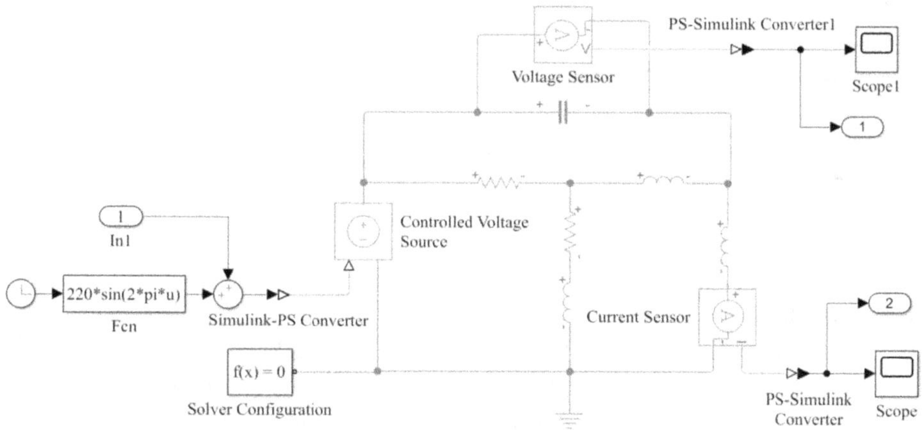

Figure 10.18: Simulation model for the circuit (file: `c10mcirc1.slx`).

Double click the capacitor block, a parameter dialog box is opened, as shown in Figure 10.19, The value of the capacitor should be specified in the **Capacitance** edit box. To simulate accurately the behavior of a real capacitor, the **Series resistance** and **Parallel conductance** values should also be specified. Normally, these values are extremely small, and they can be set to 0 here.

Figure 10.19: Capacitor parameter dialog box.

A hotlink **Source code** is provided in the dialog box. If it is clicked, a MATLAB editor is opened, and the source code in Simscape language can be opened. For instance, the source code for the capacitor is given as follows. For convenience in type setting, the format is slightly modified:

```
component capacitor < foundation.electrical.branch
   parameters
      c={1e-6,'F'}; r={1e-6,'Ohm'}; g={0,'1/Ohm'};
   end
   variables
      vc={value={0,'V'}, priority=priority.high};
   end
   equations
      assert(c>0); assert(g>=0)
      assert(isfinite(g)); assert(r>=0); assert(isfinite(r))
      v==i*r+vc; i==c*vc.der+g*vc;
end, end
```

It can be seen from the source code in Simscape that the equivalent circuit for the capacitor is as shown in Figure 10.20(a), where, r is the equivalent series resistance, g is the parallel conductance. In the ideal cases, $r = g = 0$. In certain cases, these parameters cannot be neglected. The equivalent circuit can be used to simulate accurately the practical capacitors. Besides, the equivalent of the inductor is as shown in Figure 10.20(b), also with equivalent series resistance r and parallel conductance g.

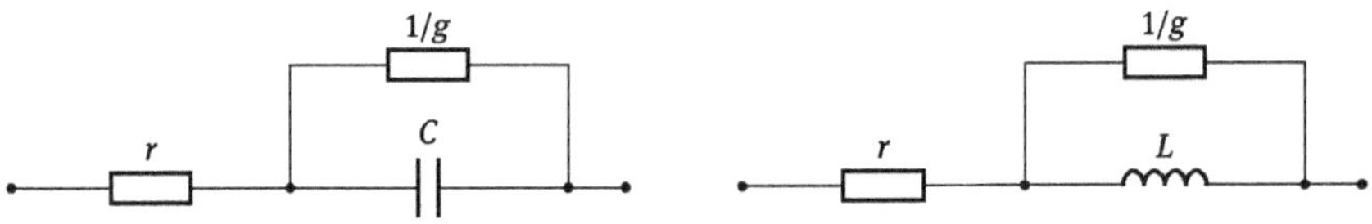

(a) equivalent capacitor (b) equivalent inductor

Figure 10.20: The equivalent circuits for capacitor and inductor.

Perform simulation for the circuit model, the simulation results in Figure 10.21 can be obtained.

With the Symbolic Math Toolbox, the following commands can be issued, to solve directly the equations in (10.1.1) to find the signals. The voltage and current obtained are exactly the same as the ones in Figure 10.21:

```
>> syms t s;
   A=[2*s+2,-(2*s+1),-1; -(2*s+1),9*s+1,-4*s; -1,-4*s,4*s+1+1/s];
   xx=inv(A)*[laplace(220*sin(2*pi*t));0;0]; U=xx(3)/s; I1=xx(2);
   u=ilaplace(U); t1=0:0.01:10; y2=double(subs(u,t,t1));
```

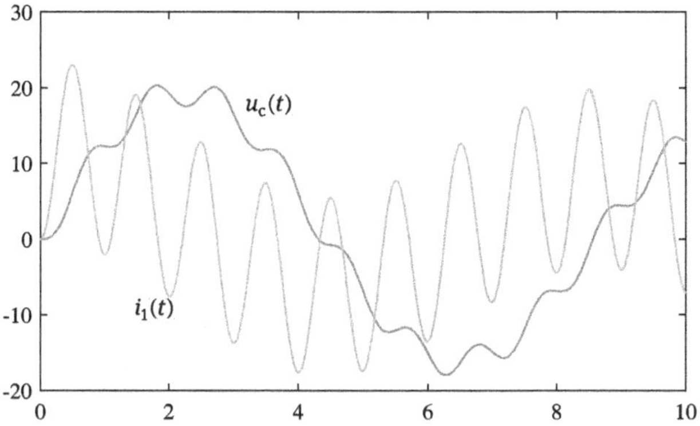

Figure 10.21: Simulation result.

```
i1=ilaplace(I1); y1=double(subs(i1,t,t1));
plot(t1,y1,t1,y2)
```

It can be seen that the solution process is rather time consuming. If the circuit has tens of loops, it is obvious that the theoretical solution method may not be feasible. Therefore, simulation method becomes the only choice.

Example 10.4. Consider the circuit in Example 10.3. Find the transfer function matrix from the input signal to the output signals.

Solutions. Since the input and output ports are reserved in the Simulation model, the following commands can be issued to carry out linearization process, and the result can be converted into zero-pole format:

```
>> G=zpk(linearize('c10mcirc1'))
```

The linearized result is

$$\boldsymbol{G}(s) = \begin{bmatrix} \dfrac{0.33333(s+0.08096)(s+1.544)}{(s+1.12)(s+0.0667)(s^2+0.06379s+0.558)} \\ \dfrac{0.33333(s+1)(s^2+0.25s+0.125)}{(s+1.12)(s+0.0667)(s^2+0.06379s+0.558)} \end{bmatrix}.$$

10.1.6 Modeling and simulation of complicated electric networks

It has been demonstrated earlier that, with the Simscape foundation library, simulation analysis can be carried out. Here, a complicated circuit is modeled with MATLAB commands, such that the modeling and simulation problems can be solved for complicated circuits.

Example 10.5. Find the equivalent resistance between points A and B in the resistance network shown in Figure 10.22.

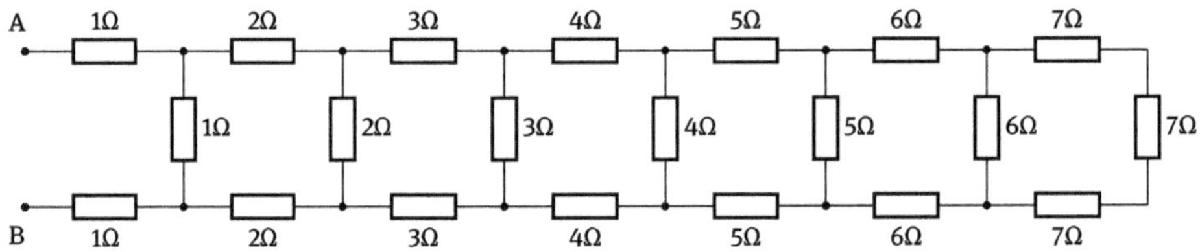

Figure 10.22: A complicated resistance network.

Solutions. If the equivalent resistance between A and B is expected, the relationship between the voltage and current should be set up first. Linearization method can be used to find the equivalent resistance. Since the model is rather complicated, the following commands in MATLAB can be issued to draw the Simulink model.

There are two difficulties in the Simscape modeling process. One is how to find the block name and property names for the resistor block, and the other is the descriptions of the ports. A simple method is that, open a blank model with `fl_lib` command, and copy to it a resistor block and save the file to a readable `mdl` file. Open the `mdl` with an editor, and it can be found that the block name is `'fl_lib/Electrical/Electrical Elements/Resistor'`, and the property name of the resistance is `R`, and its left and right port labels are respectively `LConn1` and `RConn1`. Now, let $L = 7$, the following commands can be used to draw the Simscape model for the resistance network.

```
>> L=7; M='ssss1'; new_system(M); open_system(M);
   for i=1:L, i1=int2str(i); i0=int2str(i-1);
      pos1=[100+(i-1)*70 50 125+(i-1)*70 70];
      pos2=pos1+[25 50 25 50]; pos3=pos1+[0 100 0 110];
      scr='fl_lib/Electrical/Electrical Elements/Resistor';
      add_block(scr,[M '/ra' i1],'Position',pos1,'R',i1)
      add_block(scr,[M '/rb' i1],'Position',pos2,...
                 'Orientation','down','R',i1)
      add_block(scr,[M '/rc' i1],'Position',pos3,'R',i1)
      add_line(M,['ra' i1 '/RConn1'],['rb' i1 '/LConn1'])
      add_line(M,['rb' i1 '/RConn1'],['rc' i1 '/RConn1'])
      if i>1
         add_line(M,['ra' i0 '/RConn1'],['ra' i1 '/LConn1'])
         add_line(M,['rc' i0 '/RConn1'],['rc' i1 '/LConn1'])
   end, end
```

The measurement and conversion blocks can be added and connected manually in such a model, such that a Simulink model shown in Figure 10.23 can be established. Note that, for a model containing Simscape blocks, it is better to use stiff ODE solvers, such as `ode15s`, and the relative error tolerance can be selected as a very small number, such as 3×10^{-14}. With such a model and linearization command, it is found that the equivalent resistance of the network between A and B is $R_1 = 0.351\,\Omega$, the same as the theoretical result.

```
>> R1=linearize('c10mcirc2')
```

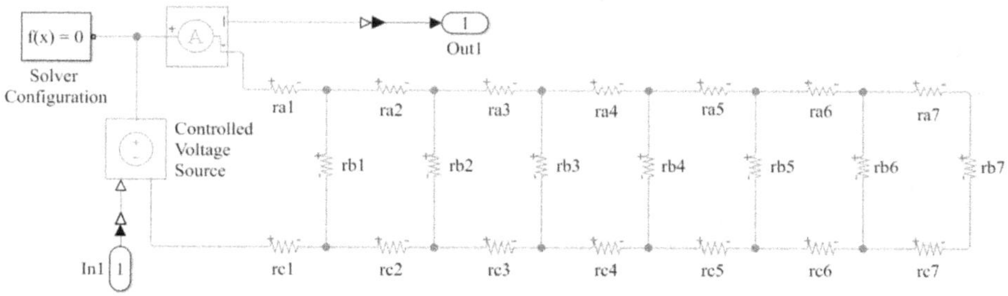

Figure 10.23: Simulink model for the resistance network (file: `c10mcirc2.slx`).

In traditional physics courses, in order to solve these types of problems, skills and tricks are recommended, while the method is always not feasible in real applications. For instance, if the resistance of rb4 block is changed from $4\,\Omega$ to $40\,\Omega$, the skill based method cannot find the equivalent resistance. With simulation method, the value of rb4 can be modified manually, and the above commands can be executed again. It is found that the new equivalent resistance is $2.849\,\Omega$, with no extra computational load added.

10.2 Introduction to electric system blockset

There are many ways in simulating the electrical systems. The foundation library of the Simscape blockset can be used, and also the Electrical blockset in Simscape can be used. The advantages of the two blocksets can be used in the modeling and simulation process. From modeling viewpoints, the Electrical blockset is more professional, and more electrical blocks and systems are provided. Simulation of circuits, power electronics and motor drive systems can be carried out.

The electrical system blockset, also known as Power Systems Blockset and SimPowerSystems, was first developed by HydroQuébec and TECSIM International companies in Canada. After several generations of evolution, the blockset is more and more powerful. It can be used easily in the simulation of circuits, power electronics, motors, and electrical transmissions. General model building facilities are equipped, as if drawing the circuits. The simulation model can be converted into state space form, such that it can be simulated in a Simulink environment. The use of the blocks in the recent versions of the electrical system blockset is rather close to those in Simscape.

Command `powerlib` can be issued to launch the electrical system blockset, however, it is recommended to start the library from the Simscape interface shown in Figure 10.3. Double click the **Electrical** icon, and the power system blockset is opened, as shown in Figure 10.24. The `ee_lib` command can also be used to launch the blockset directly. The commonly used blocks in each category are presented in this section, and illustrated with the screenshots. In different versions of the blockset, there are variations in the categories.

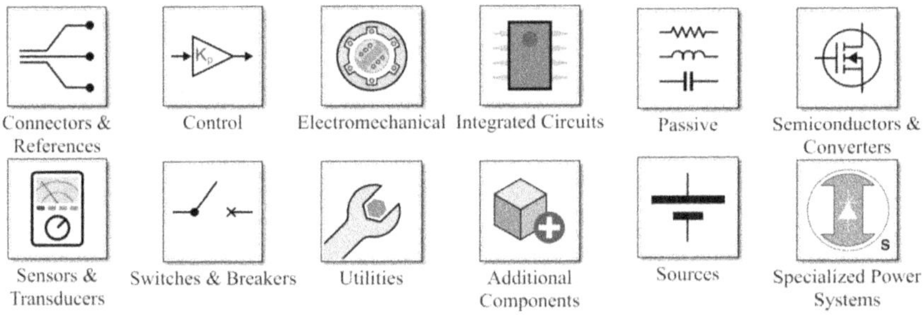

Figure 10.24: Electrical system blockset.

10.2.1 Electrical sources

Double click the **Sources** group icon, then group is opened as shown in Figure 10.25, where DC and AC sources as well as various controlled sources are provided. Also, programmable sources, three-phase sources and various batteries are provided.

10.2.2 Electrical sensors

The **Sensors & Transducers** group can be opened as shown in Figure 10.26, where many measuring blocks are provided, such as current and voltage meters and impedance meters. Other sensors are also provided. In the recent versions, the voltage and current blocks are unified with the ones in the foundation library.

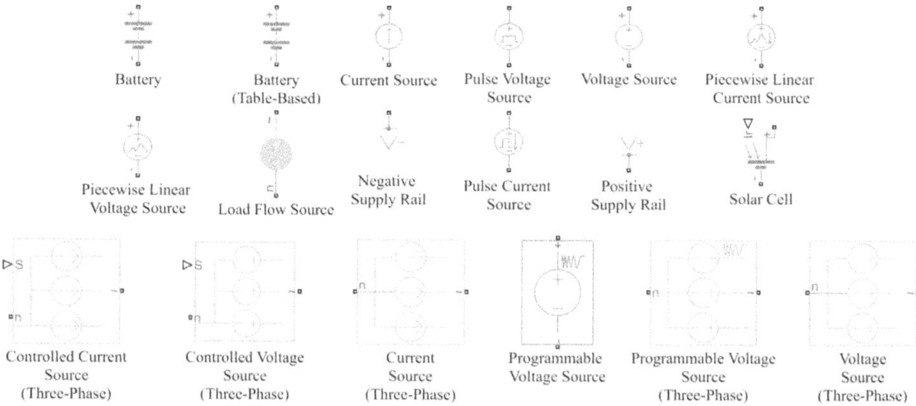

Figure 10.25: Power sources group.

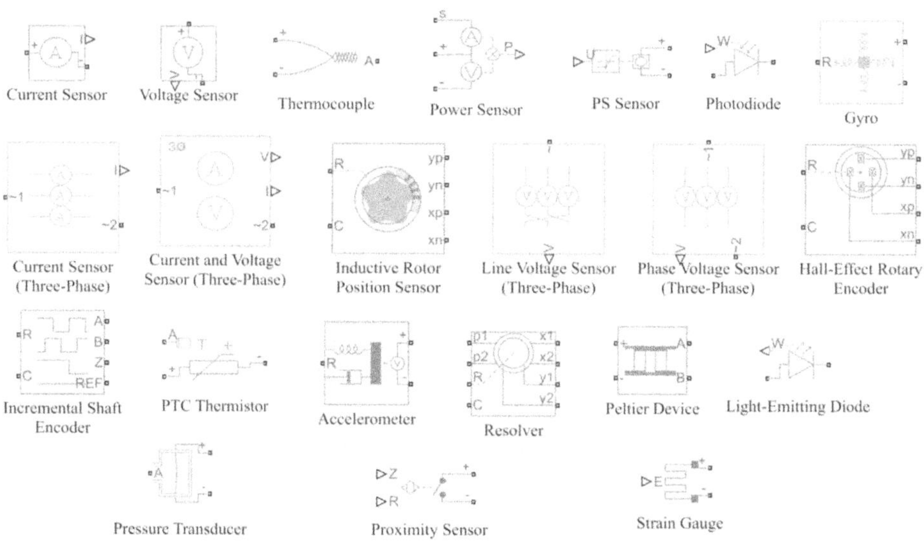

Figure 10.26: Sensors and transducer group.

10.2.3 Electric and electronic elements

A great many electrical and electronic component blocks are provided in the blockset, and they are classified into Passive, Semiconductor & Converters and Integrated Circuit groups. They are illustrated as follows:

(1) Passive group, as shown in Figure 10.27. Apart from the typical resistor, capacitor and inductor blocks, a lot of other low-level blocks and subgroups are provided.

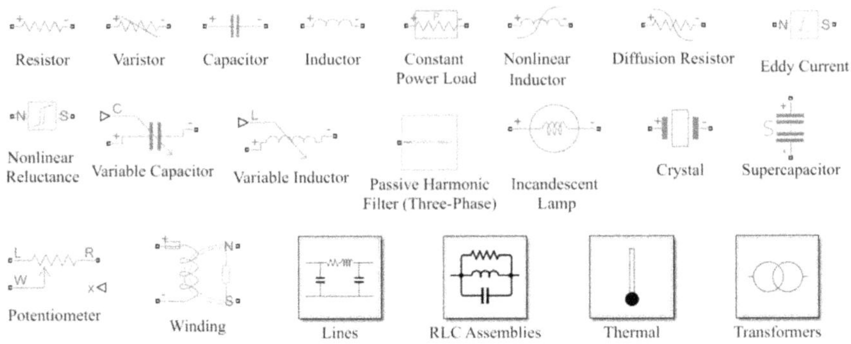

Figure 10.27: Passive element group.

(2) Semiconductor & Converters group, as shown in Figure 10.28. The commonly use analog electronic devices such as diode, transistors and power electronic components such as various FETs, IGBT and thyristors are provided. Converters is a subgroup, where different transformers, AC/DC convertors are provided.

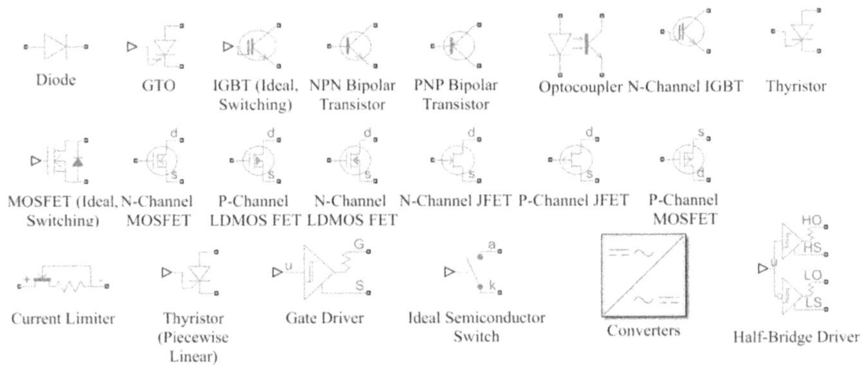

Figure 10.28: Semiconductor devices.

(3) Integrated Circuit group, as shown in Figure 10.29. Various operational amplifiers (Amps) are provided. Other integrated circuit blocks are provided, and the Logic subgroup, shown in Figure 10.30, contains various CMOS logic gates, and blocks such as S-R Latch. They can be used in modeling and simulating digital electronic circuits.

10.2.4 Electromechanical blocks

The subgroups in the Electromechanical group is shown in Figure 10.31, including various motor subgroups, such as Asynchronous, Synchronous, Brushed Motors, Perma-

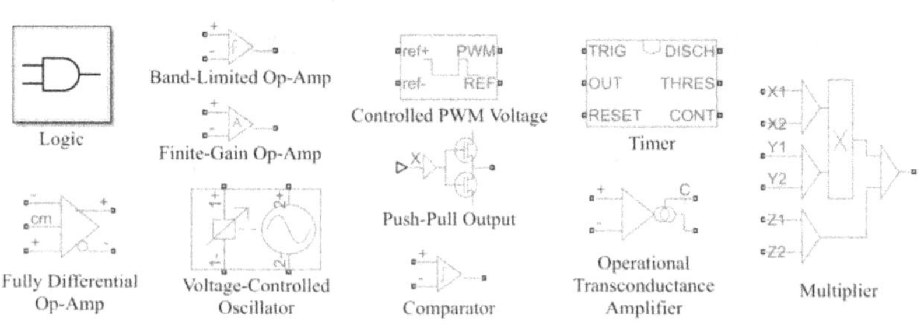

Figure 10.29: Integrated circuit group.

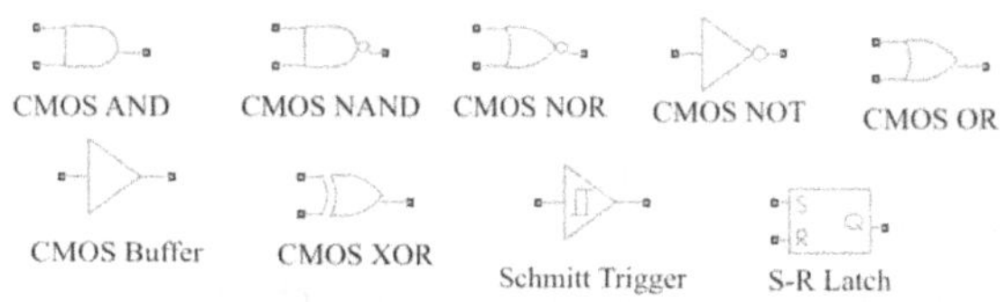

Figure 10.30: The digital logic subgroup.

Figure 10.31: Electromechanical device group.

nent Magnet, Reluctance & Stepper and so on, as well as some other electromagnetic blocks.

The motor blocks provided in the Asynchronous are shown in Figure 10.32, including Single-Phase Induction Machine, Induction Machine Squirrel Cage, Induction Machine Wound Rotor, Simplified Induction Machine and so on, and also the Induction Machine Measurement block.

Various other kinds of blocks are provided in the electrical system blockset, including different breakers, various electrical controller blocks and electric motor drive blocks.

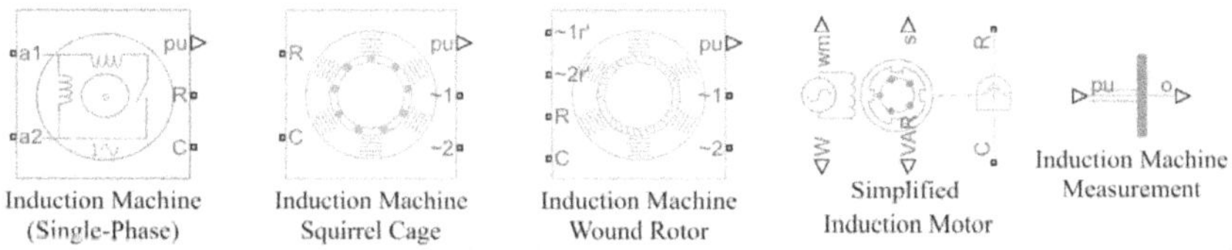

Figure 10.32: Asynchronous motor group.

10.3 Modeling and simulation of electrical systems

The contents in the electrical system blocks are illustrated in the previous section. In this section, examples are used to model and simulate analog electronic circuits and digital electronic circuits. Also, the modeling and simulation of operational amplifiers are addressed.

10.3.1 Simulation of electronic circuits

Before the appearance of the tools such as Simscape, it was extremely difficult to model and simulate electronic systems in MATLAB, since there were not even the most commonly used transistor blocks. Indirect methods must be made to simulate transistors with approximate devices. With such a blockset, the modeling of electronic systems become very simple. The physical models of the electronic circuits can be drawn, with the tools, such that the simulation model can be generated automatically for Simulink, so as to get simulation results. Examples are given next to demonstrate the modeling and simulation of simple electronic circuits.

Example 10.6. Consider the transistor amplifying circuit shown in Figure 10.33(a), where $R_0 = 10\,\mathrm{k\Omega}$, $R_1 = 1\,\mathrm{k\Omega}$, $C_1 = C_2 = 82\,\mathrm{pF}$ and $L = 120\,\mathrm{\mu H}$. Setup a simulation model for the circuit.

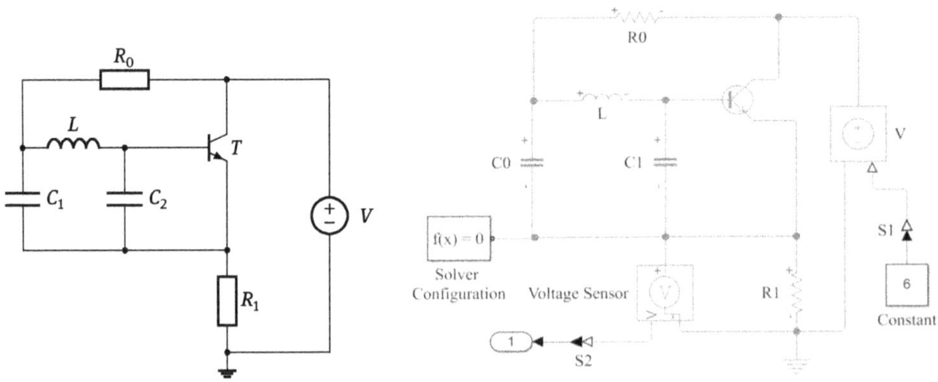

(a) transistor amplifying circuit (b) simulation model (file: `c10mtri1.slx`)

Figure 10.33: Transistor amplifying circuit.

Solutions. There is a transistor in the circuit. The NPN Bipolar Transistor block in the Semiconductor & Converters group can be used to construct the simulation model, as shown in Figure 10.33(b). Double click the transistor block, the parameter dialog box is opened as shown in Figure 10.34, where various parameters for the transistor element are provided. The users may fill in actual parameters according to the particular situation, or keep the default parameters.

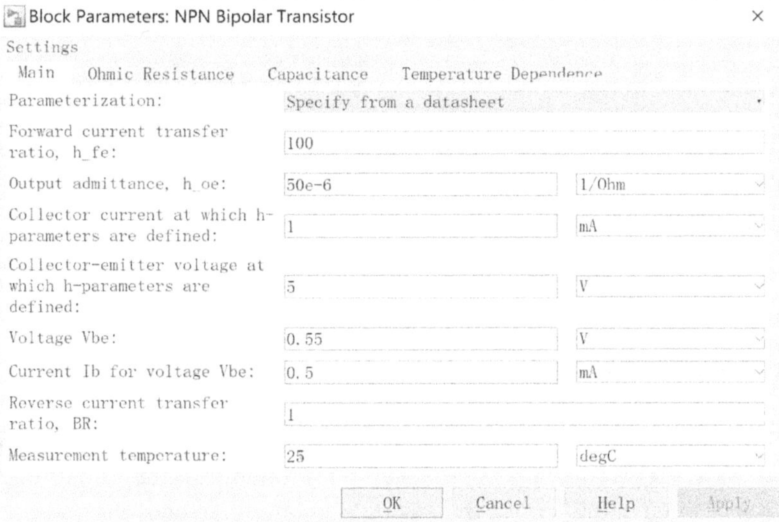

Figure 10.34: Parameter dialog box of the transistor block.

In the earlier versions of MATLAB and Simulink, the electronic devices such as transistors are not provided. Simulation via third-party tools such as Spice language must be made. With recent versions of MATLAB, Simscape can be used and better interface to Spice languages are also provided, such that they can be embedded into the Simulink model. In the next examples, transistor and operational amplifier circuits are demonstrated.

Example 10.7. Consider again the model in Example 10.6. Draw the voltage curve across the resistor R_1.

Solutions. The following commands can be used in the simulation, and the voltage curve can be obtained as shown in Figure 10.35:

```
>> [t x y]=sim('c10mtri1'); plot(t,y)
```

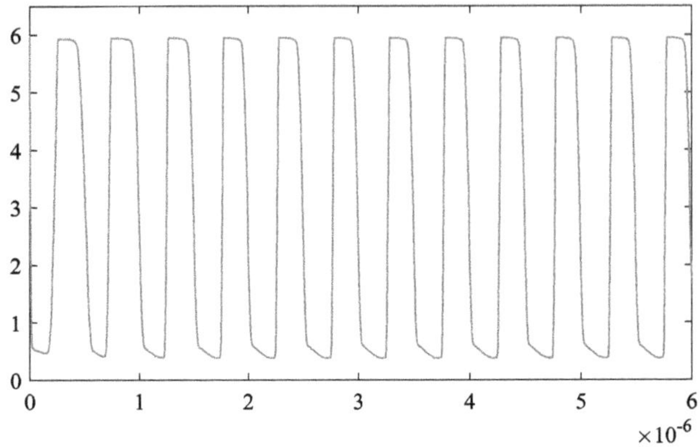

Figure 10.35: Voltage curve.

10.3.2 Simulation of circuits with operational amplifiers

Operational amplifiers are the important component in continuous controllers. The commonly used controller components in differentiators and integrators can be constructed with operational amplifiers. In the analog simulation methods described in Chapter 1, operational amplifiers were the most important components. All of the analog simulation models are established based on operational amplifiers.

In the studies such as control theory, the gains of operational amplifiers are regarded as infinity. In practice, the ideal operational amplifiers do not exist. The gains of the amplifiers are of finite values. Besides, there is a saturation effect in the output voltage signals. Therefore, there are differences between ideal operational amplifiers and practical ones. Examples are used to demonstrate simulation analysis of operational amplifiers.

Example 10.8. Consider the circuit with an operational amplifier, shown in Figure 10.36, where $R_1 = 360\text{k}\Omega$, $R_2 = 220\text{k}\Omega$, $C_1 = 5.6\,\mu\text{F}$ and $C_2 = 0.1\,\mu\text{F}$. Establish a Simulink model, and observe the impact of the magnitude of the input signal on the output signals.

Solutions. The operational amplifier block can be used as the kernel, such that the simulation model in Figure 10.37 can be established. It can be seen that the form of the model is straightforward and convenient.

Double click the operational amplifier block, and its parameter dialog box can be opened, as shown in Figure 10.38, where the gain is set to 1000, rather than the ideal ∞. Besides, there is clamping voltage of ± 15 V for the block.

If the magnitude of the input sinusoidal signal v_i is 0.3 V, the output signal v_o can be obtained with simulation method, as shown in Figure 10.39. It can be seen that

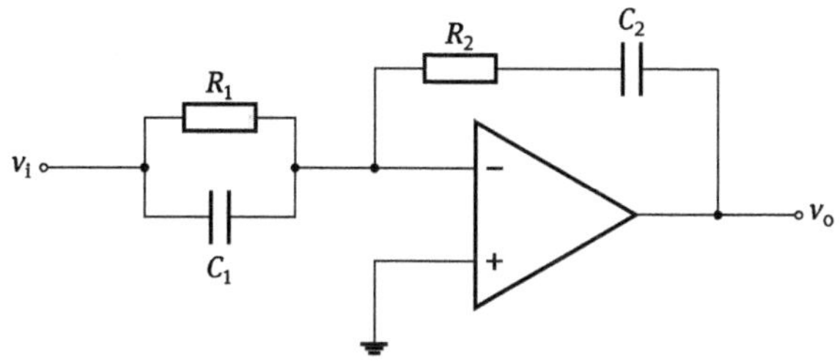

Figure 10.36: Circuit of operational amplifier.

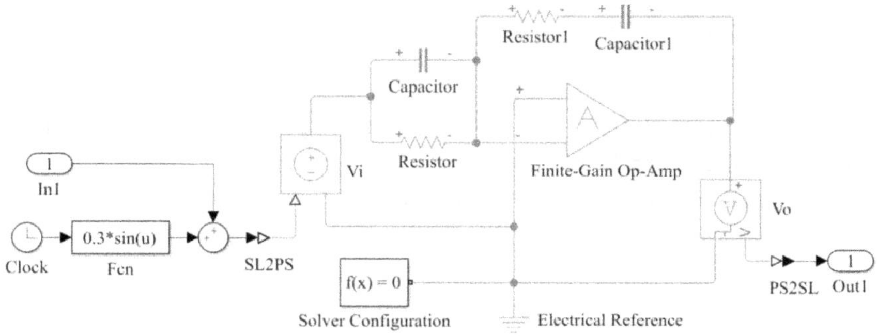

Figure 10.37: Simulation model of operational amplifiers (file: `c10moa1.slx`).

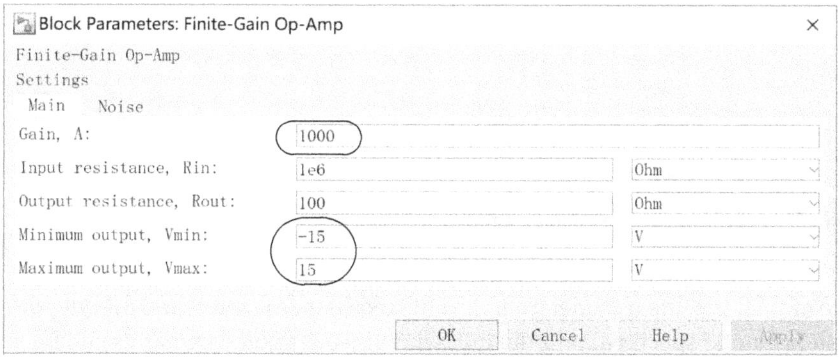

Figure 10.38: Parameter dialog box of the operational amplifiers.

there are saturation effects from the output signals. Therefore, operational amplifiers are no longer linear blocks. Its nonlinearities cannot be neglected:

```
>> [t x y]=sim('c10moa1'); plot(t,y)
```

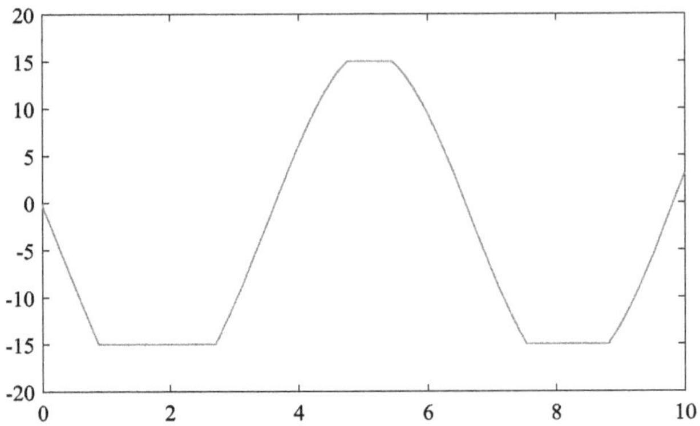

Figure 10.39: The output signal.

Further increase the magnitude of the sinusoidal input signal to 0.5 V, and the waveform of the output signal can be obtained, as shown in Figure 10.40. It can be seen that the saturation phenomenon cannot be neglected.

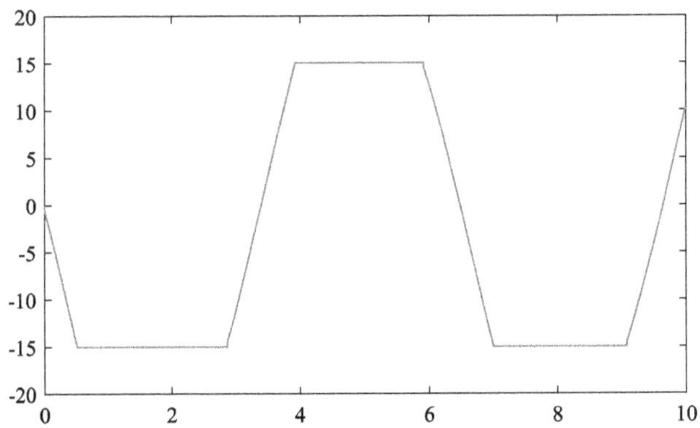

Figure 10.40: The output signal when input magnitude is increased.

In order to extract the linear model from the entire system, the input and output ports in the system can be specified, and the following linearization commands can be issued:

```
>> G=zpk(linearize('c10moa1'))
```

Therefore, after linearization computation, the equivalent transfer function model can be found

$$G(s) = \frac{-999.55(s+0.496)(s+45.45)}{(s+858.2)(s+0.03571)}.$$

10.3.3 Simulation examples in digital electronic circuits

Many digital electronic blocks are provided in the Simscape blockset, which can be used directly in simulating digital electronic circuits. The common logic gate blocks and flip-flop blocks are provided in Figure 10.30. Besides this blockset, the blocks in the Logic and Bit Operations group in Simulink can also be used to represent logic operations, such that digital electronic circuits can be modeled. In this chapter, several examples are presented for digital electronic systems.

Example 10.9. Represent the logic expression $Z = \overline{A\cdot\overline{A\cdot B} + B\cdot\overline{A\cdot B}}$ with Simulink.

Solutions. In the field of digital logic, $A\cdot B$ is referred to as the "and" operations of signals A and B, and the operation is known as the "and" gate. $A+B$ is known as "or" gate, $\overline{A}$ is the "not" gate. Besides, $\overline{A\cdot B}$ is the "nand" gate, $\overline{A+B}$ is the "nor" gate. The logic operators can be implemented directly with the gates in Figure 10.30. With the gate blocks, the Simulink model in Figure 10.41 can be established.

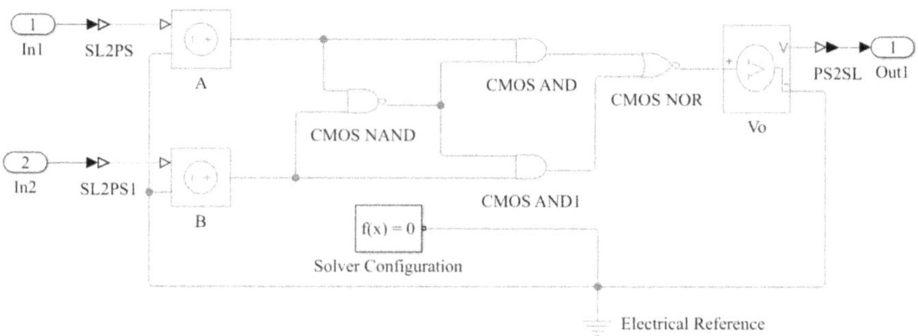

Figure 10.41: Simulink model of the digital logic expression (file: `c10mdig1.slx`).

If the ideal logic blocks are used, the simulation model established is shown in Figure 10.42. It can be seen that all the signals are Simulink signals, and no conversions and extra blocks are needed.

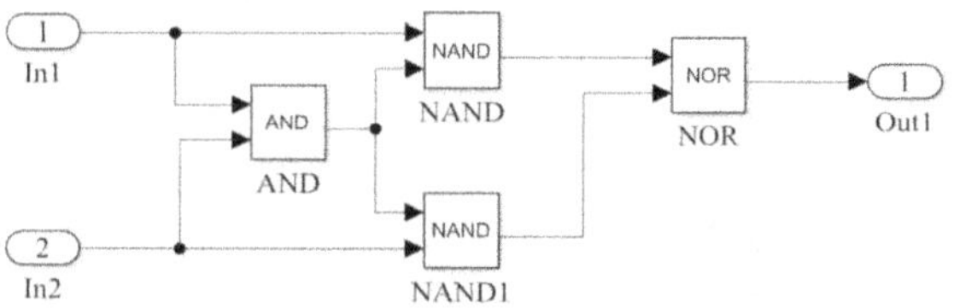

Figure 10.42: Ideal logic expression in Simulink (file: `c10mdig1a.slx`).

Truth tables are also important mathematical models in digital electronics. An example is given next to demonstrate the truth table based decoder circuit with seven-segment LED (light emitting diode) displays.

Example 10.10. Assume that there are four binary input signals A_1, A_2, A_3 and A_4. Establish a decoder circuit, such that the four signals can be used to drive seven-segment LEDs and demonstrate the binary coded decimal (BCD) decoder results.

Solutions. BCD decoder logic can be expressed with the truth table shown in Table 10.1.[20]

Table 10.1: Truth table of the BCD decoder for seven-segment LEDs.

Input signals				Decoder	Signals						
A_1	A_2	A_3	A_4	display	Y_1	Y_2	Y_3	Y_4	Y_5	Y_6	Y_7
0	0	0	0	0	1	1	1	1	1	1	0
0	0	0	1	1	0	1	1	0	0	0	0
0	0	1	0	2	1	1	0	1	1	0	1
0	0	1	1	3	1	1	1	1	0	0	1
0	1	0	0	4	0	1	1	0	0	1	1
0	1	0	1	5	1	0	1	1	0	1	1
0	1	1	0	6	0	0	1	1	1	1	1
0	1	1	1	7	1	1	1	0	0	0	0
1	0	0	0	8	1	1	1	1	1	1	1
1	0	0	1	9	1	1	1	1	0	1	1
1	0	1	0	–	0	0	0	0	0	0	1
1	0	1	1	–	0	0	0	0	0	0	1
1	1	0	0	–	0	0	0	0	0	0	1
1	1	0	1	–	0	0	0	0	0	0	1
1	1	1	0	–	0	0	0	0	0	0	1
1	1	1	1	–	0	0	0	0	0	0	1

The truth table can be described by the **Combinational Logic** block provided in the **Logic and Bit Operations** group, shown in Figure 2.37. It can be seen in Table 10.1 that the right of the truth table, $Y_1 \sim Y_7$, can be expressed by a matrix. The matrix can be specified in the parameter dialog box of the combinational logic block, to complete the decoder settings:

```
>> T=[1 1 1 1 1 1 0; 0 1 1 0 0 0 0; 1 1 0 1 1 0 1; 1 1 1 1 0 0 1;
      0 1 1 0 0 1 1; 1 0 1 1 0 1 1; 0 0 1 1 1 1 1; 1 1 1 0 0 0 0;
      1 1 1 1 1 1 1; 1 1 1 1 0 1 1; repmat([0,0,0,0,0,0,1],6,1)];
```

In earlier versions, LED blocks are provided in Simulink. while the block is no longer supported. A replacement to the block is considered. In the **Dashboard** group in Figure 2.20, a block **Multistate Image** is provided. Two image files, `red.jpg` and `white.jpg` can be created, storing red and white color boxes, respectively. They can be associated with the two states 1 and 0. Therefore, when the value to the block is 1, the red image is displayed; otherwise, the white one is displayed. In this case, the effect of a LED can be modeled. Seven such LEDs can be arranged in the shape of "8," as shown in Figure 10.43.

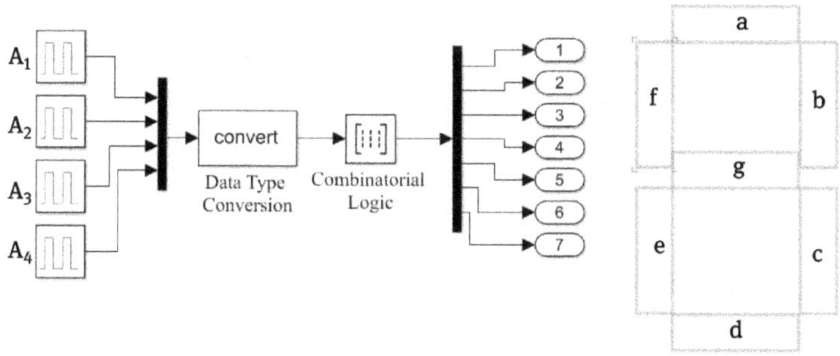

Figure 10.43: Simulink model of the decoder system (file: `c10mtru1.slx`).

The four input signals can be fed into the combinational logic block. In fact, the truth table can be understood like this: if $A_1 = 0$, $A_2 = A_3 = A_4 = 1$, then $Y_1 = Y_2 = Y_3 = 1$, and the remaining Y_i are all zeros. That is to say, in the LEDs, segments a, b and c should display red (to display number 7). The four input signals can be fed into the **Mux** block to from a vector signal. The signal cannot be connected directly into the combinational logic block, since the latter one only accepts binary signals, while the input signals are in double precision ones, therefore, they must be converted into binary ones with a **Data Type Conversion** block. The combinational logic block is a vector containing seven scalar signals, they can be fed into the **Demux** block, and then connect all the seven scalar signals to the seven **Multistate Image** blocks, and set up the associations. The BCD decoder system with LED display can be established.

The four input signals $\mathrm{A}_1 \sim \mathrm{A}_4$ can be driven by pulse signal blocks, where the period of signal A_1 is set to 1, and the periods of subsequent pulse signals can be halved. Therefore, the signals $\mathrm{A}_1 \sim \mathrm{A}_4$ may generate various combinations, so as to test the decoder display results. To better display the decoder results, a fixed-step simulation algorithm can be chosen, and the step-size can be set to 0.000001.

Example 10.11. Pseudo-random binary sequence (PRBS, or known as M-sequence) signals are widely used signals in control system identification and communication systems. The signal can be generated with the electronic circuit in Figure 10.44, where the three logic gates in the three levels are respectively exclusive or gate, or gates, and gates, respectively. Set up the Simulink model for the PRBS, and draw the output of the logic electronic circuit.

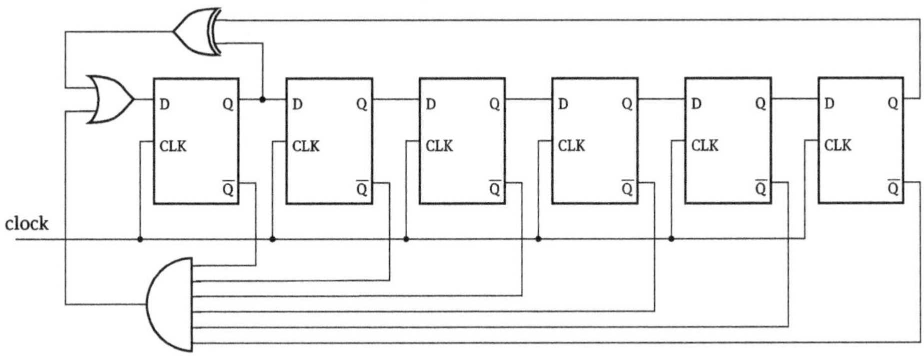

Figure 10.44: PRBS signal generator circuit.

Solutions. It can be seen from the circuit that it is rather hard to derive the mathematical model, since the mathematical model of the flip-flop blocks are not very easy to write, where truth table descriptions are needed. Therefore, a practical and feasible modeling method is to use physical modeling methods.

The kernel parts of the circuit is the flip-flop blocks. Double click the electrical system blockset in Simscape, and a flip-flop block can be found (in Figure 10.30). Unfortunately, D latch blocks are used in the circuit, not the one in the group. The flip-flop blocks in the Simulink Extra group can be used, as shown in Figure 10.45, where D Latch, D Flip-Flop, J-K Flip-Flop, S-R Flip-Flop blocks and Clock block are obtained, with the default clock period of 2 seconds.

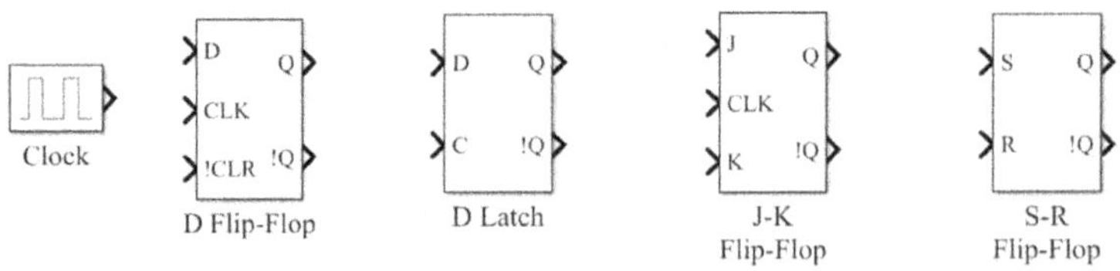

Figure 10.45: Flip-flop block group.

For this particular example, the D latch block may lead to algebraic loops, and simulation cannot be carried out. Six D Flip-Flop blocks, clock block and gates, or gates and exclusive or block can be used, to set up the simulation model in Figure 10.46. Pulse signals can be used for the clock signal, to control the six D flip-flop blocks. The difference between the model and the circuit is that for each flip-flop signal, an enable signal. This signal is connected to all the enable port !CLR such that the enable signal is always 1. The logic blocks can be built up with the blocks provided in the Logical Operator subgroup in math groups.

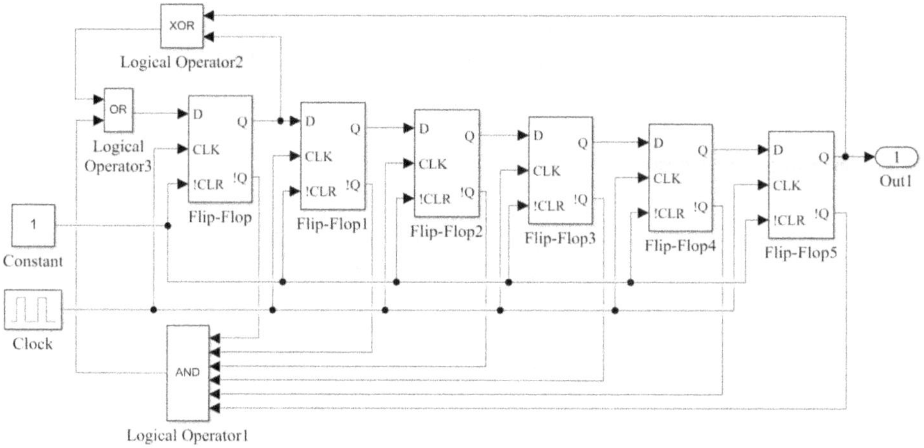

Figure 10.46: Flip-flop based PRBS sequence generator (file: `c10mflip1.slx`).

When the simulation model is constructed, simulation studies can be carried out. The simulation results obtained are as shown in Figure 10.47, where the period of the signal is 63, that is, 126 seconds. It can be seen that with the flip-flop blocks, the digital circuit with triggers can be established easily in Simulink. Scopes or output port blocks can be used to observe the signals in the system, and it can be used as a powerful tool for digital electric circuit analysis.

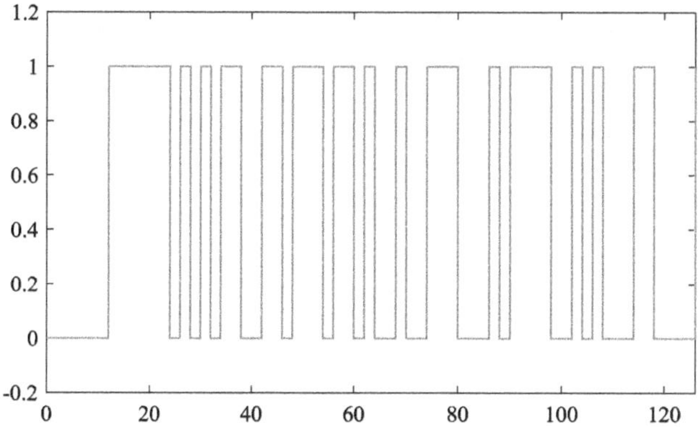

Figure 10.47: PRBS signal and the clock signal.

10.4 Modeling and simulation of mechanical systems

As it was discussed earlier, the foundation library in Simscape can be used to implement the modeling and simulation of mechanical systems. Apart from the foundation library, a more professional Multibody blockset can be used to build complicated mechanical models. A simple introduction to mechanical modeling and simulation is presented first, followed by the simulation methods with a Multibody blockset.

10.4.1 Simulation of simple mechanical systems

The blocks in the foundation library in Simscape can be used directly in the modeling and simulation of mechanical systems. The modeling and simulation of spring–damper systems are illustrated through next examples.

Example 10.12. Consider the spring–damper system in Example 10.2, if the elastic coefficient of the spring is 1000 N/m, while the damping coefficient is 100 N·s/m. The weight of the mass is $M = 1\,\mathrm{kg}$, set up the simulation model.

Solutions. Open the foundation library, and it can be seen that the Translational Spring, Translational Damper and Mass blocks can be used directly, such that the connections of them can be made according to the plot shown in Figure 10.2. The parameters can be specified to the blocks through the corresponding dialog boxes, for later use.

If an external force is added to the system, the Ideal Force Source block in the Sources group can be used. The displacement and velocity of the mass can be measured with the Translational Motion Sensor block in the Sinks group, where the P port returns the displacement signal, while V port returns the velocity signal. If the accelerator information is needed, a derivative block can be connected to the V port. As in the cases in circuits, a reference point can be added, which is the Mechanical

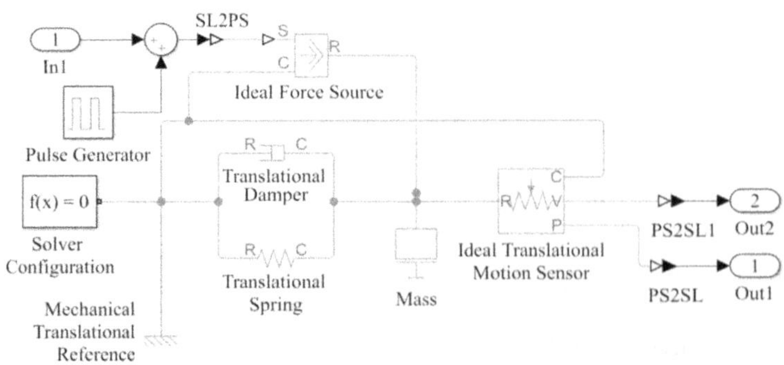

Figure 10.48: Spring–damper simulation model (file: `c10mmass1.slx`).

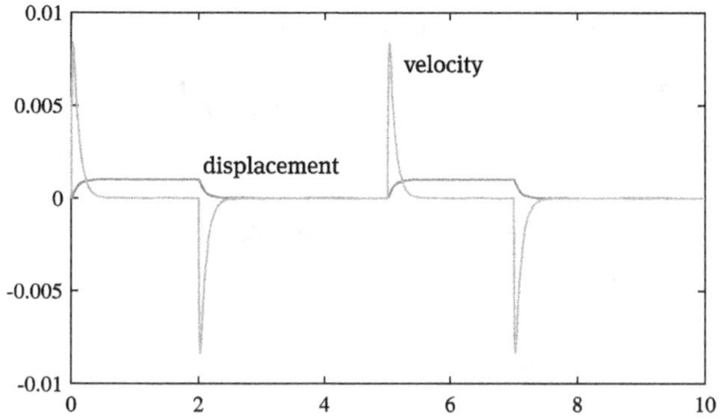

Figure 10.49: Simulation results.

Translational Reference block. Therefore, the simulation model in Figure 10.48 can be established.

Assume that the external force applied is a square wave, with magnitude of 1 N, and period of 5 seconds, with duty ratio of 40 %. The above parameters can be specified in the parameter dialog box in the Pulse Generator block. Therefore, the displacement and velocity of the mass can be obtained, as shown in Figure 10.49.

Note that the condition for the kinetic analysis is that there is no friction between the mass and the ground surface. If friction is to be considered, the problem is becoming complicated, since the direction of the mass M is often changed, and static friction forces should also be considered. A better way is to use Stateflow to describe the frictions. The stateflow and finite state machine modeling will be presented later.

Since this model is a linear one, the input and output port blocks can be added to the model, such that the transfer function matrix from the inputs to the outputs can be obtained with linearization method:

```
>> G=linearize('c10mmass1')
```

The transfer function matrix model obtained is

$$\boldsymbol{G}^{\mathrm{T}}(s) = \left[\frac{1}{s^2+100s+1000}, \frac{s}{s^2+100s+1000}\right].$$

For the modeling problem of such a simple structure, Newton's second law is sufficient to find the mathematical model. For instance, if the parameters are substituted into the equations, the transfer function model can be easily derived from the differential equation $Mx''(t) + f_{\mathrm{v}}x'(t) + Kx(t) = f(t)$.

If the system structure becomes complicated, mathematical modeling may become impossible. The physical modeling method can still be used in establishing the Simulink models.

Example 10.13. Consider the complicated spring–damper system shown in Figure 10.50. It is very complicated to derive the mathematical model from the figure. Assume that the damping coefficients of the three dampers are all $f_{v1} = f_{v2} = f_{v3} = 100$ N·s/m, and the elastic coefficients of the springs are respectively $K_1 = K_2 = 1000$ N/m, $K_3 = K_4 = 400$ N/m, with masses $M_1 = M_2 = M_3 = 1$ kg. Set up the Simulink model, and find the relationship between the external force and displacement $x_2(t)$ signal. Also, find the transfer function model.

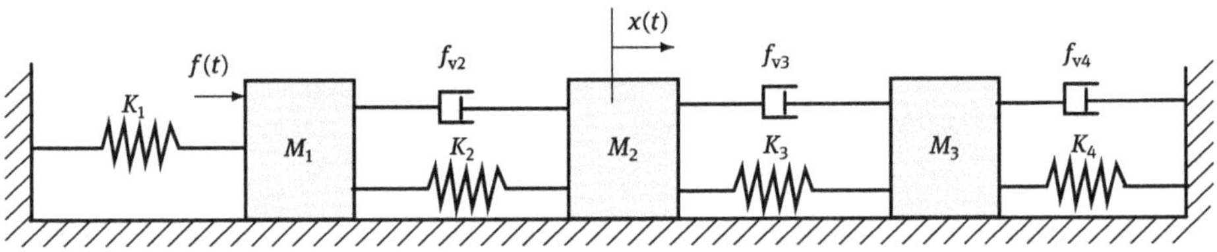

Figure 10.50: The new spring–damper system.

Solutions. Refer to the simulation model established above, the spring and damper blocks can be added to the model, and the simulation model in Figure 10.51 can be established. The parameters of the blocks can be set accordingly. It can be seen that the modeling method are simple and straightforward. It is not quite demanding to the relevant knowledge in physics. The building block method can be used directly.

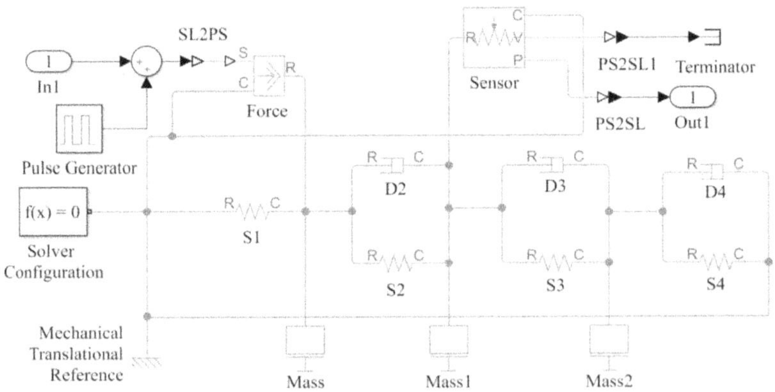

Figure 10.51: The new Simulink model (file: `c10mmass2.slx`).

If the pulling force $f(t)$ is a square wave of 4 N, the simulation results can be found. If K_2 is reduced from 1000 to 400, and simulation is carried out again, then the two results can be obtained, as shown in Figure 10.52.

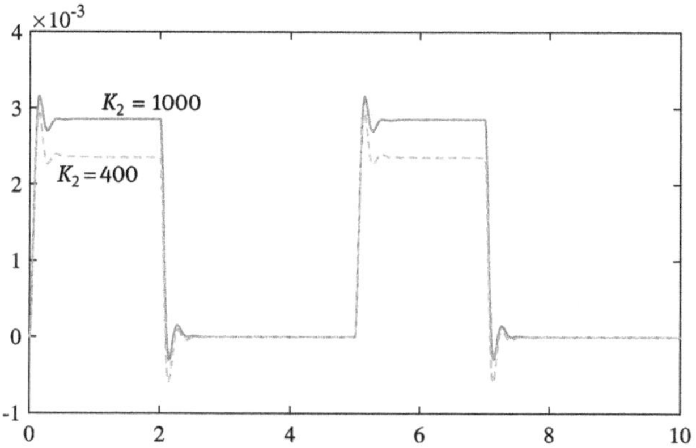

Figure 10.52: The displacement of the multiple spring–damper system.

The following commands can be issued for linearization:

```
>> G=linearize('c10mmass2'); G=minreal(zpk(G))
```

The linearized model is obtained below

$$G(s) = \frac{100(s+195.9)(s+10)(s+4.083)}{(s+318.2)(s+145.2)(s+8.486)(s+4.082)(s^2+24.01s+699.6)}.$$

10.4.2 Introduction to Multibody blockset

The former product of Multibody blockset was released in October 2001, named SimMechanics Blockset. The blockset, together with the three-dimensional animation technique can be used to set up simulation models for mechanical systems. The quantities of forces and torques in Newton's mechanics are used in the modeling of joints and bodies, so as to analyze and design mechanical systems.

The Multibody blockset can be opened from Simscape, or with command `sm_lib` in MATLAB command window. The blockset can be opened, as shown in Figure 10.53. The commonly used groups in the blockset are provided in the blockset.

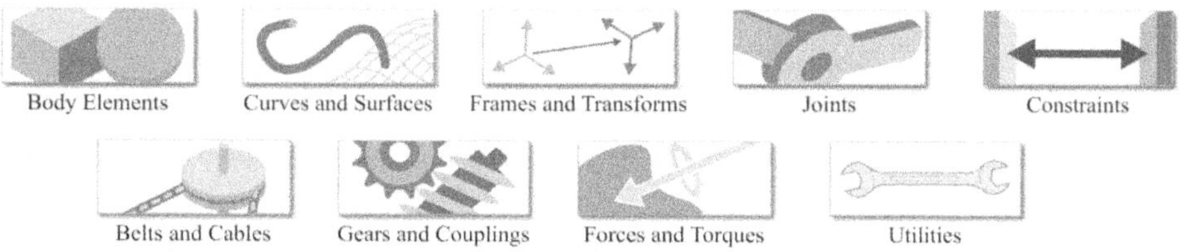

Figure 10.53: Multibody blockset.

(1) **Body Elements** group. Double click the icon, and the group is opened as shown in Figure 10.54. It can be seen that the blocks of various bodies are provided, such as the **Cylindrical Solid**, **Ellipsoidal Solid** and **Spherical Solid**. Also, **Flexible Bodies** and **Variable Mass** blocks are provided. The blocks such as **Inertia Sensor** is also included.

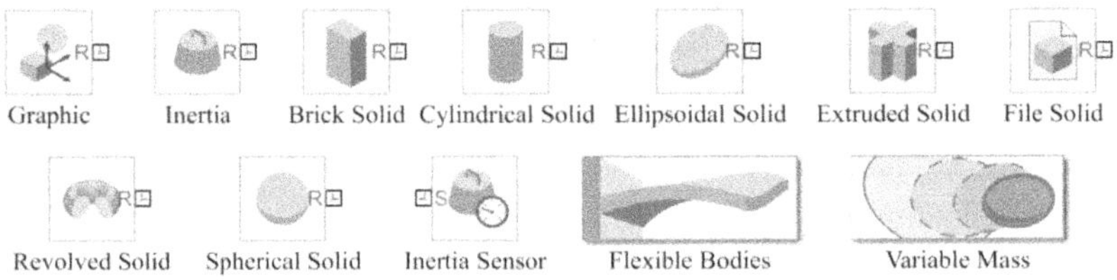

Figure 10.54: Body Element group.

(2) **Joints** group. Double click the **Joints** group icon, the group is opened as shown in Figure 10.55, where, the blocks **6-DOF Joint**, **Bearing Joint**, **Prismatic Joint**, **Planar Joint**, **Revolute Joint**, **Universal Joint** and so on are provided. Normally, the blocks are with two ports: the one labeled **B** is referred to as the base, and the **F** port is the follower. According to the system simulation concepts, **B** port can be regarded as the input, while **F** is the output.

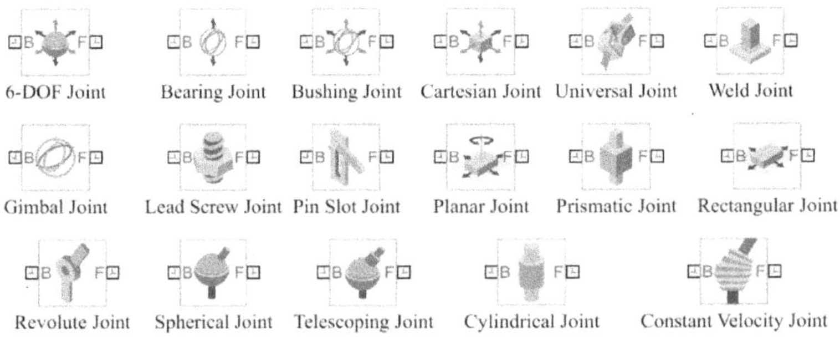

Figure 10.55: Joints group.

(3) **Force and Torque** group. Double click the **Force and Torque** group icon, and the group can be opened as shown in Figure 10.56. It can be seen that the blocks such as **External Force and Torque**, **Internal Force**, **Gravitational Field** and **Spring and Damper Force** are provided.

(4) **Constraints** group, as shown in Figure 10.57, include **Angle Constraints**, **Distance Constraints**, **Point on Curve Constraint** blocks and so on.

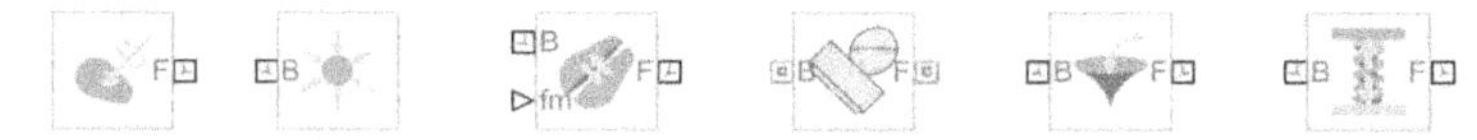

Figure 10.56: Force and Torque group.

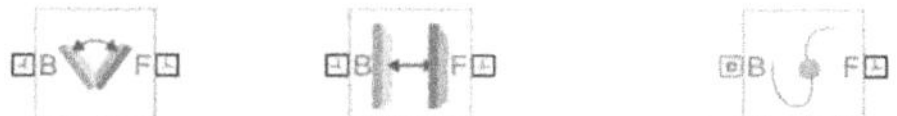

Figure 10.57: Constraints group.

(5) **Frames and Transforms** group as shown in Figure 10.58, where **Reference Frame**, **Rigid Transform**, **Transform Sensors** and **World Frame** blocks are provided.

Figure 10.58: Frames and Transforms blockset.

In fact, it can be seen from the low-level **Body Element** block that there is no base and follower ports in the block, therefore, it is not possible to connect to the joints. Some compound blocks can be used with low-level blocks. Examples are illustrated for two commonly used compound blocks.

Example 10.14. The construction of some compound blocks are illustrated in the Multibody blockset. These compound blocks are already masked into reusable ones. In Figure 10.59, two commonly used link blocks are illustrated.

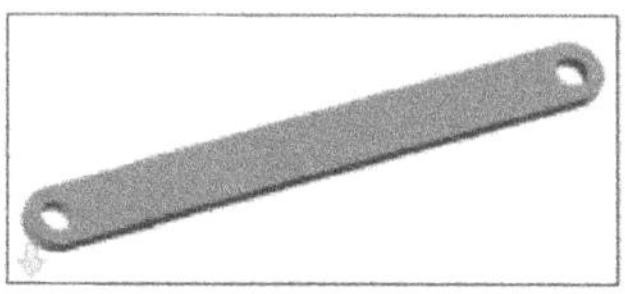

(a) compound block 1

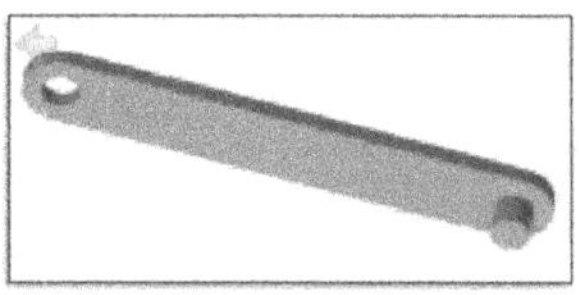

(b) compound block 2

Figure 10.59: Two compound links.

In fact, the two compound blocks are masked reusable ones, where the low-level model of compound block 2 is shown in file `sm_compound_body.slx`. The compound block is composed of three blocks, and the internal structure is shown in Figure 10.60. Note that internal coordinate transformations are used in linking the components.

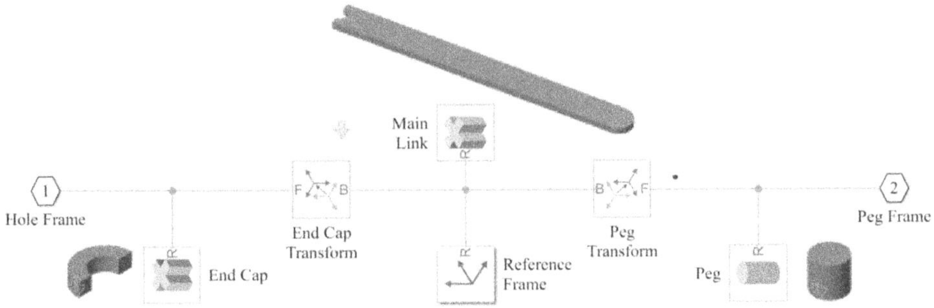

Figure 10.60: Internal structure of a compound link block.

10.4.3 Modeling and simulation of four-bar mechanisms

A planar four-bar mechanism is used as an example to demonstrate the modeling and simulation of mechanical systems with the Multibody blockset. It is expected that the reader may acquire some understanding to the modeling details, so as to solve the modeling and simulation problems of other mechanical systems with such a tool.

Example 10.15. Consider the illustration of a planar four-bar mechanism shown in Figure 10.61, with the sizes of all the links labeled in the figure. Build up a simulation model for the four-bar mechanism.

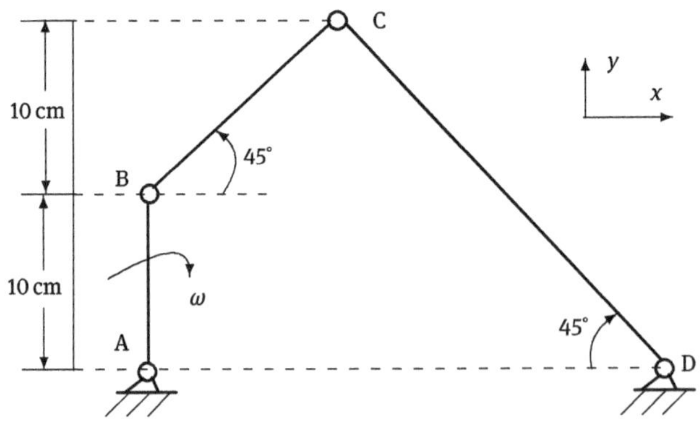

Figure 10.61: A planar four-bar mechanism.

Solutions. It can be seen that there are two fixed bases and three links in the illustration. The links and bases are joined by joints with one degree of freedom. The AB link is also known as the crank, rotating around A point in 360° rotation; BC link is known as the connector, while CD link is known as the rocker, such that the rocker rotate around joint D back and forth. The four-bar mechanism is also known as crank–rocker four-bar mechanism.

A demonstration model `sm_four_bar.slx` is provided in MATLAB, which can be used as a starting point in the simulation of four-bar mechanism. A new simulation model can be established as shown in Figure 10.61, and the new parameters can be set according to the sizes in Figure 10.62. Apart from the joints and links, two base blocks are also described.

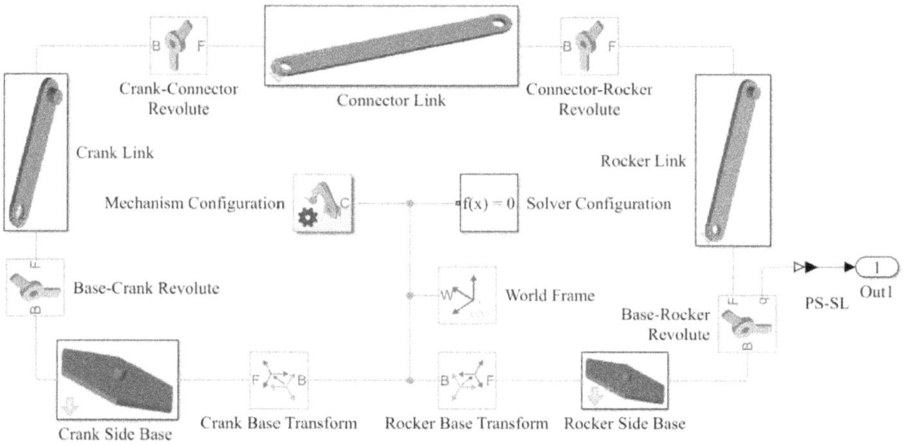

Figure 10.62: Four-bar mechanism model (file: `c10m4bar.slx`).

Double click the base block, the parameter dialog box is opened as shown in Figure 10.63. The parameters such as **Base radius**, **Base thickness**, **Peg radius** and the color of the base can be specified in the dialog box.

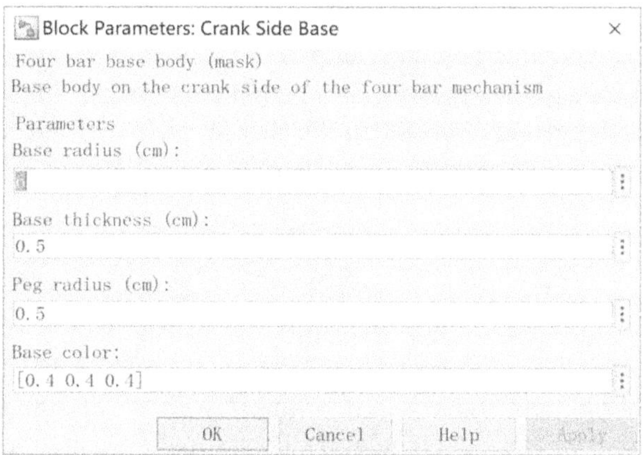

Figure 10.63: Parameter dialog box of the base.

Double click the link block, and the parameter dialog box in Figure 10.64 can be opened. The value of **Link density** can be specified. If the an iron link is used, the density is 7800 kg/m^3. The **Link length** parameters can be computed from the ones in Figure 10.61, where the length of AB link is 10, the BC length is $10\sqrt{2} \approx 14.1412$ and the CD link is $20\sqrt{2} \approx 28.2843$. Besides, the **Link Color** can also be specified. Based on the parameters specified, simulation can be carried out for the four-bar mechanism.

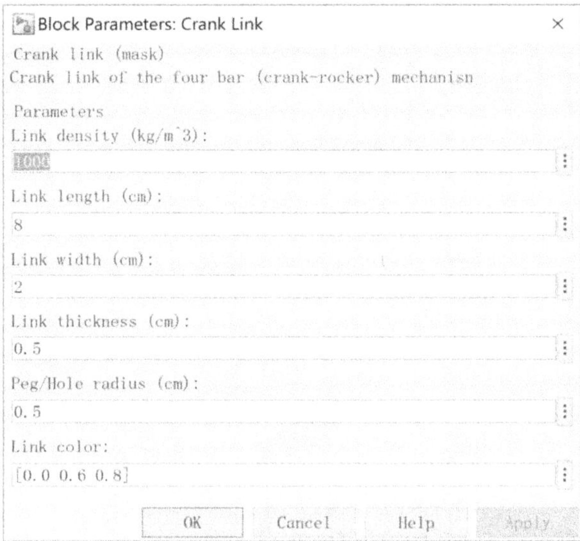

Figure 10.64: Parameter dialog box of the link.

With all the parameters specified, simulation process can be launched. In the simulation process, a Mechanics Explorer can be opened automatically in Simulink, and the three-dimensional animation of the four-bar mechanism can be played, as shown in Figure 10.65. The scroll bars can be adjusted to set the speed of the replay.

The benefit of using physical modeling and simulation is that there is no need to write out the mathematical models of the systems. The components such as links, joints and bases can be assembled together to establish the simulation model. Also, the Mechanics Explorer can be opened automatically to display three-dimensional animation of the system.

10.5 Exercises

10.1 Consider the circuit in Figure 10.66. Assume that $R_1 = R_2 = R_3 = R_4 = 10\,\Omega$, $C_1 = C_2 = C_3 = 10\,\mu\text{F}$, and the input signal is $v(t) = \sin\omega t$, with $\omega = 10$, and simulates the circuit and finds the output signal $v_c(t)$. Compare the simulation results with its analytical solution.

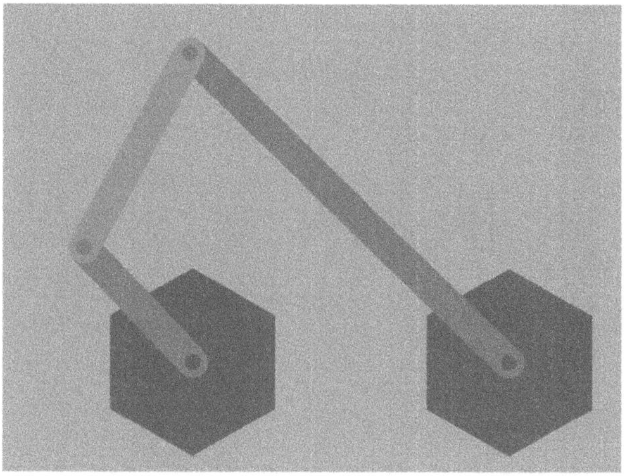

Figure 10.65: Three-dimensional animation in the Mechanics Explorer.

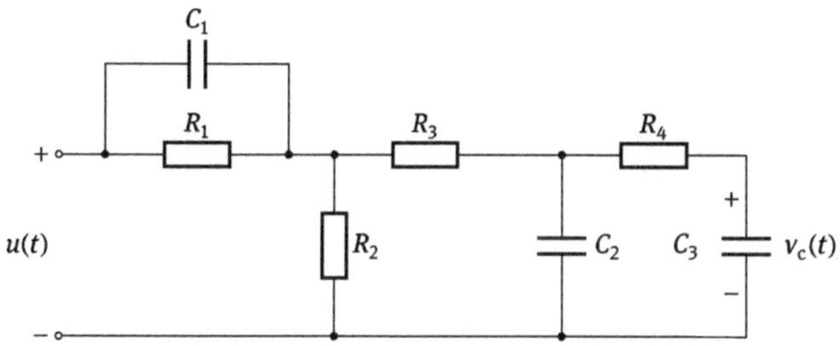

Figure 10.66: Circuit for exercise 1.

10.2 Set up the simulation models of transistor amplifier circuits in Figures 10.67(a) and (b). Draw the signals in the circuits under different control signals.

10.3 Implement the following digital logic expressions in Simulink, and show that they are equivalent:

① $Z_1 = A + B\overline{C} + D$,
② $Z_2 = AB(C + D) + D + \overline{D}(A + B)(\overline{B} + \overline{C})$.

10.4 Write out the logic expression for the circuit shown in Figure 10.68.[20] Create a Simulink model for the expression and find the simulation results.

10.5 Establish the Simulink models for the operational amplifier circuits in Figures 10.69(a), (b) and (c). Draw the output of the circuits under different input signals. Extract the transfer function models of the system. Besides, when the values of the gains in the operational amplifiers are set respectively to 10^4, 10^6 and infinity, find the differences in the output signals.

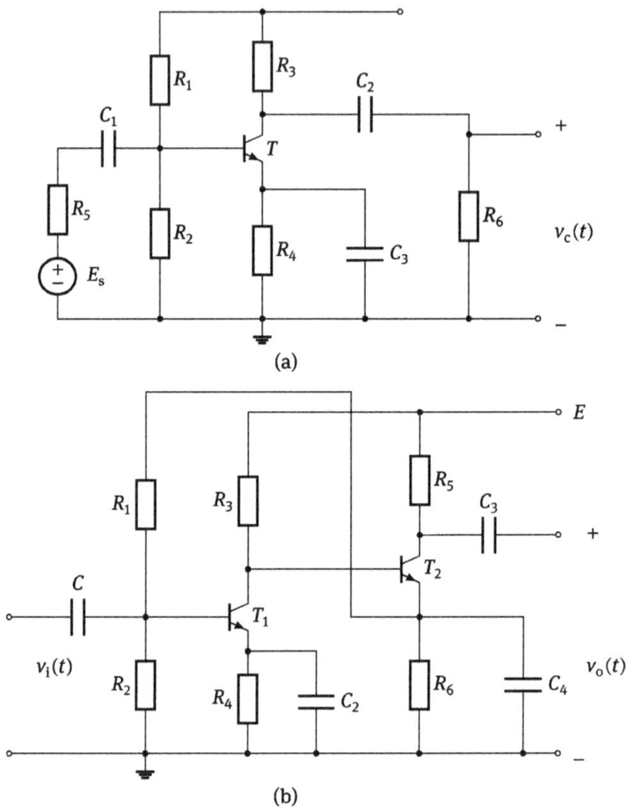

Figure 10.67: Transistor circuits for exercise 2.

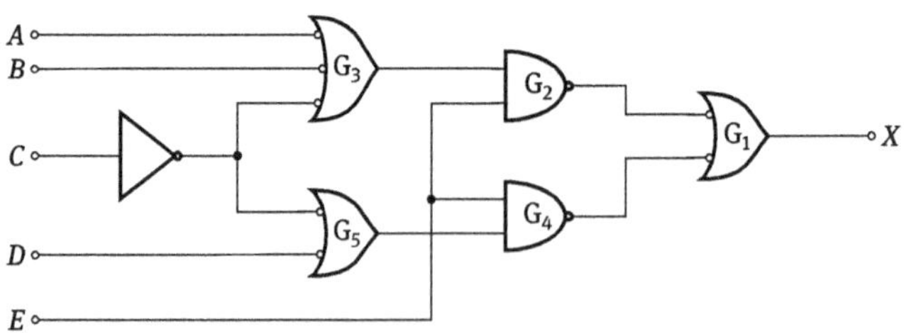

Figure 10.68: The logic circuit for exercise 4.

10.6 In the BCD decoder circuit in Example 10.10, only the digit display from 0 to 9 are supported in the truth table, while A to F cannot be displayed. Establish a complete truth table, such that the decoder can display 0 to F.

10.7 If in model in Example 10.10, the signals $A_1 = A_3 = A_4 = 0$, $A_2 = 1$, judge which segments in the LEDs are turned on, and which are turned off. If all the

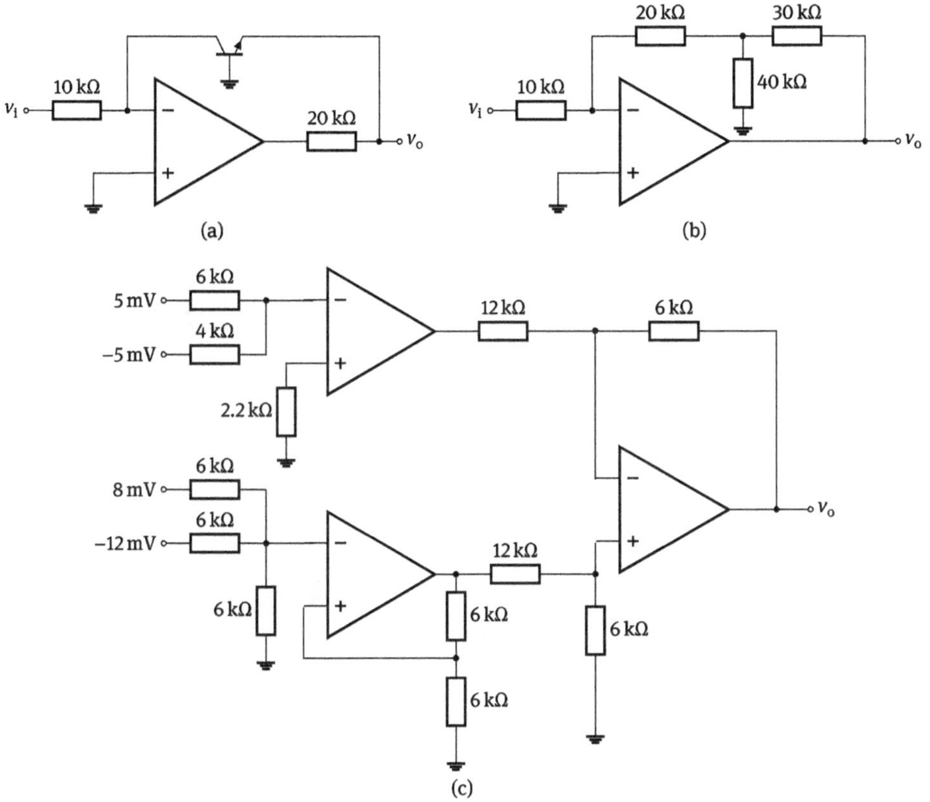

Figure 10.69: The operational amplifier circuits for exercise 5.

four signals are 1's, what will be the situation? Validate your conclusions with simulation experiment.

10.8 Assume that the J-K flip-flop circuit is Figures 10.70(a) and (b).[20] Use simulation methods to draw the time-series plot of the output signal.

10.9 Consider again the problem in Example 10.12. Assume that there is friction between the mass M and the surface, with frictional coefficient of $f_V = 0.25$. Establish a simulation model and find the equivalent linear model.

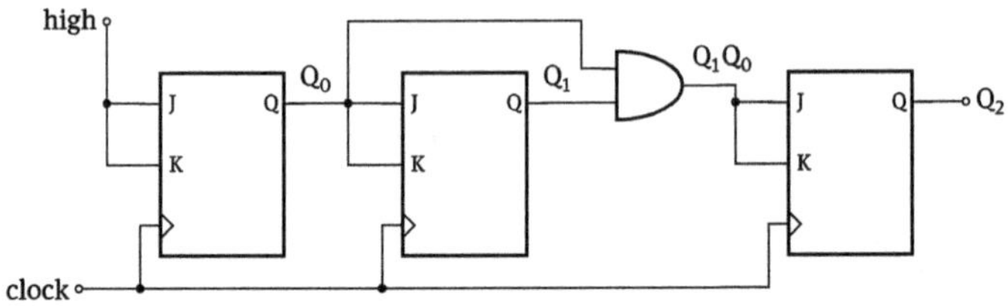

(a)

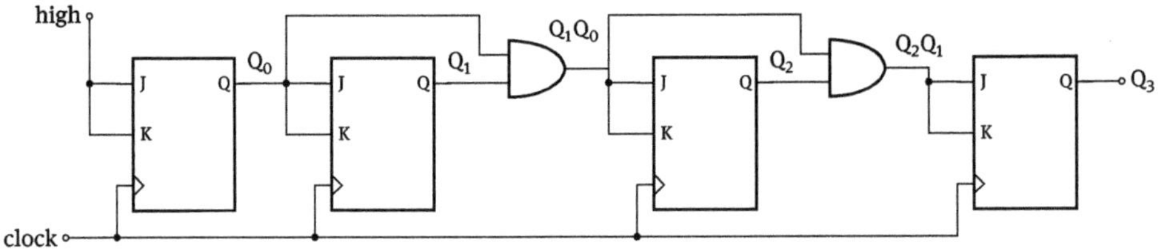

(b)

Figure 10.70: The flip-flop circuits for exercise 8.

11 Simulation of nonengineering systems

Various blocksets for engineering system modeling have been discussed earlier. The modeling and simulation tasks for engineering systems can easily be completed. For nonengineering systems, there are also toolboxes and blocksets, and also third-party tools by the researchers. These tools can be used in modeling and simulation tasks. The users may also build the blocksets in his own research area.

In this chapter, Section 11.1 introduces the modeling and simulation in pharmacokinetics. Then concentrations are made on physiological modeling and simulation methods. In Section 11.2, the concept of the finite state machine is introduced, and the modeling and simulation with Stateflow are illustrated. Stateflow can be used in the modeling and simulation of complicated supervisory control processes, where new facilities are introduced into Simulink modeling. The flow control structures such as loops and conditional ones can be implemented also in Stateflow. In Section 11.3, the modeling and simulation of discrete even systems are introduced. Examples are used to illustrate queueing systems with SimEvents blockset.

11.1 Modeling and simulation in pharmacokinetics

In this section, the basic concepts of pharmacokinetics (PK) and pharmacodynamics (PD) are introduced, and then modeling and simulation are studied. Compartment models are introduced first, then discussion on physiologically based modeling in pharmacokinetics are illustrated. Simulink models of anesthetic processes are constructed and transfer function models are obtained.

11.1.1 Introduction to pharmacokinetics

Pharmacokinetics is the study established on the theories of pharmacy, mathematics and kinetics. The study was established in 1980s, and in the study, the absorption, distribution, metabolism and excretion of drugs in human bodies are considered, when they are administrated into human bodies by intravenous (IV) or oral administration. In particular, the relationships between drug concentration in plasma or in organs are studied as a function of time. Pharmacodynamics is a discipline that studies mainly the physiological effects of drugs in human bodies. The mechanism of drug action, and the relationship of the drug effect with concentration are also studied. The relationship between pharmacokinetics and pharmacodynamics is illustrated in Figure 11.1, where $C_{\rm p}(t)$ is the drug concentration, while $C_{\rm e}(t)$ is the drug effect.

If the concept of feedback control is introduced, a controller can be designed to control the drug administration such that the best drug effect can be reached. For

https://doi.org/10.1515/9783110734959-011

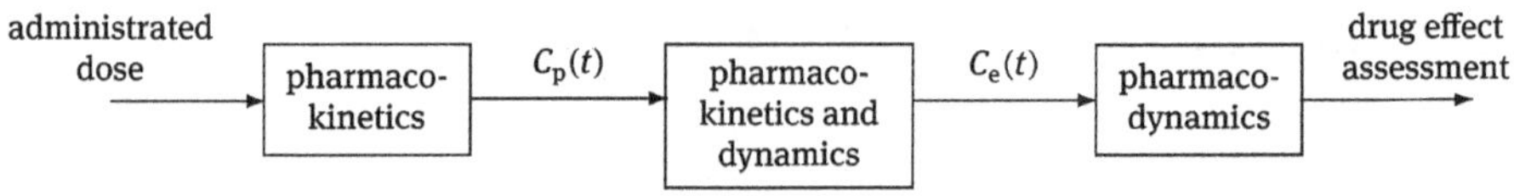

Figure 11.1: The relationship between pharmacokinetics and pharmacodynamics.

instance, the speed and dose of anesthetics can be controlled by a microinjection pump so as to achieve the best performance.

11.1.2 Compartment models in pharmacokinetics

It is known from pharmacokinetics experiments that the concentrations $c(t)$ of drugs in plasma are usually expressed as weighted sums of some exponential functions:[73]

$$c(t) = \sum_{i=1}^{n} c_i e^{-\lambda_i t}, \tag{11.1.1}$$

where c_i and λ_i are undetermined coefficients, n is the term of exponential functions. In fact, it is known from differential equation theory that if $c(t)$ can be expressed in (11.1.1), $c(t)$ function satisfies nth order homogeneous linear differential equation.

For ease of manipulation, it is not quite suitable to choose large values for n. Normally, n can be selected as 2 or 3.[16] The compartment model is a simple model to study the drug concentrations in plasma, which expresses the relationship of concentration versus time. In compartment modeling, the human body can be regarded as one or more interconnected compartments. Compartments are usually adopted as the basic unit in modeling. The interconnection of different compartments is shown in Figure 11.2, where variable k_{ij} means to flow from the jth compartment to the ith one, and subscript 0 means the association to the outside world. The entire mathematical model can be expressed as

$$\frac{dx_i}{dt} = \sum_{j=1,j\neq i}^{n} k_{ij}x_j - \sum_{j=1,j\neq i}^{m} k_{ji}x_i - k_{0i}x_i + u_i(t), \quad i = 1, 2, \cdots, n. \tag{11.1.2}$$

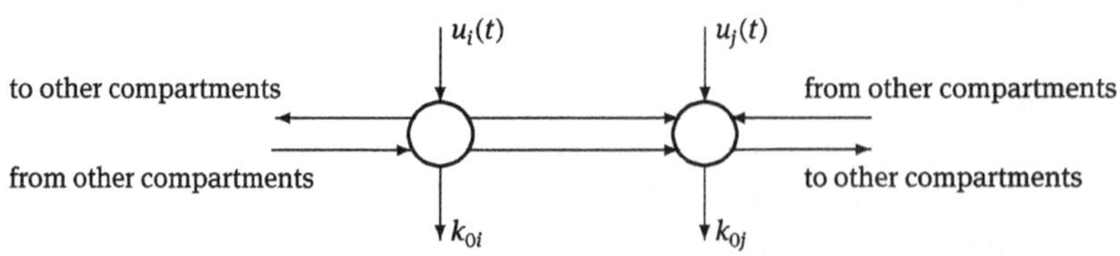

Figure 11.2: The relationship between two compartments.

Compartment model is a special case of linear constant differential equations, whose matrix form can be written as

$$\begin{bmatrix} x_1'(t) \\ x_2'(t) \\ \vdots \\ x_n'(t) \end{bmatrix} = \begin{bmatrix} a_{11} & a_{12} & \cdots & a_{1n} \\ a_{21} & a_{22} & \cdots & a_{2n} \\ \vdots & \vdots & \ddots & \vdots \\ a_{n1} & a_{n2} & \cdots & a_{nn} \end{bmatrix} \begin{bmatrix} x_1(t) \\ x_2(t) \\ \vdots \\ x_n(t) \end{bmatrix} + \boldsymbol{B}u(t), \tag{11.1.3}$$

where

$$a_{ij} = k_{ij}, \quad j \neq i, \quad \text{and} \quad a_{jj} = \sum_{i=1,j\neq j}^{n} k_{ij} - k_{0j}, \tag{11.1.4}$$

where the subscript 0 stands for the outside world, and

$$a_{jj} \leqslant 0, \quad a_{i,j} \geqslant 0 \ (i \neq j), \quad k_{0j} = 0, \quad |a_{jj}| \geqslant \sum_{i=1,i\neq j}^{n} k_{ij}. \tag{11.1.5}$$

In particular, if a model is composed of two or three compartments, the simple forms below can be used:

$$\begin{cases} \begin{bmatrix} x_1'(t) \\ x_2'(t) \end{bmatrix} = \begin{bmatrix} -(c_1+c_2) & c_3 \\ c_2 & -c_3 \end{bmatrix} \begin{bmatrix} x_1(t) \\ x_2(t) \end{bmatrix} + \begin{bmatrix} 1 \\ 0 \end{bmatrix} u(t), \\ y(t) = c_1 x_1(t), \end{cases} \tag{11.1.6}$$

$$\begin{cases} \begin{bmatrix} x_1'(t) \\ x_2'(t) \\ x_3'(t) \end{bmatrix} = \begin{bmatrix} -(c_1+c_2) & c_3 & 0 \\ c_2 & -(c_3+c_4) & c_5 \\ 0 & c_4 & -c_5 \end{bmatrix} \begin{bmatrix} x_1(t) \\ x_2(t) \\ x_3(t) \end{bmatrix} + \begin{bmatrix} 1 \\ 0 \\ 0 \end{bmatrix} u(t), \\ y(t) = c_1 x_1(t). \end{cases} \tag{11.1.7}$$

If the coefficients are known, Simulink can be used directly to express the state space equations, such that the entire model can be established. A compartment based blockset was development by Wada,[69] where the One Compartment, Two Compartment and Three Compartment blocks as shown in Figure 11.3. The Branch and Juncture blocks are also provided in the blockset. The branch method is controlled directly by the branching coefficients, which is the second input port. These blocks can be used in pharmacokinetics modeling.

Right click the Two Compartment block, and select Look Under Mask shortcut menu item, the internal structure of the model is shown in Figure 11.4(a). Double click the block icon, the parameter dialog box is shown in Figure 11.4(b). The mathematical model of the corresponding two-compartment model in Figure 11.4(a) can be written

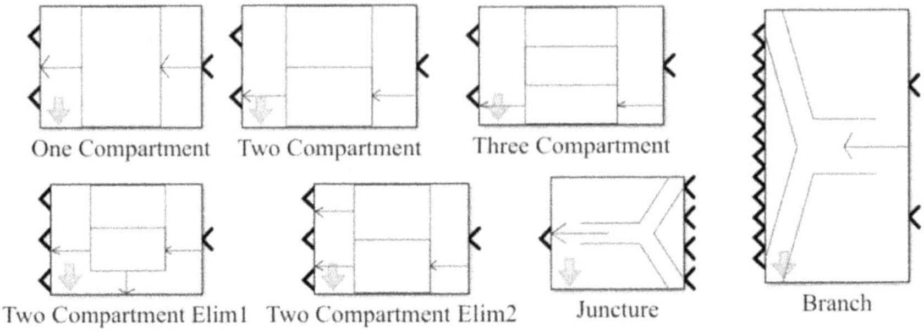

Figure 11.3: Wada's blockset (file: `wada_blocks.slx`).

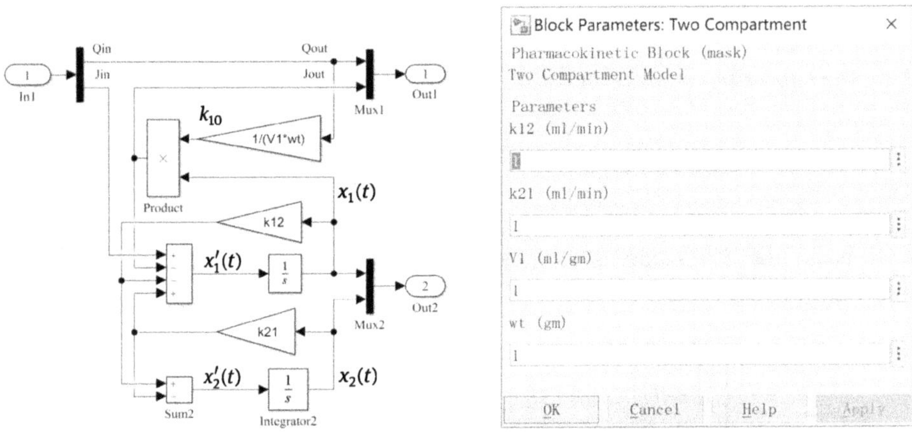

(a) internal structure (b) parameter dialog box

Figure 11.4: Two-compartment block and parameters.

as

$$\begin{cases} \begin{bmatrix} x_1'(t) \\ x_2'(t) \end{bmatrix} = \begin{bmatrix} -(k_{12}+k_{10}) & k_{21} \\ k_{12} & -k_{21} \end{bmatrix} \begin{bmatrix} x_1(t) \\ x_2(t) \end{bmatrix} + \begin{bmatrix} 1 \\ 0 \end{bmatrix} u(t), \\ \boldsymbol{y}_2(t) = \boldsymbol{x}(t), \end{cases} \tag{11.1.8}$$

where $k_{10} = 1/(Vw)$, w is the mass of the compartment and V is the volume.

Besides, the internal structures of the two compartment models are shown respectively in Figures 11.5(a) and (b), with additional output port Out3, and the mathematical models are different. They are respectively

$$\begin{cases} \begin{bmatrix} x_1'(t) \\ x_2'(t) \end{bmatrix} = \begin{bmatrix} -(k_{12}+k_{10}) & k_{21} \\ k_{12} & -k_{21} \end{bmatrix} \begin{bmatrix} x_1(t) \\ x_2(t) \end{bmatrix} + \begin{bmatrix} 1 \\ 0 \end{bmatrix} u(t), \\ \boldsymbol{y}_2(t) = \boldsymbol{x}(t), \ \ y_3(t) = k_{20}x_2(t), \end{cases} \tag{11.1.9}$$

$$\begin{cases} \begin{bmatrix} x_1'(t) \\ x_2'(t) \end{bmatrix} = \begin{bmatrix} -(k_{12}+k_{10}) & k_{21} \\ k_{12} & -k_{21} \end{bmatrix} + \begin{bmatrix} x_1(t) \\ x_2(t) \end{bmatrix} + \begin{bmatrix} 1 & k_{10} \\ 0 & 0 \end{bmatrix} \begin{bmatrix} u_1(t) \\ u_2(t) \end{bmatrix}, \\ \boldsymbol{y}_2(t) = \boldsymbol{x}(t),\ y_3(t) = k_{10}x_1(t). \end{cases} \tag{11.1.10}$$

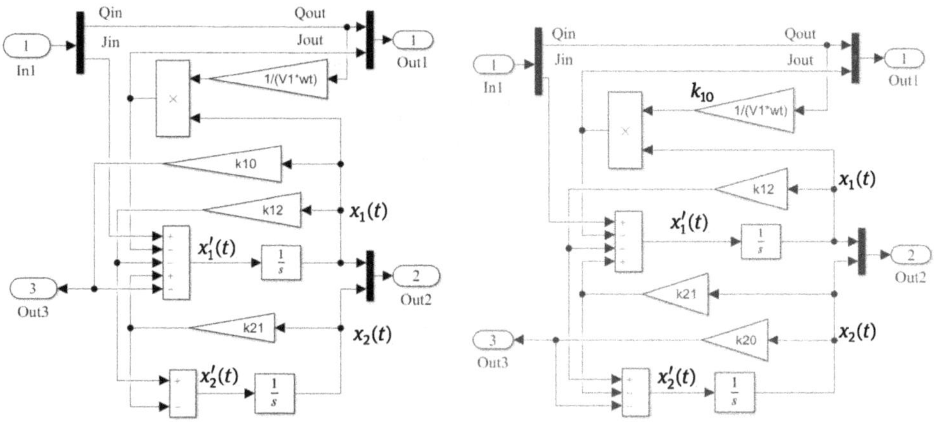

(a) two compartment model 1 (b) two compartment model 2

Figure 11.5: Two compartment model with additional metabolism output.

It is seen that the latter is with two input signals, and the definitions of metabolism and excretions are different.

If compartment models are used, and the organs and human body physiological structures are not considered, then the state variables have no physical meaning, and the modeling quality and conveniences are usually not satisfactory. Therefore, physiological modeling methods can be adopted. Besides, Wada's compartment blocks can be used to create physiological-based pharmacokinetics models, which will be further discussed later.

11.1.3 Wada models with transport delays

In ordinary compartment models, the information such as transport delay are not considered. If transport delays are considered, the mixed model in Figure 11.6 can be constructed.

It can be seen that there are two parts in the model, where the one in the forward path is referred to as the cardiopulmonary model, while the one in the feedback path is referred to as the systemic model. Select the state variables

$$\begin{cases} \boldsymbol{x}_{\rm p}(t) = [x_{\rm LH}(t), x_{\rm LB}(t), x_{\rm LT}(t), x_{\rm RH}(t)]^{\rm T}, \\ \boldsymbol{x}_{\rm S}(t) = [x_{\rm VRG}(t), x_{\rm M}(t), x_{\rm F}(t), x_{\rm R}(t)]^{\rm T}, \end{cases} \tag{11.1.11}$$

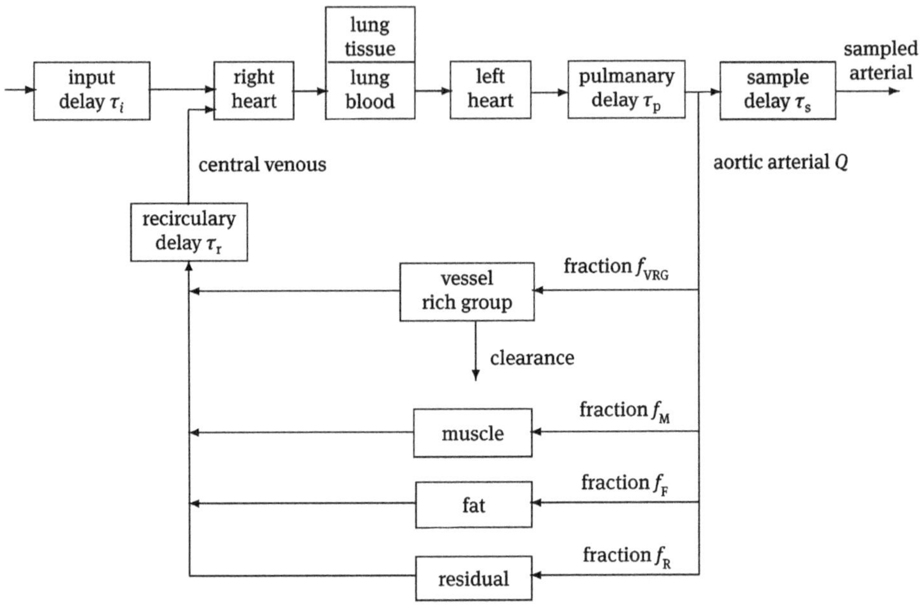

Figure 11.6: Wada's model with transport delays.

the following state space model with internal delay can be obtained:[70]

$$\begin{cases} \boldsymbol{x}_p'(t) = \boldsymbol{A}_p \boldsymbol{x}_p(t) + \boldsymbol{B}_p[Q\boldsymbol{c}_v(t) + u(t - \tau_i(t))], \\ c_a(t) = \boldsymbol{C}_p \boldsymbol{x}_p(t - \tau_p), \end{cases} \tag{11.1.12}$$

$$\begin{cases} \boldsymbol{x}_s'(t) = \boldsymbol{A}_s \boldsymbol{x}_s(t) + \boldsymbol{B}_s Q c_a(t), \\ \boldsymbol{c}_v(t) = \boldsymbol{C}_s \boldsymbol{x}_s(t - \tau_r), \end{cases} \tag{11.1.13}$$

where the state variables $\boldsymbol{x}_p(t)$ and $\boldsymbol{x}_s(t)$ are respectively the drug masses in the cardiopulmonary and systemic subsystems. The volumes of each organ can be expressed by V. The corresponding matrices in the state space expression are

$$\boldsymbol{A}_p = \begin{bmatrix} -\dfrac{Q}{V_{RH}} & 0 & 0 & 0 \\ \dfrac{Q}{V_{RH}} & -\left(k_{LB,LT} + \dfrac{Q}{V_{lung}}\right) & k_{LB,LT} & 0 \\ 0 & k_{LB,LT} & -k_{LT,LB} & 0 \\ 0 & \dfrac{Q}{V_{lung}} & 0 & -\dfrac{Q}{V_{LH}} \end{bmatrix}, \quad \boldsymbol{B}_p = \begin{bmatrix} 1 \\ 0 \\ 0 \\ 0 \end{bmatrix}, \tag{11.1.14}$$

$$\boldsymbol{C}_p = \begin{bmatrix} 0 & 0 & 0 & \dfrac{1}{V_{LH}} \end{bmatrix}, \tag{11.1.15}$$

and

$$
\boldsymbol{A}_\mathrm{S}=\begin{bmatrix} -\dfrac{Qf_\mathrm{VRG}+Cl}{V_\mathrm{VRG}} & 0 & 0 & 0\\ 0 & -\dfrac{Qf_\mathrm{M}}{V_\mathrm{M}} & 0 & 0\\ 0 & 0 & -\dfrac{Qf_\mathrm{F}}{V_\mathrm{F}} & 0\\ 0 & 0 & 0 & -\dfrac{Qf_\mathrm{R}}{V_\mathrm{R}} \end{bmatrix},\quad \boldsymbol{B}_\mathrm{S}=\begin{bmatrix} f_\mathrm{VGR}\\ f_\mathrm{M}\\ f_\mathrm{F}\\ f_\mathrm{R}\end{bmatrix}, \tag{11.1.16}
$$

$$
\boldsymbol{C}_\mathrm{S}=\begin{bmatrix} \dfrac{f_\mathrm{VGR}}{V_\mathrm{VGR}} & \dfrac{f_\mathrm{M}}{V_\mathrm{M}} & \dfrac{f_\mathrm{F}}{V_\mathrm{F}} & \dfrac{f_\mathrm{R}}{V_\mathrm{R}}\end{bmatrix}. \tag{11.1.17}
$$

Example 11.1. A set of practical model parameters are given below:[70]

```
>> T_i=5/60; T_s=5/60; T_p=5/60; T_r=30/60; Q=5.5;
   V_RH=0.26; V_lung=0.45; V_LH=0.26; k_LBLT=8.0;
   k_LTLB=2.3; Cl=0.64; rho=0.63; V_VRG=5.5;
   f_VRG=0.7; V_M=13.2; f_M=0.05; V_F=36.7; f_F=0.04;
   V_R=14.9; f_R=0.21;
   A_P=[-Q/V_RH,0,0,0; Q/V_RH,-(k_LBLT+Q/V_lung),k_LTLB,0;
        0,k_LBLT,-k_LTLB,0; 0,Q/V_lung,0,-Q/V_LH];
   B_P=[1; 0; 0; 0]; C_P=[0,0,0,1/V_LH];
   A_S=diag([-(Q*f_VRG+Cl)/V_VRG, -Q*f_M/V_M, ...
            -Q*f_F/V_F, -Q*f_R/V_R]);
   B_S=[f_VRG; f_M; f_F; f_R];
   C_S=[f_VRG/V_VRG, f_M/V_M, f_F/V_F, f_R/V_R];
   T=[0,1,1.01,1.2,1.21,5]'; U=[50,50,29,29,10,9]';
```

Set up a simulation model and analyze the drug concentration in plasma.

Solutions. From the above mathematical model, the corresponding Simulink model can be established easily, as shown in Figure 11.7. The above statements can be assigned to the PreLoadFcn property, such that each time the model is opened, and the data are loaded into MATLAB workspace automatically. Start the simulation process, the input and output signals can be obtained, as shown in Figure 11.8. For the readability of the plots, the input signal is multiplied by 0.1.

11.1.4 Pharmacokinetics toolbox and Simulink models

The pharmacokinetics toolbox designed by Wada[69] is a powerful tool in the modeling and analysis of anesthetic processes. Apart from the blockset in Figure 11.3, the physi-

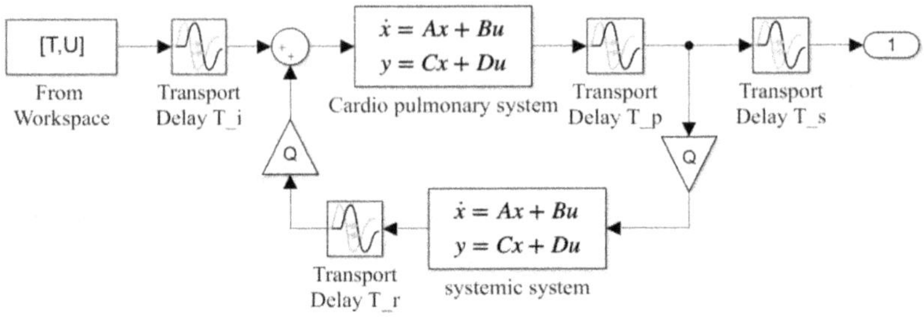

Figure 11.7: Wada's model with transport delays (file: `wada_dly.slx`).

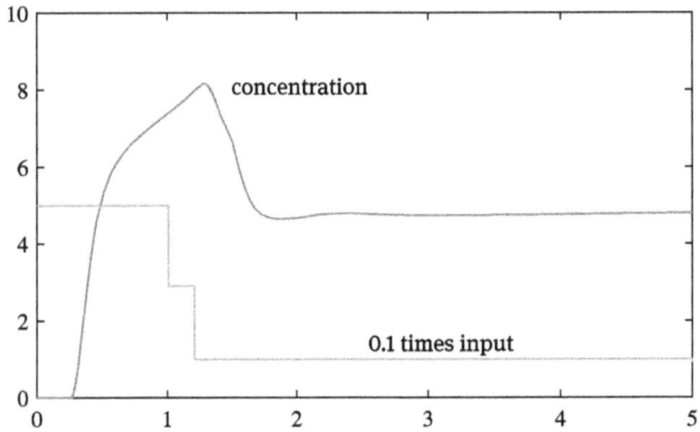

Figure 11.8: The input scheme and drug concentration.

ological based pharmacokinetics model shown in Figure 11.9 is also provided.[69] The organ blocks in the model are composed of compartment blocks and provided in the Wada's blockset. The parameters in different blocks are assigned with the programs provided in the toolbox. For instance, when fentanyl is administrated into the rat bodies, the file `fentanyl.m` can be executed. When it is administrated into human bodies, file `fent_hum.m` can be used. Open the `wada_model.slx` model, and the corresponding parameters can be assigned into Simulink model, and then the simulation process can be invoked.

11.1.5 Mapleson models and code

In this section, the effects of anesthetics upon the human body are studied, and control problems are considered. Another physiologically based pharmacokinetics model was proposed by Mapleson.[73] There are differences in Mapleson's model and Wada's model discussed earlier. In the Mapleson's model, the effect of organs upon the drug is shown in Figure 11.10, where K_x is the branching coefficient for each branch. In the kidney and liver models, excretion and metabolism are described. The Mapleson's

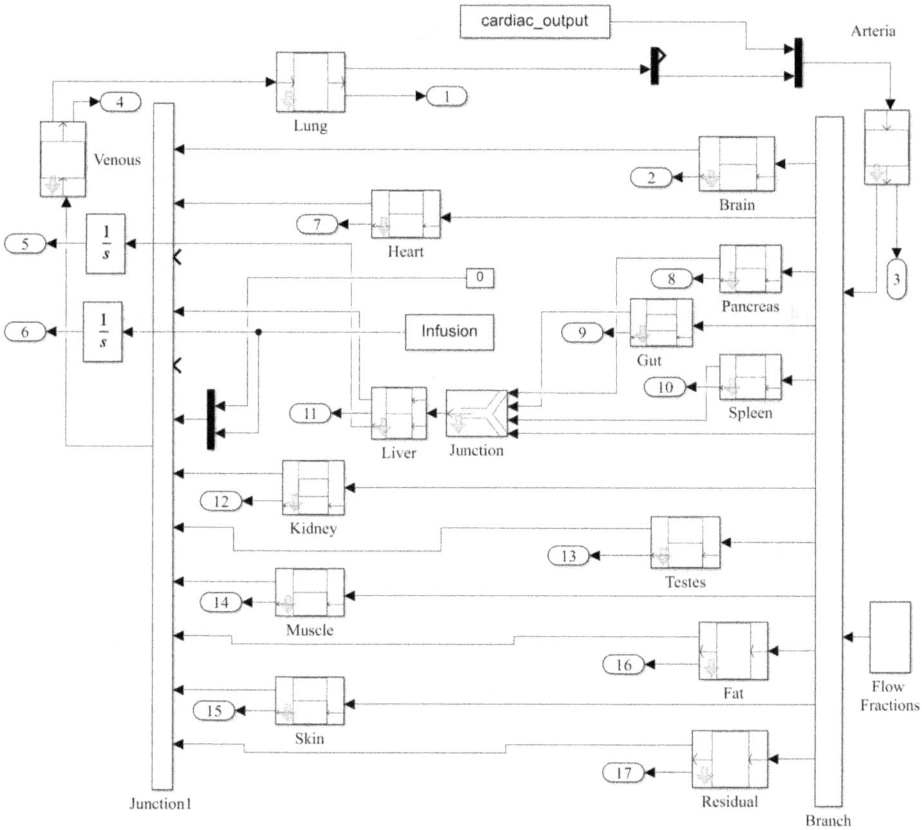

Figure 11.9: Physiological-based pharmacokinetics model (file: `wada_model.slx`).

model was extended by Higgins,[35] who wrote a program to assign data in the model when fentanyl is administrated.

It is a pity that Higgins's model was written in C. A MATLAB interface is designed for the model, such that the simulation results can loaded into MATLAB from the model. The approximate linear transfer functions are identified for each organ, such that the Simulink model in Figure 11.11 is constructed, where the coefficient 336.6 is used to convert the unit μmol in the original model into the ng unit. It can be seen from the model that, since transfer function blocks are used in the blocks, the model and parameter structures are simpler than the Wada's compartment models.

Studies show that the concentration of the drugs in various organs can be approximated in linear interpolation with the parameters such as weights, and cardiac output.[78, 46] The syntaxes to the code are given below, while the listing of the programs are not provided:

input parameters: `bd_wt, t_f, DT, card_out, FenDose, DoseDur`
`get_higgins_new` % call `higginsm.mexw64` function for simulation

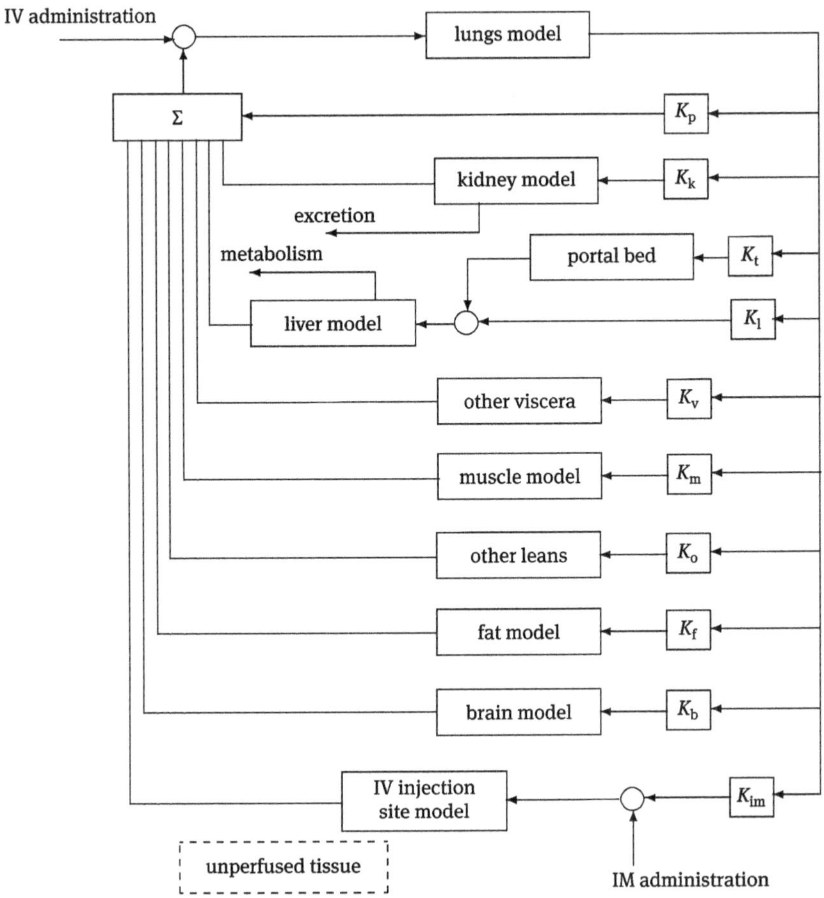

Figure 11.10: Mapleson's physiological model.

```
ident_tf              % identify all the subtransfer function models
```

where `bd_wt` is the weight in kg, `card_out` is the cardiac output in L/min, `t_f` is the stop time in seconds, `DT` is the step-size in seconds, `FenDose` and `DoseDur` are respectively the injection dose in μg and time duration in seconds.

Example 11.2. Assume that the weight of a person is 60 kg, with cardiac output of 6.8 L/min. The time duration is 3600 seconds, with step-size of 1 second. The total dose is 100 μg, and injection duration is 60 seconds, administrated evenly. Compute the drug concentration.

Solutions. The parameters can be entered first, then call the revised Higgins program, and the drug concentrations in all the organs can be found. System identification program can be called to identify the subtransfer functions of different models:

```
>> bd_wt=60; card_out=6.8; FenDose=100; DoseDur=60; t_f=3600; DT=1;
   get_higgins_new; ident_tf;
   G_Brain, G_Lungs      % extract the sub-transfer functions
```

After the function call, the identified subtransfer function models can be found, where, the brain and lungs models are

$$G_{\text{brain}}(s) = \frac{6.14\times 10^{-5}}{s+0.2988},\quad G_{\text{lungs}}(s) = \frac{0.3995s+153.8}{s^2+9.966s+31.16}.$$

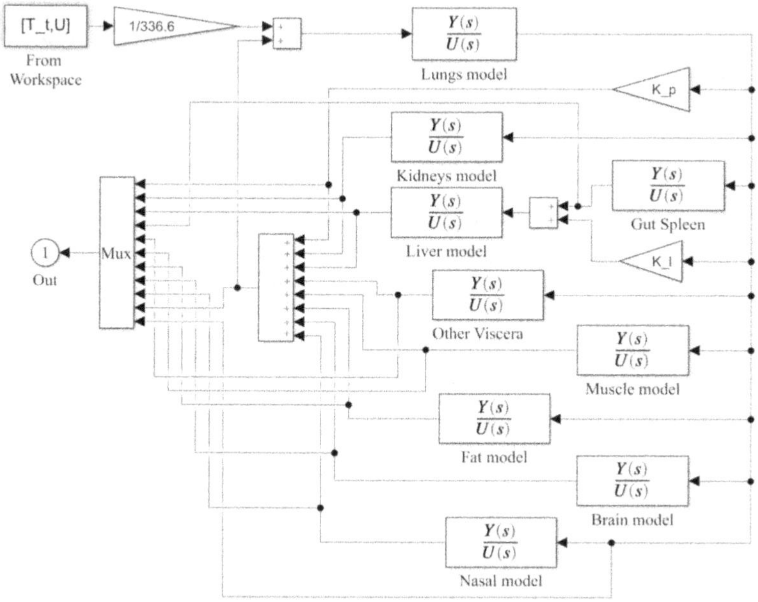

Figure 11.11: Physiological based model (file: `hig_simu0.slx`).

Apart from these subtransfer function models, the concentrations in other organs can be found. For instance, `ArtPool(:,1)` stores the aortic concentration in μmol/L. The other concentrations are also returned to MATLAB workspace, for details, the `who` command can be used.

If all the subtransfer function models are obtained, the overall transfer function from the IV input signal to the aortic concentration can be found. For instance, the equivalent transfer function model can be found with the following statements:

```
>> G_Fdbk=K_p+G_Kidneys+(G_Gut+K_l)*G_Liver+G_OtherVis+...
       G_Muscle+G_Fat+G_Nasal+G_Brain; % transfer function in feedback path
   G_Sys=zpk(feedback(G_Lungs,G_Fdbk,1))
```

where the overall model obtained is

$$G(s) = \frac{\begin{array}{c}0.39945(s+385.1)(s+1.425)(s+1.26)(s+0.8064)(s+0.8043)(s+0.2988)\\(s+0.1894)(s+0.1757)(s+0.1625)(s+0.1069)(s+0.008874)(s+0.001536)\end{array}}{\begin{array}{c}(s+1.261)(s+1.015)(s+0.81)(s+0.8063)(s+0.2984)(s+0.1927)(s+0.1626)\\(s+0.1495)(s+0.06935)(s+0.004999)(s+0.001113)(s^2+10.42s+30.86)\end{array}}.$$

Optimal reduction method can be used to find a fifth-order model.[85, 77]

```
>> G1=zpk(opt_app(G_Sys,4,5,0))
```

where `opt_app()` function[85] and other functions can be downloaded in the package for the book. The listings of the functions are not provided in the book. After optimal reduction computation, the reduced order models are

$$G_1(s) = \frac{12.973(s+0.3171)(s+0.1201)(s+0.008881)(s+0.001536)}{(s+1.956)(s+0.2878)(s+0.06987)(s+0.004935)(s+0.001098)}.$$

The concentration in aorta can be found as shown in Figure 11.12. The result from the linear model and reduced order models are also found, and it can be seen that the linear identified model is rather close to the original model, the reduced model also yields satisfactory fitting results:

```
>> u=100/60*(t<=1); y=ArtPool(:,1)*336.6; % aorta concentration in ng/ml
   y1=lsim(G_Sys,u,t); y2=lsim(G1,u,t);
   plot(t,y,t,y1,'--',t,y2,':')
```

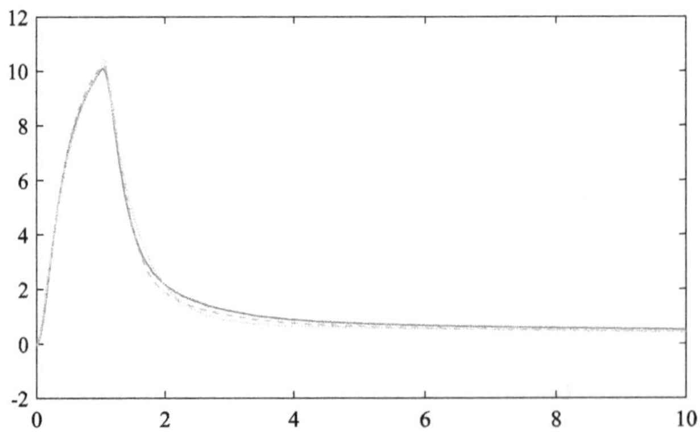

Figure 11.12: Simulation results of the aorta concentration.

It can be seen by comparing Wada's and Higgins' models that the former one can be used in block diagrams while the latter one can only be used in commands. In the former model, the parameters are fixed ones. If the parameters such as body weight or cardiac output are changed, Wada's model cannot be used, while in Higgins' code, the problem can be solved easily. The converted model from Higgins' code presented in the book can be used to solve the problems, so that a meaningful Simulink model can be constructed.

11.1.6 Pharmacodynamic modeling

Two commonly used criteria can be adopted to define the anaesthetic effectiveness:

$$C_{e_1}(t) = \frac{C_b^\gamma(t)}{\mathrm{EC}_{50}^\gamma + C_b^\gamma(t)}, \quad C_{e_2}(t) = E_0 - \frac{C_b^\gamma(t)}{\mathrm{EC}_{50}^\gamma + C_b^\gamma(t)} E_{\max}, \tag{11.1.18}$$

where EC_{50} is the concentration at 50 % of the drug effect. For different drugs, its values are different. Even for the same drug, for instance, fentanyl, the values in different papers are different. Here, the value 7.8 ng/ml is adopted. The $C_{e_1}(t)$ equation is also known as the Hill equation, where γ is known as the steepness in the Hill equation. The concentration $C_{e_1}(t)$ can be measured directly with an electroencephalogram (EEG). $C_b(t)$ is the drug concentration in the brain. The $C_{e_2}(t)$ equation is proposed in a frequency domain format, where $E_0 = 20$ Hz is a constant, and the frequency at maximum drug effect is defined as $E_{\max} = 15$ Hz. The concentration $C_{e_2}(t)$ can be measured with spectrum methods.

Based on the two pharmacodynamic criteria, the Simulink model in Figure 11.13(a) can be established. The model is masked such that the parameter dialog box shown in Figure 11.13(b) can be used to specify the parameters. This block can be used in the PK/PD modeling directly.

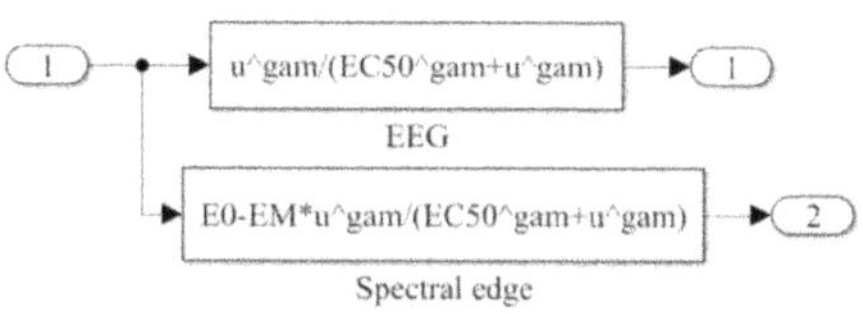

(a) pharmacodynamic model

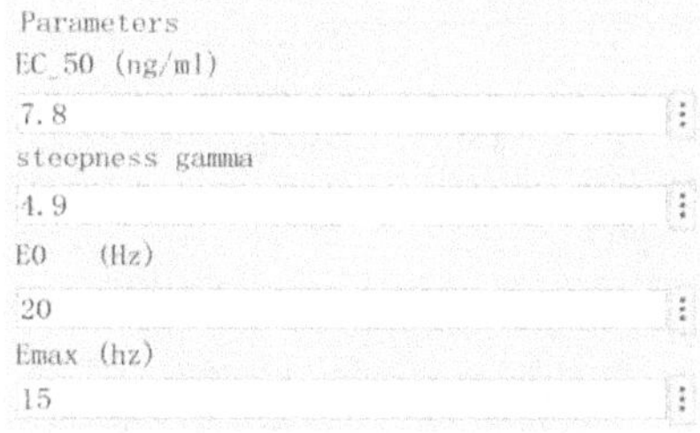

(b) parameter dialog box

Figure 11.13: Masked pharmacodynamic model (file: `c11mpdblk.slx`).

11.2 Finite state machine simulations with Stateflow

Stateflow simulation is established upon finite state machine (FSM) theory. The so-called finite state machine means that there are finite number of states in the system. Under certain events, the state can be transit from one state into another. Therefore, finite state machine system is also known as event-driven systems. In describing finite state machines, the condition for state transition is designed. In each state transition, the events are defined, such that state transition chart can be configured.

The concepts of the "states" in finite state machines are different from those in state space equations. The term "state" in this section refers to the states in finite state machine in Stateflow, while the term "states" $x_1(t)$ and $x_2(t)$ in state space models are changed into "state variables."

Stateflow is a powerful tool for describing finite state machines, which can be used in solving complicated supervisory control problems. The graphical tool can be used to implement the transitions of different states. The supervisory control logic generated in Stateflow can be embedded into ordinary Simulink models.

11.2.1 Introduction of finite state machines

A graphical interface is provided in Stateflow for designing finite state machines. The states can be defined, and the state transition conditions can be designed in a graphical way, such that the whole finite state machine can be constructed. In Stateflow, states and state transitions are the essential elements. An illustration of a finite state machine is shown in Figure 11.14. The term "finite" means that the number of states and their modes are finite. Therefore, it can be described by the illustration like this. In this chart, there are three states, and the direct transitions between any two states are conditional. It allows to transit from one state to another one, and also, "state a" in the chart also transits to itself. In the finite state machine representation, the conditions or events should be specified for the state transitions.

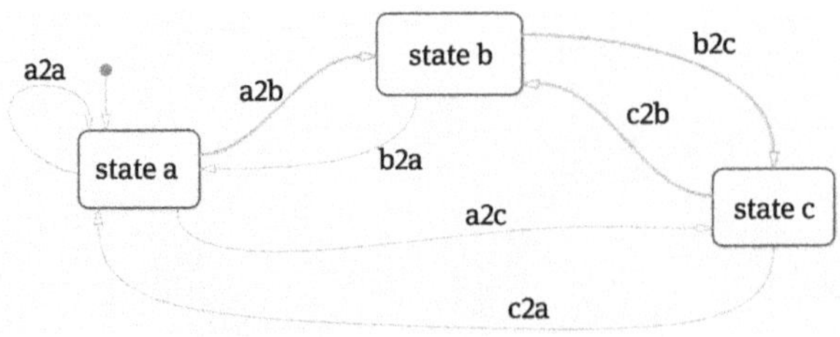

Figure 11.14: Illustration of a finite state machine.

Stateflow charts are embedded in Simulink models. Stateflow charts are event-driven, and the events may come from the same chart, or from the Simulink model.

11.2.2 Introduction to Stateflow

Type `stateflow` in MATLAB command window, a Simulink model is opened, together with the Stateflow library, as shown in Figure 11.15. In the model library, the icons **Chart**, **Truth Table**, **State Transition Table** and so on are provided. Meanwhile, in the blank Simulink model window, a **Chart** icon is given, and it is a Stateflow block. Stateflow charts can be directly embedded into the Simulink model. If command `sflib` is issued, only the Stateflow library is opened.

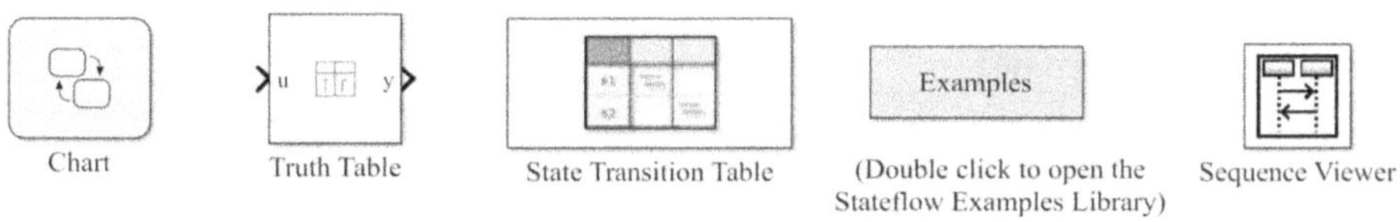

Figure 11.15: Stateflow library.

Double click **Chart** block, the editing interface in Figure 11.16 is opened. The Stateflow chart can be edited inside the interface. Powerful graphical editing facilities are provided in Stateflow, so as to describe easily the complicated logic expressions. It should be noted that the output of a Stateflow chart can be a state, or other interface variables, where the values of the state are enumerable.

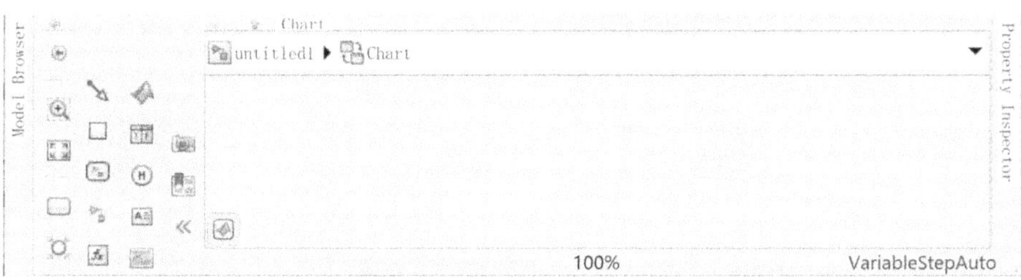

Figure 11.16: Stateflow editing interface and tools.

The toolbar on the left of Figure 11.16 is rearranged in typesetting. The commonly used tools include (add a state), (history junction), (truth table), (embed a function), (embed a MATLAB function) and so on. The tools can be dragged into the Stateflow editing interface.

Right click the blank region in Stateflow editing interface, the shortcut menu can be displayed, as shown in Figure 11.17. For typesetting purposes, the menu is divided into two parts, if the Properties menu is selected, the property dialog box is opened, as shown in Figure 11.18.

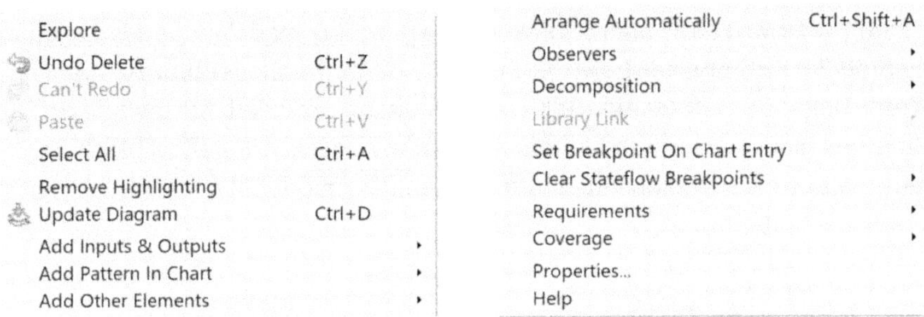

(a) upper portion (b) lower portion

Figure 11.17: Stateflow shortcut menu.

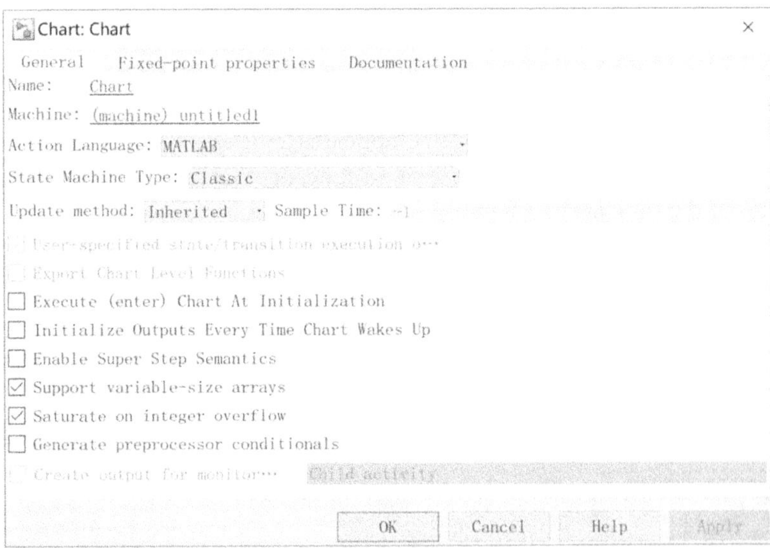

Figure 11.18: Stateflow property dialog box.

On the left of the Stateflow editing interface, some tools are provided. These tools can be used to draw Stateflow charts. The procedures in drawing Stateflow charts are introduced below:

(1) **Add a state.** The states of a chart is the executing modes in the system. Under Stateflow, there are two kinds of activities, “active” and “inactive” ones. The state tool can be used to add a state in the chart. Click the button in the toolbar, and drag it to the editing interface, and a state icon can be drawn in the chart, with a question mark on the upper-left corner of the block, indicating that the name and actions should be specified here. For instance, the name can be set to on. With such a tool, all the states can be added and named, as shown in Figure 11.19.

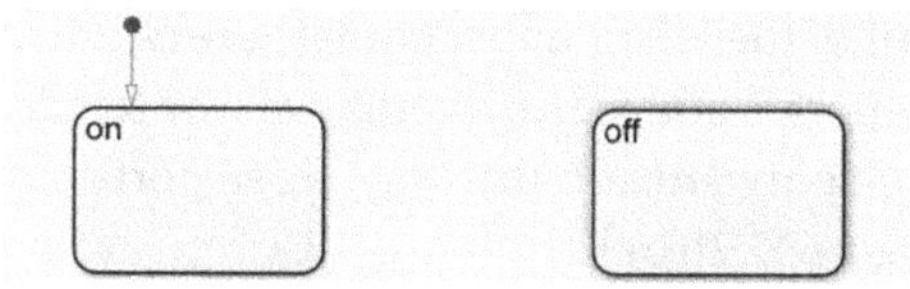

Figure 11.19: The Stateflow editing interface with two new states.

Apart from the state names, other information can be specified to the states. For instance, the command lead by `entry:` indicating the action to take in the state, while the keyword `during:` can be used to lead the command to be executed when the state is remained. Note that only one assignment statement is allowed in a command.

(2) **Set state transitions.** Click the left mouse button at the edge of a state block, drag it and release the button at another state icon, and a state transition line from one state to another is drawn. Over the state transition line, there is a ? mark, prompting the user to modify the state transition condition and the callback function. Click the right mouse button on the line, and a shortcut menu is displayed. Select the **Properties** item from the shortcut menu, and a dialog box shown in Figure 11.20 is displayed. The **Label** edit box can be used to specify the state transition expressions. In describing the state transition, a lower-level function can be called.

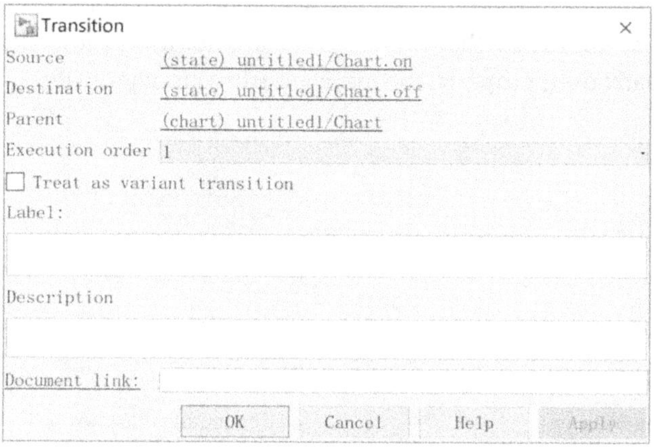

Figure 11.20: State transition in Stateflow.

(3) **Set the inputs and outputs.** Input and output ports can be added to Stateflow charts. The specific methods and procedures are demonstrated in the example next.

Example 11.3. Assign two input signals u1, u2 and an output signal y for a Stateflow chart.

Solutions. Create a Stateflow **Chart** block in Simulink model window first, as shown in Figure 11.21(a). Move the mouse button to the border of the **Chart**, the cursor is changed to ⊕, indicating a port can be added to the chart. Click the left mouse button, and a port is added automatically to the chart. By default, if the left or upper borders are clicked, an input port is added; otherwise, an output port is added. An alternative way to add a port by dragging the edge of the state block outwards. In this way, three ports are added to the Stateflow block, as shown in Figure 11.21(b).

(a) Stateflow chart (b) with ports added

Figure 11.21: Add input and output ports to Stateflow chart (file: `c11msf10.slx`).

In default cases, the name of the added ports is **data***, which is different from the ones expected in this example. The port names can be modified, if necessary. Double click the **Chart** icon, and a blank Stateflow editing interface is opened. On the right of the interface, there is a **Symbol** listbox, as shown in Figure 11.22(a). The **data** item in the listbox can be modified directly to the expected **u1**. The other two port names can also be modified accordingly, such that the modified results in Figure 11.22(b) can be found. The edited chart like this is the one expected in this example. Note that in front of the port name, there is an ⚠ icon, indicating that the variables are not used yet in the chart. If the variables are already used properly, the mark vanishes automatically.

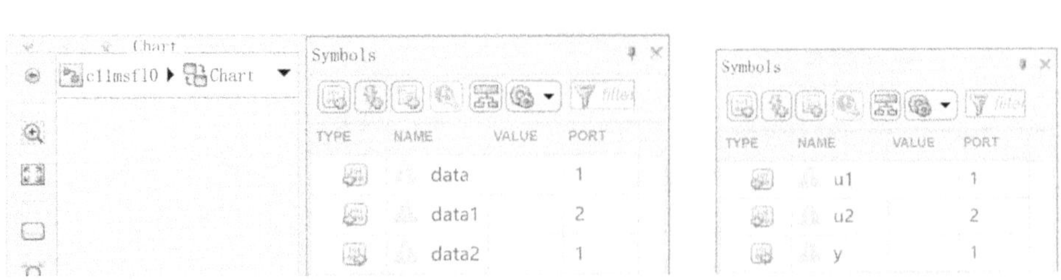

(a) Chart editing window (b) modified list box

Figure 11.22: Display and modify port name.

The input and output ports in the Stateflow chart can also be added with the shortcut menus. If the shortcut menu in Figure 11.17(a) is opened, the Add Inputs & Outputs menu item can be chosen to add input or output ports. The manipulation like this is more complicated than the ones discussed earlier, therefore, it is not presented here.

Stateflow allows the users to embed in the logic chart the MATLAB code, Stateflow subcharts, Simulink block or boxes with various contents. State transitions can be used to describe various flow controls, such as loops or conditional controls.

11.2.3 Commonly used commands in Stateflow

As introduced earlier, `stateflow` command can be used to open Stateflow main interface. The command can be used to launch Stateflow editing interface to start drawing Stateflow charts. There are other commonly used commands in using Stateflow:

(1) `sfnew` command can be used to open a new Simulink model, with a Stateflow chart icon. Command `sfnew fnm` can be used to setup a new Simulink model with the name `fnm`, where a blank Stateflow chart block is contained.

(2) `sfexit` closes all the windows with Stateflow charts, and exits Stateflow.

(3) `sfsave` command allows the user to save Stateflow charts to a file.

(4) `sfprint` command can be used to print the Stateflow chart.

11.2.4 Stateflow modeling examples

The switched differential equation models studied in Section 4.4.5 can also be modeled and solved with Stateflow models. Assume that there are two states, A and B, in the system. The state transition can be described by the switching laws. Therefore, Stateflow can be used to describe the state transition problems, and together with Simulink, the modeling and solution of switched differential equations can be completed. An example in Stateflow modeling is illustrated next in an example.

Example 11.4. Consider the switched differential equation model in Example 4.26. Build a new Simulink model with Stateflow and compare the simulation results.

Solutions. According to the simulation model in Example 4.26, the simulation model in Figure 11.23 can be created, where in the default Chart block, there are no input and output ports, and there are no connections to other blocks in Simulink. The chart is suspended. The methods in Example 11.3 can be used to add ports, such that it can be connected directly with Simulink.

Another convenient signal line drawing method is introduced here. Draw a signal line from the first output port of the Demux1 block, and settle down at the edge of the Chart block, then release the mouse button, and the first connection to the chart is established, with the name data. The port name can be used in Stateflow chart directly.

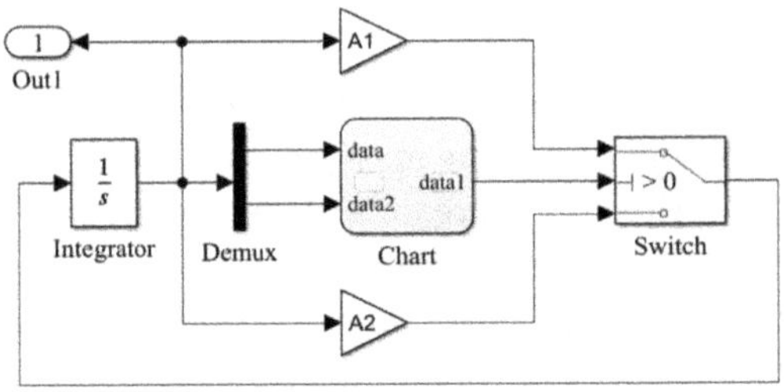

Figure 11.23: Simulink model with a Stateflow chart (file: `c11msf11.slx`).

The second input signal can be set up in the same way. If an output signal line from the **Chart** block is needed, a reverse signal line can be drawn, from the second input port of the **Switch** block, to the edge of the **Chart** block and release mouse button, the signal line can be established automatically, and the name of the port is set to **data1**. In this way the Stateflow chart is connected with other blocks in Simulink.

Double click the **Chart** block icon, and an editing interface in Figure 11.24 is opened, where the left part in the interface is left blank. Drag the ▢ icon in the toolbar to the editing interface, and a blank state block is added, with a prompt ?. Click the ? identifier, and the contents for the states can be specified, including the name and action command for the state. In this way, two state icons are designed as shown in Figure 11.24.

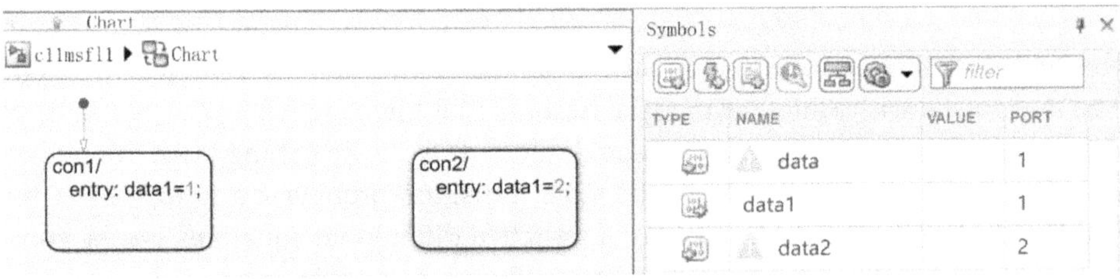

Figure 11.24: Stateflow editing interface.

Now observe the right side of the interface, the **Symbol** list. In the list, since the values of **data1** are already specified, there is no exclamation mark in the column. There are exclamation marks on the other two ports, since they are not involved yet in the chart. The name of the two ports can be renamed to **x1** and **x2**, respectively.

The state transition can now be defined. Draw a signal line from the edge of state **con1** block to **con2** and release the button on the edge of the latter state block, the transition condition, that is, the switching law, can be specified. The `[]` mark on the transition line can be modified to `[x1*x2<0]`, expressing the $x_1(t)x_2(t) < 0$ condition, which means that if the condition is satisfied, the state is changing from **con1** to **con2**.

Similarly, the other transition condition can be defined as [x1*x2>=0], as shown in Figure 11.25.

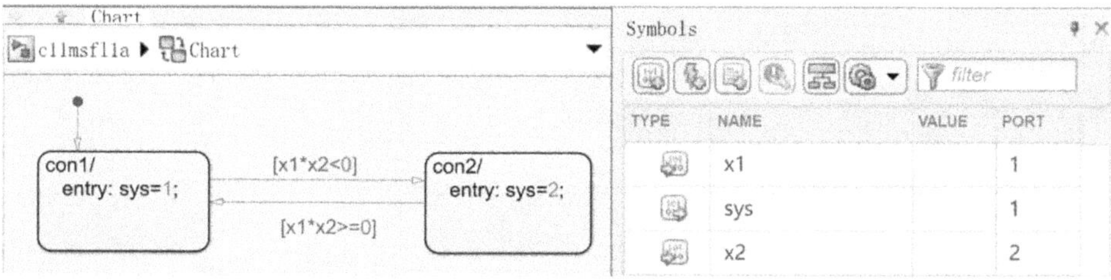

Figure 11.25: The finalized Stateflow chart model.

Now return back to the Simulink model. A virtual scope can be added at the switching signal. Since the signal switches between 1 and 2, a threshold in the Switch block can be set to 1.5. The finalized Simulink model can be constructed, as shown in Figure 11.26.

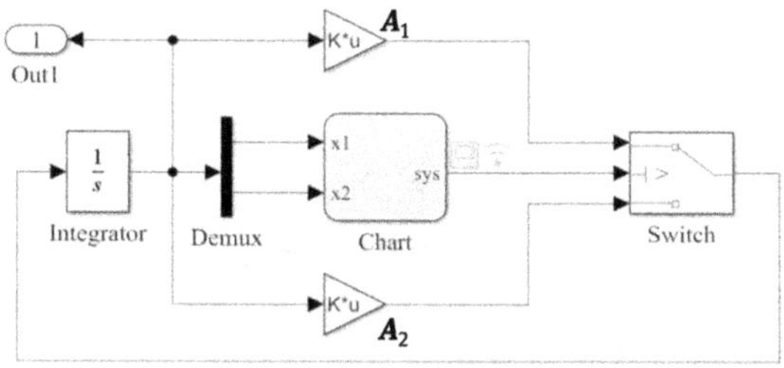

Figure 11.26: The final Simulink model (file: `c11msfl1a.slx`).

Start the simulation process in the Simulink model, and the results are shown in Figure 11.27, which are exactly the same as the one in Figure 4.26. Apart from the state variables, the $x_1(t)x_2(t)$ curve is also drawn, and the state variables are switched at the zero-crossing points of these curves.

Example 11.5. Consider the mechanical system in Figure 11.28.[47] In Example 10.2, similar problems were studied, but the assumption was made that there is no friction in the system. If frictions are considered, build up the friction model.

Solutions. In mechanical modeling, if the friction is considered, the modeling is rather complicated, since static and dynamic friction forces are to be considered, and the sizes and directions of the frictions are always changing. Several cases should be considered in real modeling process.

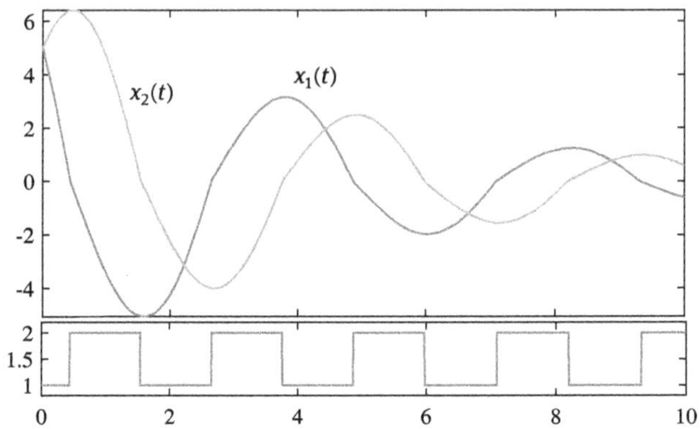

Figure 11.27: Simulation results of the switched system.

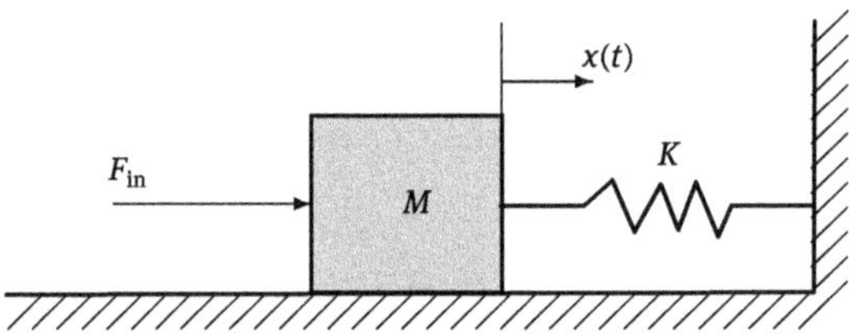

Figure 11.28: Illustration of a mechanical system.

Stateflow can be used to describe the different cases in the friction forces. In the mechanical model in Figure 11.28, the mass M is moving under the applied external force F_{in}. The total forces acting on M includes F_{in}, friction force F_f and the pulling force F_{spr} from the spring, the composite force can be found.

$$F = F_{in} - F_f - F_{spr}. \tag{11.2.1}$$

Based on Newton's second motion law, the following equation can be established:

$$Mx''(t) = F = F_{in} - F_f - F_{spr}, \tag{11.2.2}$$

where $x(t)$ is the displacement, therefore, $x''(t)$ is the acceleration. The pulling force is $F_{spr} = Kx(t)$, where K is the elastic coefficient. The friction force F_f can be computed from

$$F_f = \begin{cases} \text{sign}(x'(t))\mu F_n, & |F_{sum}| > \mu F_n, \\ F_{sum}, & x'(t) = 0 \text{ and } |F_{sum}| \leqslant \mu F_n, \end{cases} \tag{11.2.3}$$

where μ is the friction coefficient, F_n is the normal force and $x'(t)$ is the velocity of the mass. When $x'(t) = 0$, the mass is still. F_{sum} is the force in still status, satisfying $F_{sum} = F_{in} - F_{sliding}$. When the velocity is not zero, a force is imposed to make it become zero instantly. The force is usually greater than the maximum allowed force $F_{sliding}$.

When the velocity is already zero, F_{sum} force keeping the acceleration at zero. Friction force can further be classified into static and dynamic friction forces, depending on whether the mass is moving or not, that is,

$$\mu F_{\mathrm{n}} = \begin{cases} \mu_{\mathrm{static}} F_{\mathrm{n}} = F_{\mathrm{static}}, & x'(t) = 0, \\ \mu_{\mathrm{sliding}} F_{\mathrm{n}} = F_{\mathrm{sliding}}, & x'(t) \neq 0, \end{cases} \tag{11.2.4}$$

where μ_{static} and μ_{sliding} are respectively static and dynamic friction coefficients. Based on the above two conditions, the following equation can be used to evaluate the friction force:

$$F_{\mathrm{f}} = \begin{cases} \mathrm{sign}(x'(t)) F_{\mathrm{sliding}}, & x'(t) \neq 0, \\ F_{\mathrm{sum}}, & x'(t) = 0 \text{ and } |F_{\mathrm{sum}}| < F_{\mathrm{static}}, \\ \mathrm{sign}(F_{\mathrm{sum}}) F_{\mathrm{static}}, & x'(t) = 0 \text{ and } |F_{\mathrm{sum}}| \geqslant F_{\mathrm{static}}. \end{cases} \tag{11.2.5}$$

It can be seen that the mass M has two states, motion and still. An identifier variable **stuck** can be set to the state. It can be seen from (11.2.3) that, when $|F_{\mathrm{sum}}| > F_{\mathrm{static}}$, the mass is still, such that the **stuck** variable is set to 0. If the condition is not satisfied, the mass is still, but the variable **stuck** is set to 1. Therefore, the status of the friction force can be assigned. Based on the logical description, a Stateflow chart in Figure 11.29(a) can be created. In the Stateflow chart, the composite force **Fsum**, zero velocity detecting signal **novelocity** and static force **Fstatic** are the input signals from Simulink, while the identifier variable **stuck** is the output signal to Simulink.

Besides, based on (11.2.5), the subsystem for the friction force F_{f} is created as shown in Figure 11.29(b). In the subsystem, two switch blocks are used to describe the three conditions.

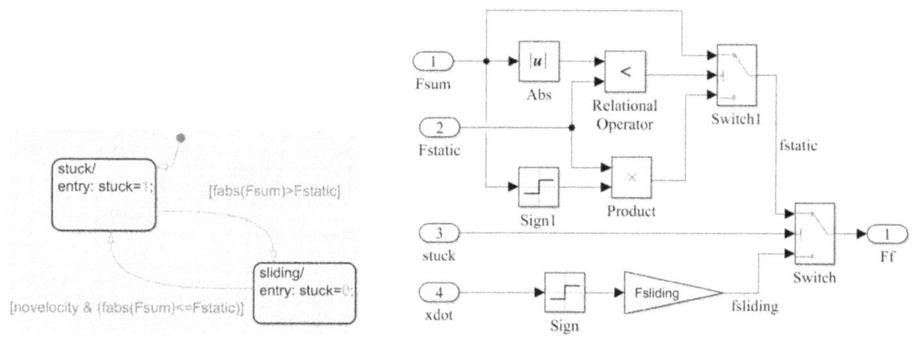

(a) Stateflow chart (b) friction force subsystem

Figure 11.29: Friction subsystem model.

With the two blocks, the Simulink model for the entire system can be established, as shown in Figure 11.30. It can be seen that in the model, the **Hit Crossing** block is used to detect the zero-crossing point of the velocity. Besides, two integrator blocks can be used to describe the mathematical model in (11.2.2). Note that the model created here is not the same as the one in [47], and the structure here is easier to understand.

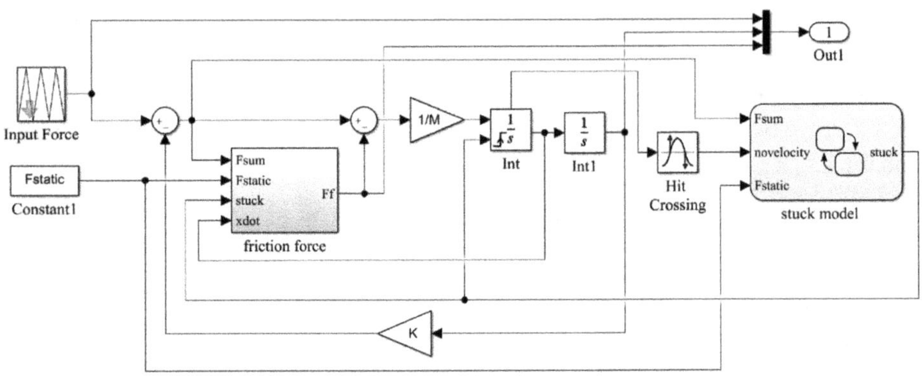

Figure 11.30: Simulink model for the friction system (file: `c11fsf12.slx`).

In the Simulink model, the velocity signal is not extracted from the output of the first integrator, but from the state variable, where the **Show state port** checkbox should be selected, to avoid algebraic loop phenomenon. Besides, **stuck** signal can be used to reset the integrator.

In the simulation system, a triangular waveform is used as the external force. The input signal is defined by the **Repeating Sequence** block in the **Sources** group. The parameter dialog box of the block is shown in Figure 11.31, where the period is set to 10 seconds, such that at time 5, the peak value of 5 is reached.

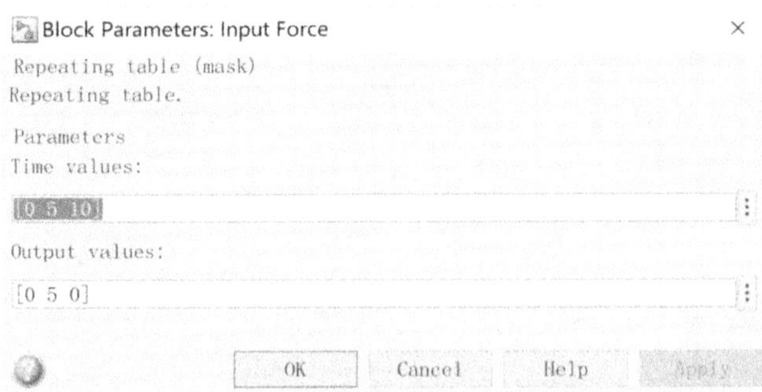

Figure 11.31: Parameter dialog box.

Since the system is a stiff one, the simulation algorithm can be set to **ode15s**, and the **Relative tolerance** is set to 10^{-6}, such that the simulation process can be executed effectively.

Before executing the simulation process, the following statements can be used to input relevant parameters into MATLAB workspace:

```
>> M=0.01; K=1; Fsliding=1; Fstatic=1;
```

Using to the parameters presented earlier, the curves of the input and output signals can be found, as shown in Figure 11.32, where the triangular waveform is the input signal.

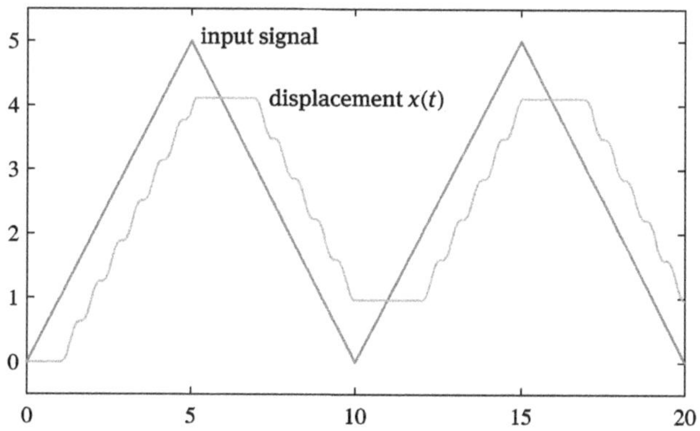

Figure 11.32: The curves of the input and output signals.

11.2.5 Draw control flow chart with Stateflow

It is easy to implement flow control with MATLAB programs, but the implementation of the controls are not so easy in Simulink, since the conditional subsystem blocks are needed. With Stateflow, the loop, conditional and switch structures can easily be constructed, by defining the state transition commands. The [transition condition] expression can be used, and it can be extended to [transition condition]{command execution} format. An example is given next to demonstrate the Stateflow modeling of flow control systems.

Example 11.6. Assume that several schemes are made to temperature control

$$y = \begin{cases} \text{scheme 1}, & u \geqslant 30\,°\text{C}, \\ \text{scheme 2}, & 22\,°\text{C} \leqslant u < 30\,°\text{C}, \\ \text{scheme 3}, & 10\,°\text{C} \leqslant u < 22\,°\text{C}, \\ \text{scheme 4}, & u < 10\,°\text{C}. \end{cases}$$

Use Stateflow to implement the schemes.

Solutions. It is complicated to model the plans with low-level blocks in Simulink. Stateflow can be used in the modeling. In fact, two virtual states can be defined, but the key point is how to specify state transitions. Assume that the input signal to a Stateflow chart is u, and output is y, the Stateflow chart in Figure 11.33(a) can be defined, to implement scheme selection. The Simulink model is shown in Figure 11.33(b). It can be seen that the Chart block has an input port of u, and an output port of y.

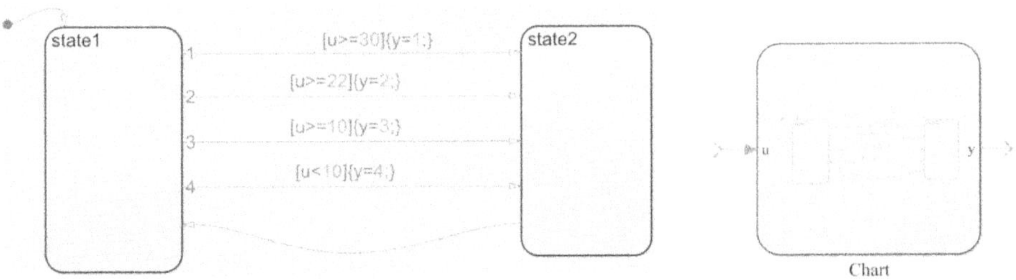

(a) Stateflow chart (b) Simulink model

Figure 11.33: Scheme selection with Stateflow (file: `c11msf13.slx`).

11.3 Simulation of discrete event systems with SimEvents

11.3.1 Fundamental concepts in discrete event dynamical systems

In everyday life, there are quite a lot of problems, which are unsuitable to be studied using continuous variable system modeling methods discussed in the previous chapters. For instance, a person enters an elevator, and presses the button of Floor 15. This is an independent event. The time and the floor happening of the event have nothing to do with other peoples' choices in the elevator. Therefore, the problem is beyond the scope of the continuous variable system descriptions. The system is an event-driven one, known as a discrete-event dynamic system (DEDS), or simply denoted as a discrete event system (DES).[88] In real world, many systems are event-driven discrete event systems, such as service systems, production scheduling systems, computer network systems and traffic management systems.

There are various ways in studying discrete event systems. The finite state machine with Stateflow tools discussed earlier can be used. Maximum algebras or Petri net models can also be used in theoretical analysis of discrete event systems. SimEvents is a simulation tool in Simulink, which can be used in studying discrete event systems, through simulation methods. In this section, some commonly used concepts in discrete event systems are presented, and an introduction to SimEvents blockset is proposed. A service queueing example is used to demonstrate simulation studies of discrete event systems.

There are two important concepts in discrete event systems: the entities and events. Entities are used to describe the objects in the system, and an event is used to describe the conditions when the states in the system have variations. Entities are usually classified as temporary entities and permanent ones. In a service queueing system, the service desk is a permanent entity, and the customers are temporary entities. The temporary entities enter the system and reside in the system for a period of time. The arrival, departure of temporary entities and also the interactions of the entities consist the internal state variations in the system.[28, 88] Each entity has its own states and properties.

Events are the conditions, which cause the internal states in the system change. For service queueing systems, the commonly used events are arrival of entities (customers), start service, complete service, the departure of the entities, and so on.

Discrete event systems can be modeled directly with Stateflow, since the states and transition of states can easily be described in Stateflow. The entire simulation model can be driven directly with the conditions such as state transitions. Discrete event systems can also be modeled with the blocks provided in the SimEvents blockset. Here, SimEvents based modeling and simulation studies of discrete event systems are demonstrated.

11.3.2 Introduction to SimEvents blockset

Double click the SimEvents icon in Simulink model library, or execute the `simevents` command, the SimEvents blockset can be opened, as shown in Figure 11.34.

It can be seen that different groups are provided in the blockset, such as Basic, Routing, Resources and so on. With the blocks provided in the blockset, simulation models for discrete event systems can be created.

It can be seen from SimEvents blockset that two kinds of signal ports are provided in SimEvents blockset. One is the standard Simulink port, and the other is labeled as ≫, which is the dedicated port in SimEvents blockset. It can be used to describe the entities, and the signals can be measured with entity detection blocks,

11.3.3 Simulation demonstration of discrete event systems

A service queueing example is given in this section to demonstrate the modeling and simulation of discrete event systems. The modeling methods with the SimEvents blockset are illustrated first, then the extension to the modeling is presented, such that useful statistical results can be obtained.

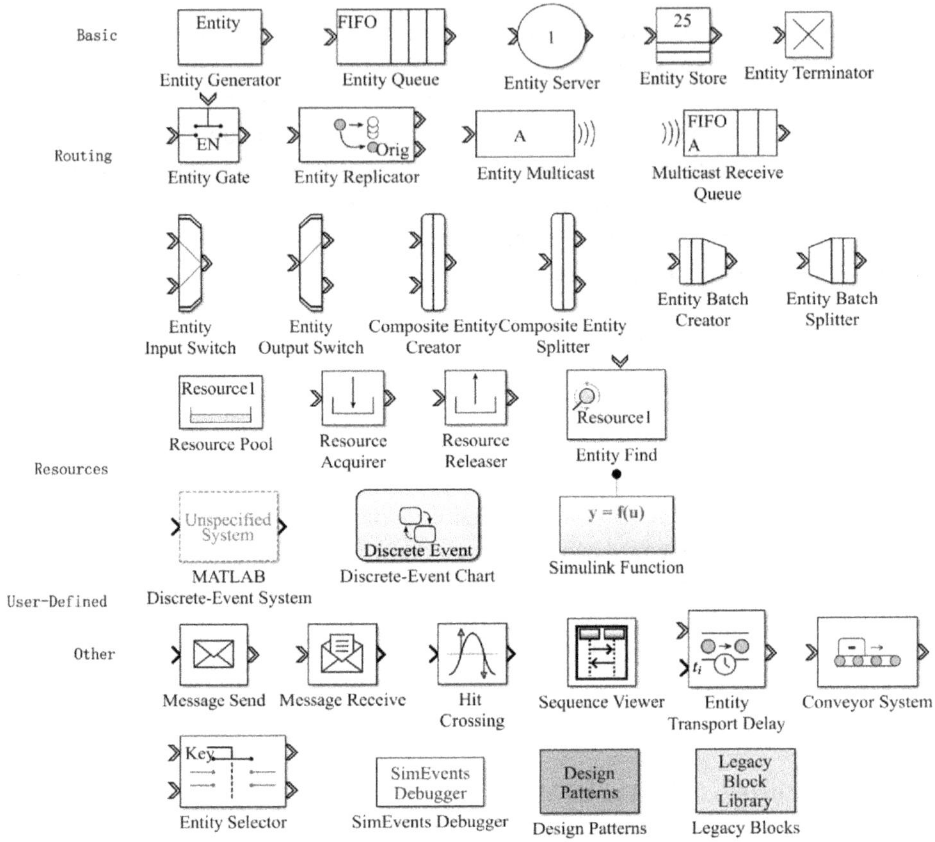

Figure 11.34: SimEvents blockset.

Example 11.7. Assume that a customer arrives in a barbershop once in every 3 minutes, and a barber completes his haircut job in 10 minutes. Check how many barbers are needed in the shop using the simulation method.

Solutions. In the SimEvents modeling, an entity (customer) arriving the barbershop can be generated by a **Entity Generator** block. When the entity arrives, he enters a first-in first-out (FIFO) queue. If the server (barber) is idle, the service begins directly; otherwise, the entity stays in the queue, waiting for service. When a service is completed, the first entity in the queue begins to be served. Based on the idea, the Simulink model shown in Figure 11.35 can be established.

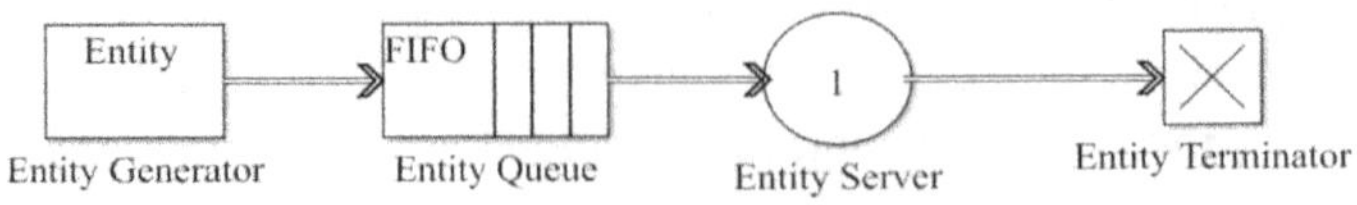

Figure 11.35: The SimEvents model (file: `c11mqueue1.slx`).

In the current model, entities are generated by the **Entity Generator** block, then enter the **Entity Queue** block, waiting for the service of the **Entity Server** block. Currently, there is only one server. When service is completed, the entity leaves and enter the **Entity Terminator** block. It can be seen that they are not any detecting blocks to observe the service status and statistical information.

To complete the simulation model, the following work should be done:

(1) Complete entity generator to describe the arrivals of entities.

(2) Describe the number of servers and service information.

(3) Monitor and display results, such as the length of the queue, the waiting time and the utilities of the barbers.

Relevant parameters in the blocks should be modified to implement the above tasks.

Double click the **Entity Generator** block, and the parameter dialog box in Figure 11.36 is opened. Each SimEvents block has its panes, and here the default pane is given. There are several items to be modified in this pane, such as the **Generation method**, with default **Time-based**, and with another choice of **Event-based**. These two options were expressed by different blocks in the old versions, and unification is made in the new versions. In the **Time source** listbox, the **Signal port** item can be chosen, such that an external signal can be used to generate arrival time, at the newly added input port, generated automatically.

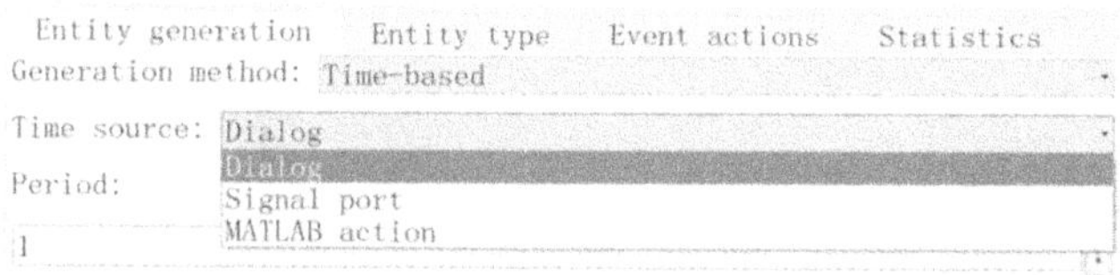

Figure 11.36: Parameter dialog box of Entity Generator block.

Double click the **Entity Queue** block, and the parameter dialog box shown in Figure 11.37 can be opened, where the information such as **Capacity** can be modified. Besides, the **Statistics** pane can be selected, as shown in Figure 11.38. Five check boxes are listed, such that the user may select which information is needed to be returned to the model. The five information items are **Number of entities departed**

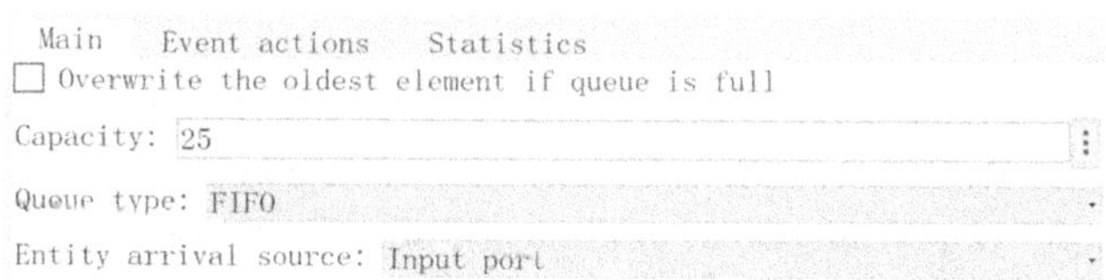

Figure 11.37: Parameter dialog box of the Entity Queue block.

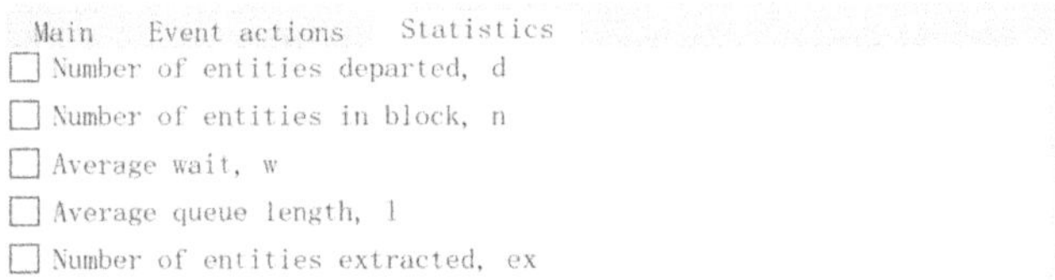

Figure 11.38: Statistics pane of the Entity Queue block.

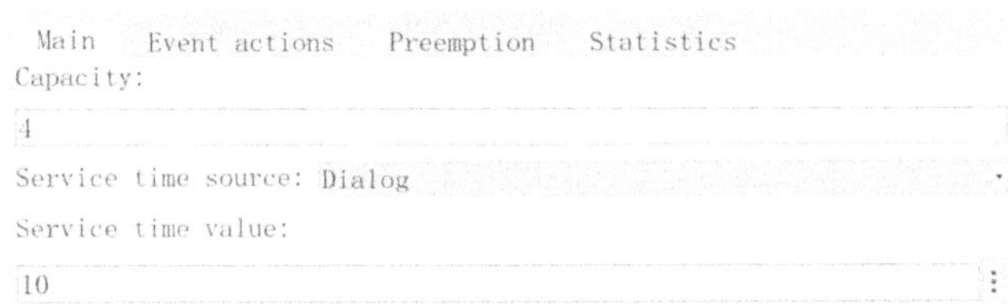

Figure 11.39: Parameter dialog box of Entity Server block.

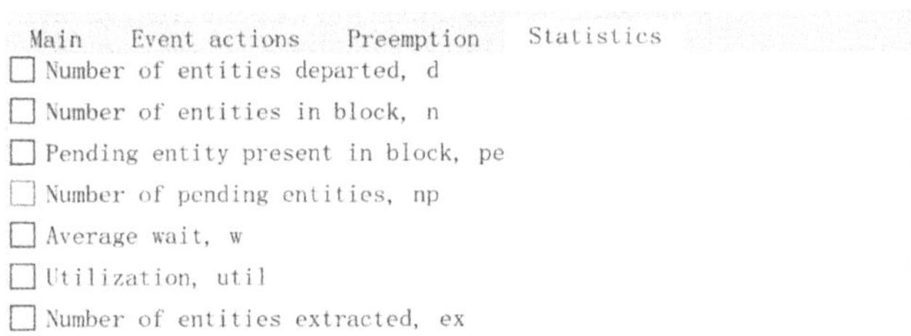

Figure 11.40: The statistics pane in the Entity Server block.

(labeled d), **Number of entities in block** (labeled n), **Average wait** (labeled w), **Queue length** (labeled l) and **Number of entities extracted** (labeled ex).

Double click the **Entity Server** block, and the parameter dialog box is opened, as shown in Figure 11.39, where the **Capacity** (i. e., the number of servers) is 4, **Service time value** is set to 10, meaning that the service time is 10 minutes. Besides, the **Statistics** pane can be opened, as shown in Figure 11.40. Seven check boxes are listed in the dialog box, allowing the user to select which information is needed. Some of them are the same as the ones in **Entity Queue** block. The different options are **Pending entity present in block** (shortened as pe), **Number of pending entities** (marked pn), **Utilization** (marked util).

If all the statistics options are checked in the two blocks, the simulation model constructed is shown in Figure 11.41. All of the ports are labeled in a shallow color. The user can select any output port from them, and extract signals to observe the simulation results. Of course, it is not necessary to select all of the signals, and some relevant signals can be selected for further analysis.

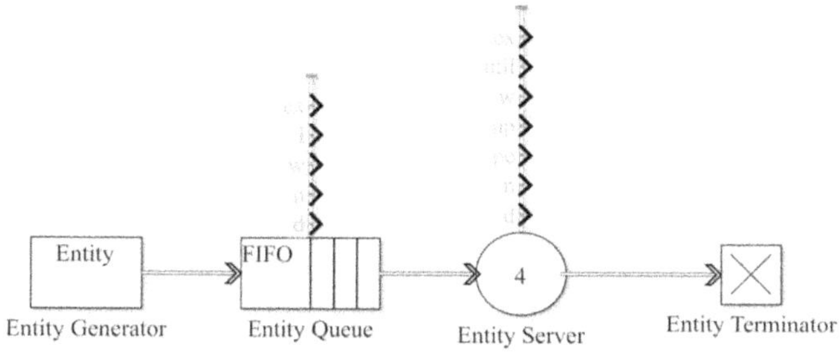

Figure 11.41: SimEvents model with all the ports (file: `c11mqueue2.slx`).

If the average number of waiting entities, and the utility of the servers are expected, the two signals can be selected, and linked to Simulink output ports, as shown in Figure 11.42. Simulate the model, and the simulation results can be obtained, as shown in Figure 11.43. It can be seen from the simulation results that since the number of servers is sufficient in this example, each entity enters the queue can be served immediately. Therefore, the number of waiting entities is zero. The utility rate of the servers is close to 80 %.

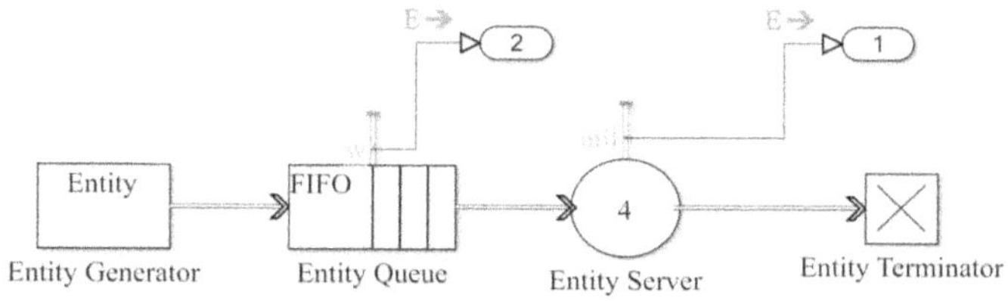

Figure 11.42: Modified SimEvents model (file: `c11mqueue3.slx`).

If the number of servers is reduced to 3, and then performs simulation, the simulation results are shown in Figure 11.44. It can be seen that the utility rate of the servers is close to 100 %. The average waiting entities has the potential of increasing, but the capacity of the queue is not exceeded. The simulation result is consistent to the directly observed results.

Example 11.8. If the arrival time of a customer is a random number uniformly distributed in the $(0, 6)$ minute interval, and the service time for each customer is a uniformly distributed random number in the $(9, 11)$ minute interval, find out the utilities of the four barbers.

Solutions. Only the fixed arrival time can be handled in the model given in Figure 11.42, so are the cases for the barbers. If the two time instances are random numbers, the two blocks should be modified. The simplest way is to double click the Entity Generator and Entity Server blocks. They all have Time source listboxes in

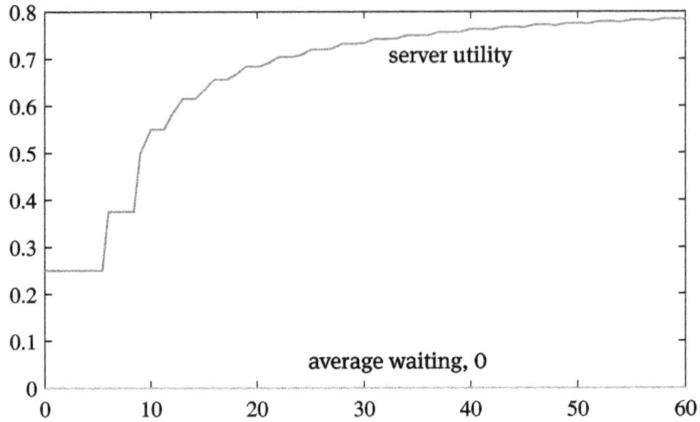

Figure 11.43: Simulation results.

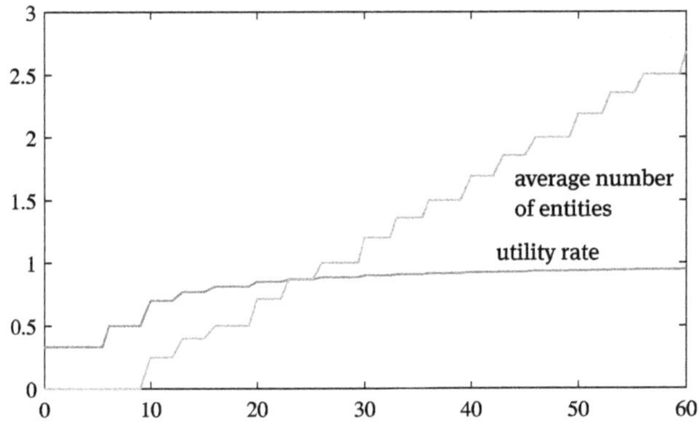

Figure 11.44: Simulation result if the number of servers is reduced to 3.

their parameter dialog boxes, and they can both be set to **Signal port**. The extra input ports are then automatically added to the two blocks. Connect the ports to **Uniform Random Number** blocks, and specify the upper and lower bounds of the random numbers, and a simulation model shown in Figure 11.45 can be established. Note that the default seeds for the two random number generator blocks are all zeros, and they should be assigned to any other numbers, as long as they are different; otherwise, the simulation results obtained may be misleading, because the two random numbers are correlated. It is recommended that both the seeds are set to pseudorandom numbers, for instance, `round(1000*rand(1))`.

Start simulation study to the model, and the result in Figure 11.46 can be found. It is worth mentioning that since the impact of random numbers, the results obtained each time are different, but the total trends are similar. For instance, the utilities of the barbers are 80 % to 85 %.

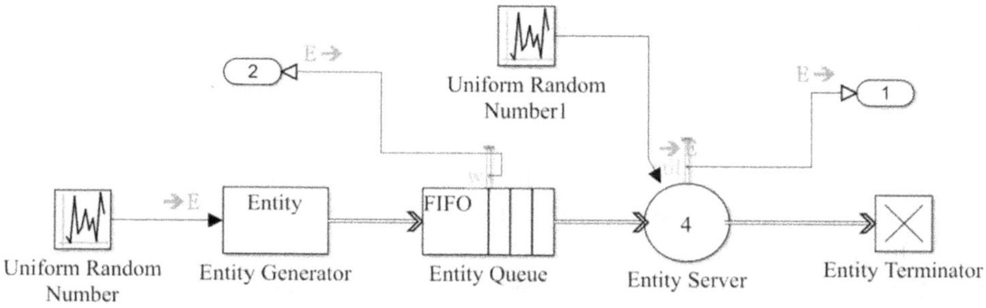

Figure 11.45: Simulation model with random input (file: `c11mqueue4.slx`).

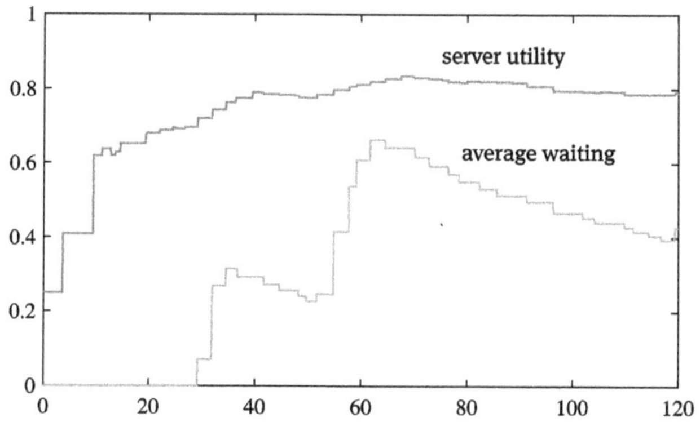

Figure 11.46: Result from the new model.

11.4 Exercises

11.1 Write out the mathematical models for two compartment and three compartment models in Wada's blockset. Also open the junction and branch blocks, and observe the internal structures.

11.2 Use Stateflow to create simulation models for loop structures and switch structures.

11.3 The decoder system with seven-segment LED display was demonstrated earlier, and the kernel part is the truth table. Use Stateflow to create a simulation model, and observe the simulation results.

11.4 In the Stateflow modeling process, the definitions of the input and output ports are very important, since the Simulink blocks can be associated to the signals outside the charts, and also provide the variables for the internal conditions and states inside the Stateflow charts. Build a blank Stateflow chart, with the clock block as its driven signal.

11.5 Create a finite-state machine, such that there are two states, On and Off. Use the manual switch block in Simulink as its input signal, and use it to turn on or off the LED block, created in Example 10.10.

11.6 Assume that the arrival time of the customers satisfies uniform random distribution in $(0, 3)$ minutes. Find out with simulation method and at least how many barbers are needed, to ensure the capacity of the queue is not exceeded within 8 hours.

11.7 Consider the simulation model in Example 11.8. Find out with simulation methods that the possible numbers of customers are served, with four barbers, within 8 hours.

11.8 If the customers entering the barbershop divided uniformly into two queues, waiting for the service of two barbers, establish the Simulink model and assess the efficiencies of the barbers.

Bibliography

[1] Anderson B D O, Moore J B. Linear optimal control. Englewood Cliffs: Prentice-Hall, 1971.

[2] Ascher U M, Mattheij R M M, Russel R D. Numerical solution of boundary value problems for ordinary differential equations. Philadelphia: SIAM Press, 1995.

[3] Ascher U M, Petzold L R. Computer methods for ordinary differential equations and differential–algebraic equations. Philadelphia: SIAM Press, 1998.

[4] Åström K J. Computer aided tools for control system design. In: Jamshidi M, Herget C J (eds). Computer-aided control systems engineering. Amsterdam: Elsevier Science Publishers B V, 1985.

[5] Åström K J, Hägglund T. PID controllers: theory, design and tuning. Research Triangle Park: Instrument Society of America, 1995.

[6] Åström K J, Hang C C, Persson P, Ho W K. Towards intellegient PID control. Automatica, 1992, 28(1): 1–9.

[7] Asuni N, Giachetti A. TESTIMAGES — Free collection of digital images for testing. https: //testimages.org.

[8] Atherton D P. Nonlinear control engineering — Describing function analysis and design. London: Van Nostrand Reinhold, 1975.

[9] Bellen A, Zennaro M. Numerical methods for delay differential equations. Oxford: Oxford University Press, 2003.

[10] Bennett S. Development of the PID controllers. IEEE Control Systems Magazine, 1993, 13(6): 58–65.

[11] Brenann R D, Linebarger R N. A survey of digital simulation: Digital analog simulator programs. Simulation, 1964, 3(6): 22–36.

[12] Brosilow C, Joseph B. Techniques of model-based control. Englewood Cliffs: Prentice Hall, 2002.

[13] Cai S F. Automatic control principles. Beijing: Machine Industry Press, 1980 (in Chinese).

[14] Caputo M, Mainardi F. A new dissipation model based on memory mechanism. Pure and Applied Geophysics, 1971, 91(8): 134–147.

[15] Dai L Y. Singular control systems. In: Thoma M, Wyner A (eds). Lecture Notes in Control and Information Sciences. Berlin: Springer-Verlag, 1989.

[16] Davis N R, Mapleson W W. A physiological model for the distribution of injected agents, with special reference to pethidine. British Journal of Anaesthsia, 1993, 70: 248–258.

[17] Diethelm K. The analysis of fractional differential equations: an application-oriented exposition using differential operators of Caputo type. New York: Springer, 2010.

[18] Dorf R C, Bishop R H. Modern control systems. 9th Ed. Upper Saddle River: Prentice-Hall, 2001.

[19] Felhberg E. Low-order classical Runge–Kutta formulas with stepsize control and their applications to some heat transfer problems. Technical report, Washington DC: NASA Technical Report TR R-315, 1969.

[20] Floyd T L. Digital fundamentals. 11th Ed. Harlow: Pearson Education Limited, 2015.

[21] Forsythe G E, Malcolm M A, Moler C B. Computer methods for mathematical computations. Englewood Cliffs: Prentice-Hall, 1977.

[22] Fortuna L, Frasca M. Chua's circuit implementations — Yesterday, today and tomorrow. Singapore: World Scientific, 2009.

[23] Frederick D K, Rimer M. Benchmark problem for CACSD packages. In: Abstracts of the Second IEEE Symposium on Computer-aided Dontrol System Design, Santa Barbara, 1985.

[24] Gaskill R A, Harris J W, McKnight A L. DAS: A digital analog simulator. In: Proceedings of Spring Joint Computer Conference, 1963, 83–90.

https://doi.org/10.1515/9783110734959-012

[25] Gladwell I. The development of the boundary-value codes in the ordinary differential equations — Chapter of the NAG library. In: Childs B eds. Codes for boundary-value problems in ordinary differential equations. Berlin: Springer-Verlag, 1979.
[26] Gongzalez R C, Woods R E. Digital image processing. 2nd Ed. Englewood Cliffs: Prentice-Hall, 2002.
[27] Green W L, Speckhart F H. CSMP. Simulation, 1980, 4: 131–133.
[28] Gu Q T. Modeling and simulation of discrete event systems. Beijing: Tsinghua University Press, 1999 (in Chinese).
[29] Hairer E, Nørsett S P, Wanner G. Solving ordinary differential equations I: Nonstiff problems. 2nd Ed. Berlin: Springer-Verlag, 1993.
[30] Hairer E, Wanner G. Solving ordinary differential equations: Stiff and differential–algebraic problems. 2nd Ed. Berlin: Springer-Verlag, 1996.
[31] Han J Q. Active disturbance rejection control technique – the technique foe estimating and compensating the uncertainties. Beijing: National Defense Industry Press, 2008 (in Chinese).
[32] Han J Q, Yuan L L. Discrete form of tracking and differentiator. System Science and Mathematics, 1999, 19(3): 268–273 (in Chinese).
[33] Hartung F, Krisztin T, Walther H-O, Wu J. Functional differential equations with state-dependent delays: Theory and applications. In: Cañada A, Drábek P and Fonda A (eds). Handbook of differential equations — Ordinary differential equations, Volume 3. Amsterdam: Elsevier, 2006.
[34] Hellerstein J L, Diao Y, Parekh S, Tilbury D M. Feedback control of computing systems. Hoboken: IEEE Press and John Wiley & Sons Inc, 2004.
[35] Higgins M J. Clinical and theoretical studies with opioid analgestic fentanyl. Master's thesis, University of Glasgow, 1990.
[36] Hilfer R. Applications of fractional calculus in physics. Singapore: World Scientific, 2000.
[37] Houbak N. SIL — A simulation language: User's guide. Berlin: Springer-Verlag, 1990.
[38] Johnson M A, Moradi M H. PID control — new identification and design methods. London: Springer, 2005.
[39] Kuehni R G. Color space and its divisions — color order from antiquity to the present. Hoboken: Wiley-Interscience, 2003.
[40] Laub A J, Little J N. Control system toolbox: A uses' guide – for use with MATLAB. The Mathworks Inc, 1986.
[41] Li B H. System simulation may become the third means in investigating the world. Scientific Chinese, 2012, 1(5): 47 (in Chinese).
[42] Li Z G, Soh Y C, Wen C Y. Switched and impulsive systems — Analysis, design, and applications. Berlin: Springer, 2005.
[43] Little J N, Emami-Naeini A, Bangert S N. CTRL-C and matrix environments for the computer-aided design of control systems. In: Jamshidi M, Herget C J (eds). Computer-aided control systems engineering. Amsterdam: Elsevier Science Publishers B V, 1985.
[44] Liu B. Modern control theory. Beijing: Machine Industry Press, 1980 (in Chinese).
[45] Liu D G, Fei J G. Numerical simulation algorithms for dynamical systems. Beijing: Science Press, 2001 (in Chinese).
[46] Mahfouf M, Linkens D A, Xue D Y. A new generic approach to model reduction for complex physiologically based drug models. Control Engineering Practice, 2002, 10(1): 67–82.
[47] MathWorks. Simulink/Stateflow technical examples — Using Simulink and Stateflow in automotive applications. Technical Report, 1998.
[48] MathWorks. Audio System Toolbox documentation, 2016.
[49] Mattsson S E, Elmqvist H. Modelica — An international effort to design the next generation modeling language. In: Proceedings of the 7th IFAC Symposium on Computer-aided Control System Design. Gent, Belgium, 1997.

[50] Mitchell E E L, Gauthier J S. Advanced continuous simulation language (ACSL): user's guide and reference manual. 3rd Ed. Mitchell & Gauthier Associates, 1981.
[51] Munro N. Multivariable control 1: the inverse Nyquist array design method. In: Lecture Notes of SERC Vacation School on Control System Design. Manchester: UMIST, 1989.
[52] Oustaloup A, Levron F, Nanot F, Mathieu B. Frequency band complex non integer differentiator: characterization and synthesis. IEEE Transactions on Circuits and Systems I: Fundamental Theory and Applications, 2000, 47(1): 25–40.
[53] Peterson H E, Sansom F J. MIMIC: A digital simulator program. SESCA Internal Memo 65-12, Wright Patterson AFB, Dayton, Ohio, 1965.
[54] Polyanin A D, Zaitsev V F. Handbook of ordinary differential equations — Exact solutions, methods and problems. Boca Raton: CRC Press, 2018.
[55] Ren X Q. Computer simulation of control systems. Beijing: Machine Industry Press, 1988 (in Chinese).
[56] Scott M R, Watts H A. A systematized collection of codes for solving two-point boundary-value problems. In: Lapidus L and Schiesser W E eds. Numerical methods for differential systems — Recent developments in algorithms, software, and applications. New York: Academic Press, 1976.
[57] Shampine L F, Gladwell I, Thompson S. Solving ODEs with MATLAB. Cambridge: Cambridge University Press, 2003.
[58] The MathWorks Inc. Simulink user's manual, 2007.
[59] Speckhart F H, Green W L. A guide to using CSMP. Englewood Cliffs: Prentice-Hall, 1976.
[60] Sun G J, Li R H. Computer simulation technique. Beijing: National Defence Press, 1979 (in Chinese).
[61] Sun Z D, Ge S S. Stability theory of switched dynamical systems. London: Springer, 2011.
[62] Syn W M, Linebarger R N. DSL/90: A digital simulation program for continuous system modeling. In: Proceedings of Spring Joint Computer Conference, 1965.
[63] Tao Y H, Yin Y X, Ge L S. New types of PID control and their applications. Beijing: Machine Industry Press, 2001 (in Chinese).
[64] Tarr E. Hack audio — An introduction to computer programming and digital signal processing in MATLAB. London: Routledge, 2019.
[65] The Modelica Association. https://www.modelica.org/.
[66] The SCi. Continuous system simulation language (CSSL). Simulation, 1967, 9: 282–303.
[67] Tuinenga P W. Spice — A guide to circuit simulation and analysis with PSpice. Englewood Cliffs: Prentice-Hall, 1992.
[68] Van der Pol B, Van der Mark J. Frequency demultiplication. Nature, 1927, 120: 363–364.
[69] Wada D R, Stanski D R, Ebling W F. A PC-based graphical simulator for physiological pharmacokinetic models. Computer Methods and Programs in Biomedicine, 1995, 46: 245–255.
[70] Wada D R, Ward D S. Open-loop control of multiple drug effects in anaesthesia. IEEE Transaction on Biomedical Engineering, 1995, 42(7): 666–677.
[71] Wang W L. Automatic control principles. Beijing: Science Press, 2001 (in Chinese).
[72] Wang Z B, Han J Q. CADCSC software system – Control system computer-aided design. Beijing: Science Press, 1997 (in Chinese).
[73] Waters D J, Mapleson W W. Exponentials and the anaesthetist. Anaesthesia, 1964, 19: 274–293.
[74] Xie X K. Fundamentals of modern control theory. Shenyang: Liaoning People's Press, 1980 (in Chinese).
[75] Xu X H. Model reference adaptive control systems. In: Lecture notes in Northeast University of Technology, Shenyang, 1982 (in Chinese).
[76] Xue D. Analysis and computer aided design of nonlinear systems with Gaussian inputs. Ph. D. thesis, Sussex University, U. K., 1992.

[77] Xue D, Atherton D P. A suboptimal reduction algorithm for linear systems with a time delay. International Journal of Control, 1994, 60(2): 181–196.

[78] Xue D Y. Experimentation of linear dynamic fitting analysis for the Higgins model with fentanyl drug administrations. Technical report, Department of Automatic Control and Systems Engineering, University of Sheffield, 1999.

[79] Xue D Y. Fractional-order control systems — Fundamentals and numerical implementations. Berlin: de Gruyter, 2017.

[80] Xue D Y. Fractional calculus and fractional-order control. Beijing: Science Press, 2018 (in Chinese).

[81] Xue D Y. Differential equation solutions with MATLAB. Berlin: De Gruyter, 2020.

[82] Xue D Y. Calculus computation with MATLAB. Berlin: De Gruyter, 2020.

[83] Xue D Y. Solving optimization problems with MATLAB. Berlin: De Gruyter, 2020.

[84] Xue D Y. MATLAB programming – mathematical problem solutions. Berlin: De Gruyter, 2020.

[85] Xue D Y, Chen Y Q. Modeling, analysis and design of control systems with MATLAB and Simulink. Singapore: World Scientific, 2014.

[86] Yang H, Jiang B, Cocquempot V. Stabilization of switched nonlinear systems with unstable modes. Switzerland: Springer, 2014.

[87] Zhao X D, Kao Y G, Niu B, Wu T T. Control synthesis of switched systems. Switzerland: Springer, 2017.

[88] Zheng D Z, Zhao Q C. Discrete event dynamic systems. Beijing: Tsinghua University Press, 2001 (in Chinese).

MATLAB function index

Bold page numbers indicate where to find the syntax explanation of the function. The function or model name marked by * are the ones developed by the authors. The items marked with ‡ are those downloadable freely from internet.

Index

www.ingramcontent.com/pod-product-compliance
Lightning Source LLC
Chambersburg PA
CBHW080535250926
51846CB00016B/991
* 9 7 8 3 1 1 0 7 3 9 0 4 6 *